Encyclopaedia
of
STATISTICAL THEORY

Encyclopaedia
of
STATISTICAL THEORY

Volume 1

Dr. Rakesh Kumar Pandey

ANMOL PUBLICATIONS PVT. LTD.
NEW DELHI - 110 002 (INDIA)

ANMOL PUBLICATIONS PVT. LTD.
Regd. Office: 4360/4, Ansari Road, Daryaganj,
New Delhi-110 002 (India)
Ph.: 23278000, 23261597
Branch Office: No. 1015, Ist Main Road, BSK IIIrd Stage
IIIrd Phase, IIIrd Block,
Bangalore-560 085 (India)
Tel.: 080-41723429
Visit us at: www.anmolpublications.com

Encyclopaedia of Statistical Theory

First Edition, 2009

ISBN 978-81-261-4109-8 (Set)

PRINTED IN INDIA

Printed at Mehra Offset Press, Delhi.

Contents

Preface

Encyclopaedia of Statistical theory defines a statistic as a function of a sample where the function itself is independent of the sample's distribution: the term is used both for the function and for the value of the function on a given sample. A statistic is distinct from an unknown statistical parameter, which is not computable from a sample. A key use of statistics is as estimators in statistical inference, to estimate parameters of a distribution given a sample. For instance, the *sample mean* is a statistic, while the *population mean* is a parameter.

Statisticians often contemplate a parameterized family of probability distributions, any member of which could be the distribution of some measurable aspect of each member of a population, from which a sample is drawn randomly. Statistics are used to describe the basic features of the data gathered from an experimental study in various ways. A descriptive Statistics is distinguished from inductive statistics. They provide simple summaries about the sample and the measures. Together with simple graphics analysis, they form the basis of virtually every quantitative analysis of data. In addition to showing how to apply statistical procedures, the authors explain how to meaningfully describe real sets of data, what the statistical tests mean in terms of their practical applications, how to evaluate the validity of the assumptions behind statistical tests, and what to do when statistical assumptions have been violated.

Author

List of Symbols

$\pm \quad \not\subset \quad \subset \quad \supset \quad \subseteq \quad \in \quad \notin \quad \partial$

$\neq \quad \equiv \quad \approx \quad \angle \quad \nabla \quad \Sigma \quad \Pi \quad \Omega$

$\phi \quad \alpha \quad \beta \quad \chi \quad \varepsilon \quad \gamma \quad \eta \quad \lambda$

$\mu \quad \nu \quad \perp \quad \pi \quad \theta \quad \rho \quad \sigma \quad \tau$

$\Re \quad \omega \quad \xi \quad \psi \quad \zeta \quad \Upsilon \quad \exists \quad \int$

$\iint \quad \iiint \quad \oint \quad \times \quad \sqrt{\ } \quad \therefore \quad \because \quad \Delta$

$\geq \quad \leq \quad \Rightarrow \quad \Downarrow \quad \Leftrightarrow \quad \infty \quad \propto \quad \mathbb{R}$

$\mathbb{Z} \quad \mathbb{N} \quad \mathbb{C} \quad \mathbb{Q}$

Chapter 1

Probability Sampling Distributions

In the historical development of probability theory, the first step in the streamlining occurred in connection with what are known as binomial probabilities. Although the term is perhaps new to you, the concept it describes is simply an extension of the matters.

A binomial probability situation is one, such as the mere-chance occurrence of heads when tossing coins, or the mere-chance recovery of patients from a disease, for which you can specify

p	the probability that the event or outcome in question will occur in any particular instance	E.g., p=.5, the probability that any particular tossed coin will come up as a head; p=.4, the probability that any particular patient will spontaneously recover.
q	the probability that the event or outcome in question will not occur in any particular instance	E.g., q=.5, the probability that any particular tossed coin will not come up as a head; q=.6, the probability that any particular patient will not spontaneously recover.
N	the number of instances in which the event or outcome has the opportunity to occur	

To show how the streamlining came about, we will begin by considering the very simple binomial probability situation in which N is equal to 2, and then work our way up through some more complex binomial situations.

If you are tossing 2 coins with elemental probabilities of p=.5 for the outcome of getting a head on any particular one of the coins (labeled below as "H") and q=.5 for the outcome of not getting a head (labeled as "–"), the possible conjunctive outcomes and their associated probabilities are

Coin A	Coin B	*Sub-pathway Probability*	*Number of Heads*	*Main Pathway Probability*	
–	–	.5x.5=.25	0	.25	(25%)
–	H	.5x.5=.25	1	.50	(50%)
H	–	.5x.5=.25			
H	H	.5x.5=.25	2	.25	(25%)
			Totals	1.0	(100%)

Similarly, if you are considering 2 randomly selected patients with elemental probabilities of p=.4 for any particular one of the patients spontaneously recovering (labeled below as "*R*") and q=.6 for not spontaneously recovering (labeled as "–"), the possible conjunctive outcomes and their associated probabilities are

Patient A	Patient B	*Sub-pathway Probability*	*Number of Recoveries*	*Main Pathway Probability*	
–	–	.6x.6=.36	0	.36	(36%)
–	*R*	.6x.4=.24	1	.48	(48%)
R	–	.4x.6=.24			
R	*R*	.4x.4=.16	2	.16	(16%)
			Totals	1.0	(100%)

In Figure we display these two sets of probabilities side-by-side in the form of two histograms. In each case, you can think of the total area of the histogram as representing 100% of the total probability that applies to that particular situation. Thus, for both examples there is a 100% chance that the outcome will include either zero or 1 or 2 of the particular events in question—heads or patient recoveries. All that differs is the way in which the 100% total is divided up by the 3 possible outcomes: 25%|50%|25% for the coin-toss example versus 36%|48%|16% for the patient-recovery example.

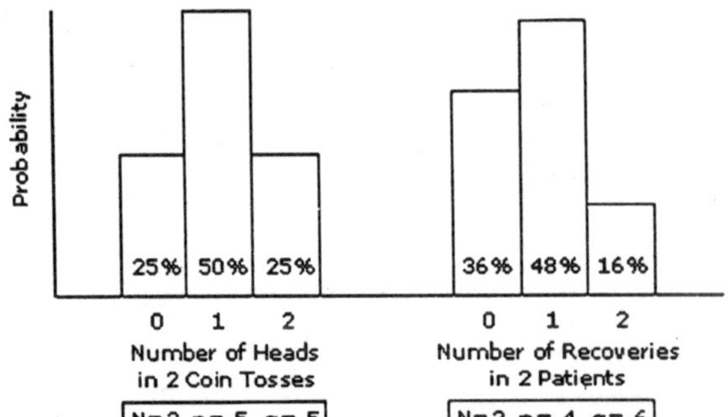

Fig. Two Binomial Sampling Distributions for the Case where N=2

These two sets of probabilities are illustrative of a large class of theoretical structures collectively known as probability sampling distributions, so named because they describe the probabilities for the entire range of outcomes that are possible in the particular situations that are under consideration.

Thus, the sampling distribution for the coin-toss situation (N=2, p=.5, q=.5) specifies that any particular randomly selected sample of 2 tossed coins has a 25% chance of including zero heads, a 50% chance of including exactly 1 head, and a 25% chance of including 2 heads. The sampling distribution for the patient-recovery situation (N=2, p=.4, q=.6) specifies that any particular sample of 2 randomly selected patients who have come down with this disease has a 36% chance of ending up with zero recoveries, a 48% chance of ending up with exactly 1 recovery, and a 16% chance of ending up with 2 recoveries. By extension, these two sampling distributions would also specify that any particular randomly selected sample of 2 tossed coins has a 50%+25%=75% chance of including at least 1 head, and that any particular sample of 2 randomly selected patients has a 48%+16%=64% chance of ending up with at least one recovery.

Another way of interpreting a probability sampling distribution would be to say that it describes the manner in which the outcomes of any large number of randomly selected samples will tend to be distributed. Thus, a large number of randomly selected samples of 2 tossed coins would tend to have 25% of the samples including zero heads, 50% including exactly 1 head, and 25% including 2 heads. A large number of randomly selected 2-patient samples would tend to have 36% of the samples with zero recoveries, 48% with exactly 1 recovery, and 16% with 2 recoveries.

The interpretation of central tendency and variability for a sampling distribution is essentially the same as for any other distribution. The central tendency of the distribution describes the average of all the outcomes, and its variability is the measure of the tendency of individual outcomes to be dispersed away from that average. For the particular case of a binomial probability sampling distribution, these parameters can be easily determined according to the formulas given below. [Note the new symbols used here. When referring to an entire population of potential outcomes, the convention is to use (lower-case Greek letter "mu") for the mean and (lower-case Greek letter "sigma") for the variance and standard deviation.]

mean:	$\mu = Np$
variance:	$\sigma^2 = Npq$
standard deviation:	$\sigma =$ sqrt[Npq]

Thus, for the coin-toss example, with N=2, p=.5, and q=.5, you have

mean:	$\mu = 2\times.5 = 1.0$
variance:	$\sigma^2 = 2\times.5\times.5 = 0.5$
standard deviation:	$\sigma =$ sqrt[0.5] $= \pm 0.71$

And for the patient-recovery example, with N=2, p=.4, and q=.6, you have

mean:	$m = 2\times.4 = 0.8$
variance:	$\sigma^2 = 2\times.4\times.6 = 0.48$
standard deviation:	$\sigma =$ sqrt[0.48] $= \pm 0.69$

Although we are illustrating these various points about probability sampling distributions with the specific examples of coin tosses and patient recoveries, please keep in mind as we proceed that these distributions are abstract and general, in the sense that they apply to any probability situation at all that has certain defining properties. Thus, the sampling distribution shown in the left-hand histogram of Figure is not limited to the example of tossing two coins. It applies to any situation at all in which the probability of a certain event is p =.5, the complementary probability of its non-occurrence is q = .5, and there are N = 2 opportunities for it to occur. Similarly, the sampling distribution shown in the right-hand histogram of Figure applies to any situation at all in which the defining properties are p = .4, q = .6, and N = 2.

At any rate, in examining Figure you can hardly fail to notice that the two sampling distributions have rather different shapes. Both are unimodal, but whereas the one for the coin-toss situation (p=.5, q=.5) is symmetrical, the one for the patient-recovery example (p=.4, q=.6) has a pronounced positive skew.

But now see what happens when the size of the sample is increased from N=2 to N=10. The specific probabilities for the outcomes depicted by the histogram columns in Figure have been calculated using the factorial and exponential formula,

$$P_{(k \text{ out of } N)} = \frac{N!}{k!(N-k)!} \times p^k \times q^{N-k} T$$

substituting N=10 along with the appropriate values of p, q, and k.

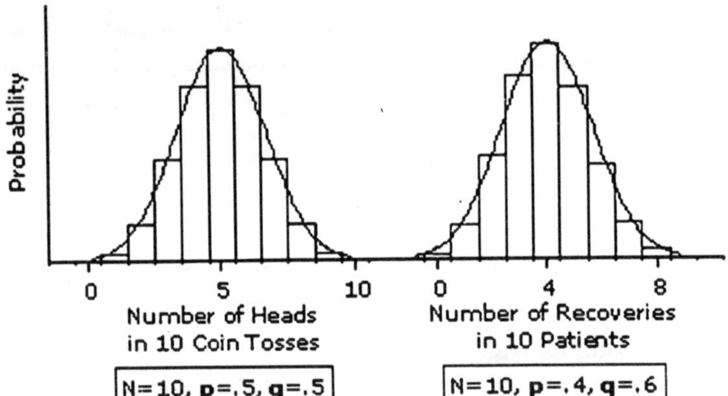

Fig. 6.2. Two Binomial Sampling Distributions for the Case where N=10

Please take a moment to compare the outlines of Figure with those of Figure, for the transformation is quite remarkable. As a result of increasing the size of the sample to N=10, both sampling distributions have grown smoother, and the one for the patient-recovery example has in addition grown much more symmetrical. You will also surely recognize the smooth curve that has been superimposed upon the two sampling distributions in Figure as the by-now familiar outline of the normal distribution.

In Figure we increase the size of the sample to N=20, and there you can see these trends carried even further. The size of N increases, the shape of a binomial sampling distribution comes closer and closer to the shape of the normal distribution. Eventually it comes so close as to be equivalent to the normal distribution, for all practical purposes—and that is the point where all the jagged complexity that we have been describing suddenly gives way to a smooth, streamlined simplicity. The reason for this remarkable convergence is explained in SideTrip.

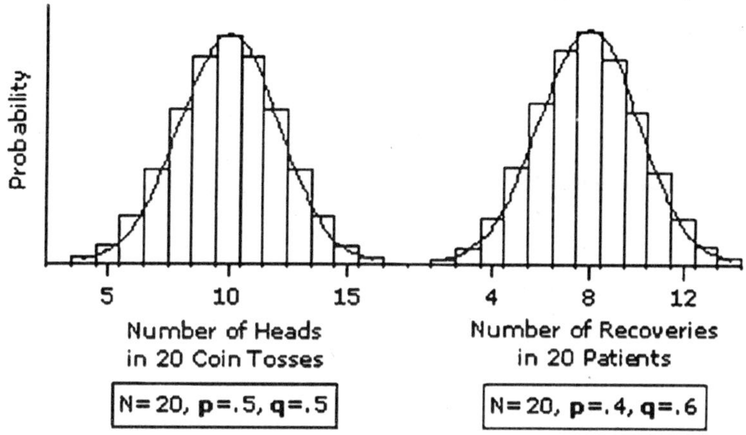

Fig. Two Binomial Sampling Distributions for the Case where N=20

If the elemental probabilities, p and q, are both equal to .5, as in the coin-toss example, a sufficiently close approximation to the normal distribution is reached at the point where N is equal to 10. If p and q are anything other than .5, it is reached at some point beyond N=10. In general, a binomial sampling distribution may be regarded as a sufficiently close approximation to the normal distribution if the products of Np and Nq are both equal to or greater than 5. That is

$Np \geq 5$ and $Nq \geq 5$ Criterion for determining that a binomial sampling distribution is a sufficiently close approximation of the normal distribution.

For elemental probabilities of .4 and .6, as in our patient-recovery example, the point of sufficiently close approximation is reached at N=13; for .3 and .7, it is reached at N=17; for .2 and .8, it is reached at N=25; and so on. The simple way of determining these points is to divide 5 by either p or q, whichever is smaller, rounding the result up to the next higher integer if it comes out with a decimal fraction. Thus, for .4 and .6 you have 5/.4=12.5, which rounds up to N=13; for .3 and .7 it is 5/.3=16.67, which rounds up to N=17; and so on.

The graph of the normal distribution that appears in Figure will help you see where all this is leading. Once you know that a distribution is normal, or at least a close approximation of the normal, you are then in a position to specify the proportion of the distribution that falls to the left or right of any particular point along the horizontal axis, or alternatively the proportion that falls in-between any two points along the horizontal axis.

And specifying the proportion is tantamount to specifying the probability. For the standardized normal distribution, these points along the horizontal axis are marked out in units of z, with each unit of z being equal to 1 unit of standard deviation, σ.

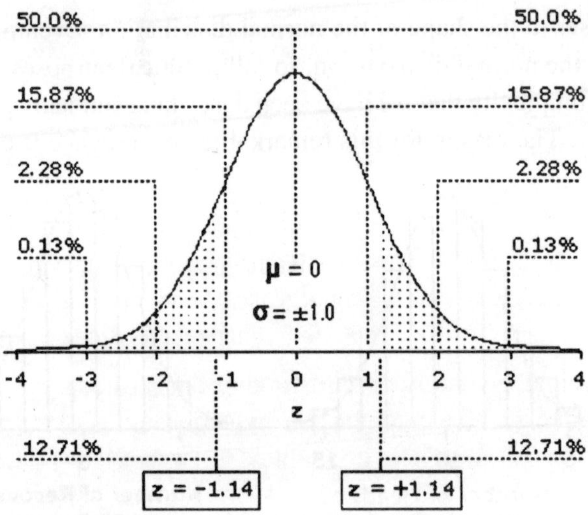

Fig. The Unit Normal Distribution

Most of the percentage values that appear in the body of the graph will already be familiar to you. To the right of z=+1 falls 15.87% of the total distribution; to the right of z=+2 falls 2.28% of the total distribution; and so on.

And since the normal distribution is precisely symmetrical, the same percentages also fall to the left of corresponding negative values of z. For reasons that will be evident in a moment, Figure also shows shaded areas corresponding to the proportions of the normal distribution that fall to the left of z=–1.14 and to the right of z=+1.14–in each case, 12.71%.

Table Proportions of the normal distribution falling to the left of negative values of z or to the right of positive values of z. Bold-face entries pertain to examples discussed in the text.

$\pm z$	*Area Beyond* $\pm z$	$\pm z$	*Area Beyond* $\pm z$
0.00	.5000	1.80	.0359
0.20	.4702	1.90	.0287
0.40	.3446	2.00	.0228
0.60	.2743	2.20	.0139
0.80	.2119	2.40	.0082
1.00	.1587	2.60	.0047
1.14	.1271	2.80	.0026
1.20	.1151	3.00	.0013
1.40	.0808	3.50	.0002
1.60	.0548	4.00	.00003

Now turn to Figure where you will see this same standardized normal distribution superimposed on the binomial sampling distribution that applies to our patient-recovery example with a sample size of 20. The defining properties of this sampling distribution are N=20, p=.4, q=.6. Hence *mean:* m = 20x.4 = 8.0

standard deviation: σ = sqrt[20×.4×.6] = ±2.19

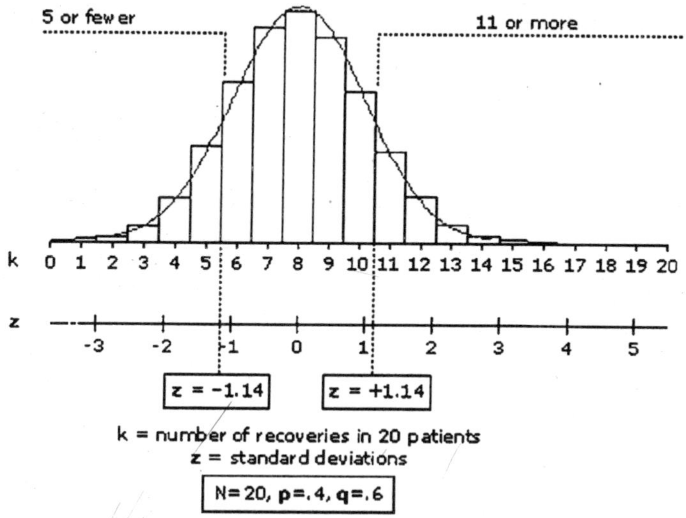

Fig. Binomial Sampling Distribution for N=20, p=.4, q=.6

You will note that there are two scales along the horizontal axis of Figure. The first, labeled k, delineates "number of recoveries in 20 patients"; and the second, labeled z, is a direct translation of this binomial k scale into units of standard deviation, with each unit of standard deviation equal to ±2.19 units of k. Any particular one of the discrete values on the k scale–0, 1, 2, ... , 18, 19, 20–can be directly translated into its corresponding value on the z scale by application of the formula

---

$$z = \frac{(k - \mu) \pm .5}{\sigma}$$

Note 1. Only the positive value of is used in the denominator of this expression.

Note 2. The component ±.5 ('plus or minus .5') in the numerator is a 'correction for continuity', aimed at transforming the zig-zag angularity of the binomial distribution into the smooth curve of the normal distribution. It achieves this effect by reducing the distance between k and the mean by one-half unit. Thus, when k is smaller than the mean you add half a unit (+.5), and when k is larger than the mean you subtract half a unit (–.5).

Thus, for k=5 we have

$$z = \frac{(5-8)+.5}{2.19} = \frac{-2.5}{2.19} = -1.14$$

and for k=11:

$$z = \frac{(11-8)-.5}{2.19} = \frac{+2.5}{2.19} = +1.14.$$

With the calculation of z for a binomial outcome of the general type "*k* or fewer" or "*k* or more," you are then in a position to use the normal distribution for assessing the probability associated with that outcome. You simply treat your calculated value of z as though it belonged to the z scale of the normal distribution–and read off the corresponding probability value from the table of the normal distribution. The two particular values of z that we have just calculated are $z = -1.14$ for "5 or fewer" recoveries and z=+1.14 for "11 or more" recoveries. As we saw a moment ago in Figure and Table, the proportions of the normal distribution falling to the left of $z = -1.14$ and to the right of z=+1.14 are in each case equal to .1271. Hence, on the basis of the normal distribution we would judge that with a random sample of 20 patients there is a 12.71% chance that as few as 5 will spontaneously recover, and an equal 12.71% chance that as many as 11 will spontaneously recover. If you work out the exact binomial probabilities for these two outcomes using the factorial and exponential formula given earlier, you will find that they come out to 12.56% for "5 or fewer" and 12.75% for "11 or more." The 12.71% probability values arrived at via the normal distribution do not hit these targets precisely, but for all practical purposes they are close enough. The difference between .1271 and .1256 is only .0015; between .1271 and .1275 it is only .0004.

To appreciate just how useful and labor-saving this streamlining can be, return for a moment to the second experiment of our medical researchers, the one in which there were 430 recoveries out of 1,000 patients. The question is: In a random sample of 1,000 patients, how likely is it that as many as 430 (k>430) would recover by mere chance, if the plant extract had no effectiveness whatsoever? First try to imagine how much it would cost you in time and patience to perform the requisite 571 exact binomial calculations according to the factorial and exponential formula, assuming you had access to a calculator powerful enough to perform huge operations such as 1,000! and $.4^{430}$x$.6^{570}$–and then reflect on how very much easier it is to reach the same goal on the basis of the normal distribution.

First establish the defining properties of the relevant binomial sampling distribution, making sure that the products of Np and Nq are both equal to or greater than 5:

N=1,000, p=.4, q=.6. Hence

mean: $\mu = 1{,}000 \times .4 = 400$

standard deviation: $\sigma =$ sqrt[1,000 × .4 ×.6] = ±15.49

Then perform the simple calculation that translates the details of the binomial situation into the language of the normal distribution:

$$z = \frac{(430 - 400) - .5}{15.49} = \frac{+29.5}{15.49} = +1.90.$$

And then refer your calculated value of z to the table of the normal distribution. For the moment you need only refer to the abbreviated version shown in Table 6.1, where you will find that the proportion of the normal distribution falling to the right of z=+1.9 is equal to .0287.

The meaning of this proportion is illustrated in Figure. Of all the possible spontaneous-recovery outcomes that could have occurred within this sample–zero recoveries, 1 recovery,

2 recoveries, and so on–only 2.87% would include as many as 430 recoveries. The mere-chance probability of the observed result is therefore P=.0287, which of course clears the conventional cut-off point for statistical significance (P<.05) by a considerable margin.

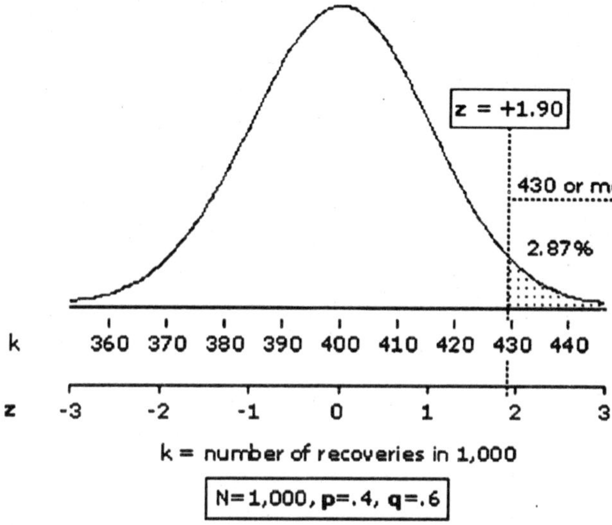

Fig. Location of z = +1.90 within the Unit Normal Distribution

Moving beyond the rather rigid mechanics of the .05 criterion of statistical significance, you can also think of this P=.0287 probability value in terms of confidence. There is only about a 3% likelihood that the observed outcome would have occurred by mere chance coincidence. Hence, you can be about 97% confident that it is the result of something other than mere chance coincidence–presumably some active ingredient of the plant extract. The researchers can now go on to the costly task of isolating and refining that ingredient, with a high degree of confidence that they are on the right track.

The streamlining of binomial probabilities was only the first historical step. In the course of time there came the discovery and detailed analysis of several other families of sampling distributions, each providing a relatively simple streamlined method for answering inferential-statistical questions of the general type: How likely is it that such-and-such might occur by mere chance coincidence? In the remaining chapters of this text we will be making extensive use of some of these families of sampling distributions, in particular, several that are variations on the normal distribution,.plus some others that are known as t-distributions, F-distributions, and chi-square distributions.

MEASURE AND PROBABILITY THEORY

The evolution of probability theory was based more on intuition rather than mathematical axioms during its early development. In 1933, A. N. Kolmogorov provided an axiomatic basis for probability theory and it is now the universally accepted model. There are certain 'non commutative' versions that have their origins in quantum mechanics, see for instance K. R. Parthasarathy, that are generalizations of the Kolmogorov Model. We shall however use exclusively Kolmogorov's framework.

The basic intuition in probability theory is the notion of randomness. There are experiments whose results are not predicatable and can be determined only after performing it and then observing the outcome. The simplest familiar examples are, the tossing of a fair coin, or the throwing of a balanced die. In the first experiment the result could be either a head or a tail and the throwing of a die could result in a score of any integer from 1 through 6. These are experiments with only a finite number of alternate outcomes. It is not difficult to imagine experiments that have countably or even uncountably many alternatives as possible outcomes.

Abstractly then, there is a space Ω of all possible outcomes and each individual outcome is represented as a point ω in that space Ω. Subsets of Ω are called events and each of them corresponds to a collection of outcomes. If the outcome ω is in the subset A, then the event A is said to have occurred. For example in the case of a die the set $A = \{1, 3, 5\} \subset \Omega$ corresponds to the event 'an odd number shows up'. With this terminology it is clear that union of sets corresponds to 'or', intersection to 'and', and complementation to 'negation'.

One would expect that probabilities should be associated with each outcome and there should be a 'Probability Function' $f(\omega)$ which is the probabilty that ω occurs. In the case of coin tossing we may expect $\Omega = \{H, T\}$ and

$$f(T) = f(H) = \frac{1}{2}.$$

Or in the case of a die

$$f(1) = f(2) = ... = f(6) = \frac{1}{6}.$$

Since 'Probability' is normalized so that certainty corresponds to a Probability of 1, one expects

$$\sum_{\omega \in \Omega} f(\omega) = 1.$$

If Ω is uncountable this is a mess. There is no reasonable way of adding up an uncountable set of numbers each of which is 0. This suggests that it may not be possible to start with probabilities associated with individual outcomes and build a meaningful theory. The next best thing is to start with the notion that probabilities are already defined for events. In such a case, $P(A)$ is defined for a class B of subsets $A \subset \Omega$. The question that arises naturally is what should B be and what properties should $P(\cdot)$ defined on B have? It is natural to demand that the class B of sets for which probabilities are to be defined satisfy the following properties:

The whole space Ω and the empty set Φ are in B. For any two sets A and B in B, the sets $A \cup B$ and $A \cap B$ are again in B. If $A \in B$, then the complement A^c is again in B. Any class of sets satisfying these properties is called a *field*.

Definition: A 'probability' or more precisely 'a finitely additive probability measure' is a nonnegative set function $P(\cdot)$ defined for sets $A \subset B$ that satisfies the following properties:

$$P(A) \geq 0 \text{ for all } A \in B,$$

$$P(\Omega) = 1 \text{ and } P(\Phi) = 0.$$

If $A \in B$ and $B \in B$ are disjoint then

$$P(A \cup B) = P(A) + P(B).$$

In particular

$$P(A^c) = 1 - P(A)$$

for all $A \in B$.

A condition which is some what more technical, but important from a mathematical viewpoint is that of countable additivity. The class B, in addition to being a field is assumed to be closed under countable union (or equivalently, countable intersection); i.e. if $A_n \in B$ for every n, then $A = \bigcup_n A_n \in B$. Such a class is called a σ-field. The 'probability' itself is presumed to be defined on a σ-field B.

Definition: A set function P defined on a σ-field is called a 'countably additive probability measure' if in addition to satsfying equations, it satisfies the following countable additivity property: for any sequence of pairwise disjoint sets A_n with $A = \bigcup_n A_n$

$$P(A) = \sum_n P(A_n).$$

Definition: A probability measure P on a field F is said to be countably additive on F, if for any sequence $A_n \in F$ with $A_n \downarrow \Phi$, we have $P(A_n) \downarrow 0$.

Definition: The σ-field in the above exercise is called the σ-field generated by F.

CONSTRUCTION OF MEASURES

The following theorem is important for the construction of countably additive probability measures. A detailed proof of this theorem, as well as other results on measure and integration, can be found or in any one of the many texts on real variables. In an effort to be complete we will sketch the standard proof.

Caratheodory Extension Theorem

Theorem: Any countably additive probabilty measure P on a field F extends uniquely as a countably additive probability measure to the σ-field B generated by F.

Proof: The proof proceeds along the following steps:

Step 1. Define an object P^* called the *outer measure* for all sets A.

$$P^*(A) = \inf_{\cup_j A_j \supset A} \sum_j P(A_j)$$

where the infimum is taken over all countable collections $\{A_j\}$ of sets from F that cover A. Without loss of generality we can assume that $\{A_j\}$ are disjoint. (Replace A_j by $\left(\bigcap_{i=1}^{j-1} A_i^c\right) \cap A_j$).

Step 2. Show that P^* has the following properties:

1. P^* is countably sub-additive, i.e.

$$P^*\left(\bigcup_j A_j\right) \leq \sum_j P^* A_j$$

2. For $A \in F$, $P^*(A) \leq P(A)$. (Trivial)

3. For $A \in F$, $P^*(A) \geq P(A)$. (Need to use the countable additivity of P on F)

Step 3. Define a set E to be measurable if

$$P^*(A) \geq P^*(A \cap E) + P^*(A \cap E^c)$$

holds for all sets A, and establish the following properties for the class M of measurable sets. The class of measurable sets M is a σ-field and P^* is a countably additive measure on it.

Step 4. Finally show that $M \supset F$. This implies that $M \supset B$ and P^* is an extension of P from F to B.

Uniqueness is quite simple. Let P_1 and P_2 be two countably additive probability measures on a σ-field B that agree on a field $F \subset B$. Let us define $A = \{A: P_1(A) = P_2(A)\}$. Then A is a monotone class i.e., if $A_n \in A$ is increasing (decreasing) then $\bigcup_n A_n \left(\bigcap_n A_n\right) \in A$. Uniqueness will follow from the following fact left as an excercise.

Exercise: The smallest monotone class generated by a field is the same as the σ-field generated by the field.

It now follows that A must contain the σ-field generated by F and that proves uniqueness.

The extension Theorem does not quite solve the problem of constructing countably additive probability measures. It reduces it to constructing them on fields. The following theorem is important in the theory of Lebesgue integrals and is very useful for the construction of countably additive probability measures on the real line. The proof will again be only sketched. The natural σ-field on which to define a probability measure on the line is the Borel σ-field. This is defined as the smallest σ-field containing all intervals and includes in particular all open sets.

Let us consider the class of subsets of the real numbers, $I = \{I_{a,b}: -\infty \leq a < b \leq \infty\}$ where $I_{a,b} = \{x: a < x \leq b\}$ if $b < \infty$, and $I_{a,\infty} = \{x: a < x < \infty\}$. In other words I is the collection of intervals that are left-open and right-closed. The class of sets that are finite disjoint unions of members of I is a field F, if the empty set is added to the class. If we are given a function $F(x)$ on the real line which is nondecreasing, continuous from the right and satisfies

$$\lim_{x \to -\infty} F(x) = 0 \text{ and } \lim_{x \to \infty} F(x) = 1,$$

we can define a finitely additive probability measure P by first defining

$$P(I_{a,b}) = F(b) - F(a)$$

for intervals and then extending it to F by defining it as the sum for disjoint unions from I. Let us note that the Borel σ-field B on the real line is the σ-field generated by F.

Lebesgue

Theorem: P is countably additive on F if and only if $F(x)$ is a right continuous function of x. Therefore for each right continuous nondecreasing function $F(x)$ with $F(-\infty) = 0$ and $F(\infty) = 1$ there is a unique probability measure P on the Borel subsets of the line, such that $F(x) = P(I_{-\infty,x})$. Conversely every countably additive probability measure P on the Borel subsets of the line comes from some F. The correspondence between P and F is one-to-one.

Proof: The only difficult part is to establish the countable additivity of P on F from the right continuity of $F(\cdot)$. Let $A_j \in F$ and $A_j \downarrow \Phi$, the empty set.

Let us assume, $P(A_j) \geq \delta > 0$, for all j and then establish a contradiction.

Step 1. We take a large interval $[-\ell, \ell]$ and replace A_j by $Bj = A_j \cap [-\ell, \ell]$. Since $|P(A_j) - P(B_j)| \leq 1 - F(\ell) + F(-\ell)$, we can make the choice of ℓ large enough that $P(B_j) \geq \frac{\delta}{2}$. In other words we can assumes without loss of generality that $P(A_j) \geq \frac{\delta}{2}$ and $A_j \subset [-\ell, \ell]$ for some fixed $\ell < \infty$.

Step 2. If

$$A_j = \bigcup_{i=1}^{k_j} I_{a_{j,i}, b_{j,i}}$$

use the right continuity of F to replace A_j by B_j which is again a union of left open right closed intervals with the same right end points, but with left end points moved ever so slightly to the right. Achieve this in such a way that

$$P(A_j - B_j) \leq \frac{\delta}{10.2^j}$$

for all j.

Step 3. Define C_j to be the closure of B_j, obtained by adding to it the left end points of the intervals making up B_j. Let $E_j = \bigcap_{i=1}^{j} B_i$ and $D_j = \bigcap_{i=1}^{j} C_i$. Then, (i) the sequence D_j of sets is decreasing, (ii) each D_j is a closed bounded set, (iii) since $A_j \supset D_j$ and $A_j \downarrow \Phi$, it follows that $D_j \downarrow \Phi$. Because $D_j \supset E_j$ and $P(E_j) \geq \frac{\delta}{2} - \sum_j P(A_j - B_j) \geq \frac{4}{10\delta}$, each D_j is nonempty and this violates the finite intersection property that every decreasing sequence of bounded nonempty closed sets on the real line has a nonempty intersection, i.e. has at least one common point.

The rest of the proof is left as an exercise. The function F is called the distribution function corresponding to the probability measure P.

Example: Suppose $x_1, x_2, \ldots, x_n \ldots$ is a sequence of points and we have probabilities p_n at these points then for the discrete measure

$$P(A) = \sum_{n: x_n \in A} p_n$$

we have the distribution function

$$F(x) = \sum_{n: x_n \leq x} p_n$$

that only increases by jumps, the jump at x_n being p_n. The points $\{x_n\}$ themselves can be discrete like integers or dense like the rationals.

Example: If $f(x)$ is a nonnegative integrable function with integral 1, i.e. $\int_{-\infty}^{\infty} f(y)dy = 1$ then $F(x) = \int_{-\infty}^{x} f(y)dy$ is a distribution function which is continuous. In this case f is the density of the measure P and can be calculated as $f(x) = F'(x)$.

There are (messy) examples of F that are continuous, but do not come from any density. More on this later.

Exercise: Let us try to construct the Lebesgue measure on the rationals $Q \subset [0, 1]$. We would like to have

$$P[I_{a,b}] = b - a$$

for all rational $0 \leq a \leq b \leq 1$. Show that it is impossible by showing that $P[\{q\}] = 0$ for the set $\{q\}$ containing the single rational q while $P[Q] = P\left[\bigcup_{q \in Q} \{q\}\right] = 1$. Where does the earlier proof break down?

Once we have a countably additive probability measure P on a space (Ω, Σ), we will call the triple (Ω, Σ, P) a probabilty space.

INTEGRATION

An important notion is that of a random variable or a measurable function.

Definition: A random variable or measurable function is map $f: \Omega \to R$, i.e. a real valued function $f(\omega)$ on Ω such that for every Borel set $B \subset R$, $f^{-1}(B) = \{\omega: f(\omega) \in B\}$ is a measurable subset of Ω, i.e $f^{-1}(B) \in \Sigma$.

A function that is measurable and satisfies $|f(\omega)| \leq M$ all $\omega \in \Omega$ for some finite M is called a bounded measurable function.

The following statements are the essential steps in developing an integration theory. Details can be found in any book on real variables.

1. If $A \in \Sigma$, the indicator function A, defined as

$$1_A(\omega) = \begin{cases} 1 & if \omega \in A \\ 0 & if \omega \notin A \end{cases}$$

   is bounded and measurable.

2. Sums, products, limits, compositions and reasonable elementary operations like min and max performed on measurable functions lead to measurable functions.

3. If $\{A_j: 1 \leq j \leq n\}$ is a finite disjoint partition of into measurable sets, the function $f(\omega) = \sum_j c_j 1_{A_j}(\omega)$ is a measurable function and is referred to as a 'simple' function.

4. Any bounded measurable function f is a uniform limit of simple functions. To see this, if f is bounded by M, divide $[-M, M]$ into n subintervals I_j of length $\frac{2M}{n}$ with midpoints c_j. Let

$$A_j = f^{-1}(I_j) = \{\omega: f(\omega) \in I_j\}$$

   and

$$f_n = \sum_{j=1}^{n} c_j 1_{A_j}.$$

   Clearly f_n is simple, $\sup_\omega |f_n(\omega) - f(\omega)| — \frac{M}{n}$, and we are done.

5. For simple functions $f = \sum c_j 1_{A_j}$ the integral $\int f(\omega) dP$ is defined to be $\sum_j c_j P(A_j)$. It enjoys the following properties:

   (a) If f and g are simple, so is any linear combination $af + bg$ for real constants a and b and

$$\int (af + bg) dP = a \int f dP + b \int g dP.$$

   (b) If f is simple so is $|f|$ and $\left| \int f dP \right| \leq \int |f| dP \leq \sup_\omega |f(\omega)|$

6. If f_n is a sequence of simple functions converging to f uniformly, then $a_n = \int f_n dP$ is a Cauchy sequence of real numbers and therefore has a limit a as $n \to \infty$. The integral $\int f dP$ of f is defined to be this limit a. One can verify that a depends only on f and not on the sequence f_n chosen to approximate f.

7. Now the integral is defined for all bounded measurable functions and enjoys the following properties.
   (a) If f and g are bounded measurable functions and a, b are real constants then the linear combination $af+bg$ is again a bounded measurable function, and
   $$\int (af+bg)dP = a\int fdP + b\int gdP.$$
   (b) If f is a bounded measurable function so is $|f|$ and $\left|\int fdP\right| \leq \int |f|\, dP \leq$ $\sup_\omega |f(\omega)|$.
   (c) In fact a slightly stronger inequality is true. For any bounded measurable f,
   $$\int |f|\, dP \leq P(\{\omega : |f(\omega)| > 0\}) \sup_\omega |f(\omega)|$$
   (d) If f is a bounded measurable function and A is a measurable set one defines
   $$\int_A f(\omega)dP = \int 1_A(\omega) f(\omega)dP$$
   and we can write for any measurable set A,
   $$\int fdP = \int_A fdP + \int_{A^c} fdP$$

In addition to uniform convergence there are other weaker notions of convergence.

Definition: A sequence f_n functions is said to converge to a function f everywhere or pointwise if

$$\lim_{n\to\infty} f_n(\omega) = f(\omega)$$

for every $\omega \in \Omega$.

In dealing with sequences of functions on a space that has a measure defined on it, often it does not matter if the sequence fails to converge on a set of points that is insignificant. For example if we are dealing with the Lebesgue measure on the interval [0, 1] and $f_n(x) = x_n$ then $f_n(x) \to 0$ for all x except $x = 1$. A single point, being an interval of length 0 should be insignificant for the Lebesgue measure.

Definition: A sequence f_n of measurable functions is said to converge to a measurable function f almost everywhere or almost surely (usually abbreviated as a.e.) if there exists a measurable set N with $P(N) = 0$ such that

$$\lim_{n\to\infty} f_n(\omega) = f(\omega)$$

for every $\omega \in N^c$.

Note that almost everywhere convergence is always relative to a probability measure.

Another notion of convergence is the following:

Definition: A sequence f_n of measurable functions is said to converge to a measurable function f 'in measure' or 'in probability' if

$$\lim_{n\to\infty} P[\omega : | f_n(\omega) - f(\omega) | \geq \varepsilon] = 0$$

for every $\varepsilon > 0$.

Let us examine these notions in the context of indicator functions of sets $f_n(\omega) = 1_{A_n}(\omega)$. As soon as $A \neq B$, $\sup_\omega | 1_A(\omega) - 1_B(\omega) | = 1$, so that uniform convergence never really takes place. On the other hand one can verify that $1_{A_n}(\omega) \to 1_A(\omega)$ for every ω if and only if the two sets

$$A_n = \bigcap_n \bigcup_{m\geq n} A_m$$

and

$$\liminf_n A_n = \bigcup_n \bigcap_{m\geq n} A_m$$

both coincide with A. Finally $1_{A_n}(\omega) \to 1_A(\omega)$ in measure if and only if

$$\lim_{n\to\infty} P(A_n \Delta A) = 0$$

where for any two sets A and B the symmetric difference $A\Delta B$ is defined as $A\Delta B = (A \cap B^c) \cup (A^c \cap B) = A \cup B \cap (A \cap B)^c$. It is the set of points that belong to either set but not to both. For instance $1_{A_n} \to \omega$ in measure if and only if $P(A_n) \to 0$.

Lemma: If $f_n \to f$ almost everywhere then $f_n \to f$ in measure.

Proof: $f_n \to f$ outside N is equivalent to

$$\bigcap_n \bigcup_{m\geq n} [\omega : | f_m(\omega) - f(\omega) | \geq \varepsilon] \subset N$$

for every $\varepsilon > 0$. In particular by countable additivity

$$P[\omega : | f_n(\omega) - f(\omega) | \geq \varepsilon] \leq P\left[\bigcup_{m\geq n} [\omega : | f_m(\omega) - f(\omega) | \geq \varepsilon\right] \to 0$$

as $n \to \infty$ and we are done.

Bounded Convergence Theorem

Theorem: If the sequence $\{f_n\}$ of measurable functions is uniformly bounded and if $f_n \to f$ in measure as $n \to \infty$, then $\lim_{n\to\infty} \int f_n dP = \int f dP$.

Proof: Since

$$\left| \int f_n dP - \int f dP \right| = \left| \int (f_n - f) dP \right| \leq \int | f_n - f | dP$$

we need only prove that if $f_n \to 0$ in measure and $|f_n| \leq M$ then $\int |f_n| dP \to 0$. To see this

$$\int |f_n| dP = \int_{|f_n| \leq \varepsilon} |f_n| dP + \int_{|f_n| > \varepsilon} |f_n| dP \leq \varepsilon + MP[\omega : |f_n(\omega) > \varepsilon]$$

and taking limits

$$\limsup_{n \to \infty} \int |f_n| dP \leq \varepsilon$$

and since $\varepsilon > 0$ is arbitrary we are done.

The bounded convergence theorem is the essence of countable additivity. Let us look at the example of $f_n(x) = x^n$ on $0 \leq x \leq 1$ with Lebesgue measure. Clearly $f_n(x) \to 0$ a.e. and therefore in measure. While the convergence is not uniform, $0 \leq x^n \leq 1$ for all n and x and so the bounded convergence theorem applies. In fact

$$\int_0^1 x^n dx = \frac{1}{n+1} \to 0.$$

However if we replace x^n by nx^n, $f_n(x)$ still goes to 0 a.e., but the sequence is no longer uniformly bounded and the integral does not go to 0.

We now proceed to define integrals of nonnegative measurable functions.

Definition: If f is a nonnegative measurable function we define

$$\int f dP = \left\{\sup \int g dP : g \text{ bounded}, 0 \leq g \leq f\right\}.$$

Fatou's Lemma

Theorem: If for each $n \geq 1, f_n \geq 0$ is measurable and $f_n \to f$ in measure as $n \to \infty$ then

$$\int f dP \leq \liminf_{n \to \infty} \int f_n dP.$$

Proof: Suppose g is bounded and satisfies $0 \leq g \leq f$. Then the sequence $h_n = f_n \wedge g$ is uniformly bounded and

$$h_n \to h = f \wedge g = g.$$

Therefore, by the bounded convergence theorem,

$$\int g dP = \liminf_{n \to \infty} \int h_n dP$$

Since $\int h_n dP \leq \int f_n dP$ for every n it follows that

$$\int g dP \leq \liminf_{n \to \infty} \int f_n dP.$$

As g satisfying $0 \leq g \leq f$ is arbitrary and we are done.

Monotone Convergence Theorem

Corollary: If for a sequence $\{f_n\}$ of nonnegative functions, we have $f_n \uparrow f$ monotonically then

$$\int f_n dP \to \int f dP \text{ as } n \to \infty.$$

Proof: Obviously $\int f_n dP \leq \int f dP$ and the other half follows from Fatou's lemma.

Now we try to define integrals of arbitrary measurable functions. A nonnegative measurable function is said to be *integrable* if $\int f dP < \infty$. A measurable function f is said to be integrable if $|f|$ is integrable and we define $\int f dP = \int f^+ dP - \int f^- dP$ where $f^+ = f \vee 0$ and $f^- = -f \wedge 0$ are the positive and negative parts of f. The integral has the following properties.

1. It is linear. If f and g are integrable so is $af + bg$ for any two real constants and $\int (af + bg) dP = a\int f dP + b\int g dP$.
2. $\left|\int f dP\right| \leq \int |f| dP$ for every integrable f.
3. If $f = 0$ except on a set N of measure 0, then f is integrable and $\int f dP = 0$. In particular if $f = g$ almost everywhere then $\int f dP = \int g dP$.

Theorem: If $\phi(x)$ is a convex function of x and $f(\omega)$ and $\phi(f(\omega))$ are integrable then

$$\int \phi(f(\omega)) dP \geq \phi\left(\int f(\omega) dP\right)$$

Proof: We have seen the inequlity already for $\phi(x) = |x|$. The proof is quite simple. We note that any convex function can be represented as the supremum of a collection of affine linear functions.

$$\phi(x) = \sup_{(a,b)\in E} \{ax + b\}.$$

It is clear that if $(a, b) \in E$, then $a f(\omega) + b \leq (f(\omega))$ and on integration this yields $am + b \leq E[(f(\omega))]$ where $m = E[f(\omega)]$. Since this is true for every $(a, b) \in E$, in view of the represenattion, our theorem follows.

The Dominated Convergence Theorem

Theorem: If for some sequence $\{f_n\}$ of measurable functions we have $f_n \to f$ in measure and $|f_n(\omega)| \leq g(\omega)$ for all n and ω for some integrable function g, then $\int f_n dP \to \int f dP$ as $n \to \infty$.

Proof: $g + f_n$ and $g - f_n$ are nonnegative and converge in measure to $g + f$ and $g - f$ respectively. By Fatou's lemma

$$\liminf_{n\to\infty} \int (g+f_n)dP \le \int (g+f)dP$$

Since $\int gdP$ is finite we can subtract it from both sides and get

$$\liminf_{n\to\infty} \int f_n dP \ge \int fdP .$$

Working the same way with $g - f_n$ yields

$$\liminf_{n\to\infty} \int f_n dP \le \int fdP$$

and we are done.

Exercise: Take the unit interval with the Lebesgue measure and define $f_n(x) = n^{\alpha} 1_{\left[0,\frac{1}{n}\right]}(x).$ Clearly $f_n(x) \to 0$ for $x \ne 0$. On the other hand $\int f_n(x)dx = n^{\alpha-1} \to 0$ if and only if $\alpha < 1$. What is $g(x) = \sup_n f_n(x)$ and when is g integrable ?

If $h(\omega) = f(\omega) + ig(\omega)$ is a complex valued measurable function with real and imaginary parts $f(\omega)$ and $g(\omega)$ that are integrable we define

$$\int h(\omega)dP = \int f(\omega)dP + i\int g(\omega)dP$$

TRANSFORMATIONS

A measurable space (Ω, B) is a set Ω together with a σ-field B of subsets of Ω.

Definition: Given two measurable spaces (Ω_1, B_1) and (Ω_2, B_2), a mapping or a transformation from $T: \Omega_1 \to 2$, i.e. a function $\omega_2 = T(\omega_1)$ that assigns for each point $\omega_1 \in \Omega_1$ a point $\omega_2 = T(\omega_1) \in \Omega_2$, is said to be measurable if for every measurable set $A \in B_2$, the inverse image

$$T^{-1}(A) = \{\omega_1: T(\omega_1) \in A\} \in B_1.$$

Theorem: If $f: \Omega_2 \to R$ is a real valued measurable function on Ω_2, then $g(\omega_1) = f(T(\omega_1))$ is a measurable real valued function on (Ω_1, B_1). Moreover g is integrable with respect to P if and only if f is integrable with respect to Q, and

$$\int_{\Omega_2} f(\omega_2)dQ = \int_{\Omega_1} g(\omega_1)dP$$

Proof: If $f(\omega_2) = 1_A(\omega_2)$ is the indicator function of a set $A \in B_2$, the claim in equation is in fact the definition of measurability and the induced measure. We see, by linearity, that the claim extends easily from indicator functions to simple functions. By uniform limits, the claim can now be extended to bounded measurable functions. Monotone limits then extend it to nonnegative functions. By considering the positive and negative parts separately we are done.

A measurable trnasformation is just a generalization of the conçept of a random variable. We can either think of a random variable as special case of a measurable transformation where the target space is the real line or think of a measurable transformation as a random variable with values in an arbitrary target space.

The induced measure $Q = PT^{-1}$ is called the *distribution* of the random variable F under P. In particular, if T takes real values, Q is a probability distribution on R.

PRODUCT SPACES

Given two sets Ω_1 and Ω_2 the Cartesian product $\Omega = \Omega_1 X \Omega_2$ is the set of pairs (ω_1, ω_2) with $\omega_1 \in \Omega_1$ and $\omega_2 \in \Omega_2$. If Ω_1 and Ω_2 come with σ-fields B_1 and B_2 respectively, we can define a natural σ-field B on Ω as the σ-field generated by sets (measurable rectangles) of the form $A_1 X A_2$ with $A_1 \in B_1$ and $A_2 \in B_2$. This σ-field will be called the product σ-field.

Definition: The product -field B is the σ-field generaated by the field F.

Given two probability measures P_1 and P_2 on (Ω_1, B_1) and (Ω_2, B_2) respectively we try to define on the product space (Ω, B) a probability measure P by defining for a measurable rectangle $A = A_1 X A_2$

$$P(A_1 X A_2) = P_1(A_1) X P_2(A_2)$$

and extending it to the field F of finite disjoint unions of measurable rectangles as the obvious sum.

Lemma: The measure P is countably additive on the field F.

Proof: For any set $E \in F$ let us define the section E_{ω_2} as

$$E_{\omega_2} = \{\omega_1 : (\omega_1, \omega_2) \in E\}.$$

Then $P_1(E_{\omega_2})$ is a measurable function of ω_2 (is in fact a simple function) and

$$P(E) = \int_{\Omega_2} P_1(E_{\omega_2}) dP_2.$$

Now let $E_n \in F \downarrow \Phi$, the empty set. Then it is easy to verify that E_{n,ω_2} defined by

$$E_{n,\omega_2} = \{\omega_1 : (\omega_1, \omega_2) \in E_n\}$$

satisfies $E_{n,\omega_2} \downarrow \Phi$ for each $\omega_2 \in \Omega_2$. From the countable additivity of P_1 we conclude that $P_1(E_{n,\omega_2}) \to 0$ for each $\omega_2 \in \Omega_2$ and since, $0 \leq P_1(E_{n,\omega_2}) \leq 1$ for $n \geq 1$, it follows from equation and the bounded convergence theorem that

$$P(E_n) = \int_{\Omega_2} P_1(E_{n,\omega_2}) dP_2 \to 0$$

establishing the countable additivity of P on F.

By an application of the Caratheodory extension theorem we conclude that P extends uniquely as a countably additive measure to the σ-field B (product σ-field) generated by F. We will call this the Product Measure P.

Corollary: For any $A \in B$ if we denote by A_{ω_1} and A_{ω_2} the respective sections

$$A_{\omega_1} = \{\omega_2: (\omega_1, \omega_2) \in A\}$$

and

$$A_{\omega_2} = \{\omega_1: (\omega_1, \omega_2) \in A\}$$

then the functions $P_1(A_{\omega_2})$ and $P_2(A_{\omega_1})$ are measurable and

$$P(A) = \int P_1(A_{\omega_2})dP_2 = \int P_2(A_{\omega_1})\, dP_1.$$

In particular for a measurable set A, $P(A) = 0$ if and only if for almost all ω_1 with respect to P_1, the sections A_{ω_1} have measure 0 or equivalently for almost all ω_2 with respect to P_2, the sections A_{ω_2} have measure 0.

Proof: The assertion is clearly valid if A is rectangle of the form $A_1 X A_2$ with $A_1 \in B_1$ and $A_2 \in B_2$. If $A \in F$, then it is a finite disjoint union of such rectangles and the assertion is extended to such a set by simple addition. Clearly, by the monotone convergence theorem, the class of sets for which the assertion is valid is a monotone class and since it contains the field F it also contains the σ-field B generated by the field F.

Warning. It is possible that a set A may not be *measurable* with respect to the product σ-field, but nevertheless the sections A_{ω_1} and A_{ω_2} are all measurable, $P_2(A_{\omega_1})$ and $P_1(A_{\omega_2})$ are measurable functions, but

$$\int P_1(A_{\omega_2})dP_2 \neq \int P_2(A_{\omega_1})\, dP_1.$$

In fact there is a rather nasty example where $P_1(A_{\omega_2})$ is identically 1 whereas $P_2(A_{\omega_1})$ is identically 0. The next result concerns the equality of the double integral, (i.e. the integral with respect to the product measure) and the repeated integrals in any order.

Fubini's Theorem

Theorem: Let $f(\omega) = f(\omega_1, \omega_2)$ be a measurable function of ω on (Ω, B). Then f can be considered as a function of ω_2 for each fixed ω_1 or the other way around. The functions $g_{\omega_1}(\cdot)$ and $h_{\omega_2}(\cdot)$ defined respectively on Ω_2 and Ω_1 by

$$g_{\omega_1}(\omega_2) = h_{\omega_2}(\omega_1) = f(\omega_1, \omega_2)$$

are measurable for each ω_1 and ω_2. If f is integrable then the functions $g_{\omega_1}(\omega_2)$ and $h_{\omega_2}(\omega_1)$ are integrable for almost all ω_1 and ω_2 respectively. Their integrals

$$G(\omega_1) = \int_{\Omega_2} g_{\omega_1}(\omega_2)dP_2$$

and

$$H(\omega_1) = \int_{\Omega_1} h_{\omega_2}(\omega_2)dP_1$$

are measurable, finite almost everywhere and integrable with repect to P_1 and P_2 respectively. Finally

$$\int f(\omega_1, \omega_2)dP = \int g(\omega_1)dP_1 = \int H(\omega_2)dP_2$$

Conversely for a nonnegative measurable function f if either G are H, which are always measurable, has a finite integral so does the other and f is integrable with its integral being equal to either of the repeated integrals, namely integrals of G and H.

Proof: The proof follows the standard pattern. It is a restatement of the earlier corollary if f is the indicator function of a measurable set A. By linearity it is true for simple functions and by passing to uniform limits, it is true for bounded measurable functions f. By monotone limits it is true for nonnegative functions and finally by taking the positive and negative parts seperately it is true for any arbitrary integrable function f.

Warning. The following could happen. f is a measurable function that takes both positive and negative values that is not integrable. Both the repeated integrals exist and are unequal. The example is not hard.

DISTRIBUTIONS AND EXPECTATIONS

Let us recall that a triplet (Ω, B, P) is a Probability Space, if Ω is a set, B is a σ-field of subsets of Ω and P is a (countably additive) probability measure on B. A random variable X is a real valued measurable function on (Ω, B). Given such a function X it induces a probability distribution on the Borel subsets of the line $\alpha = PX^{-1}$. The distribution function $F(x)$ corresponding to is obviously

$$F(x) = ((-\infty, x]) = P[\omega: X(\omega) \leq x].$$

The measure α is called the distibution of X and $F(x)$ is called the distribution function of X. If g is a measurable function of the real variable x, then $Y(\omega) = g(X(\omega))$ is again a random variable and its distribution $\beta = PY^{-1}$ can be obtained as $\beta = \alpha g^{-1}$ from α. The *Expectation* or *mean* of a random variable is defined if it is integrable and

$$E[X] = E^P[X] = \int X(\omega)dP.$$

By the change of variables formula it can be obtained directly from α as

$$E[X] = \int x d\alpha.$$

Here we are taking advantage of the fact that on the real line x is a very special real valued function. The value of the integral in this context is referred to as the expectation or mean of α. Of course it exists if and only if

$$\int |x|\, d\alpha < \infty$$

and

$$\left|\int x d\alpha\right| \leq \int |x|\, d\alpha.$$

Similarly

$$E(g(X)) = \int g(X(\omega))dP = \int g(x)d\alpha$$

and anything concerning X can be calculated from. The statement X is a random variable with distribution α has to be interpreted in the sense that somewhere in the background there is a Probability Space and a random variable X on it, which has for its distribution. Usually only α matters and the underlying (Ω, B, P) never emerges from the background and in a pinch we can always say Ω is the real line, B are the Borel sets, P is nothing but α and the random variable $X(x) = x$.

Some other related quantities are

$$Var(X) = \sigma^2(X) = E[X^2] - [E[X]]^2.$$

$Var(X)$ is called the variance of X.

MEASURE THEORY PROBABILITY

The Riemann Integral the integral that is treated in freshman calculus applies to continuous functions. It can be extended a little beyond the class of continuous functions, but not very far. It can be used to define the lengths, areas, and volumes of sets in R, R^2, and R^3, provided those sets are reasonably nice, in particular not too irregularly shaped. In R^2, the Riemann Integral defines the area under the graph of a function by dividing the x-axis into a collection of small intervals.

On each of these small intervals, two rectangles are erected: one lies entirely inside the area under the graph of the function, while the other rectangle lies entirely outside the graph. The function is Riemann integrable (and its integral equals the area under its graph) if, by making the intervals sufficiently small, it is possible to make the sum of the areas of the outside rectangles arbitrarily close to the sum of the areas of the inside rectangles.

Measure theory provides a way to extend our notions of length, area, volume etc. to a much larger class of sets than can be treated using the Riemann Integral. It also provides a way to extend the Riemann Integral to Lebesgue integrable functions, a much larger class of functions than the continuous functions.

The fundamental conceptual difference between the Riemann and Lebesgue integrals is the way in which the partitioning is done. As noted above, the Riemann Integral partitions the domain of the function into small intervals.

By contrast, the Lebesgue Integral partitions the *range* of the function into small intervals, then considers the set of points in the domain on which the value of the function falls into one of these intervals. Let $f: [0, 1] \to R$.

Given an interval $[a, b) \subseteq R, f^{-1}([a, b))$ may be a very messy set. However, as long as we can assign a "length" or "measure" $\mu(f^{-1}([a, b)))$ to this set, we know that the contribution of this set to the integral of f should be between $a\mu\,(f^{-1}([a, b)))$ and $b\mu\,(f^{-1}([a, b]))$. By making the partition of the range finer and finer, we can determine the integral of the function.

Clearly, the key to extending the Lebesgue Integral to as wide a class of functions as possible is to define the notion of "measure" on as wide a class of sets as possible. In an ideal world, we would be able to define the measure of *every* set; if we could do this, we could then define the Lebesgue integral of *every* function. Unfortunately, as we shall see, it is not possible to define a measure with nice properties on every subset of R.

Measure theory is thus a *second best* exercise. We try to extend the notion of measure from our intuitive notions of length, area and volume to as large a class of *measurable* subsets of R, R^2, and R^3 as possible. In order to be able to make use of measures and integrals, we need to know that the class of measurable sets is closed under certain types of operations. If we can assign a sensible notion of measure to a set, we ought to be able to assign a sensible notion to its complement.

Probability and statistics focus on questions about convergence of sequences of random variables. In order to talk about convergence, we need to be able to assign measures to countable unions and countable intersections of measurable sets. Thus, we would like the collection of measurable sets to be a σ-algebra:

Definition: A *measure space* is a triple (Ω, B, μ), where

1. Ω is a set
2. B is a σ-algebra of subsets of Ω, i.e.
   (a) $B \subset 2^{\Omega}$, i.e. B is a collection of subsets of Ω
   (b) $\phi, \Omega \in B$
   (c) $B_n \in B, n \in N \Rightarrow \bigcup_{n\in N} \in B$
   (d) $B \in B \Rightarrow \backslash B \in B$
3. μ is a nonnegative, countably additive set function on B, i.e.
   (a) $\mu: B \to R^+ \cup \{\infty\}$
   (b) $B_n \in B,\ n \in N,\ B_n \cap B_m = \phi$ if $n \neq m \Rightarrow \mu\left(\bigcup_{n\in N} B\right) - \sum_{n\in N} \mu(B_n)$
   and $\mu(\phi) = 0$.

Remark: The definition of a σ-algebra is closely related to the properties of open sets in a metric space. Recall that the collection of open sets is closed under (1) arbitrary unions and (2) finite intersections; by contrast, a σ-algebra is closed under (1) countable unions and (2) countable intersections. Notice also that σ-algebras are closed under complements; the complement of an open set is closed, and generally not open, so closure under taking complements is not a property of the collection of open sets. The analogy between the properties of a σ-algebra and the properties of open sets in a metric space will be very useful in developing the Lebesgue integral. Recall that a function $f: X \rightarrow Y$ is continuous if and only if $f^{-1}(U)$ is open in X for every open set U in Y.

Recall from the earlier discussion that the Lebesgue integral of a function f is defined by partitioning the range of the fuction f into small intervals, and summing up numbers of the form $a\mu\ (f^{-1}([a, b)))$; thus, we will need to know that $f^{-1}([a, b)) \in B$. We will see in a while that a function $f: (\Omega, B, \mu) \rightarrow (\Omega', B')$ is said to be measurable if $f^{-1}(B) \in B$ for every $B' \in B'$. Thus, there is a close analogy between measurable functions and continuous functions.

As you know from calculus, continuous functions on a closed interval can be integrated using the so-called Riemann integral; the Lebesgue integral extends the Riemann integral to all bounded measurable functions (and many unbounded measurable functions).

Remark: Countable additivity implies $\mu(\phi) = 0$ provided there is some set B with $\mu(B) < \infty$; thus, the requirement $\mu(\phi) = 0$ is imposed to rule out the pathological case in which $\mu(B) = \infty$ for all $B \in B$.

Remark: If we have a finite collection $B_1,..., B_k \in B$ with $B_n \cap B_m = \phi$ if $n \neq m$, we can write $B_n = \phi$ for $n > k$, and obtain $\mu(B_1 \cup ... \cup B_k) = \mu\left(\bigcup_{n \in N} B_n\right) = \sum_{n \in N} \mu(B_n)$ $= \sum_{n=1}^{k} \mu(B_k) + \sum_{n=k+1}^{\infty} \mu(\phi) = \sum_{n=1}^{k} \mu(B_k)$, so the measure is additive over finite collections of disjoint measurable sets.

Example: Suppose that we are given a partition $\{\Omega_\lambda : \lambda \in \Lambda\}$ of a set Ω, i.e. $\Omega = \bigcup_{\lambda \in \Lambda} \Omega_\lambda$ and $\lambda \neq \lambda' \Rightarrow \Omega_\lambda \bigcap \Omega_{\lambda'} \phi$. Then we can form a σ-algebra as follows: Let

$$B_\Lambda = \left\{\bigcup_{\lambda \in C} \Omega_\lambda : C \subseteq \Lambda\right\}$$

In other words, B_Λ is the collection of all subsets of Ω which can be formed by taking unions of partition sets. B_Λ is closed under complements, as well as arbitrary (not just countable) unions and intersections. Suppose the partition is finite, i.e. Λ is finite, say $\Lambda = \{1,..., n\}$. Then B_Λ is finite; it has exactly 2^n elements, each corresponding to a subset of Λ. Suppose now that Λ is countably infinite; since every subset $C \subseteq \Lambda$ determines a different set $B \in B_\Lambda$, B_Λ is uncountable. Suppose finally that Λ is uncountable. For concreteness, let $\Omega = R$, $\Lambda = R$, and $\Omega_\lambda = \{\lambda\}$, i.e. each set in the partition consists of a single real number. There are many Γ-algebras containing this partition. As before, we can take

$$B_\Lambda = \left\{\bigcup_{\lambda \in C} \Omega_\lambda : C \subseteq \Lambda\right\} = \{C: C \subseteq R\} = 2^R$$

This is a perfectly good σ-algebra; however, as we indicated above (and as we shall show) it is not possible to define a measure with nice properties on this σ-algebra. The smallest σ-algebra containing the partition is

$$B0 = \{C: C \subseteq R,\ C \text{ countable or } R\setminus C \text{ countable}\}$$

This σ-algebra is too small to allow us to develop probability theory. We want a σ-algebra on R which lies between B_0 and B_Λ.

Definition: The Borel σ-*algebra* on R is the smallest σ-algebra containing all open sets in R. In other words, it is the σ-algebra

$$B = \cap\ \{C: C \text{ is a σ-algebra}, U \text{ open} \Rightarrow U \in C\}$$

i.e. it is the intersection of the class of all σ-algebras that contain all the open sets in R. A set is called *Borel* if it belongs to B.

Remark: Obviously, if U is an open set in R, then $U \in B$. If C is a closed set in R, then $R\setminus C$ is open, so $R\setminus C \in B$, so $C \in B$; thus, every closed set is a Borel set. Every countable intersection of open sets, and every countable union of closed sets, is Borel.

Every countable set is Borel (exercise), and every set whose complement is countable is Borel. If $\{U_{nm}: n, m \in N\}$ is a collection of open sets, then $\bigcup_n \left(\bigcap_m U_{nm}\right)$ is a Borel set. Thus, Borel sets can be quite complicated.

The most important example of a measure space is the Lebesgue measure space, which comes in two flavors. The first flavor is (R, B, μ), where B is the Borel σ-algebra and μ is Lebesgue measure, the measure defined in the following theorem; the second flavor is (R, C, μ), where C is the σ-algebra defined in the proof sketch of the following theorem.

Theorem: If B is the Borel σ-algebra on R, there is a unique measure μ (called Lebesgue measure) defined on B such that $\mu((a, b)) = b - a$ provided that $b > a$. For every Borel set B,

$$\begin{aligned}\mu(B) &= \sup\{\mu(K): K \text{ compact}, K \subseteq B\}\\ &= \inf\{\mu(U): U \text{ open}, U \supseteq B\}\end{aligned}$$

The proof works by gradually extending μ from the open intervals to the Borel σ-algebra. First, one shows one can extend μ to bounded open sets; it follows that one can extend it to compact sets. Then let $C_n = \{C \subset [-n, n]: \sup\{\mu(K): K \text{ compact}, K \subset C\} = \inf\{\mu(U): U \text{ open}, C \subset U\}\}$, and $C = \{C \subset \mathrm{R}: C \cap [-n, n] \in C_n \text{ for all } n\}$. If $C \in C$, we define $\mu(C) = \sup\{\mu(K): K \text{ compact}, K \subset C\}$. One can verify that C is a σ-algebra containing every open set, hence $C \supset B$.

Definition: A measure space (Ω, B, μ) is *complete* if $B\ \mu\ B, A \subseteq B, \mu(B) = 0 \Rightarrow A \in B$. The *completion* of a measure space (Ω, B, μ) is the measure space $(\Omega, \overline{B}, \overline{\mu})$, where

$$\overline{B} = \{B \subseteq \Omega : \exists_{D,D\in B} C \subseteq B \subseteq D, \mu(D\setminus C) = 0\}$$

and

$$\overline{\mu}(B) = \sup\{\mu(C): C \in B, C \subseteq B\} = \inf\{\mu(D): D \in B, B \subseteq D\}$$

for $B \in \overline{B}$.

It is easy to verify that the Lebesgue measure space (R, C, μ) is complete, and is the completion of (R, B, μ), where B is the Borel σ-algebra.

Definition: Suppose μ is a measure on the Lebesgue σ-algebra C on R. μ is *translation-invariant* if, for all $x \in R$ and all $C \in C$, $\mu(C + x) = \mu(C)$.

Theorem: The Lebesgue measure space (R, C, μ) is translation-invariant.

The theorem follows readily from the construction of Lebesgue measure, since translation doesn't change the length of intervals.

Observe if $x \in R$, $\mu(\{x\}) \leq \mu((x - \varepsilon/2, x + \varepsilon/2)) = \varepsilon$ for every positive ε, so $\mu(\{x\}) = 0$.

As we have already indicated, it is not possible to extend Lebesgue measure to every subset of R, at least in the conventional set-theoretic foundations of mathematics.

Theorem: There is a set $D \subset R$ which is not Lebesgue measurable.

Proof: We actually prove the following stronger statement: there is no translation-invariant measure μ defined on all subsets of R such that $0 < \mu([0, 1]) < \infty$.

Let μ be a translation-invariant measure defined on all subsets of R. Define an equivalence relation $\sim$ on R by

$$x \sim y \Leftrightarrow x - y \in Q$$

Let D be formed by choosing exactly one element from each equivalence class, and such that all the chosen elements lie in $[0, 1)$. Thus,

$$\forall_{x \in R} \exists_{d \in D} d - x \in Q$$

$$\forall_{d_1, d_2 \in D} d_1 - d_2 \notin Q$$

Given $x, y \in [0, 1)$, define

$$x +' y = \begin{cases} x + y & \text{if } x + y \in [0,1) \\ x + y - 1 & \text{if } x + y \in [1,2) \end{cases}$$

The operation +' is *addition modulo one*. It is easy to check that, given any $C \subseteq [0, 1)$ and $y \in [0, 1)$,

$$\mu(C + y) = \mu(C)$$

i.e. μ is translation-invariant with respect to translation using the operation +'. Then

$$[0, 1) = \bigcup_{q \in Q \cap [0,1)} D +' q$$

so

$$\mu([0, 1)) = \sum_{q \in Q \cap [0,1)} \mu(D +' q) = \sum_{q \in Q \cap [0,1)} \mu(D)$$

Then either $\mu(D) = 0$, in which case $\mu([0, 1)) = 0$, so $\mu([0, 1]) = 0$, or $\mu(D) > 0$, in which case $\mu([0, 1)]) \geq \mu([0, 1)) = \infty$.

Definition: A measure space (Ω, B, μ) is a *probability space* if $\mu(\Omega) = 1$.

From now on, we will restrict attention to probability spaces.

Example: By abuse of notation, let C denote the collection of Lebesgue measurable sets which are subsets of $[0, 1]$, and let μ be the restriction of

Lebesgue measure to this σ-algebra. $([0, 1], C, \mu)$ is called the Lebesgue probability space.

Theorem: Let (Ω, B, μ) be a probability space. Suppose $B_1 \supseteq B_2 \supseteq \ldots \supseteq B_n \supseteq \ldots$ with $B_n \in B$ for all n. Then

$$\lim_{n\to\infty} \mu(B_n) = \mu\left(\bigcap_{n\in N} B_n\right)$$

In particular, if $\bigcap_{n\in N} B_n = \phi$, then

$$\lim_{n\to\infty} \mu(B_n) = 0$$

Proof: Let $C_m = B_m \setminus B_{m+1}$. Then $C_n \cap C_m = \phi$ if $n \neq m$, and

$$B_1 = \left(\bigcap_{n\in N} B_n\right) \cup \left(\bigcup_{m\in N} C_m\right)$$

so

$$\sum_{m\in N} \mu(C_m) = \mu(B_1) - \mu\left(\bigcap_{m\in N} B_n\right) \leq \mu(B_1) < \infty$$

so $\sum_{m\in N} \mu(C_m)$ converges and hence

$$\sum_{m=n}^{\infty} \mu(C_m) \to 0 \text{ as } n \to \infty$$

$$\mu(B_n) = \mu\left(\bigcap_{m\in N} B_m\right) + \sum_{m=n}^{\infty} \mu(C_m)$$

$$\to \mu\left(\bigcap_{m\in N} B_m\right)$$

We now turn to the definition of a *random variable*. We think of Ω as the set of all possible states of the world tomorrow. Exactly one state will occur tomorrow. Tomorrow, we will be able to observe some function of the state which occurred. We do not know today which state will occur, and hence what value of the function will occur. However, we do know the probability that any given state will occur, and we know the mapping from possible states into values of the function, so we know the probability that the function will take on any given possible value. Thus, viewed from today, the value the function will take on tomorrow is a random variable; we know its probability distribution, but not its exact value. Tomorrow, we will observe the value which is realized.

Definition: Let (Ω, B, P) be a probability space. A random variable on Ω is a function X: $\Omega \to R \cup \{-\infty, \infty\}$ satisfying

$$P(\{\omega \in \Omega: X(\omega) \in \{-\infty, \infty\}) = 0$$

which is measurable, i.e. for every Borel set $B \subset R$, $X^{-1}(B) \in B$.

Observe the analogy beteen the definition of a measurable function and the characterization of a continuous function in terms of open sets: f is continuous if and only if $f^{-1}(U)$ is open for every open set U.

Lemma: Let (Ω, B, P) be a probability space. A function X: $\Omega \to \mathrm{R} \cup \{-\infty, \infty\}$ satisfying

$$P(\{\omega \in \Omega: X(\omega) \in \{-\infty, \infty\}\}) = 0$$

is a random variable if and only if for every open interval (a, b), $X^{-1}((a, b)) \in B$.

Proof: Since every open interval is a Borel set, if X is a random variable, then $X^{-1}((a, b)) \in B$.

Now suppose that $X^{-1}((a, b)) \in B$ for every open interval (a, b). Consider

$$M = \{B \subset R : X^{-1}(B) \in B\}$$

We claim that M is a σ-algebra. Note that $X^{-1}(\phi) = \phi \in B$, so $\phi \in M$. $X^{-1}(R) = \Omega \backslash X^{-1}(\{-\infty, \infty\})$. Since $P(X^{-1}(\{-\infty, \infty\}))$ is zero (and hence defined), $X^{-1}(\{-\infty, \infty\}) \in B$; since B is a σ-algebra, it is closed under complements, so X–1 (R) Ω B, and RΩ M.

If $B \in M$, $X^{-1}(B) \in B$, so $X^{-1}(R\backslash B) = X^{-1}(R) \backslash X^{-1}(B) \in B$ because $X^{-1}(R) \in B$ and $X^{-1}(B) \in B$. Therefore, $R\backslash B \in M$, so M is closed under complements.

Now suppose $B_n \in M$, $n \in N$. Then $X^{-1}(B_n) \in B$ for all n, so

$$X^{-1}\left(\bigcup_{n \in N} B_n\right) = \bigcup_{n \in N} X^{-1}(B_n) \in B$$

and hence M is closed under countable unions.

Therefore, M is a Ω-algebra. Since every open set is a countable union of open intervals, M contains every open set; since the Borel Ω-algebra is the smallest Ω-algebra containing the open sets in R, M contains every Borel set, so X is measurable, and hence a random variable.

Theorem: If f: $[0, 1] \to R$ is continuous, then f is a random variable on the Lebesgue probability space $([0, 1], C, \mu)$.

Proof: Since f takes values in R, $f^{-1}(\{-\infty, \infty\}) = \phi$. If (a, b) is an open interval, then since f is continuous, $f^{-1}((a, b))$ is an open subset of $[0, 1]$, so $f^{-1}((a, b)) \in C$. By Lemma, f is a random variable.

Example: A random variable f: $[0, 1] \to R$ can be discontinuous everywhere. Consider $f(x) = 1$ if x is rational, 0 if x is irrational. f is discontinuous everywhere. Suppose B is a Borel set. There are four cases to consider:

1. If $0, 1 \notin B$ $f^{-1}(B) = \phi \in C$.

2. If $0 \in B$, $1 \notin B, f^{-1}(B) = [0, 1] \setminus Q \in C$.
3. If $0 \notin B$, $1 \in B, f^{-1}(B) = [0, 1] \cap Q \in C$.
4. If $0, 1 \in B, f^{-1}(B) = [0, 1] \in C$.

Thus, f is measurable.

In elementary probability theory, random variables are not rigorously defined. Usually, they are described only by specifying their cumulative distribution functions. Continuous and discrete distributions often seem like entirely unconnected notions, and the formulation of mixed distributions (which have both continuous and discrete parts) can be problematic. Measure theory gives us a way to deal simultaneously with continuous, discrete, and mixed distibutions in a unified way.

We shall first define the cumulative distribution function of a random variable, and establish that it satisfies the defining properties of a cumulative distribution functions, as defined in elementary probability theory. We will then show (first in examples, then in a general theorem) that given any function F satisfying the defining properties of a cumulative distribution function, there is in fact a random variable defined on the Lebesgue probability space whose cumulative distribution function is F.

Definition: Given a random variable $X: (\Omega, B, P) \to R \cup \{-\infty, \infty\}$, the *cumulative distribution function* of X is the function $F: R \to [0, 1]$ defined by

$$F(t) = P(\{\omega \in \Omega: X(\omega) \leq t\})$$

Theorem: If X is a random variable, its cumulative distribution function F satisfies the following properties:

1. $\lim_{t \to -\infty} F(t) = 0$

   this is often abbreviated as $F(-\infty) = 0$.

2. $\lim_{t \to -\infty} F(t) = 1$

   this is often abbreviated as $F(\infty) = 1$.

3. F is increasing, i.e.

   $s < t \Rightarrow F(t) \geq F(s)$

4. F is right-continuous, i.e. for all t,

   $\lim_{s \searrow t} F(s) = F(t)$

Proof: We prove only right-continuity, leaving the rest as an exercise. Since F is increasing, it is enough to show that (as an exercise, think through carefully why this *is* enough)

$$\lim_{n\to\infty} F\left(t+\frac{1}{n}\right) = F(t)$$

$$F\left(t+\frac{1}{n}\right) - F(t)$$

$$= P\left(\left\{\omega \in \Omega : X(\omega) \le t+\frac{1}{n}\right\}\right) - P(\{\omega \in \Omega : X(\omega) \le t\})$$

$$= P\left(X^{-1}\left(-\infty, t+\frac{1}{n}\right)\right) - P(\{X^{-1}((-\infty, t]))$$

$$= P\left(x^{-1}\left(t, t+\frac{1}{n}\right)\right) \to P\left(\bigcap_{n\in N} X^{-1}\left(t, t+\frac{1}{n}\right]\right)$$

$$= P\left(X^{-1}\left(\bigcap_{n\in N} t, t, +\frac{1}{n}\right]\right)$$

$$= P(X^{-1}(\phi))$$

$$= P(\phi) = 0$$

Example: The *uniform distribution* on [0, 1] is the cumulative distribution function

$$F(t) = \begin{cases} 0 & \text{if} \quad t < 0 \\ t & \text{if} \quad t \in [0,1] \\ 1 & \text{if} \quad t > 1 \end{cases}$$

Consider the random variable X defined on the Lebesgue probability space X: ([0, 1], C, μ) $\to R$, $X(t) = t$. Observe that

$$\mu(\{\omega \in [0, 1]: X(\omega) \le t\}) = \mu([0, t]) = t$$

Thus, X has the uniform distribution on [0, 1]. Notice also that F is strictly increasing and hence one-to-one on [0, 1]. Thus, $F|_{[0,1]}$ has an inverse function $\left(F|_{[0,1]}\right)^{-1} : [0,1] \to [0,1]$ In fact, $X = \left(F|_{[0,1]}\right)^{-1}$.

Example: The *standard normal distribution* has the cumulative distribution function

$$f(t) = \frac{1}{\sqrt{2\pi}} \int_{-\infty}^{t} e^{-x^2/2} dx$$

Notice that F is strictly increasing, hence one-to-one, and the range of F is (0, 1), so F has an inverse function X: (0, 1) $\to R$; extend X to [0, 1] by defining $X(0) = -\infty$ and $X(1) - \infty$. One can show that the inverse of a strictly increasing, continuous function is strictly increasing and continuous. Since $X|_{(0,1)}$ is continuous, it is measurable on the Lebesgue probability space. Observe that

$$\mu(\{\omega \in [0, 1]: X(\omega) \leq t\}) = \mu X^{-1}((-\infty, t]) = F(t)$$

so X has the standard normal distribution.

Example: Consider the cumulative distribution function

$$F(t) = \begin{cases} 0 & \text{if} \quad t < -1 \\ \frac{1}{2} & \text{if} \quad t \in [-1, 1] \\ 1 & \text{if} \quad t \geq 1 \end{cases}$$

F is the cumulative distribution function of a random variable which takes on the values -1 and 1, each with probability $\frac{1}{2}$. We define a random variable on the Lebesgue probability space by

$$X(\omega) = \begin{cases} -\infty & \text{if} \quad \omega = 0 \\ -1 & \text{if} \quad \omega \in \left(0, \frac{1}{2}\right) \\ 1 & \text{if} \quad \omega \in \left(\frac{1}{2}, 1\right] \\ \infty & \text{if} \quad \omega = 1 \end{cases}$$

Clearly, the cumulative distribution function of X is F. Notice that F is weakly but not strictly increasing, so it is not one-to-one and hence does not have an inverse function. However, notice that X satisfies the following weakening of the definition of an inverse:

$$X(\omega) = \inf \{t: F(t) \geq \omega\}$$

Theorem: Let $F: R \to [0, 1]$ be an arbitrary function satisfying conclusions. There is a random variable X defined on the Lebesgue probability space whose cumulative distribution function is F.

Proof: Let

$$X(\omega) = \inf \{t: F(t) \geq \omega\}$$

Since $\omega' > \omega$

$$\Rightarrow \{t: F(t) \geq \omega'\} \subseteq \{t: F(t) \geq \omega\}$$

$$\Rightarrow \inf\{t: F(t) \geq \omega'\} \geq \inf \{t: F(t) \geq \omega\}$$

$$\Rightarrow X(\omega') e'' X(\omega)$$

X is increasing (not necessarily strictly). Since $\lim_{\sigma\to-\infty} F(s) = 0$ and $\lim_{t\to\infty} F(t) = 1$, if X^{-1} (0, 1), then there exist $s, t \in R$ such that $F(s) < \omega < F(t)$; since F is increasing, $-\infty < X < X(\omega) < \infty$. Thus,

$$\mu(\{\omega: X(\omega) \in \{-\infty, \infty\}\}) \leq \mu\{0, 1\}) = 0$$

Since X is increasing, if (a, b) is an open interval in R, $X^{-1}((a, b))$ is an interval (not necessarily open) in [0, 1], and hence $X^{-1}((a, b)) \in C$. By Lemma, X is a random variable.

$$X(\omega) \leq t \Leftrightarrow \inf\{s: F(s) \geq \omega\} \leq t$$

$$\Leftrightarrow \exists_{s \leq t} F(s) \geq \omega$$

$$\Leftrightarrow F(t) \geq \omega$$

so

$$\mu(\{\omega: X(\omega) \leq t\}) = \mu(\{\omega: \omega \leq F(t)\})$$

$$= \mu([0, F(t)])$$

$$= F(t)$$

In probability theory, the choice of a particular probability space is usually considered arbitrary and unimportant; you should choose a probability space which is convenient for the particular problem at hand, in particular one on which you can easily write down random variables with the desired joint distribution. In what follows, we will allow (Ω, B, P) to be an arbitrary probability space, not necessarily the Lebesgue probability space.

Next, we develop a notion of integration.

Definition: A simple function is a function of the form

$$f = \sum_{i=1}^{n} \alpha_i \chi_{B_i}$$

where each B_i is a measurable set in Ω and χ_{B_i} is the characteristic function of B_i:

$$\chi_{B_i}(\omega) = \begin{cases} 1 & \text{if} \quad \omega \in B_i \\ 0 & \text{if} \quad \omega \notin B_i \end{cases}$$

We define the integral of f by

$$\int_{\Omega} f dP = \sum_{i=1}^{n} \alpha_i P(B_i)$$

A given simple function may be expressed in the form of Equation in more than one way; one can show that one gets the same value of the integral from all of these different expressions.

Given a nonnegative random variable $f: (\Omega, B, P) \to R_+ \cup \{\infty\}$, we can construct a simple function which closely approximates it from below. Choose constants

$$0 = \alpha_1 < \alpha_2 < .... < \alpha_n < \alpha_{n+1} = \infty$$

Let
$$f_n = \sum_{i=1}^{n} \alpha_i \chi_{f^{-1}}([\alpha_i, \alpha_{i+1}))$$

Thus,

$$f_n(\omega) = \alpha_i \to f(\omega) \in [\alpha_i, \omega_{i+1})$$

Because f is a random variable, $f^{-1}([\alpha_i, \omega_{i+1})) \in B$, so f_n is a simple function.

Definition: If f: $(\Omega, B, P) \to R_+ \cup \{\infty\}$ is a random variable, define

$$\int_\Omega f dP = \sup\left\{\int_\Omega g dP : 0 \le g \le f, g \text{ simple}\right\}$$

The value of the integral may be ∞. We say that f is integrable if $\int_\Omega f dP < \infty$.

Example: Consider the function $f(x) = \dfrac{1}{x}$ on the Lebesgue probability space $([0, 1], C, \mu)$. Let

$$f_n = \sum_{i=1}^{n} i\chi_{f^{-1}([i,i+1))} = \sum_{i=1}^{n} i\chi_{\left(\left(\frac{1}{i+1}, \frac{1}{i}\right]\right)}$$

f_n is a simple function and $0 \le f_n \le f$.

$$\int f_n d\mu = \sum_{i=1}^{n} i\mu\left(\left(\frac{1}{i+1}, \frac{1}{i}\right]\right)$$

$$= \sum_{i=1}^{n} i\left(\frac{1}{i} - \frac{1}{i+1}\right) = \sum_{i=1}^{n} \frac{1}{i+1} \to \infty \text{ as } n \to \infty$$

Thus, $\int_{[0,1]} f d\mu = \infty$.

Definition: If f: $(\Omega, B, P) \to R \cup \{-\infty, \infty\}$ is a random variable, define $f_+(\omega) - \max\{f(\omega), 0\}$, $f_-(\omega) = -\min\{f(\omega), 0\}$. Note that $f = f_+ - f_-$ and $|f| = f_+ + f_-$. We say f is *integrable* if f_+ and f_- are both integrable, and we define

$$\int_\Omega f dP = \int_\Omega f_+ dP - \int_\Omega f_- dP$$

The following theorem, which we shall not prove, shows that the Lebesgue and Riemann integrals coincide when the Riemann integral is defined.

Theorem: Suppose that f: $[a, b] \to R$ is Riemann integrable (in particular, this is the case if f is continuous). Then f is Lebesgue integrable and

$$\int_{[a,b]} f d\mu = \int_a^b f(t)dt$$

In other words, the Lebesgue integral of f equals the Riemann integral of f.

Definition: We say that a property holds *almost everywhere* (abbreviated a.e.) or *almost surely* (abbreviated a.s.) if the set of ω for which it does *not* hold is a set of measure zero.

Definition: Suppose f: (Ω, B, P) → R is a random variable. If $B \in B$, we define

$$\int_B f d\mu = \int_\Omega f \chi_B dP$$

where χ_B is the characteristic function of B.

Theorem: Suppose f, g: (Ω, B, P) → R are integrable, $A, B \in B$, $A \cap B = \phi$, and $c \in R$. Then

1. $\int_B CfdP = c\int_B fdP$
2. $\int_B f + gdP = \int_B fdP + \int_B gdP$
3. $f \leq g$ a.e. on $B \Rightarrow fdP \leq \int_B gdP$
4. $\int_{A\cup B} fdP = \int_A fdP + \int_B fdP$

Definition: Suppose f, g: (Ω, B, P) → R are random variables. Say $f \sim g$ if $f(\omega) = g(\omega)$ a.e. Let $[f]$ denote the equivalence class of f under the equivalence relation ~, i.e. $[f] = \{g: (\Omega, B, P) \to R$: g is measurable, $g \sim f\}$. Define

$L^1(\Omega, B, P) = \{[f]: f: \Omega \to R, f$ a random variable, f integrable$\}$

$L^2(\Omega, B, P) = [f]: f: \Omega \to R, f$ a random variable, f^2 integrable$\}$

Theorem: If (Ω, B, P) is a probability space,

$$L^2(\Omega, B, P) \subseteq L^1(\Omega, B, P)$$

Definition: Given a metric space (X, d), the *completion* of (X, d) is a metric space $(\bar{X}, \bar{d})$ such that $(\bar{X}, \bar{d})$ is complete, $\bar{d}\,|_X = d$, and X is dense in $(\bar{X}, \bar{d})$. We are justified in calling $(\bar{X}, \bar{d})$ *the* completion of (X, d) because if $(\hat{X}, \hat{y})$ is another completion of (X, d), then there is an isomorphism $\phi: (\bar{X}, \bar{d}) \to (\hat{X}, \hat{y})$ such that $\phi(x) = x$ for every $x \in X$, and $\hat{d}(\phi(\bar{x}), \phi(\hat{y})) = \bar{d}(\bar{x}, \bar{y})$ every $\bar{x}, \bar{y} \in \bar{X}$.

Theorem: $L^1(\Omega, B, P)$ and $L^2(\Omega, B, P)$ are Banach spaces under the respective norms

$$\| f \|_1 = \int_\Omega | f(\omega) | dP$$

$$\| f \|_2 = \sqrt{\int_\Omega f^2(\omega) dP}$$

For the Lebesgue probability space, $L^1([0, 1], C, \mu)$ and $L^2([0, 1], C, \mu)$ are the completions of the normed space $C([0, 1])$, with respect to the norms $\|.\|_1$ and $\|\cdot\|_2$.

Example: Recall that $C([0, 1])$ is a Banach space with the sup norm $f = \sup\{|f(x)| : x \in [0, 1]\}$. $(C[0, 1], \|\cdot\|_1)$ and $(C[0, 1], \|\cdot\|_2)$ are normed spaces, but they are not complete in these norms, hence are not Banach spaces. To see this, let $f_n(x)$ be the function which is zero for $x \in \left[0, \frac{1}{2} - \frac{1}{n}\right]$, one for $x \in \left[\frac{1}{2} + \frac{1}{n}, 1\right]$ and linear for $x \in \left[\frac{1}{2} - \frac{1}{n}, \frac{1}{2} + \frac{1}{n}\right]$. f_n is not Cauchy with respect to $\|\cdot\|_\infty$. However, f_n *is* Cauchy and has a natural limit with respect to $\|\cdot\|_1$. Indeed, let

$$f(x) = \begin{cases} 0 & \text{if} \quad x < \dfrac{1}{2} \\ 1 & \text{if} \quad x \geq \dfrac{1}{2} \end{cases}$$

Then

$$\int_{[0,1]} |f_n - f| \, d\mu \leq \frac{1}{2} \times \frac{2}{n} = \frac{1}{n} \to 0$$

so f_n converges to f in $L^1([0, 1], C, \mu)$, and hence must be Cauchy with respect to $\|\cdot\|_1$. Notice that this limit does not belong to $C([0, 1])$.

Definition: For $f \in L^1(\Omega, B, P)$, define the *mean* or *expectation* of f by

$$E(f) = \int_\Omega f dP$$

For $f \in L^2(\Omega, B, P)$, define the variance of f by

$$\text{Var}(f) \int_\Omega (f(\omega) - E(f)^2 dP$$

Definition: If $f, g \in L^2(\Omega, B, P)$, define the *inner product* of f and g by

$$f.g = \int_\Omega f(\omega) g(\omega) dP$$

The properties of the inner product are closely analogous to those of the dot product of vectors in Euclidean space R^n. In particular, they determine a geometry, including lengths $\left(\|f\|_2 = \sqrt{f.f}\right)$ and angles. The most basic property of the dot product, the Cauchy-Schwarz inequality, extends to the inner product.

Cauchy-Schwarz Inequality

Theorem: If $f, g \in L^2(\Omega, B, P)$,

$$|f \cdot g| \leq \|f\|_2 \|g\|_2$$

Definition: If $f, g \in L^2(\Omega, B, P)$, define the *covariance* of f and g by

$$\text{Covar}(f, g) = E((f - E(f))(g - E(g)))$$

Proposition: If $f, g - L^2(\Omega, B, P)$, then

$$\text{Covar}(f, g) = E(fg) - E(f) E(g)$$

Observe from the definition that Covar (f, g) = Covar (g, f). Thus, given $f_1,\ldots, f_n - L^2(\Omega, B, P)$, the covariance matrix C whose (i, j) entry is $c_{ij} = \text{Covar}(f_i, f_j)$ is a symmetric matrix. Hence, there is an orthonormal basis of R^n composed of eigenvectors of C; expressed in this basis, the matrix becomes diagonal. Observe also that

$$\text{Covar}\left(\sum_{i=1}^{n}\alpha_i f_i, \sum_{j=1}^{n}\beta_j f_j,\right) = (\alpha_1,\ldots,\alpha_n)C\begin{pmatrix}\beta_1\\ \vdots\\ \beta_n\end{pmatrix}$$

is a quadratic form of the type we studied earlier.

One very useful consequence of the inner product structure on L^2 is the existence of orthogonal projections. Given a linearly independent family

$f_1,\ldots, f_n - L^2(\Omega, B, P)$, let V be the vector space spanned by $f_1,\ldots, f_n$. Then any $g \in L^2$ can be written in a unique way as $g = \pi(g) + \omega$, where $\pi(v) = \sum_{i=1}^{n}\alpha_i f_i \in V$ and $\omega \cdot v = 0$ for all $v \in V$ (in particular, $\omega \cdot f_i = 0$ for each i). $\pi(g)$ is also characterized as the point in V closest to g. The coefficients $\alpha_1,\ldots, \alpha_n$ are the regression coefficients of g on $f_1,\ldots, f_n$. If $f_1,\ldots, f_n$ are orthonormal (i.e. $f_i \cdot f_j = 1$ if $i = j$ and 0 if $i \neq j$), $\alpha_i = g \cdot f_i$ for each i.

One frequently encounters two measures living on a single probability space. For example, Lebesgue measure is one measure on the real line R; any random variable determines another measure on R by its distribution.

Definition: Let μ and v be two measures on the same measurable space (X, A), i.e. (X, A, μ) and (X, A, v) are measure spaces. We say v is *absolutely continuous with respect to* μ if

$$B \in A,\ \mu(B) = 0 \Rightarrow v(B) = 0$$

We say μ is σ-finite if there exist $B_1, B_2,\ldots \in A$ such that $X = \bigcup_{n\in N} B_n$ and $\mu(B_n) < \infty$ for all n.

Radon-Nikodym

Theorem: Suppose that μ and v are two σ-finite measures on the same measurable space (X, A). Then v is absolutely continuous with respect to μ if and only if there exists $f \in L^1(X, A, \mu)$, $f \geq 0$ almost everywhere, such that

$$v(B) = \int_B f d\mu$$

for all $B \in A$.

The Radon-Nikodym theorem tells us that if a random variable Z has a continuous distribution (in other words, the measure v determined by its cumulative distribution function F is absolutely continuous with respect to Lebesgues measure), then Z has a density function f, namely f is the Radon-Nikodym derivative of v with respect to Lebesgue measure.

We now turn to products of probability spaces (X, A, P) and (Y, B, Q). If $A \in A$ and $B \in B$, it is natural to define

$$(P X Q)(A X B) = P(A)Q(B)$$

However, most of the subsets of $X X Y$ we are interested in are not of the form $A X B$ with $A \in A$ and $B \in B$. For example, if $X = Y = [0, 1]$, the diagonal $\{(x, y): x = y\}$ is *not* of that form. However, it is possible to write the diagonal in the form

$$\bigcap_{m \in N} \bigcup_{n=1}^{i_m} A_{mn} \times B_{mn}$$

with $A_{mn} \in A$ and $B_{mn} \in B$. This suggests we should extend $P X Q$ to the smallest σ-algebra containing $\{A X B: A \in A, B \in B\}$.

Definition: Suppose (X, A, P) and (Y, B, Q) are probability spaces. The product probability space is $(X X Y, A X B, P X Q)$, where

1. $A \times' B$ is the smallest σ-algebra containing all sets of the form $A \times B$, where $A \in A$ and $B \in B$
2. $P \times' Q$ is the unique measure on $A \times' B$ that takes the values $P(A)Q(B)$ on sets of the form $A \times B$, where $A \in A$ and $B \in B$.
3. $(X \times Y, A \times B, P \times Q)$ is the completion of $(X \times Y, A \times' B, P \times' Q)$

The existence of the product measure $P X Q$ is proven by extending the definition of $P X Q$ for sets of the form $A X B$ to $A X B$, in a manner similar to the process by which Lebesgue measure is obtained from lengths of intervals.

Fubini

Theorem: Suppose (X, A, P) and (Y, B, Q) are complete probability spaces. If $f: X X Y \to R$ is $A X B$ measurable and integrable, then

1. For almost all $x \in X$, the function defined by $f_x(y) = f(x, y)$ is integrable in Y;
2. For almost all $y \in Y$, the function defined by $f_y(x) = f(x, y)$ is integrable in X;
3. $\int_Y f(x, y)dQ(y)$ is an integrable function of $x \in X$;
4. $\int_X f(x, y)dP(x)$ is an integrable function of $y \in Y$;
5. $$\int_X \left(\int_Y f(x,y)dQ(y)\right)dP(x) = \int_Y \left(\int_X f(x,y)dP(x)\right)dQ(y)$$

$$\int_{X \times Y} f(x,y)d(P \times Q)(x,y)$$

There are many notions of convergence of functions, and results showing that the integral of a limit of functions is the limit of the integrals.

Definition: Suppose $f_n, f: (\Omega, B, P) \to R$ are random variables.

1. We say f_n converges to f in probability or in measure if

$$\forall_{\varepsilon>0}\exists_N n > N \Rightarrow P(\{\omega : | f_n(\omega) | > \varepsilon\}) < \varepsilon$$

2. We say f_n converges to f almost everywhere (abbreviated a.e.) or *almost surely* (abbreviated *a.s.*) if

$$P(\{\omega: f_n(\omega) \to f(\omega)\}) = 1$$

Theorem: If f_n converges to f almost surely, then f_n converges to f in probability.

In the lecture, I gave an example of a sequence of functions which converges in probability, but not almost surely. The example is much easier to describe with pictures than with words, so I won't give the details here. The idea is to construct a sequence of functions f_n: [0, 1] $\to R$ so that $\mu(\{x: f_n(x) \neq 0\}) \to 0$; hence f_n converges in probability to the identically zero function f. However, we can move the set on which f_n is not zero around so that, for every $x \in [0, 1]$, $f_n(x) = 1$ for infinitely many n, so almost surely, $f_n(x)$ fails to converge to $f(x)$.

Example: Although convergence almost surely is a strong notion of convergence of functions, it is not sufficient for convergence of the integrals of the functions to the integral of the limit. For example, consider f_n: [0, 1] $\to R$ defined by $f_n = n\chi_{[0,1/n]}$, which takes the value n on [0, 1/ n] and 0 everywhere else. f_n converges almost surely to f, the function which is identically zero, but if P is Lebesgue measure, [0 1]$f_n dP = n X \frac{1}{n} = 1$, while $\int_{[0,1]} f dP = 0$. Roughly speaking, f_n is converging to a function which is ∞ with probabilty 0, and in the Lebesgue integral, 0 X ∞ = 0; some of the mass in f_n is lost in the limit operation. The next result tells us that, for nonnegative functions, this is the only thing that can go wrong, and the limit of the integrals is always at least as big as the integral of the limit function.

Chapter 2

Chi-Square Procedures

The binomial procedures to situations where there are exactly two mutually exclusive. categories into which observations might fall–female/male, head/tail, recovery/non-recovery, and so on. The family of inferential statistical procedures known as chi-square (pronounced 'kai' to rhyme with 'sky') extends the logic of binomial procedures to cover situations where there are more than two categories of possible outcome; for example, students categorized according to academic class (freshman, sophomore, junior, senior) or patients with a certain disease categorized according to whether their condition, following an experimental treatment, is improved, worsened, or unchanged. Chi-square procedures also extend this logic to cover situations where there is more than one dimension of classification; for example, students categorized according to whether they describe themselves as "conservative" or "liberal," as well as by their academic class as freshman, sophomore, junior, or senior. When the observed items–students or whatever else they might be–are categorized in this fashion according to two or more separate dimensions of classification concurrently, they are said to be cross-categorized. We will begin with the simpler situation in which there is only one dimension of categorization.

CHI-SQUARE PROCEDURES FOR ONE DIMENSION OF CATEGORIZATION

The underlying logic of chi-square would be best illustrated with the example of tossing a three-sided coin, with each of the sides having an outcome probability of exactly one-third. But as a coin with these specifications is difficult to imagine, we will have to improvise with an analogy.

For more than a century, the three species of large fish–gumpies, sticklebarbs, and spotheads–that are native to a certain river have been observed to co-exist in equal proportions of one-third each. But now a random sample of 300 large fish drawn from a standard fish-sampling location has turned up numbers and proportions suggesting that something has occurred to upset the natural ecology of the river. If the three fish species still inhabited the river in equal proportions, we would expect to find about 100 instances of each in a sample of size N=300; whereas what we actually observe are 89 gumpies, 120 sticklebarbs, and 91 spotheads. Here is an overview of the observed counts and percentages for the three fish-species categories, each in comparison with the corresponding mean chance expected values (MCE).

	gumpies	*sticklebarbs*	*spotheads*	*Totals*
Observed frequency of cases	89 (29.7%)	120 (40.0%)	91 (30.3%)	300
Expected frequency of cases (MCE)	100 (33.3%)	100 (33.3%)	100 (33.3%)	300

The question of statistical significance in this situation is of the same form we have already encountered, except that here what is at issue is the difference between two *patterns,* namely, the observed frequency pattern of 89/120/91 versus the MCE frequency pattern of 100/100/100. And the first step in answering the question is to devise a way for measuring the degree to which the two patterns differ from each other in the aggregate. A straightforward way of going about this would be to take, for each of the three categories, the difference between the observed frequency and the expected frequency, and then divide that difference by the expected frequency.

$$\frac{\text{observed frequency} - \text{expected frequency}}{\text{expected frequency}}$$

The outcome of this operation would be a measure of the proportionate amount by which each observed frequency deviates from its corresponding expected frequency. Thus

$$\text{gumpies: } \frac{(89-100)}{100} = -.11$$

$$\text{sticklebarbs: } \frac{(120-100)}{100} = +.20$$

$$\text{spotheads: } \frac{(91-100)}{100} = -.09.$$

In percentage terms, the observed number of gumpies in the sample is 11 percent smaller than the corresponding MCE frequency, the number of sticklebarbs is 20 percent greater, and the number of spotheads is 9 percent smaller.

The advantage of this procedure is that you will find it intuitively obvious just what is being measured. Its limitation is that the proportionate differences measured for the several categories–in the present example, –.11, +.20, and –.09–will always sum to zero and thus will not be able to provide a measure of how much the observed and expected patterns of frequencies differ from each other overall. That is, instead of taking just the difference between an observed frequency and its corresponding expected frequency, we take the squared difference.

$$\frac{(\text{observed frequency} - \text{expected frequency})^2}{\text{expected frequency}}.$$

The effect of this operation will be to get rid of the minus signs and thus provide a set of measures whose sum will reflect the aggregate degree of difference that actually exists between the observed and expected patterns of frequencies. For the present example, where the patterns are 89/120/91 and 100/100/100, it comes out as

$$\text{gumpies: } \frac{(89-100)^2}{100} = 1.21 \text{ sum}$$

$$\text{sticklebarbs: } \frac{(120-100)^2}{100} = 4.0 = 6.02$$

$$\text{spotheads: } \frac{(91-100)^2}{100} = .81.$$

This, in brief, is the measure known as chi-square, conventionally rendered in symbolic notation as χ^2, which is simply the lower-case Greek letter 'chi' ('kai') with a squaring sign appended. For the sake of simplicity, we will henceforth represent an observed frequency as O and its corresponding MCE expected frequency as E. The only other item of symbolic notation is the summation sign, Σ, with which you are already familiar. Henceforth we will also be describing each category (e.g., gumpies, sticklebarbs, spotheads) as a cell; this is merely a linguistic convention reflecting the box-like tabular format usually employed for representing categorical frequency data within the context of chi-square procedures.

Here, then, is the simple two-step operation for calculating χ^2 for the general case where there is one dimension of categorization. Keep in mind that what we are measuring through this procedure is the degree to which an observed pattern of frequencies differs, overall, from an MCE expected pattern.

First, for each cell, calculate the component value

$$\frac{(\text{observed frequency} - \text{expected frequency})^2}{\text{expected frequency}} = \frac{(\text{O}-\text{E})^2}{\text{E}}.$$

which, as we have just seen, yields

gumpies: $(89–100)^2/100 = 1.21$

sticklebarbs: $(120–100)^2/100 = 4.0$

spotheads: $(91–100)^2/100 = .81$

and then: sum up these several component values to get the overall value of chi-square, according to the formula$_T$

$$\chi^2 = \Sigma\frac{(\text{O}-\text{E})^2}{\text{E}}$$

$$= 1.21 + 4.0 + .81 = 6.02.$$

If we were building the inferential apparatus of chi-square from scratch, our next task would be to figure out the properties of the sampling distribution to which this or any other calculated value of χ^2 belongs. Fortunately, that particular wheel has already been invented, and we do not need to try to reinvent it. For the sake of illustration, however, suppose for just a moment that we were starting from scratch and had no idea at all of the sampling distribution to which our calculated value of =6.02 belongs. The specific probability question for which we would need to figure out the answer is this:

If the proportions of the three fish species in the river were still actually one-third each, how likely is it that this or any other random sample of size *N*=300 might end up with a discrepancy between the observed and expected frequency patterns this large or larger; that is, with a calculated chi-square value equal to or greater than 6.02?

Until fairly recently it would have required a quite high level of mathematical expertise and acumen to figure out the answer to this question from scratch. The only alternative method would have been to construct the sampling distribution of chi-square empirically. That is, start out with a river that you know in advance contains vast numbers of gumpies, sticklebarbs, and spotheads in exactly equal proportions of one-third each; draw a random sample of size *N*=300; calculate and record the resulting value of chi-square–and then repeat that operation again and again, many times over. After a very large number of such samples, we would then take the proportion of cases in which the calculated chi-square value was equal to or greater than 6.02–and that would be our probability value, or at least a quite close estimate of it.

The limitations of this empirical method will surely be obvious. First we would have to find a river with those precise specifications; and then we would have to devote quite a lot of time, energy, and expense to the task of drawing a multitude of samples from it. Until quite recently it would have been difficult if not impossible to construct sampling distributions in this fashion. Nowadays, on the other hand, it can be done quite easily–or at least simulated–within the virtual reality of a computer. Using a versatile, programmable spreadsheet application known as Wingz, I instructed the desktop computer in my office to perform exactly the empirical sampling procedure just described–the only material difference being that what the computer was pulling up in its nets was not random gumpies, sticklebarbs, and spotheads, but random a's, b's, and c's. The programming instructions were of course given in a language the computer understands. Here is the English translation:

Generate a series of *a*'s, *b*'s, and *c*'s, such that the probability of each is exactly one-third.

Label each instance of a as a "gumpie," each instance of b as a "sticklebarb," and each instance of c as a "spothead."

Count up the respective frequencies of the three species within the sample of 300 and calculate and record the value of chi-square, using the expected frequency of E=100 for each cell.

Go back to the beginning and repeat this operation for a total of 10,000 times.

Figure shows the distribution of the 10,000 chi-square values that resulted from this simulation. The format of the graph is that of a histogram, with relative frequencies expressed in terms of percentages. Thus, the first column on the left represents the percentage of chi-square values that fell between zero and .99; the second column shows the percentage that fell between 1.0 and 1.99; and so on. As indicated on the right-hand side of the graph, the 10,000 sample values of chi-square included only 4.99 percent that were equal to or greater than our originally calculated value of =6.02—from which we would infer that the mere chance probability of the outcome observed in our fish example is somewhere in the vicinity of P=.0499.

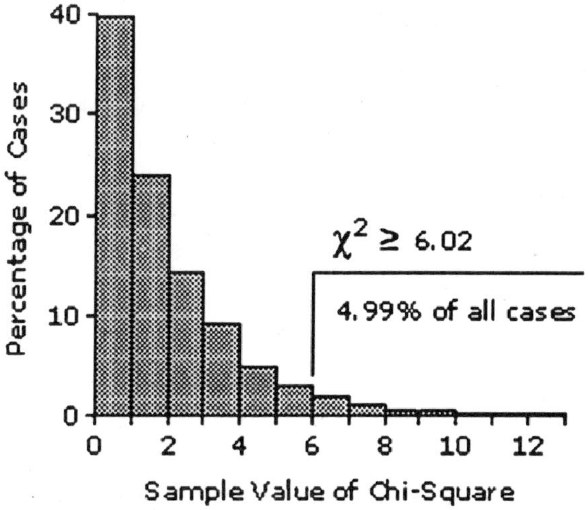

Fig. Empirical Approximation of a Chi-Square Sampling Distribution

Figure shows the "official" abstract theoretical sampling distribution of chi-square that would apply to our fish example, and you will certainly see that the correspondence between the two is very close. In the theoretical distribution, the critical value of chi-square for significance at the P=.05 level is =5.99. That is, of all the possible values of chi-square that might have resulted in this situation–on the null-hypothesis assumption that the three species of fish still inhabit the river in identical proportions of one-third each–only 5 percent would have been equal to or greater than 5.99.

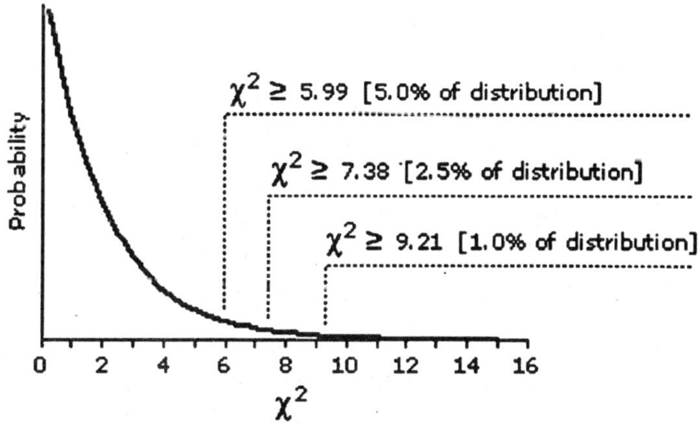

Fig. Theoretical Sampling Distribution of Chi-Square (df=2)

Hence, an observed chi-square value precisely equal to 5.99 could be said to be significant *at* the .05 level. As our calculated value of χ^2 =6.02 is slightly larger than this 5.99 critical value, we can infer that its probability is a shade smaller than .05. Alternatively, we could say that it is significant slightly *beyond* the .05 level. In any case, you will recall that our computer simulation of this same sampling distribution yielded an estimated probability of P=.0499 for the calculated value of χ^2 = 6.02. Figure also shows the critical values of chi-square for significance at the .025 (2.5 percent) and .01 (1.0 percent) levels.

Although this next point will be obvious from the way in which chi-square values are calculated—with the difference between O and E squared to get rid of the minus signs—do take a moment nonetheless to notice that there are no negative chi-square values. All possible values of χ^2 in this or any other chi-square sampling distribution are either equal to or greater than zero. The immediate implication of this fact is that a chi-square test of statistical significance is intrinsically non-directional. In our fish example there are several possible directions in which the results might have gone. What we in fact ended up with was more observed sticklebarbs than expected and fewer gumpies and spotheads:

	gumpies	*sticklebarbs*	*spotheads*
O	89	120	91
E	100	100	100

But we might also have ended up with a preponderance of gumpies or spotheads; or alternatively, with a preponderance of any two of the species along with relatively small numbers of the third. Any one of these possibilities, providing it involved a sufficiently large discrepancy between the patterns of O and E, would have resulted in a chi-square value equal or greater than the value of χ^2 = 6.02 calculated for our example. The following two possibilities, among others, would each have resulted in chi-square values exactly equal to 6.02.

	gumpies	*sticklebarbs*	*spotheads*
O	120	89	91
O	89	91	120

We will see later that there are certain circumstances in which the apparatus of chi-square can be adjusted to provide a directional test of significance; but until then, the basic precept is that a chi-square test of statistical significance is intrinsically non-directional.

The essential difference between the graphs in Figures 8.1 and 8.2 is not that one is a histogram and the other is a smoothed polygon. It is rather that one is fairly specific and the other broadly general. Strictly speaking, the simulated sampling distribution shown in Figure applies only to situations that correspond to the stipulated details of the simulation; namely, that there are three categories, the size of the sample is *N*=300, and the pattern of MCE frequencies for the three categories is 100/100/100. The theoretical distribution of Figure, on the other hand, includes cases of this particular type, but also extends more generally beyond them to all other three-category situations, irrespective of the size of the sample or the particular values of the expected frequencies; for example, a sample of size *N*=100 with expected frequencies of 25/50/25, or of size *N*=62 with expected frequencies of 13/22/27.

If our sample of fish had been sorted into four categories, or five, or six, on the other hand, we would need to refer to a different member of the family of chi-square sampling distributions. So long as we are dealing with only one dimension of categorization, the sampling distribution of chi-square that is appropriate for any particular case is determined quite simply by the number of categories–3, 4, 5, 9, 12, or whatever it might be. With two or more dimensions of categorization it becomes a bit more complex. In both kinds of cases, the appropriate sampling distribution of chi-square is determined not by the number of categories as such, but rather by a more general property spoken of as degrees of freedom, symbolized as df. In working to understand the meaning of this concept, your first step should be to disentangle it entirely from most of the meanings that you normally associate with the word "freedom," since in this context it has nothing at all to do with freedom of the will, freedom of conscience, or anything of the sort. Degrees of freedom, df, is simply an index of the amount of random variability, mere chance coincidence, that can be present in a particular situation. Its closest literal translation would be something along the line of "degrees of arbitiariness."

Here is a working definition of the concept that should prove sufficient for our immediate purposes. Suppose you have two cells such as the ones shown below, and you are free to plug any integer numbers that you want into them, subject only to the stipulation that the sum of the two numbers must be equal to a certain specified quantity. For purposes of illustration we will set the sum at 20, though it could actually be any positive integer value.

$$a \quad b$$
$$+ \quad = 20.$$

It will be obvious at a glance that your "freedom" in this case is limited by the fixed sum of 20. If you start by plugging the integer 8 into cell a, the number that goes into b is then inescapably fixed as 20–8=12. Start by plugging 16 into cell b and the number that goes into a is then rigidly fixed as 20–16=4. So in this two-cell situation, only one of the cells is "free" to vary arbitrarily–which is to say, there is only one degree of freedom. If you have three cells subject to a fixed sum of 20

$$a \quad b \quad c$$
$$+ \quad + \quad = 20$$

you can plug numbers arbitrarily into any two of the cells, but once those cells are plugged the value of the third is rigidly fixed. Thus, plug 6 into any one cell and 8 into either of the remaining two, and the value of the third is then fixed as 20–6–8=6. So here, with three cells, your degrees of freedom would be equal to two. The same logic then extends to cases where the number of cells is four, five, six, and so on. When applying chi-square procedures to situations in which there is only one dimension of categorization, the general principle for determining degrees of freedom is

$$\text{df} = (\text{number of cells}) - 1.$$

In our fish example the number of categorical cells is three; hence df=2. Similarly, the chi-square sampling distribution shown in Figure applies to the general case where df=2.

Figure shows the same df=2 sampling distribution that appears in Figure, along with two other members of the family of chi-square sampling distributions, the ones for df=3 and df=4. We need not be distracted just now by the fact that these three distributions have somewhat different shapes on the left-hand side of the scale. What is of immediate relevance are the similarities and differences that appear on the right-hand side of the scale, in the tails of the distributions. The similarity among all three distributions is that values of chi-square in the tail regions become less probable, hence more significant, as they increase in size. And the difference is that the mere chance probability of any particular value of chi-square is greater for a larger value of df than for a smaller value. Thus, for all three distributions a value falling at or beyond $\chi^2=6$ is less probable than one falling at or beyond $\chi^2=5$; however, the probability for both of these cases is larger for df=4 than for df=3, and larger for df=3 than for df=2. This point will be most readily visible if you draw a vertical line in Figure straight up from the point on the scale at which $\chi^2=6.0$. For df=2, the portion of the distribution falling to the right of your line will be about 5 percent, whereas for the df=3 and df=4 distributions it will be about 10 percent and 18 percent, respectively. Thus, for df=2 an observed chi-square value of 6.0 would be significant at the conventional .05 level, while for df=3 and df=4 it would not be. Although we have made these observations with particular reference to the distributions for df=2, 3, and 4, they extend in general to the whole family of chi-square sampling distributions.

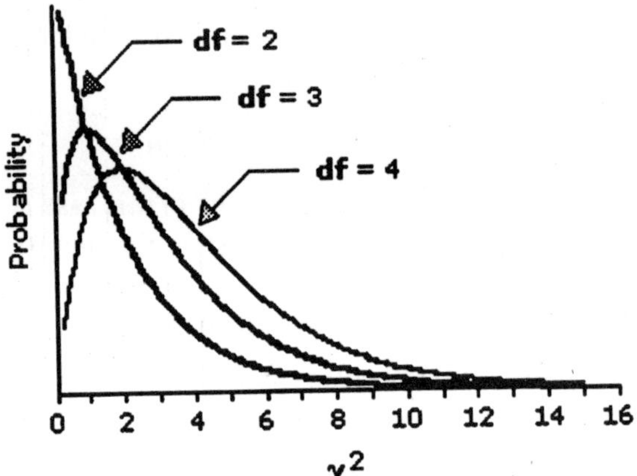

Fig. 8.3. Chi-Square Sampling Distributions for df=2, 3, and 4

In the practical application of chi-square procedures, you of course do not need to examine every detail of the relevant sampling distribution; it will be sufficient just to know the critical value that a calculated value of chi-square must equal or exceed in order to be judged statistically significant. A full set of such critical values is listed in Appendix B, and a somewhat abbreviated version of the same set is shown here in Table. For the moment we will confine our attention to Table. The first column on the left in this table, labeled df, lists various values of degrees of freedom; the row across the top lists various levels of significance (P=.05, .025, etc.); and each of the other entries indicates the critical value of chi-square that an observed value must meet or exceed in order to be judged significant at a given level of significance for a given value of df. Thus, for df=2 the minimum value of chi-square required for significance at the basic P=.05 level is 5.99; for df=3 it is 7.81; for df=4 it is 9.49; and so on. For the more stringent significance level of P=.025, the required value of chi-square is 7.38 for df=2; 9.35 for df=3; 11.14 for df=4; and so on. In applying chi-square and other statistical procedures, it is conventional to speak of a given result as being either non-significant, or significant *at* a certain level, or significant *beyond* a certain level. The illustration at the bottom of Table will give you an idea of what these three phrases mean and how they are used.

Table. Partial Table of Critical Values of Chi-Square

Each entry indicates the critical value that an observed value of chi-square must meet or exceed in order to be judged significant at a given level of significance for a given value of df.

Level of Significance (non-directional test)

df	.05	.025	.01	.005	.001
1	3.84	5.02	6.63	7.88	10.83
2	5.99	7.38	9.21	10.60	13.82
3	7.81	9.35	11.34	12.84	16.27
4	9.49	11.14	13.28	14.86	18.47
5	11.07	12.83	15.09	16.75	20.52
–	–	–	–	–	–
10	18.31	20.48	23.21	25.19	29.59
11	19.68	21.92	24.73	26.76	31.26
–	–	–	–	–	–

Illustration for df=2

If the observed value of chi-square is:	Then it is:
smaller than 5.99	non-significant
equal to 5.99	significant at the .05 level
greater than 5.99	significant beyond the .05 level
equal to 7.38	significant at the .025 level
greater than 7.38	significant beyond the .025 level
equal to 9.21	significant at the .01 level
greater than 9.21	significant beyond the .01 level
etc.	etc.

Here is one other example to illustrate this simpler one-dimensional version of chi-square. Suppose that a questionnaire administered to a large national sample of college students included an item aimed at measuring conservativism versus liberalism on a certain issue of social/political relevance. The item took the form of a statement, and the response categories were "strongly disagree," "moderately disagree," "undecided," "moderately agree," and "strongly agree." I will leave it to your imagination to fill in the blanks concerning what the statement was and which end of the response scale was taken to reflect "conservative" or "liberal" attitudes. Suffice it to say that the percentages of response within each category were as follows:

strongly disagree	moderately disagree	undecided	moderately agree	strongly agree
9.4%	15.6%	34.3%	27.5%	13.2%

Professor H, upon reading the results of this survey, suspects that the students at her particular college are considerably more polarized into "conservative" and "liberal" camps, in comparison with the more general population of students studied in the national sample. To test this hypothesis, she administers the same question to a random sample of 204 students at her college and then compares the pattern of responses to the pattern of the national sample.

The null hypothesis in this situation is that the responses of Professor H's 204 respondents should not differ significantly from the pattern of the national survey. Thus, the MCE expected frequency of response in the "strongly disagree" category would be 9.4% of the 204 subjects: .094x204=19.2. For the "moderately disagree" category, it would be 15.6% of the subjects: .156x204=31.8. And so on.

What Professor H actually found, however, was something that seemed to fit rather well with her suspicion concerning polarization. All that remained was to determine whether the difference between the two patterns was significant. Here is an overview of the observed counts and percentages for the five response categories, each in comparison with the corresponding MCE expected values, along with the steps required for calculating chi-square.

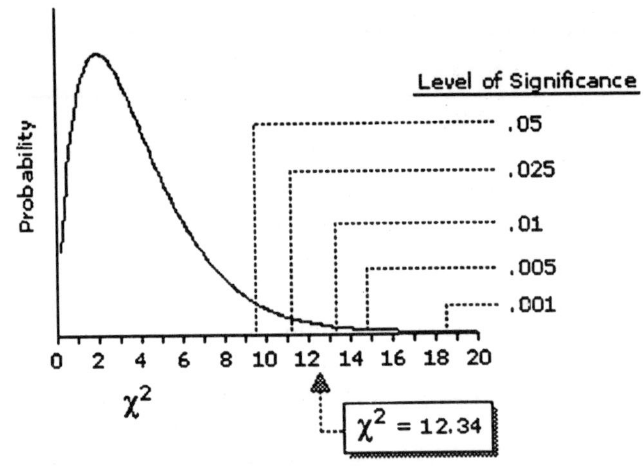

Level of Significance (non-directional test)

df	.05	.025	.010	.005	.001
4	9.49	11.14	13.28	14.86	18.47

critical values of chi-square for **df** = 4

Fig. Chi-Square Sampling Distribution for df=4

	strongly disagree	*moderately disagree*	*undecided*	*moderately agree*	*strongly agree*	*Total*
O	28	34	50	57	35	204
	(13.7%)	(16.7%)	(24.5%)	(27.9%)	(17.2%)	
E	19.2	31.8	70.0	56.1	26.9	204
	(9.4%)	(15.6%)	(34.3%)	(27.5%)	(13.2%)	
$(O-E)^2$	$(28-19.2)^2$	$(34-31.8)^2$	$(50-70)^2$	$(57-56.1)^2$	$(35-26.9)^2$	
E	19.2	31.8	70	56.1	26.9	$\chi^2 =$
	=4.03	=0.15	=5.71	=0.01	=2.44	12.34
						df=4

The top part of Figure shows a graph of the sampling distribution of chi-square for the case of df=4, while the bottom part reproduces the portion of the table of critical values of chi-square that pertains to this distribution. As indicated in both the graphic and tabular parts, our observed value of χ^2=12.34 is significant at the minimal .05 level, but also at and beyond the more stringent .025 level. In brief, we can be about 97.5 percent confident that the difference between the observed and MCE expected patterns of frequencies does not result from mere random variability.

Do keep in mind, however, that the chi-square test we have just performed is intrinsically non-directional. In and of itself, the significant chi-square value says nothing at all about the particular texture or direction of the difference. Examine the details of Figure, however, and you will see that the texture of the difference *is* consistent with Professor H's hypothesis concerning greater conservative-liberal polarization among the students at her college. In particular, there was a smaller proportion of respondents in the "undecided" category than would have been expected on the null hypothesis, and greater proportions in the "strongly agree" and "strongly disagree" categories. These three categories, in fact, accounted for all but a small fraction of the calculated chi-square value of 12.34.

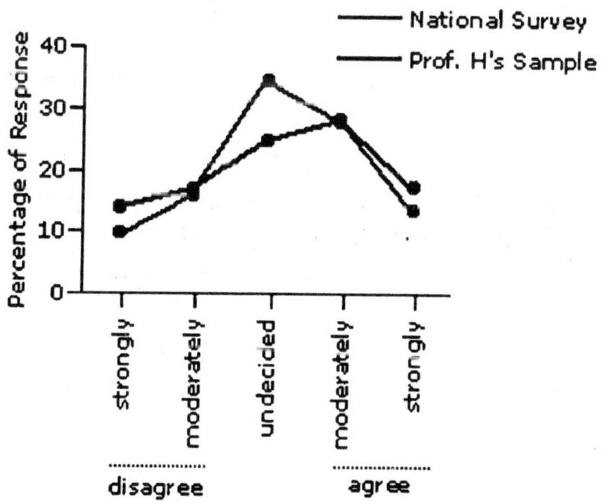

Fig. Professor H's Sample Versus the Results of the National Survey

CHI-SQUARE PROCEDURES FOR TWO DIMENSIONS OF CATEGORIZATION

When observed items are sorted according to two or more separate dimensions of classification concurrently, they are said to be cross-categorized. The most efficient way to represent cross-categorized frequency data is with what is known as a contingency table. Suppose, for example, that there is a certain University–we will call it University Z, or UZ for short–which has several graduate professional schools. An investigator is interested in determining whether the students at two of these entities, the L-school and the B-school, differ in their respective proportions of liberals and conservatives. To this end, the investigator takes a random sample of students from each of the two schools and asks them which of the two labels they prefer as a self-description, "liberal" or "conservative." She then cross-categorizes the subjects according to;

- . Their school affiliation, L- or B;
- Their preferred self-description as either "liberal" or "conservative." The resulting contingency table would have the following form, except that the various verbal descriptions beginning with the phrases "number of" and "total" would of course be replaced with specific numerical values.

		Preferred Self-Description		
		Liberal	*Conservative*	
School	L-school	number of L-school liberals	number of L-school conservatives	total L-school subjects
	B-school	number of B-school liberals	number of B-school conservatives	total B-school subjects
		total liberals	total conservatives	total subjects

When chi-square procedures are applied to a contingency table of this sort, it is typically with the aim of determining whether the two categorical variables are associated; hence the name of this version, the chi-square test of association. If L-school and B-school subjects divided themselves as "liberal" and "conservative" in equal proportions, as in the following example

	Liberal	*Conservative*
L-school	55% of L-school subjects	45% of L-school subjects
B-school	55% of B-school subjects	45% of B-school subjects

there would clearly be no evidence of association between the two variables. If on the other hand they divided themselves up in different proportions, for example

	Liberal	*Conservative*
L-school	55% of L-school subjects	45% of L-school subjects
B-school	45% of B-school subjects	55% of B-school subjects

then there would be at least the suggestion of an association between the two variables. In this particular instance, the null hypothesis would be that there is no overall difference between L-school and B-school students at UZ with respect to how they divide themselves up on the liberal/conservative self-description. The chi-square test of association permits one to determine how likely it is that random samples of L-school and B-school students could have produced a liberal/conservative difference this large or larger, by mere chance coincidence, if the null hypothesis were true.

Here is a concrete example of cross-categorized frequency data, adapted from a medical study that sought to determine whether estrogen supplementation might delay or prevent the onset of Alzheimer's disease in postmenopausal women.

M-*X*. Tang, D. Jacobs, *Y*. Stern, et al., "Effect of oestrogen during menopause on risk and age at onset of Alzheimer's disease." *Lancet* 1996; 348, 429-32.

The subjects in the study were 1,124 postmenopausal women who were initially free of Alzheimer's disease, Parkinson's disease, and stroke, and who were taking part in a longitudinal study of ageing and health in a New York City community. Of these women, 156 had received estrogen supplementation following the onset of menopause, while the remaining 968 had not. The following table breaks these two groups down according to whether the women did or did not show signs of Alzheimer's disease onset during a five-year follow-up period, as judged by annual clinical assessments and criterion-based diagnoses.

		Alzheimer's onset during 5-year period		
		No	Yes	
received	Yes	147	9	156
estrogen	No	810	158	968
		957	167	1,124

In case the bare numbers in this table do not jump right out at you, here is the bottom line in percentage terms: Of the women who did not receive estrogen supplementation, 16.3% (158/968) showed signs of Alzheimer's disease onset during the five-year period; whereas, of the women who did receive estrogen supplementation, only 5.8% (9/156) showed signs of disease onset. It is not difficult to adduce reasons that might account for such a difference. As the authors of the study note, estrogen "promotes the growth of cholinergic neurons, stimulates the secretase metabolism of the amyloid precursor protein, and may interact with apolipoprotein E. All these factors could affect the risk of Alzheimer's disease." But of course, none of these possibilities amounts to anything at all unless we can establish that the difference between the observed percentages–16.3% versus 5.8%–reflects anything other than mere random variability. Once again it is the question of statistical significance: How likely is it that a difference this large or larger could have occurred by mere chance coincidence? The procedures for applying chi-square to a two-dimensional situation of this sort are the same as we saw for the one-dimensional situation, except for two small modifications. The first has to do with what we take to be the MCE expected cell frequencies, and the second pertains to how we calculate the appropriate value for degrees of freedom. In both of these modifications the underlying logic is the same; it is only the details that differ.

In the one-dimensional chi-square test the values of E are simply stipulated in advance. In the fish example they were given as 100/100/100, because the sample was to be compared with the proportions of the three fish species that had been recorded over the preceding years. In the student survey example they were set to match the proportions of responses in the various categories that had been found in the large national survey.

In the two-dimensional situation, on the other hand, the expected cell frequencies are not given in advance, nor are they intuitively obvious. We will illustrate the logic of the point with a simple if somewhat fanciful example.

Two friends, A and B, believe their friendship to be so deep as to produce a remarkable pattern of correspondences. Often they find that they are thinking the same thing at the same time. Often, even when separated, they find that they are doing the same thing at the same time. To put their faith to the test, they each toss a coin 100 times in succession, recording on each occasion the head/tail outcome of A's toss and the corresponding head/tail outcome of B's toss. Their hypothesis is that when A gets a head, B will also tend to get a head; and that when A gets a tail, B will also tend to get a tail.

They of course do not suppose that even their relationship is so deep as to ensure these correspondences in 100 percent of the tosses, though they do believe that the pattern of such correspondences will significantly exceed what would be expected on the basis of mere chance coincidence.

For the sake of discussion, suppose that A and B each end up with exactly 50 heads and 50 tails. In that case the contingency table would have the following marginal totals, irrespective of how much or little the heads and tails outcomes of A and B might correspond.

		Outcomes for B		
		Tails	*Heads*	
Outcomes for A	Heads	[cell a]	[cell b]	50
	Tails	[cell c]	[cell d]	50
	50	50	100	

Actually, this is one of the rare scenarios for which the values of E would be fairly intuitively obvious. If you were asked to guess the values of the MCE expected frequencies for cells a, b, c, and d, I expect you would answer 25/25/25/25–and this would be quite correct. Now, if only you can make explicit the hidden logic that leads you to this answer, you will have the procedure for figuring out the values of E for two-dimensional chi-square situations in general. I suspect the core of your implicit logic runs something like this: If A gets 50 percent heads and B gets 50 percent tails; and if nothing other than mere chance coincidence is operating in the situation; then the (conjunctive) probability that any particular one of the 100 paired tosses will include a head for A and a tail for B is .5x.5=.25. Thus, the expected frequency for cell a is 25 percent of the total number of paired tosses: E_a=.25x100=25. The same logic would also lead you to E_b=25, E_c=25, and E_d=25.

Now see how the same logic can be extended to situations that are not intuitively obvious. When A and B perform their series of 100 paired tosses, it is actually not very likely that they would both end up with exactly 50 percent heads and 50 percent tails. It is much more likely that they each would end up with something slightly different from an exact 50/50 split. Suppose that A comes out with 46 heads and 54 tails, while B ends up with 48 heads and 52 tails. In this case the marginal totals would be distributed as follows

		Outcomes for B		
		Tails	*Heads*	
Outcomes	Heads	[cell a]	[cell b]	46
for A	Tails	[cell c]	[cell d]	54
	52	48	100	

and the logic would run like this. It is the same logic as outlined above, but now we will lay it out more formulaically. The proportion of paired tosses that include a head for person A is 46/100=.46; the proportion that include a tail for person B is 52/100. If nothing other than mere chance coincidence is operating in the situation, the probability that any particular one of the 100 paired tosses will include a head for A and a tail for B is therefore .46x.52=.2392; and the mean chance expected number of such conjunctions in 100 paired tosses is accordingly

$$\mathrm{E}_a = \frac{46}{100} \times \frac{52}{100} \times 100 = 23.92.$$

This same structure can now be generalized to any situation at all where we have frequency data arranged in a rows by columns matrix. Defining "*R*" as the marginal total of the row to which the cell belongs, "*C*" as the marginal total of the column to which the cell belongs, and "*N*" as the total number of cross-categorized observations, we can write it as

$$E_{\text{cell}} = \frac{R}{N} \times \frac{C}{N} \times N.$$

which reduces algebraically to the simpler structure

$$E_{\text{cell}} = \frac{R \times C}{N}.$$

That is: For each cell, multiply the marginal total for the row to which the cell belongs by the marginal total for the column to which the cell belongs, and then divide the result by the total number of cross-categorized observations.

The following illustration shows how this simple calculation works out for each of the cells of the present coin-toss example.

		Outcomes for B		
		Tails	*Heads*	
Outcomes for A	Heads	$E_a = \frac{46\times52}{100} = 23.92$	$E_b = \frac{46\times48}{100} = 46\times48$	46
	Tails	$E_c = \frac{54\times52}{100} = 22.08$	$E_a = \frac{54\times48}{100}$ 54×52	54

And here is the same procedure applied to our estrogen/Alzheimer's example.

		Alzheimer's onset during 5-year period		
		No	*Yes*	
receivedestrogen	Yes	$E_a = \frac{156\times957}{1124} = 132.82$	$E_b = \frac{156\times167}{1124} = 23.18$	156
	No	$E_c = \frac{968\times957}{1124} = 824.18$	$E_d = \frac{968\times167}{167} = 143.82$	968
		957	167	1,124

Translation: The null hypothesis holds that estrogen supplementation in postmenopausal women makes no difference with respect to Alzheimer's onset. If this null hypothesis were true, then these are the frequencies we would have expected for the various cells of the table, given the marginal totals that appear along the right edge and the bottom of the table.

In case all this calculation and its accompanying jargon are proving a bit confusing, step back for a moment and think of it this way. On the null hypothesis, we would have expected the two categories of subjects–those who received estrogen supplementation and those who did not–to show approximately the same proportionate incidence of Alzheimer's onset during the five-year period. And that is exactly what the calculated values of *E* are specifying. Thus, of the 156 subjects who did receive estrogen supplementation, we would have expected a 23.18/156=14.9% rate of onset; and for the 968 subjects who did not receive supplementation, we would have expected an identical 143.82/968=14.9% rate. The same holds true for the expected rate of non-onset: 132.82/156=85.1% for those who received supplementation, and 824.18/968=85.1% for those who did not receive supplementation.

The next table shows these values of *E*, now listed in red, in comparison with the original values of O. Notice, incidentally, that the values of O and *E*, when summed across the rows and the columns, both add up to the same marginal totals.

		Alzheimer's onset during 5-year period		
		No	*Yes*	
received estrogen	Yes	147	9	
		132.82	23.18	156
	No	810	158	
		824.18	143.82	968
		957	167	1,124

Beyond this point, the calculation of chi-square for the two-dimensional situation depends on the particular numbers of rows and columns. If there are *more than two rows or more than two columns,* it proceeds exactly as it does in the one-dimensional case, in accordance with the formula

$$\chi^2 = \Sigma\frac{(O-E)^2}{E}.$$

When there are *exactly two rows and two columns,* the calculation requires a *correction for continuity* analogous to the correction for binomial probability situations:

$$\chi^2 = \Sigma\frac{(|O-E|-.5)^2}{E}.$$

I.e., for each cell, subtract one-half unit from the absolute (unsigned) difference between O and E before squaring that difference.

As our current example has exactly two rows and two columns, the appropriate procedure is the latter. The following table shows the specific calculations for each of the four cells.

		Alzheimer's onset during 5-year period		
		No	*Yes*	
receivedestrogen	Yes	$\frac{(\lvert 147-132.82\rvert-.5)^2}{132.82}=1.41$	$\frac{(\lvert 9-23.18\rvert-.5)^2}{23.18}=8.07$	
	No	$\frac{(\lvert 810-824.18\rvert-.5)^2}{824.18}=0.23$	$\frac{(\lvert 158-143.82\rvert-.5)^2}{143.82}=1.3$	sum:
				χ^2= 11.01

The next step is to refer our calculated value of chi-square to the appropriate sampling distribution, which you will recall is defined by the applicable number of degrees of freedom. When applying chi-square procedures to situations in which there is only one dimension of categorization, the general principle for determining degrees of freedom is

$$df = (\text{number of cells}) - 1$$

With two dimensions of classification, it is a different formula but the same basic logic. Suppose you have two rows and two columns of cells, as shown below, and you are free to plug any integer numbers that you want into them, subject only to the stipulation that the cell values summed across rows, down columns, and overall, must add up to the fixed row, column, and overall sums that appear along the right edge and across the bottom.

cell a = ?	cell b = ?	50
cell c = ?	cell d = ?	40
48	42	90

Here again your "freedom" is limited by the fixed sums. Arbitrarily plug 10 into cell a and the remaining three cells are instantly fixed at b=40, c=38, and d=2. Plugging 20 into cell d fixes the other cells at b=22, c=20, and a=28; and so on. For chi-square situations involving two rows and two columns of cells, degrees of freedom will in every instance be equal to one.

If you have two rows and three columns, your "freedom" is increased a bit, though still limited by the fixed sums.

cell a = ?	cell b = ?	cell c = ?	50
cell d = ?	cell e = ?	cell f = ?	40
25	40	25	90

If you plug 10 arbitrarily into cell a, cell d becomes fixed at 15, but the other four cells remain "free" to vary. Plug some other number into any of these remaining four cells, however, and everything else becomes instantly fixed. For chi-square situations involving two rows and three columns (or three rows and two columns), degrees of freedom is in every case equal to two. More generally, when applying chi-square procedures to situations in which there are two dimensions of categorization, the general principle for determining degrees of freedom is

$$df = (r-1)(c-1) \frac{r = \text{number of rows}}{c = \text{number of columns}}.$$

For our estrogen/Alzheimer's example, there are 2 rows and 2 columns. Hence

$$df = (2-1)(2-1) = 1.$$

Figure shows the sampling distribution of chi-square for df=1, along with the critical values of chi-square for several levels of significance. As indicated, a chi-square value equal to or greater than 3.84 would be significant at or beyond the .05 level; a chi-square value equal to or greater than 5.02 would be significant at or beyond the .025 level; and so on.

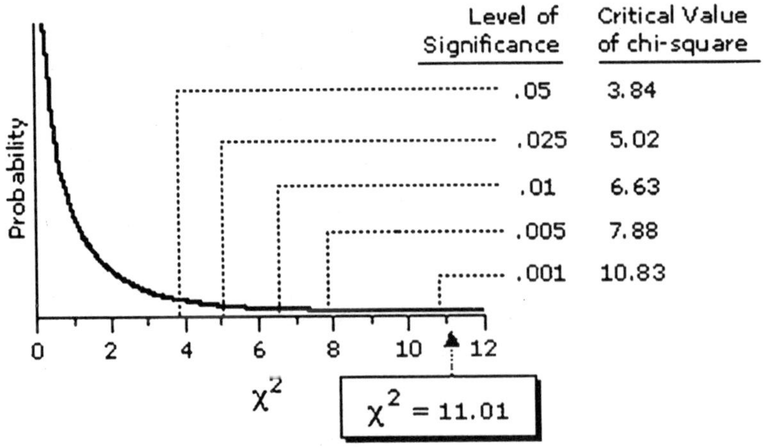

Fig. 8.6. Sampling Distribution of Chi-Square for df=1

Our calculated value of =11.01 exceeds the value of chi-square (10.83) required for significance at the .001 level; hence we can say that the observed result is significant *beyond* the .001 level. That is: If the null hypothesis were true–if estrogen supplementation were unrelated to the onset of Alzheimer's disease in postmenopausal women–the likelihood of obtaining a differential onset rate as great as the one observed (16.3% versus 5.8%), by mere chance coincidence, would be smaller than one-tenth of one percent. Our investigators can therefore reject the null hypothesis with a high degree of confidence.

Chi-square procedures for two dimensions of categorization are not limited to the case where there are just two rows and two columns. In principle, they can be applied to contingency tables involving any number of rows and columns, although in practice you will not often find more than three or four of each. We will illustrate this type of application with an example involving two rows and three columns.

A team of clinical researchers is interested in assessing the effectiveness of a certain newly developed form of behavior therapy for the treatment of chronic anxiety conditions. To this end, they collect a sample of 150 persons who, although diagnosed with chronic anxiety, are currently receiving no treatment for their condition. Of these subjects, 60 are randomly selected to receive a two-month program of the behavior therapy, while the remaining 90 continue to receive no treatment. At the end of the two-month period, all 150 subjects are assessed with respect to whether their anxiety condition has improved, grown worse, or remained essentially unchanged. The following contingency table shows the results of this study, cross-categorized according to the two variables;

- Therapy vs. no therapy,
- Improved, worse, or no change (*N*/C). Also shown here, in parentheses, are the percentage figures within each of the two rows.

	Worse	*N/C*	*Improved*	
Therapy	24	11	25	
	(40.0%)	(18.3%)	(41.7%)	60
No Therapy	30	31	29	
	(33.3%)	(34.4%)	(32.2%)	90
	54	42	54	150

And here is a graph of the outcome, portrayed in terms of the percentages of subjects within each of the two groups who ended up being assessed as improved, worse, or no change.

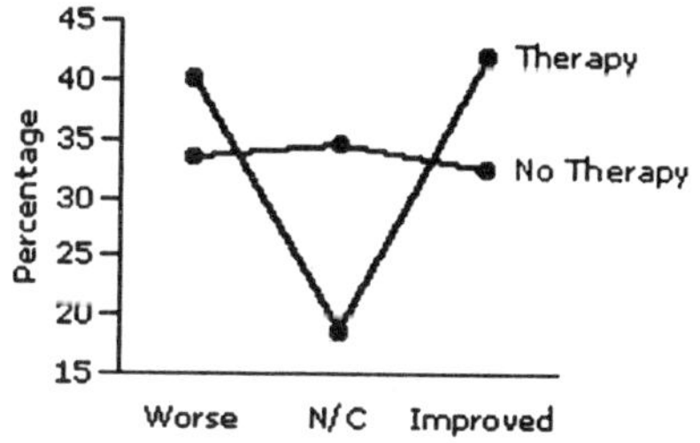

So the investigators do appear to have found a difference between the two groups, though it is not the simple difference they might have expected. Sure enough, the percentage of those assessed as "improved" is greater within the therapy group than within the no-therapy group

(41.7% vs. 32.2%); but at the same time, the percentage judged "worse" within the therapy group is also greater (40.0% vs. 33.3%). Might it be that this particular form of therapy has positive effects for some chronic-anxiety subjects and negative effects for others? If so, the researchers would certainly want to investigate further, to determine just what spells the difference between those cases where the effects are positive and those where they are negative.

But of course there is no point in doing anything at all until it can be established that the observed difference between the therapy and no-therapy groups reflects anything other than mere chance coincidence. The first of the following two tables shows the calculations for expected cell frequencies and the second details the calculations for the cell components of chi-square.

Expected Cell Frequencies

	Worse	*N/C*	*Improved*	
Therapy	$\frac{60\times54}{150}=21.6$	$\frac{60\times42}{150}=16.8$	$\frac{60\times54}{150}=21.6$	60
No Therapy	$\frac{90\times54}{150}=32.4$	$\frac{90\times42}{150}=25.2$	$\frac{90\times54}{150}=32.4$	90
	54	42	54	150

Recall that the calculation of chi-square requires a correction for continuity only when there are exactly two rows and two columns. For the present example we accordingly use the simpler chi-square formula that does not involve a correction for continuity:

$$\chi^2 = \Sigma\frac{(O-E)^2}{E}.$$

Thus:

Cell Components of Chi-Square

	Worse	*N/C*	*Improved*	
Therapy	$\frac{(24-21.6)^2}{21.6}=0.27$	$\frac{(11-16.8)^2}{16.8}=2.00$	$\frac{(25-21.6)^2}{21.6}=0.54$	
No Therapy	$\frac{(30-32.4)^2}{32.4}=0.18$	$\frac{(31-25.2)^2}{25.2}=1.33$	$\frac{(29-32.4)^2}{32.4}=0.36$	sum:
				$\chi^2=4.68$

With two rows and three columns, degrees of freedom is equal to 2, so that is the sampling distribution of chi-square to which this value gets referred. Figure shows the shape of the sampling distribution for df=2, along with the values of chi-square required for several levels of significance.

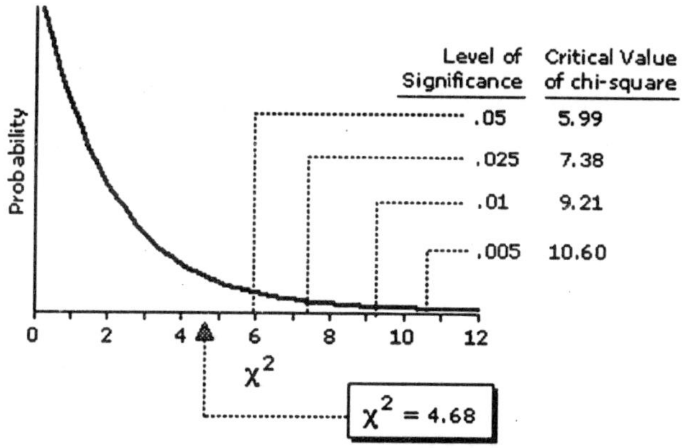

Fig. Sampling Distribution of Chi Square for df = 2.

As you can see, the calculated value of =4.68 falls well short of the 5.99 that would be required for significance at the minimal .05 level. Hence, although the observed difference between the therapy and no-therapy groups is impressive to the naked eye, it is non-significant. That is, by the accepted criterion of statistical inference, it cannot be taken as signifying anything other than mere random variability. So our investigators, unable to reject the null hypothesis, cannot confidently conclude anything at all about the effects of this form of therapy, one way or the other. All that work for nothing!

Ah! But suppose, as shown below, that they had started out with samples twice the size and ended up with the same proportionate results.

	Worse	*N/C*	*Improved*	
Therapy	48	22	50	
	(40.0%)	(18.3%)	(41.7%)	120
No Therapy	60	62	58	
	(33.3%)	(34.4%)	(32.2%)	180
	108	84	108	300

Exactly the same proportions. Exactly the same graph.

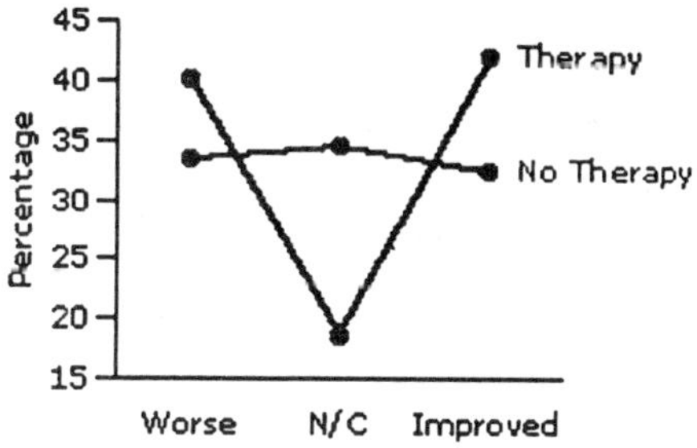

But now they would end up with = 9.35, which, as you can see in Figure, is significant beyond the .01 level.

In this second scenario the researchers could reject the null hypothesis with a high degree of confidence, and could accordingly infer that the therapy does have the effect suggested by the observed differences between the therapy and no-therapy groups. While laying out some basic concepts of probability, we noted that the mere-chance likelihood of getting as many as 70% heads in N tosses of a coin is fairly large (.17) when N=10 and very tiny (.00005) when N=100. It is the same principle at work here in these two chi-square scenarios. For any particular proportionate difference between an observed value and the corresponding mean chance expected value: the larger the sample, the less likely it is that a difference that large or larger will occur by mere chance coincidence; hence the greater will be the statistical significance of such a difference.

AVOIDING AN EGREGIOUS LOGICAL FALLACY:

Interlude on What NOT to do with a Non-Significant Chi-Square Value

There is an important implication here that we must take a moment to draw out. By way of analogy, suppose you were planning to hike through a forest; but before you hike very far, you want to determine whether the forest includes among its denizens a certain species of large, irritable bear. So you cautiously sample a limited area of the forest to see whether there are any signs of the bear. Your null hypothesis is that the bear is not present in the forest. And here are the two possible outcomes of your investigation: Either you find indications of the bear's presence–its growlings, droppings, perhaps even a visual sighting–or you do not. In the first case, you could reject the null hypothesis with a certain degree of confidence and conclude that the bear probably does abide there, in which event you might wish to choose a different venue for your hike. This much is fairly obvious. But what can you you conclude in the second case, when the outcome is negative? You have looked for the bear and found no sign of it. Does that permit you to conclude that the species is probably not to be found in the forest?

The short answer is that this second outcome would allow you to conclude nothing at all with any degree of confidence. Of course, it might be that the bear truly is absent from the forest. But then again, it might be that you have simply not sampled enough of the forest. There is a certain tendency of the human mind to assume that if we look for something and do not find it, then it does not exist. It is a fundamental error of logic, obviously to be avoided in the case of irritable bears, and no less to be eschewed with tests of statistical significance.

There is a certain application of the chi-square procedures for a one-dimensional situation, known as the chi-square goodness-of-fit test, that rests in my view on this same basic logical fallacy. The following example is rather fanciful, but not at all far-fetched.

Suppose a researcher frames the hypothesis that four variations on a certain genetic characteristic, A, B, C, and D, exist within the muglout species in the proportions: A=20%, B=30%, C=30%, and D=20%. To test his hypothesis, he examines N=60 randomly selected muglouts and sorts each according to whether it displays variation A, B, C, or D. The following table shows the results he obtained (O), laid out in comparison with the results that would have been predicted (E) on the basis of his genetic hypothesis.

	A	*B*	*C*	*D*	*Totals*
O	15	15	15	15	
	(25.0%)	(25.0%)	(25.0%)	(25.0%)	60
E	12	18	18	12	
	(20.0%)	(30.0%)	(30.0%)	(20.0%)	60

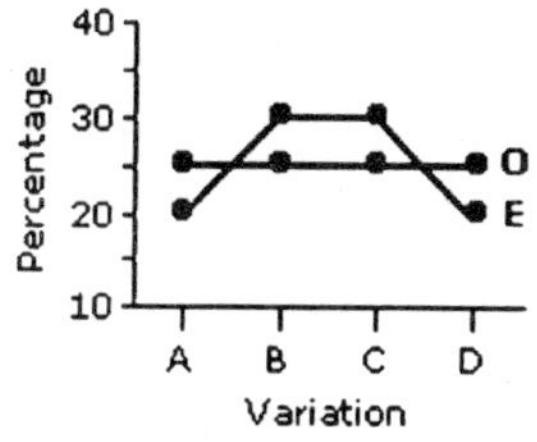

Performing the requisite calculations, our investigator finds a non-signigicant chi-square value of 2.5 (for df=3 a value of 7.81 would be required for significance at the minimal .05 level) and joyfully exclaims: "Aha! Although the observed pattern of frequencies does differ a bit from the pattern stipulated by my hypothesis, it does not differ significantly. Therefore–it's a fit! My hypothesis is supported by the data. Tenure Heaven, here I come!"

The logical fallacy in this scenario is contained in the phrase "Therefore–it's a fit!" The non-significant test result in this case would allow the investigator to conclude that this particular sample provides no evidence contradicting his hypothesis. But the absence of evidence contradicting an hypothesis is not at all the same thing as positive evidence in support of an hypothesis. Indeed, the pattern of frequencies and percentages observed in this particular sample would be consistent with (would not significantly differ from the pattern stipulated by) any number of other hypotheses that the investigator might have framed. The pattern predicted by the investigator's hypothesis was

[E1] A=20%, B=30%, C=30%, D=20%.

Suppose that he had instead predicted quite the opposite:

[E2] A=30%, B=20%, C=20%, D=30%.

As suggested by the adjacent graph, this mirror-image null hypothesis would have yielded precisely the same non-significant chi-square value of 2.5. So here as well our investigator might have proclaimed, enthusiastically but quite wrongly: "It's a fit! My hypothesis is supported by the data!"

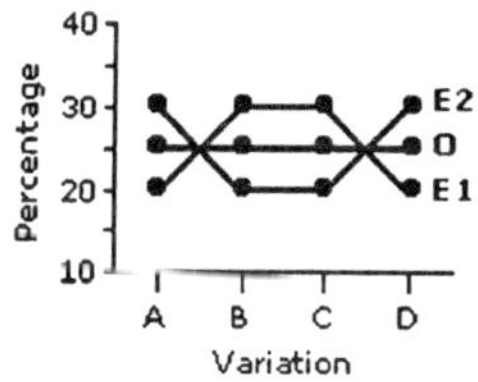

Similarly, if the same proportionate outcome had been found in a sample of N=200 muglouts, the resulting chi-square value of 8.33 *would* be significant beyond the .05 level, and our investigator's ascension to Tenure Heaven would accordingly have to be postponed.

	A	*B*	*C*	*D*	*Totals*
O	50	50	50	50	
	(25.0%)	(25.0%)	(25.0%)	(25.0%)	200
E	40	60	60	60	
	(20.0%)	(30.0%)	(30.0%)	(20.0%)	200

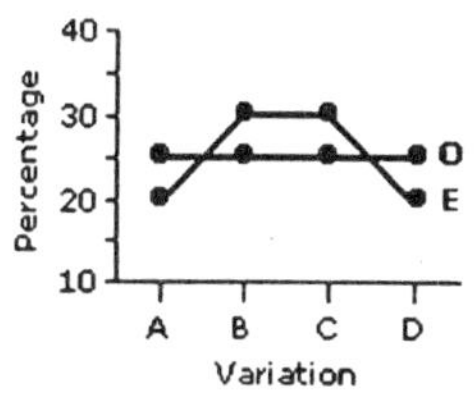

The principle at stake here is one that applies not just to chi-square, but to statistical tests in general. It is that, while a significant test result permits you to *reject* the null hypothesis with a certain degree of confidence, a non-significant result does not allow you to *accept* the null hypothesis.

TWO LIMITATIONS ON THE USE OF CHI-SQUARE PROCEDURES

Although chi-square procedures are computationally simple, they rest upon a rather complex logical substructure; and this, in turn, imposes certain limitations on how they can be applied. Two of these in particular are worth mentioning at the present time.

Restriction 1. Chi-square procedures can be legitimately applied only if the categories into which the *N* observations are sorted are independent of each other; that is, only if the placement of each observation into a particular category does not in any way depend on the placement of any of the other observations. For the beginning student of the subject, this restriction is best observed by ensuring that the categories are both exhaustive and mutually exclusive, such that each observation fits into one or another of the categories and no observation fits into more than one.

Restriction 2. The logical validity of the chi-square test is greatest when the values of E, the mean chance expected frequencies within the cells, are fairly large, and decreases as these values of E become smaller. Although the statistical cognoscenti do not always agree on just where to draw the line between "large enough" and "too small," the beginning student can take this as a practical rule of thumb: Chi-square procedures can be legitimately applied only if all values of E are equal to or greater than 5. For the special case of two rows by two columns, this limitation can usually be circumvented through application of the Fisher Exact Probability Test.

THE FISHER EXACT PROBABILITY TEST

Chi-square procedures can be legitimately applied only if all values of E are equal to or greater than 5. The best way to avoid running afoul of this limitation is of course to base your observations on relatively large samples. But alas in real life there are many situations where all we can muster is a relatively small sample, and we must make the best of it. For the special case of two rows by two columns, a useful alternative to chi-square in situations of this sort is the Fisher Exact Probability Test. There is, in fact, one respect in which the Fisher test is plainly superior to a chi-square test, even in those cases where a 2x2 chi-square might be legitimately employed. It is that chi-square is intrinsically non-directional, whereas the Fisher procedure is capable of being applied as either a directional test or a non-directional test.

The bad news about the Fisher test is that the calculations required for it can be quite laborious and about as pleasant as an afternoon of root-canal surgery. The good news is that the computer on which you are now working will happily do all the tedious number-crunching for you, leaving your mind free to wrap itself around the underlying concepts.

First the concepts. To illustrate, suppose that 19 subjects who display a certain medical condition are assessed with respect to whether they do or do not possess two other characteristics, *X* and *Y*. On the basis of earlier informal observations on patients of this particular type, the investigators are aiming to test the directional hypothesis that subjects who possess characteristic *X* will tend also to show characteristic *Y*, and that those who do not show characteristic *X* will also tend not to show characteristic *Y*.

		X		
		No	*Yes*	
Y	Yes	2	7	9
	No	8	2	10
		10	9	19

The results of the assessment are clearly in line with the hypothesis. Of the 9 subjects who show characteristic *X*, all but 2 also show characteristic *Y*; and of the 10 subjects who do not show characteristic *X*, all but 2 also do not show characteristic *Y*. Alternatively, you can say that in 15 of the 19 subjects the presence or absence of characteristic *X* corresponds with the presence or absence, respectively, of characteristic *Y*. All that remains is to determine whether this ostensible association between *X* and *Y* reflects anything more than mere chance coincidence. In case it is not obvious from the bare numbers in the above table, please take a moment to note that any attempt to perform a chi-square test on this array of frequencies would come to a grinding halt as soon as we calculated the values of E, since three of these are smaller than the required minimal value of 5. The next table shows the specific values of E for each cell, as calculated.

		X		
		No	*Yes*	
Y	Yes	E=4.7	E=4.3	9
	No	E=5.3	E=4.7	10
		10	9	19

The basic precept:

Chi-square procedures can be legitimately applied only if all values of E are equal to or greater than 5.

THE LOGIC OF THE FISHER TEST

		X		
		No	*Yes*	
Y	Yes	2	7	9
	No	8	2	10
		10	9	19

The null hypothesis in our illustrative example is that there is no association between *X* and *Y*. And the question of statistical significance is accordingly this: If the null hypothesis were true–if any ostensible association between characteristics *X* and *Y* were the result of nothing more than mere chance coincidence–how likely is it that we might end up with a result this large or larger?

To develop the logic of the Fisher test, suppose that our investigators have performed their initial assessment and counted up the number of subjects who do and do not show characteristics *X* and *Y*, but have not yet sorted their subjects according to the correspondences of *X* and *Y*. In this case, all they would have would be the marginal totals shown in the following table.

		X		
		No	*Yes*	
Y	Yes			9
	No			10
		10	9	19

Given these marginal totals, there are 10 possible ways in which the specific correspondences between *X* and *Y* might sort themselves out by mere chance. We will label these as Ô1, Ô2, Ô3, and so on, with "Ô" standing in as an abbreviation for "outcome."

Ô1	Ô2	Ô3	Ô4	Ô5	Ô6	Ô7	Ô8	Ô9	Ô10
9 0	8 1	7 2	6 3	5 4	4 5	3 6	2 7	1 8	0 9
1 9	2 8	3 7	4 6	5 5	6 4	7 3	8 2	9 1	10 0

"this large or larger"

Outcomes falling toward the left end of this array would betoken ostensible negative association between *X* and *Y*, those falling toward the right end would betoken ostensible positive association, and those lying toward the middle would approximate zero association. I expect it is intuitively obvious to you that the mere-chance probability of an outcome is greatest toward the middle of the range and decreases sharply as you go out toward either extreme. The particular outcome observed by the investigators is T8, so the question of statistical significance in this case takes the specific form: If the null hypothesis were true, how likely is it that we could end up with either Ô8 or Ô9 or Ô10?

Ô8	Ô9	Ô10
2 7	1 8	0 9
8 2	9 1	10 0

In principle, the application of the Fisher test to a situation of this sort is quite simple: figure out the exact probability for each possible outcome that is "this large or larger," and then add up these separate disjunctive probabilities to get the answer.

There are several paths we could follow at this point, all of which would eventually lead to the same destination. I choose the one most likely to make sense to the beginning student and least likely to bring the non-mathematical reader to the brink of tears. The mere-chance probability of any particular outcome is fundamentally a proportion of the general form

$$P_{(outcome)} = \frac{\text{number of possibilities favorable to the occurrence of the outcome}}{\text{total number of pertinent possibilities}}.$$

We begin by figuring out the value of the denominator of the expression, which is the total number of pertinent possibilities. If you were to toss 19 coins, the number of possible combinations that would yield exactly 9 heads out of those 19 tosses would be

$$\frac{N!}{k!(N-k)!} = \frac{19!}{9!10!} = 92{,}378.$$

		X		
		No	*Yes*	
Y	Yes			9
	No			10
		10	9	19

The same concept applies when you are assessing 19 subjects with respect to the possession or non-possession of characteristic *X*. The total number of combinations that would yield exactly 9 subjects displaying *X* and 10 subjects not displaying it would be

$$\frac{19!}{9!10!} = 92{,}378.$$

		X		
		No	*Yes*	
Y	Yes			9
	No			10
		10	9	19

And similarly for characteristic *Y*. The total number of combinations that would involve exactly 10 subjects displaying *Y* and 9 subjects not displaying it would be

$$\frac{19!}{9!10!} = 92{,}378.$$

The total number of possible outcomes in which you might have exactly 9 out of 19 subjects displaying *X* and exactly 10 out of the same 19 subjects displaying *Y* is accordingly

$$92{,}378 \times 92{,}378 = 8{,}533{,}694{,}884$$

So here is how our probability calculation would begin to shape up for any particular one of the 10 possible outcomes enumerated above, T1 through T10.

$$P_{(outcome)} = \frac{\text{number of possibilities favorable to the occurrence of the outcome}}{\text{total number of pertinent possibilities}}.$$

$$P_{(outcome)} = \frac{?}{8{,}533{,}694{,}884}.$$

- All we now have to do for each particular outcome is replace the question mark with a specific numerical value.

		X		
		No	*Yes*	
Y	Yes	0	9	9
	No	10	0	10
		10	9	19

We begin with T10, which is the most extreme of the possible "this large or larger" outcomes and by far the easiest to examine. For each of the 92,378 possible combinations that would yield exactly 9 subjects displaying *X* and 10 subjects not displaying it, there is exactly one of the *Y* combinations that would involve every instance of *X* being associated with an instance of *Y* and every instance of not-*X* being associated with an instance of not-*Y*. So the probability of this particular outcome would be

$$P_{(\hat{O}10)} = \frac{92{,}378}{8{,}533{,}694{,}884} = \frac{1}{92{,}378}.$$

		X		
		No	*Yes*	
Y	Yes	1	8	9
	No	9	1	10
		10	9	19

The logic is the same with the less extreme outcomes, although rather more complicated in computational detail. Here, to illustrate, is the next less extreme outcome, T9. For each of the 92,378 possible combinations that would yield exactly 9 subjects displaying *X* and 10 subjects not displaying it, the number of combinations that would produce the particular correspondences of *X* and *Y* shown in the first column of the above table (*X*="No") would be

$$\frac{10!}{1!9!} = 10.$$

and the number of combinations that would produce the particular correspondences shown in the second column (*X*="Yes") would be

$$\frac{9!}{8!1!} = 9.$$

So the number of combinations that would produce the overall array of frequencies in the gray cells of the table would be

$$10 \times 9 \times 92{,}378 = 8{,}314{,}020$$

and the probability associated with this outcome is accordingly

$$P_{(\hat{O}9)} = \frac{8{,}314{,}020}{8{,}533{,}694{,}884} = \frac{90}{92{,}378}.$$

		X		
		No	*Yes*	
Y	Yes	2	7	9
	No	8	2	10
		10	9	19

One more turn of this wheel and we are finished, at least for the present example. Here is the table for the actually observed outcome, T8. For each of the 92,378 possible combinations that would yield exactly 9 subjects displaying *X* and 10 subjects not displaying it, the number of combinations that would produce the particular correspondences of *X* and *Y* shown in the first column of the table would be

$$\frac{10!}{2!8!} = 45$$

and the number of combinations that would produce the particular correspondences shown in the second column would be

$$\frac{9!}{7!2!} = 36.$$

So the number of combinations that would produce the overall array of frequencies in the gray cells of this table would be

$$45 \times 36 \times 92{,}378 \;=\; 149{,}652{,}360$$

and the probability associated with this outcome is accordingly

$$P_{(\hat{O}8)} = \frac{149{,}652{,}360}{8{,}533{,}694{,}884} = \frac{1620}{92{,}378}.$$

The rest of the ride is a smooth one along a familiar path. The probability of getting a result "this large or larger" is the sum of the separate probabilities for Ô8, Ô9, and Ô10:

2 7	1 8	0 9
8 2	9 1	10 0

$$\frac{1620}{92{,}378} + \frac{90}{92{,}378} + \frac{1}{92{,}378} = \frac{1711}{92{,}378} = .0185.$$

And that, in a nutshell, is the probability that our investigators can take to the printing press. If the null hypothesis were true, the exact probability of finding a positive association between *X* and *Y* as large as the one observed would be a scant P=.0185. The investigators can therefore reject the null hypothesis with a comfortable degree of confidence and conclude that characteristics *X* and *Y* do tend to be associated for this particular type of subject.

A Simpler Formulaic Approach

Once you have the concepts, the actual calculations for the Fisher test can be performed by way of a relatively simple formulaic structure. For any 2 × 2 contingency table, designate the cell frequencies as a, b, c, and d, and the marginal totals as a+b, c+d, etc., with *N* equal to the sum of the frequencies in the four cells.

a	b	$a+b$
c	d	$c+d$
$a+c$	$b+d$	N

Given this notation, the exact probability of any particular outcome can then be calculated according to the formula

$$P_{(\text{outcome})} = \frac{(a+b)!\,(c+d)!\,(a+c)!\,(b+d)!}{N!\,a!\,b!\,c!\,d!}$$

In performing factorial operations recall that 0!=1 and 1!=1.

0	9	9
10	0	10
10	9	19

Thus, for outcome 10 in the above example:

$$P_{(\hat{O}10)} = \frac{9!\ 10!\ 10!\ 9!}{19!\ 0!\ 9!\ 10!\ 0!} = .000010825.$$

Note that this rounded decimal value is equivalent to the ratio 1/92,378 calculated above.

1	8	9
9	1	10
10	9	19

For outcome 9:

$$P_{(\hat{O}9)} = \frac{9!\ 10!\ 10!\ 9!}{19!\ 1!\ 8!\ 9!\ 1!} = .000974258.$$

Equivalent to 90/92,378

2	7	9
8	2	10
10	9	19

And for T8, the observed outcome:

$$P_{(\hat{O}8)} = \frac{9!\ 10!\ 10!\ 9!}{19!\ 2!\ 7!\ 8!\ 2!} = .017536642.$$

Equivalent to 1620/92,378.

THE FISHER TEST FOR DIRECTIONAL AND NON-DIRECTIONAL HYPOTHESES [H]

		9
		10
10	9	19

If you were to perform these calculations for each of the 10 possible outcomes in the example, the composite result would constitute the sampling distribution of outcomes for the general case where you have a 2x2 contingency table with these particular marginal totals. The following graph shows the outlines of this sampling distribution. As usual in such depictions, the vertical axis represents relative frequency, which is of course the same thing as probability.

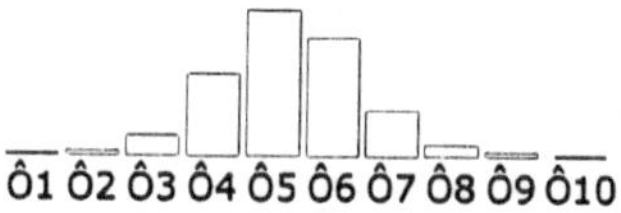

As our investigators begin with a directional hypothesis, their probability assessment needs to refer to only one of the two tails of this distribution, namely, the "this large or larger" portion on the right that includes the probabilities for T8, T9, and T10. If they had instead begun with a non-directional hypothesis–in effect, "Let's do this and see if there's an association one way or the other"–they would also have to refer to the portion of the left tail that includes the possible outcomes that are "this large or larger" in the opposite direction.

Ô1	Ô5	Ô10
9 0	5 4	0 9
1 9	5 5	10 0

Within this context, the concept of "this large or larger" is defined by the degree to which the frequencies within the cells are arrayed disproportionately, so as to fall predominantly along the upward diagonal, as in T10, or along the downward diagonal as in T1. The degree of disproportion within any array of cell frequencies–in effect, the degree of ostensible association in either direction–can be measured by the absolute difference

$$\text{disproportion} = \left|\frac{a}{a+b} - \frac{c}{c+d}\right|.$$

The following table shows this value as calculated for each of the 10 possible outcomes. The disproportion measure for T8, the observed outcome, is indicated in red, and the measures for the other outcomes that are "this large or larger," in either direction, are shown in boldface.

Ô1	Ô2	Ô3	Ô4	Ô5	Ô6	Ô7	Ô8	Ô9	Ô10
9 0	8 1	7 2	6 3	5 4	4 5	3 6	2 7	1 8	0 9
1 9	2 8	3 7	4 6	5 5	6 4	7 3	8 2	9 1	10 0
0.90	0.69	0.48	0.27	0.06	0.16	0.37	0.58	0.79	1.00

The short of it is that the test of a non-directional hypothesis would also have to fold in the probabilities for T1 and T2, which you would find upon calculation to be

$$P_{(\hat{O}1)} = .000108251 \quad \text{and } P_{(\hat{O}2)} = .004384161$$

Adding these values to the right-tail probability (.0185) already calculated yields a non-directional two-tailed probability of P=.023.

A HANDY NUMBER-CRUNCHER

Although the concepts are relatively simple, the calculations, as I warned at the outset, can be rather laborious. I am therefore providing you with this handy calculator, which will perform the Fisher exact probability test for any particular 2x2 table of cross-classified frequency data, providing that the numbers are not too large. Just for practice, and to assure yourself that the calculator actually works, you might want to try entering some of the 2x2 frequency arrays for the various outcomes, Ô1, Ô2, etc., of the foregoing example. You need not enter anything into the cells that contain "", as these marginal totals will be calculated automatically.

Chapter 3

Procedures Involving Sample Means

The example of a "certain disease that has an abrupt and unmistakable onset, and for which there is currently no effective treatment." In brief, a team of investigators administered a certain plant extract to a certain number of randomly selected patients, beginning for each patient immediately after the onset of the disease and continuing for two months. At the end of a two-month period, they sorted the patients into two outcome categories, "recovered" and "not recovered," and then employed binomial probability procedures to determine whether the number of patients in the "recovered" category was significantly greater than the 40% recovery rate that would have been expected by mere chance coincidence.

It perhaps occurred to you at the time that sorting patients into the two simple categories "recovered" and "not recovered" is rather like a shoe manufacturer producing shoes in only two sizes, "large" and "small." It makes no allowance for the possibility of gradations in-between. Up to a point you could allow for this possibility by sorting the outcomes into a larger number of categories and applying chi-square procedures: for example, "not improved," "somewhat improved," "much improved," and "completely recovered." Even this, however, would potentially leave out much of the fine-grained detail of the reality you are investigating.

Of course there are some distinctions, such as "female" and "male," that are intrinsically categorical and cannot be measured any other way. For others, such as "conscious,' "semi-conscious," and "unconsciousness," we might have a range of continuous gradations, but with no meaningful way of measuring them. Still, there is a wide range of phenomena for which the finer-grained detail of things can be measured.

Suppose, for example, that a certain normally present component of the patient's blood—component *Z*–systematically decreases in proportion to the degree of the disease. In this case you could take the concentration of component *Z* as an inverse measure of the degree of the disease in each particular patient. Providing these measures derived from an equal-interval scale of measurement, you could then also calculate their mean and (pending some other matters yet to be covered) end up being able to say something along the lines of: "On average, patients who received the experimental treatment showed greater improvement than they would have shown if they had not received the treatment." Notice that this is potentially saying quite a lot more than could be said if the outcomes were measured simply as "Recovered" and "Not Recovered." For you can easily imagine a scenario in which patients receiving the experimental treatment show greater improvement, on average, even though the number of full recoveries among them does not exceed the 40% rate to be expected on the basis of mere chance.

The key word in all of this is the phrase "on average." There is a very wide range of research situations where the focus is on the mean of a sample or on the means of two or more samples. This is not simply because science loves averages or the fine-grained measures on an equal-interval scale that averages require. It is because the statistical procedures applicable to this form of measurement are considerably more powerful and versatile.It is, however, very important theory, because it forms the logical groundwork for the eminently useful procedures that we will examine in the chapters following this one. The transition from pure theory to practical application will begin ; but first we must start at the beginning and build up the logic of it step-by-step.

SOME THINGS YOU ALREADY KNOW OR COULD READILY FIGURE OUT

Somewhere deep within the electronic workings of your computer I have created a vast population of X_i values. To prove to you that this population really exists, I have constructed a kind of window through which you can inspect some of its individual constituents. Each time you click the button labeled "*N*=1," your computer will reach into this population and draw one value of X_i at random. I will ask you to begin by clicking the button twenty or thirty times to gain an intuitive sense of how this population is put together.

The population from which you have just drawn your random values of X_i is shown below in Figure. As indicated, it is normally distributed with a mean of $\mu = 18$ and a standard deviation of $\sigma = \pm 3$. The first scale along the horizontal axis of the graph is framed in terms of z, with each unit of z representing a unit of standard deviation, while the scale below it is delineated in terms of the variable *X*. We will refer to this distribution henceforth as our "reference source population."

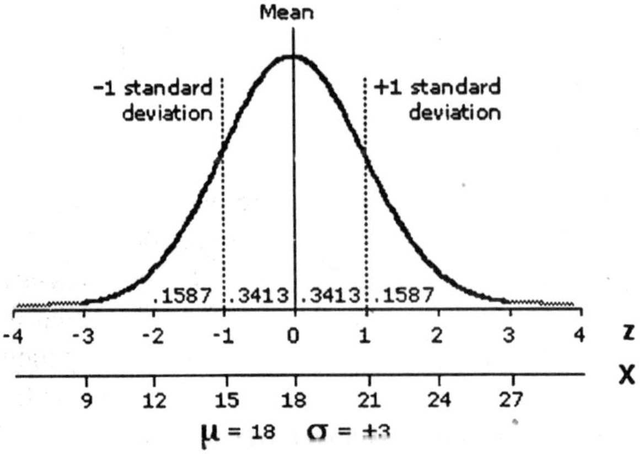

Fig. Reference Source Population, Normally Distributed with m =18 and s = ±3

Given that this source population is normally distributed, you know in advance certain facts about how the individual values of X_i sort themselves out. From the proportions listed in the graph (.1587, .3413, etc.), you know at a glance that 34.13% of all instances of X_i fall between the mean and +1 standard deviation; symmetrically, that 34.13% of all instances fall between the mean and −1 standard deviation; and hence, that 68.26% of all values of X_i fall between −1 and +1 standard deviation. Similarly, you know that 15.87% of all instances fall to the left of −1 standard deviation, that an equal 15.87% fall to the right of +1 standard deviation, and so on. Consult the table of the standard normal distribution and you will be able to comb it out in much finer-grained detail. Here again is an abbreviated version of that table, just to remind you of how it is laid out.

Table. Proportions of the Normal Distribution Falling to the Left of Negative Values of z or to the Right of Positive Values of z.

±z	Area Beyond ±z	±z	Area Beyond ±z
0.00	.5000	1.80	.0359
0.20	.4702	2.00	.0228
0.40	.3446	2.20	.0139
0.60	.2743	2.40	.0082
0.80	.2119	2.60	.0047
1.00	.1587	2.80	.0026
1.20	.1151	3.00	.0013
1.40	.0808	3.50	.0002
1.60	.0548	4.00	.00003

This detailed knowledge of the properties of a normal distribution puts us in a position to answer all sorts of questions concerning matters of probability. For example: If you were to reach into the source population and select one instance of X_i at random, what is the probability that its value would fall between 18 and 21. The answer practically jumps off the page. Within the source population 34.13% of all instances of X_i fall between the mean (18) and +1 standard deviation (21). Hence: P=.3413.

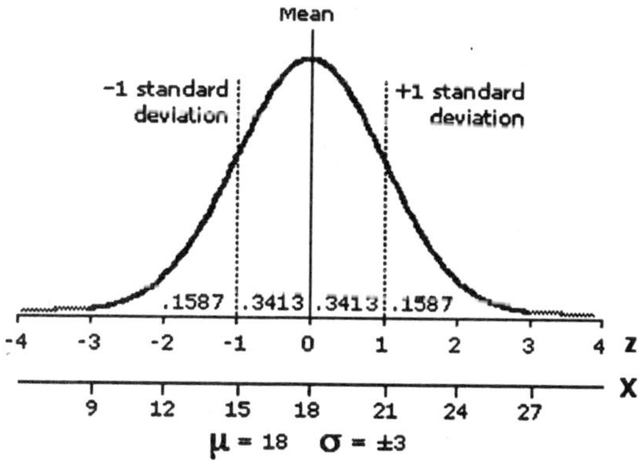

Similarly for this one: If you were to reach into the source population and select one instance of X_i at random, what is the probability that its value would be equal to or greater than 21? ~~ Within the source population 15.87% of all instances of X_i fall at or beyond +1 standard deviation (21). Hence: P=.1587.

Or for this "two-tailed" variation on the same theme: If you were to reach into the source population and select one instance of X_i at random, what is the probability that its value would be at least 3 points distant from the mean of the population? [That is: $X_i \leq 15$ or $X_i \geq 21$.] ~~ Within the source population 15.87% of all instances of X_i fall to the left of —1 standard deviation (15), and an equal 15.87% fall to the right of +1 standard deviation (21). Hence: P=.1587+.1587=.3174.

In fact, you could answer any probability question of this general type through the simple strategy of converting the relevant information into units of z, which is to say, into units of standard deviation. The version of z introduced *N* pertains only to the special case of binomial probability situations. The basic concept is much more general and can be refitted to apply to a wide range of probability situations that are not binomial in nature. Recall that the essential meaning of z is simply "one standard deviation away from the mean."

$$z = \frac{(k - \mu) \pm .5}{|\sigma|}$$

k = the observed value

μ = the mean of the binomial sampling distribution

σ = the standard deviation of th binomial sampling distribution

Thus "+1z" signifies "one standard deviation above the mean"; "–2.5z" says "two and one-half standard deviations below the mean"; and so forth. The general, abstract formulaic structure is simply this:

$$z = \frac{\text{(observed value)} - \text{(mean of the relevant sampling distribution)}}{\text{standard deviation of the relevant sampling distribution}}.$$

For the present type of situation where we are randomly drawing one value of X_i from a normally distributed source population, it is the source population itself that constitutes the sampling distribution. Hence the general structure translates into

$$z = \frac{X_i}{|\sigma|}$$

X_i = the observed or stipulated value

μ = the mean of the source population

σ = the standard deviation of the source population.

Note that there is no "±.5" correction for continuity in this formulation. A z-ratio normally requires a correction for continuity only in the special case of binomial probabilities.

Thus, a value of X_i=22 drawn from our reference population would be correspond to a z-ratio of

$$z \quad \frac{22-18}{3} = +1.33\cdots$$

For X_i=12 it would be

$$z = \frac{12-18}{3} = -2.0.$$

And so on. From the table of the normal distribution will see that z=+1.33 marks the point beyond which (away from the mean, to the right) fall 9.18% of all possible instances of X_i, while z=–2.0 marks the point beyond which (away from the mean, to the left) fall only a scant 2.28% of all possible instances of X_i. Thus, the probability that a randomly drawn value of X_i will be equal to or greater than 22 is P=.0918, while the probability that it will be equal to or smaller than 12 is .0228.

THE SAMPLING DISTRIBUTION OF SAMPLE MEANS

Among the remarkable properties of a normal distribution is the fact that it is capable of generating other distributions, which are also normal, and which can in turn be used to make probability assessments of a more complex and sophisticated kind. For example, imagine you were to draw a vast number of random samples from our source population, each sample of size *N*=10. If you were then to take the means of all those samples and lay them out in the form of a frequency distribution, the result would be a very close approximation of the theoretical sampling distribution shown in Figure. As this is a distribution composed of sample means (M), we label its mean as μ_M and its standard deviation as σ_M.

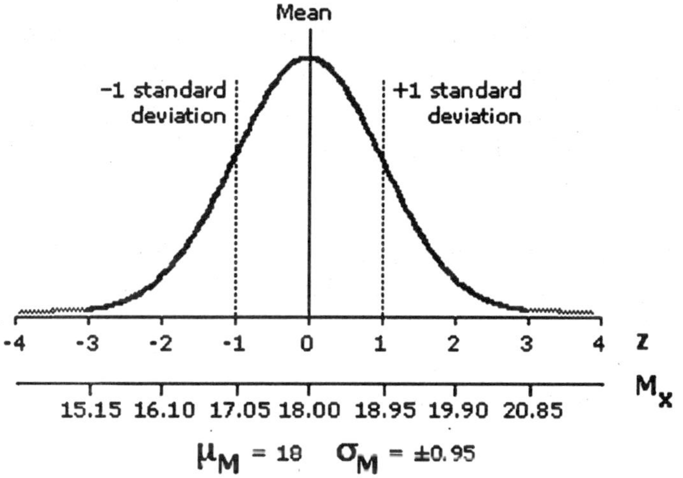

Fig. Sampling Distribution of the Means of Samples of Size *N*=10 Randomly Drawn from a Normally Distributed Source Population with μ =18 and σ = ±3

The fact that this sampling distribution of sample means is normally distributed is only the first part of the story. Equally important is that the two main parameters of the sampling distribution, its central tendency and variability, will be systematically related to those of the source population. Indeed, the means of the two distributions will be identical. As illustrated by the present example, the mean of the source population is 18.0, and so too is the mean of the sampling distribution of means. For this or any other situation where samples of size *N* are randomly drawn from a normally distributed source population, the relationship is

$$\mu_M = \mu_{source}$$

μ_M = mean of the sampling distribution of sample means

source = mean of the source population.

The reason for this identity is simply that both distributions are constructed of the same basic elements. In drawing random samples of size *N* from the source population, all we are really doing is randomly sorting the individual values of X_i from the source population into numerous groups, each of size *N*. The overall mean of all these items will be the same no matter how we sort them.

The respective variabilities of the two distributions are also precisely related, although here the relationship is somewhat more complex since it must take into account not only the variability of the source population but also the *size* of the samples that are being drawn from it. The basic logic of this point is that the properties of the source population will tend to be more accurately reflected by a larger sample than by a smaller one. In particular, the larger the sample, the greater the likelihood that its mean will approximate the mean of the population from which it is drawn. The result is that the means of larger samples will tend to cluster more tightly around the mean of the source population, while those of smaller samples will cluster more loosely and disperse more widely.

To illustrate this point more concretely, suppose we were drawing samples of different sizes from our reference source population. For samples of relatively small size, a single extreme X_i value such as 13 or 24 would have a relatively large effect on the mean of the sample. With samples of size *N*=2, for instance, we could quite easily get two sample values such as 18 and 24, which would yield a sample mean of 21; or such as 18 and 12, which would produce a sample mean of 15. With samples of larger size, however, occasional extreme values such as 12 and 24 will have a proportionately smaller effect upon sample means because they are now being averaged in with a larger number of more moderate values such as 17, 18, and 19.

The precise form of the relationship can be expressed in terms of either variance (σ^2) or standard deviation (σ). When drawing random samples of size *N* from a normally distributed source population:

The variance of the sampling distribution of sample means is equal to the variance of the source population divided by *N*

$$\mathrm{x}\sigma^2_M = \frac{\mathrm{x}^2_{\text{source}}}{N} \qquad \begin{aligned} \mathrm{x}\,\sigma^2_M &= \text{variance of sampling distribution} \\ \mathrm{x}\,\sigma^2_{\text{source}} &= \text{variance of source population} \end{aligned}$$

and the standard deviation of the sampling distribution of sample means is equal to the standard deviation of the source population divided by the square root of *N*

$$\mathrm{x}\sigma_M = \frac{\mathrm{x}\sigma^2_{\text{source}}}{\mathrm{sqrt}[N]} \qquad \begin{aligned} \mathrm{x}\,\sigma_M &= \text{standard deviation of sampling distribution} \\ \mathrm{x}\,\sigma_{\text{source}} &= \text{standard deviation of source population.} \end{aligned}$$

Thus for samples of size *N*=4 drawn from our reference source population, the standard deviation of the sampling distribution of means would be

$$\mathrm{x}\sigma_M = \frac{\pm 3}{\mathrm{sqrt}[4]} = \pm 1.5.$$

For samples of size N=10 it would be

$$x\sigma_M = \frac{\pm 3}{\text{sqrt}[10]} = \pm 0.95.$$

And so forth. In each of these two cases, 68.26% of all sample means would fall within the range bounded by —1 and +1 standard deviation, that is, between $-1\sigma_M$ and $+1\sigma_M$. However, for the N=4 sampling distribution this interval is relatively wide (±1.5), whereas for the N=10 sampling distribution it is relatively narrow (±0.95). In the N=4 distribution you would find 68.26% of all sample means falling between 16.5 and 19.5, while in the N=10 distribution you would find the same proportion clustering more compactly between 17.05 and 18.95. For samples of size N=20 (σ_M = ±0.67) the middle 68.26% of the sampling distribution would cluster even more compactly between 17.33 and 18.67; for samples of size N=40 (σ_M= ±0.47), still more compactly between 17.53 and 18.47; and so on.

The upshot of all this is that if you know a source population to be normally distributed, and if you happen also to know the values of its mean and standard deviation, you are then in a position to answer all sorts of probability questions concerning the means of such samples as might be drawn from that source population. For example: If you were to draw from our reference source population one random sample of size N=10, what is the probability that the mean of the sample would be equal to or greater than 20? Here again is Figure, which shows the sampling distribution that would apply to this scenario.

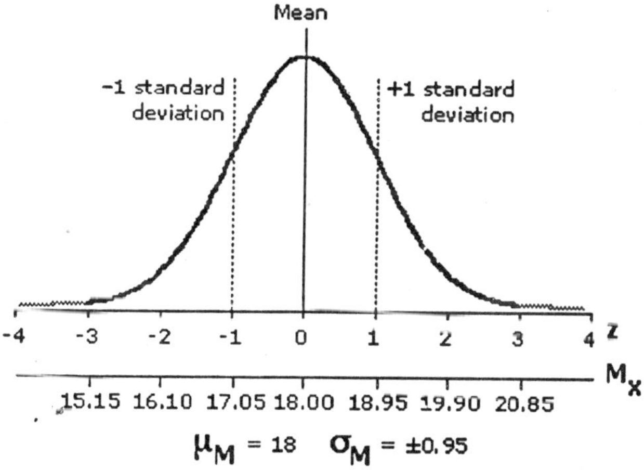

Fig. Sampling Distribution of the Means of Samples of Size N=10 Randomly Drawn from a Normally Distributed Source Population with μ =18 and σ = ±3

It will be obvious at a glance that a sample mean as large as M_X=20 would be fairly unlikely to occur through mere chance coincidence, falling as it does at about two standard deviations away from the mean of the sampling distribution. The general structure of a z-ratio

$$z = \frac{\text{(observed value)} - \text{(mean of the relevant sampling distribution)}}{\text{standard deviation of the relevant sampling distribution}}$$

can be tailored to apply to a variety of situations. In the present case, the tailoring takes the form

$$z = \frac{M_X - \mu_M}{\sigma_M} = \frac{20-18}{0.95} == +2.11.$$

M_X = observed sample mean

μ_M = mean of the sampling distribution of sample means

σ_M = standard deviation of the sampling distribution of sample means

From the table of the normal distribution you will see that z=+2.11 marks the point beyond which (away from the mean, to the right) fall 1.74% of all the possible sample means that might result from a random sample of size N=10. Thus, the probability that the mean of any particular sample of size N=10 will be equal to or greater than 20 is a scant 0.0174. Falling at the same distance at the opposite extreme of the sampling distribution (z=–2.11), hence equally unlikely, would be a sample mean equal to or smaller than 16. The probability of ending up with a sample mean as small as 16 ($M_X \leq 16$) or as large as 20 ($M_X \geq 20$) is therefore 0.0174+0.0174=0.0348.

Here again is an opportunity to click a button. Each time you click the one below, labeled "N=10," your computer will reach into the reference source population and draw a sample of 10 random values of X_i. It will also calculate and display the means (M_X) of your samples, both separately and cumulatively. Please click the button twenty or thirty times, taking particular note of the following as your sample means accumulate:

Within each sample, occasional extreme values of X_i will tend to be balanced off by the other values of X_i in the sample. In consequence, the means of the samples will tend to fall fairly close to the sampling distribution mean of μ_M=18.

Few if any of your sample means will be as small as 16 or as large as 20. If you were to continue clicking the "N=10" button several hundred times, you would find the proportions of such extreme sample means coming closer and closer to the theoretical proportions of 0.0174 (for $M_X \leq 16$) and 0.0174 (for $M_X \geq 20$).

THE SAMPLING DISTRIBUTION OF SAMPLE-MEAN DIFFERENCES [H]

Now for a somewhat more complex variation on the same theme. Suppose we were to reach into our reference source population and draw a vast number of pairs of samples. We will designate the first sample in each pair as A, the second as B, and stipulate that their respective sizes are N_a and N_b. For each pair we take the difference between the mean of sample A and the mean of sample B, according to the formula

$$\text{difference} = M_{Xa} - M_{Xb}$$

and then at the end we sort out all of these sample-mean differences in the form of a frequency distribution. Here as well we would end up with a sampling distribution that is normal in form and whose properties would be systematically related to those of the source population from which the samples are drawn. The average of all sample means of type-A would be equal to the mean of the source population (μ_{source}=18), and so too would be the average of all sample means of type-B. The average difference between type-A and type-B sample means would therefore be zero, and the mean of this sampling distribution of sample-mean differences (subscripted as "*M–M*") would accordingly be

$$\mu_{M-M} = 0.$$

The measures of variability for this sampling distribution are determined on the one hand by the amount of variability that exists within the source population and on the other by the sizes of the two types of samples, N_a and N_b. When drawing pairs of random samples of sizes N_a and N_b from a normally distributed source population:

The variance of the sampling distribution of sample-mean differences (again subscripted as "M-M") is equal to

the variance of the source population divided by N_a

plus$_a$

the variance of the source population divided by N_b

$$x\sigma^2_{M-M} = \frac{x\sigma^2_{source}}{N_a} + \frac{x\sigma^2_{source}}{N_b} \qquad \begin{array}{l} x\sigma^2_{M\text{-}M} = \text{variance of sampling distribution} \\ x\sigma^2_{source} = \text{variance of source population} \end{array}$$

and the standard deviation of the sampling distribution of sample-mean differences is equal to the square root of this quantity

$$x_{M-M} = \text{sqrt}\left[\left|\frac{x^2_{source}}{N_a}\right| + \left|\frac{x^2_{source}}{N_b}\right|\right].$$

Although we have not yet explicitly specified the variance of our source population, you will know that it must be σ^2_{source}=9, since its standard deviation, which is simply the square root of the variance, is σ_{source}= ±3. Thus for the case where we are drawing random samples of sizes N_a=10 and N_b=10 from our source population, the standard deviation of the relevant sampling distribution would be

$$x\sigma_{M-M} = \text{sqrt}\left[\left|\frac{9}{10}\right| + \left|\frac{9}{10}\right|\right] = \pm 1.34.$$

Figure shows the outlines of the sampling distribution of sample-mean differences for the case where the source population is normally distributed with a mean of 18 and a standard deviation of ±3, and where the samples of type-A and type-B are of sizes N_a=10 and N_b=10. The same general outlines would appear for any other values of N_a and N_b. The only difference would be in the value of σ_{M-M}, the standard deviation of the sampling distribution.

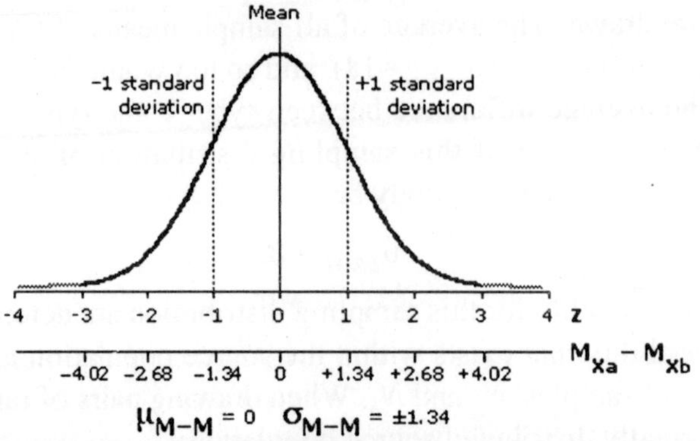

Fig. Sampling Distribution of Sample-Mean Differences, for Samples of Sizes N_a=10 and N_b=10 Randomly Drawn from a Normally Distributed Source Population with μ =18 and σ = ±3

You will certainly be able to guess where we go next. Suppose you were to reach into our reference source population and draw two random samples, A and B, each of size *N*=10. Three questions:

Question 1	Question 2	Question 3
What is the probability that the mean of sample A will be 2 or more points greater than the mean of sample B?	What is the probability that the mean of sample B will be 2 or more points greater than the mean of sample A?	What is the probability that there will be at least a 2-point difference between the means of the two samples, in either direction?
$[M_{Xa}-M_{Xb}] \geq 2$	$[M_{Xa}-M_{Xb}] \leq -2$	$[M_{Xa}-M_{Xb}] \geq 2$ or $[M_{Xa}-M_{Xb}] \leq -2$

To answer questions of this type, we once again take the general structure of a z-ratio

$$z = \frac{\text{(observed value)} - \text{(mean of the relevant sampling distribution)}}{\text{standard deviation of the relevant sampling distribution}}$$

and refit it to apply to the current situation. Now the re-tailoring takes the form

$$z = \frac{[M_{Xa} - M_{Xb}] - m_{M-M}}{s_{M-M}}$$

M_{Xa} = mean of sample A
M_{Xb} = mean of sample B
$\mu_{M\text{-}M}$ = mean of the sampling distribution
$\sigma_{M\text{-}M}$ = standard deviation of the sampling distribution.

As indicated above, the mean of the sampling distribution will always be $_{M-M}$=0 (except for certain specialized applications that we need not go into just now). So it can simply be dropped out of the equation to yield

$$z = \frac{M_{Xa} - M_{Xb}}{\sigma_{M-M}}.$$

Thus for Question 1, the probability that $[M_{Xa} - M_{Xb}] \geq 2$:

$$z = \frac{M_{Xa} - M_{Xb}}{\sigma_{M-M}} = \frac{2}{1.34} = +1.49.$$

And for Question 2, the probability that $[M_{Xa} - M_{Xb}] \leq -2$:

$$z = \frac{M_{Xa} - M_{Xb}}{\sigma_{M-M}} = \frac{-2}{1.34} = -1.49.$$

From the table of the normal distribution you will see that z=+1.49 marks the point beyond which (away from the mean, to the right) fall 6.81% of all the possible differences between M_{Xa} and M_{Xb} in this particular scenario. The probability (Question 1) that the mean of sample A will be 2 or more points greater than the mean of sample B in any randomly drawn pair of samples is therefore 0.0681. The same is true of z=–1.49, though in the opposite direction. So 0.0681 is also the probability (Question 2) that the mean of sample B will be 2 or more points greater than the mean of sample A in any randomly drawn pair of samples. The sum of these two will give the disjunctive probability (Question 3) of finding at least a 2-point difference between the means of sample A and sample B, in either direction: 0.0681+0.0681=0.1362.

One more button-click exercise, and then we will move on to see how all this abstract theory can be adapted to purposes of practical application. Below is a button labeled "Two Samples." Each time you click it, your computer will draw two random samples from our normally distributed reference population, each of size *N*=10. If you have the patience to click the button several hundred times, you will find the proportions of the types of outcomes stipulated in Questions 1, 2, and 3 approaching closer and closer to the theoretical proportions of 0.0681, 0.0681, and 0.1362.

Vocabulary Note: "standard deviation of the sampling distribution of sample means"	You will often find this quantity spoken of as the "standard error of the mean."
"standard deviation of the sampling distribution of sample-mean differences"	By extension, you could speak of this quantity as the "standard error of sample-mean differences." In general, the "standard error" of a statistic (sample means, sample-mean differences, etc.) is the standard deviation of the sampling distribution of that statistic.

Oh, the joys of creation, especially when it can be done fairly simply with a few lines of programming code. Once again, somewhere deep within the electronic workings of your computer I have created a vast population of X_i values. This time, however, all I will tell you about the population is that it is normally distributed. The rest of its main properties—mean, variance, and standard deviation–you will have to figure out for yourself, starting from scratch. There are many real-life research situations where we can reasonably suppose that a source population is normally distributed, but rarely if ever do we know in advance the true values of its central tendency and variability. We will begin with the easiest part of the task, which is to figure out the mean of the population.

ESTIMATING THE MEAN OF A NORMALLY DISTRIBUTED SOURCE POPULATION

A random sample will tend to reflect the properties of the population from which it is drawn. Among these reflected properties is the population's central tendency. So here is the simple, straightforward way of estimating the mean of our current source population. Draw from it a random sample of size *N*, and then let the mean of that sample serve as your estimate of the mean of the population.

The principle underlying such a process is that the mean of any particular sample can be taken as an unbiased estimate of the mean of the population from which the sample is drawn. In general, a biased estimate is one that will systematically underestimate the true value, or systematically overestimate it, while an unbiased estimate is one that avoids this tendency. The unbiased estimate might prove to be either under or over the true value in particular cases, but it will not move in either of these directions systematically. It is roughly analogous to shooting arrows at a target. The archer who tends to hit below the bull's eye is systematically biased in one direction, while the archer who tends to hit above it is systematically biased in the other. An archer without such a systematic bias will hit below and above the bull's eye in equal measure, and occasionally she will even hit it dead center.

But of course, even an unbiased archer is not necessarily a contender for the world title. The shots of one might tend to cluster within two or three inches of dead center, while those of another might scatter out over a distance of two or three feet. Similarly, some sample means can be regarded as very close estimates of the population mean, while others can be taken only as loose ballpark estimates.

In the following table is a button labeled "Sample." Each time you click it, you computer will draw a random sample of whatever size is indicated in the cell labeled "sample size." The default value of "sample size" is 5, though you can reset it to any positive integer value you might wish. I suggest you begin by drawing a few samples of size N=5, then a few of size N=10, and so on until you have worked your way up to some fairly large sample sizes. By the time you are drawing samples of size N=100, you will begin to get a pretty close fix on the mean of the source population. With samples of larger sizes, it will be closer still. (Note, however, that large sample sizes will take longer to run, especially if your browser is Internet Explorer.) To help you avoid getting lost in the fractional numerical details of your samples, each sample mean is also shown rounded to the nearest integer value.

If you have taken the time to perform the above exercise, you have almost certainly arrived at the conclusion that the mean of our current source population is somewhere in the vicinity of 15. Perhaps a shade more than 15, perhaps a shade less, but in any event somewhere in that vicinity.

Just how narrowly or broadly you can set the boundaries of "vicinity" depends on the size of the sample. The fact that the mean of our current source population lies somewhere in the neighborhood of 15 might not have been very obvious when you were drawing small samples, but it must surely have become so as the sizes of your samples increased. In general, a larger sample sets the boundaries of "vicinity" more narrowly, while a smaller sample sets them more broadly. From what we observed in Part 1 concerning the sampling distribution of sample means, you will know that these boundaries are also determined by the amount of variability that exists within the population from which the sample is drawn. A relatively small amount of variability within the source population would define "vicinity" fairly narrowly, while a relatively large amount would define it fairly broadly. Either way, the general structure of this process of estimation is

$$\text{estimated } \mu_{\text{source}} = M_X \pm \text{[definition of "vicinity"]}$$

The mean of the source population from which you were drawing your samples is precisely

$$\mu_{\text{source}} = 15.0$$

ESTIMATING THE VARIABILITY OF A NORMALLY DISTRIBUTED SOURCE POPULATION

The principle here is akin to the one just discussed in connection with the central tendencies of samples and populations, thought with one very important difference. The variability that is observed to exist within a sample can be taken as an estimate of the variability of the population, but not as an unbiased estimate. In general, the variability that appears within samples will tend to be smaller than the variability that exists within the entire population. In a relatively small percentage of cases it will be larger, and occasionally it might even hit the variability of the population dead center; but overall there is a strong bias in favor of the observed variability of a sample coming out as an underestimate of the variability of the population.

Here as well, the basic concept is that a random sample will tend to reflect the properties of the population from which it is drawn. Samples drawn from a highly variable source population will tend to contain relatively large amounts of variability, while those drawn from a fairly homogeneous population will tend to contain smaller amounts of variability. Moreover, the larger the size of the sample, the closer the reflection will tend to be.

The raw measure of variability within a set of numerical values—X_1, X_2, X_3, etc.—is the sum of squared deviates, *SS*, for which the formulas are

$$\text{conceptual: } SS = \text{S}(X_i - M_X)^2$$

$$\text{computational: } SS = \Sigma X_i^2 - \frac{(\Sigma X_i)^2}{{}_2N_2} \quad \text{From Ch. 2, Pt. 2.}$$

The variance of the set of X_i values is then the average of these squared deviates:

$$\text{variance} = \frac{SS}{N}$$

and the standard deviation is the square root of that average:

$$\text{standard deviation} = \text{sqrt}\left[\left|\frac{SS}{N}\right|\right] \quad \text{From Ch. 2, Pt. 2.}$$

The following demonstration focuses on the variance. Each time you click one of the buttons below, your computer will reach into our current source population and draw 10 random samples of the indicated size, either N=5 or N=20. It will also calculate and display the variance of each sample along with the average variance of the whole set of 10 samples and the cumulative average variance of all the samples that you draw through repeated clicking. Please click each of the buttons for as many times as you can muster the patience, noting in particular the values that appear toward the bottom of each text box under the heading of "cumulative average variance." If you click each button 30 or 40 times, so as to accumulate 300 to 400 samples of each size, you will almost certainly find that the cumulative average variance for samples of size N=20 will be larger than for samples of size N=5.

If you really want to get a hands-on idea of what is going on here, keep clicking the two buttons again and again, non-stop over the next several weeks, so as to accumulate a vast number of samples of each size. I do not imagine anyone will actually do this. But if you were to do it, here is what you would find. For your zillion random samples of size N=5, the cumulative average sample variance would very closely approximate the value of 6.0; and for your zillion random samples of size N=20, it would very closely approximate the value of 7.125.

There are two reasons why I am able to make this claim. The first is merely adventitious, occasioned by the fact that I needed to construct a specific source population in order to illustrate these principles. As the designer of this population I happen to know, and will now share the fact with you, that its variance is exactly

$$:\sigma^2_{\text{source}} = 7.5 \qquad [\text{hence: } \sigma_{\text{source}} = \pm 2.74].$$

The second reason is one of principle and obtains irrespective of whether you know the variability of the source population in advance. When drawing random samples of size N from a normally distributed source population, the average variance of those samples will end up in the long haul being equal to a certain proportion of the variance of the population. That proportion is determined by the size of the samples; its precise value is given by the ratio $(N–1)/N$. For samples of size N=5 the proportion is 4/5=0.80, so here we would expect the average sample variance to end up as 0.80x7.5=6.0. For samples of size N=20 it is 19/20=0.95, so we would expect 0.95x7.5=7.125. And so on. For all cases where samples of size N are randomly drawn from a normally distributed source population, the form of the relationship is

$$\text{mean sample variance} =: \sigma^2_{\text{source}} \times \frac{N-1}{N}.$$

Conversely, the variance of the source population would be equal to the average sample variance multiplied by $N/(N – 1)$:

$$:: \sigma^2_{\text{source}} - \text{mean sample variance} \times \frac{N}{N-1}.$$

It is this latter version of the relationship that permits us to estimate the variability of a source population – in those more realistic cases where we do not know it in advance – on the basis of a single sample.

For even though the variability contained within individual samples might not fall precisely at the average, it will nonetheless tend to fall somewhere *near* the average. In effect, therefore, the observed variance of a single sample can be taken as an estimate of "mean sample variance"; and that, in turn, can lead us to an estimate of the variance of the source population. What we end up with, of course, is still only an estimate. It is not precise, it does not pretend to be precise, and that fact will eventually have to be taken into account. But more of this later. Our immediate concern is with how the estimate can be obtained, and what can be done with it once we have it.

The general precept is this: If a random sample of size N is drawn from a normally distributed source population, a useful estimate of the source population's variance ($i^2\sigma_{source}$) can be obtained through multiplying the observed variance of the sample (s^2) by the ratio $N/(N-1)$. Thus,

$$\text{estimated:}\sigma^2\text{source} = s^2 \times \frac{N}{N-1}.$$

For practical computational purposes, this estimate can be reached in a somewhat more streamlined fashion. As we reminded ourselves just a moment ago, the variance of an observed set of X_i values is simply the average of the sum of squared deviates:

$$s^2 = \frac{SS}{N}.$$

Multiply that average by $N/(N-1)$ and you end up with

$$\frac{SS}{N} \times \frac{N}{N-1} = \frac{SS}{N-1}.$$

We will symbolize this modified version of a sample's variance as "$\{s^2\}$" in order to distinguish it from the plain-vanilla version; similarly, we will use "$\{s\}$" to denote the modified standard deviation that would result from taking the square root of $\{s^2\}$. Thus, on the basis of any particular sample randomly drawn from a population that we can reasonably suppose to be normally distributed, the variance and standard deviation of the population can be estimated as

$$\{s^2\} = \frac{SS}{N-1} \left[= \text{estimate of: } \sigma^2_{source}\right]$$

and

$$\{s\} = \text{sqrt}\left[\frac{SS}{N-1}\right] \left[= \text{estimate of: } \sigma^2_{source}\right].$$

ESTIMATING THE STANDARD DEVIATION OF THE SAMPLING DISTRIBUTION OF SAMPLE MEANS

The variance of the sampling distribution of sample means is equal to the variance of the source population divided by *N*.

$$:\sigma^2_M = \frac{:^2_{\text{source}}}{N}.$$

When $:\sigma^2_{\text{source}}$ is unknown, as it usually is, the value of $:\sigma^2_M$ can be estimated by substituting $\{s^2\}$, which is the estimate of $:\sigma^2_{\text{source}}$ obtained through the procedure described above. Thus

$$\text{estimated}:\sigma^2_M = \frac{\{s^2\}}{N} \quad \text{Recall that } \{s^2\} = SS/(N-1).$$

This in turn would allow you to estimate the standard deviation of the sampling distribution of sample means ("standard error of the mean") as

$$\text{estimated: } \sigma_M = \text{sqrt}\left[\frac{\{s^2\}}{N}\right].$$

To give you an idea where this is headed, suppose you have reason to believe that a certain source population is normally distributed, but you have no precise knowledge of its central tendency or variability. On the basis of certain theoretical considerations, however, you do have reason to suspect that the mean of the population is $\mu_{\text{source}}=50.0$. To test this hypothesis, you draw a random sample of size $N=25$ from the population. Your reasoning for the test is straightforward: If the mean of the population is 50.0, then the mean of any particular sample randomly drawn from the population should fall somewhere in the vicinity of 50.0. So you draw your sample and find its mean to be $M_X=53.0$, as compared with your hypothetical population mean of $\mu_{\text{source}}=50.0$.

Well, yes, 53.0 is "somewhere in the vicinity" of 50.0 – although, depending on your scale of measurement, that could be a bit like saying that Glasgow is somewhere in the vicinity of London. The observed sample mean of 53.0 clearly differs from 50.0. The question is, does it differ significantly? That is: If the mean of the population truly were 50.0, how likely would it be, by mere chance coincidence, that the mean of a sample randomly drawn from the population would fall 3 or more points distant from 50.0?

If only you knew the variability of the source population, you could directly calculate $:\sigma^2_M$, the standard deviation of the relevant sampling distribution of sample means. And with that value in hand you could then plug your numbers into the appropriate version of the sample-mean z-ratios:

$$z = \frac{M_X - \mu_M}{:\sigma_M} = \frac{53-50}{\sigma:_M} \quad \text{From Ch. 9, Pt. 1.}$$

As it happens, you do not know the variability of the source population, hence cannot directly calculate either $:\sigma^2_M$ or z. But as Moliere had his character Tartuffe say on one occasion:

Though Heaven forbids certain gratifications, there are nonetheless "ways and means of accommodation." In the present case, the accommodation comes about by way of systematic estimates. Suppose that you observe within your sample a sum of squared deviates of *SS*=625. On this basis you could estimate the variance of the population as

$$\{s^2\} = \frac{SS}{N-1} = \frac{625}{24} = 26.04.$$

And that in turn would allow you to estimate ('est.") the standard deviation of the sampling distribution as

$$\text{est.:s}_M = \text{sqrt}\left[\frac{\{s^2\}}{N}\right]$$
$$= \text{sqrt}\left[\frac{26.04}{25}\right] = \pm 1.02.$$

The next step will be obvious. With your estimated value of $:\sigma_M$ in hand, you can now go on to calculate what is essentially an estimate of the z-ratio examined a moment ago. To make it clear that what we are now calculating is grounded on several layers of estimation, the convention is to label it with the letter t rather than z.

$$t = \frac{M_X - \mu_M}{\text{est.:}\sigma_M}$$
$$= \frac{53-50}{1.02} = +2.94.$$

ESTIMATING THE STANDARD DEVIATION OF THE SAMPLING DISTRIBUTION OF SAMPLE-MEAN DIFFERENCES

There is a certain species of small furry animal known as the golliwump, of which half are green and the other half are blue. An investigator has framed the hypothesis that green golliwumps are on average smarter than blue golliwumps. To test this hypothesis he draws one random sample of N_a=20 greens and another of N_b=20 blues. He then administers a standard test of golliwump intelligence to each of the individual subjects in the two samples, finding that the mean score of the greens is M_{Xa}=105.6, while that of the blues is only M_{Xb}=101.3.

Here again is one of those "if only" scenarios. If only our investigator knew the variability of test scores within the entire population of golliwumps, he would be able to calculate directly the standard deviation of the relevant sampling distribution,

$$i\sigma_{M-M} = \text{sqrt}\left[\frac{i\sigma^2_{\text{source}}}{N_a}\right] + \left[\frac{i\sigma^2_{\text{source}}}{N_b}\right] \quad \text{From Ch. 9, Pt. 1}$$

which would in turn permit the calculation of the two-sample z-ratio described in Part 1:

$$z = \frac{Mx_a}{i\sigma_{M-M}} = \frac{105.6-101.3}{i\sigma_{M-M}} \quad \text{From Ch. 9, Pt. 1.}$$

The logic of estimation in this case is analogous to what we examined above for the one-sample situation. Suppose that the values for sum of squared deviates within the two samples were SS_a=4321 and SS_b=4563. If you were to estimate the variance of the source population on the basis of sample A separately, it would be

$$\{s^2_a\} = \frac{SS_a}{N_a - 1} = \frac{4321}{19} = 227.42.$$

Estimated on the basis of sample B separately, it would be

$$\{s^2_b\} = \frac{SS_b}{N_b - 1} = \frac{4563}{19} = 240.16.$$

Blending these two separate variance estimates together in just the right way will give you a composite estimate known as the pooled variance. (Note the subscript "*p*" to indicate "pooled.")

$$\{s^2_p\} = \frac{SS_a + SS_b}{(N_a - 1) + (N_b - 1)}$$

$$= \frac{4321 + 4563}{19 + 19} = 233.79.$$

Returning now to the formula for the direct calculation of the standard deviation of the sampling distribution,

$$i\sigma_{M\text{-}M} = \text{sqrt}\left[\frac{i\sigma^2_{\text{source}}}{N_a} + \frac{i\sigma^2_{\text{source}}}{N_b}\right] \quad \text{From Ch. 9, Pt. 1}$$

we substitute $\{s^2_p\}$ for $i\sigma_{\text{source}}$ and end up with

$$\text{est.}is_{M\text{–}M} = \text{sqrt}\left[\frac{\{s^2_p\}}{N_a} + \frac{\{s^2_p\}}{N_b}\right]$$

$$= \text{sqrt}\left[\frac{233.79}{20} + \frac{233.79}{20}\right] = \pm 4.84.$$

The next step is then to calculate a t-ratio on analogy with the two-sample z-ratio described above:

$$t = \frac{M_{Xa} - M_{Xb}}{\text{est.}i\sigma_{M\text{-}M}}$$

$$= \frac{105.6 - 101.3}{4.84} = +0.89.$$

In effect, we are estimating that the true value of the sampling distribution's standard deviation is somewhere in the vicinity of ±4.84, and accordingly that the true value of z is somewhere in the vicinity of +0.89.

So here it stands, estimate piled on estimate, and the whole structure hedged in with an escape clause that reads "somewhere in the vicinity." At first glance you might think it hardly possible to squeeze from these rather spongy estimates anything even remotely resembling a precise probability assessment. Indeed, were it not for the work of the statistician W.S. Gosset (who wrote under the pseudonym of "Student"), we might have to conclude at this point that the first-hand impression is accurate. In the final portion of this chapter you will catch your first glimpse of the extraordinarily useful inferential tool that Gosset's work created.

t-ratio for comparing the mean of a sample with the mean (either actual or hypothetical) of a population:

$$t = \frac{M_X - \mu_M}{\text{est.:}\sigma_M}$$

t-ratio for comparing the means of a two samples:

$$t = \frac{M_{Xa} - M_{Xb}}{\text{est.:}\,\sigma_{M\text{-}M}}.$$

William S. Gosset's notable achievement was the mathematical discovery that, even though t-ratios are based on estimates, they nonetheless behave in a precise, orderly fashion. Specifically, what he discovered was that when t-ratios are considered en masse they form relative-frequency distributions whose outlines are capable of being precisely known; and further, that this precise knowledge of the outlines can then be used as a basis for making precise probability judgments, in much the same way that precise knowledge of the properties of the normal distribution permits us to make precise probability assessments in connection with z-ratios.

To illustrate this point, let us return for a moment to the reference source population.

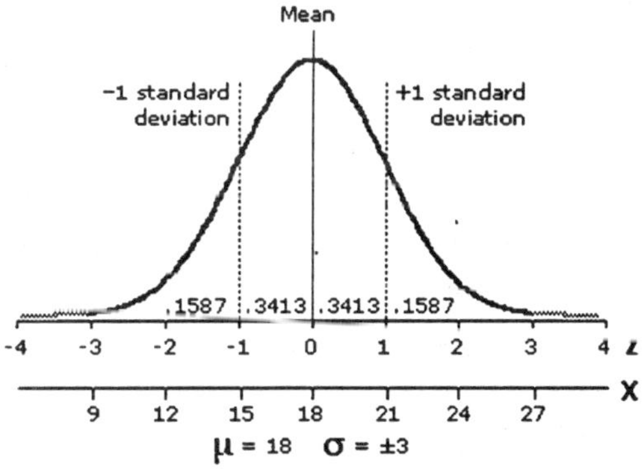

Fig. Reference Source Population, Normally Distributed with $\mu = 18$ and $\sigma = \pm 3$

Suppose we were to draw a vast number of pairs of random samples from this population, each sample of size N=10. Since this is a source population for which we know the true values of $i\,\sigma^2$ and $i\,\sigma$, we could easily calculate a proper z-ratio for each pair. If we were then to take all these z-ratios and lay them out in the form of a frequency distribution, we would find that it very closely approximates the shape of the unit normal distribution. That, in fact, is precisely why it is meaningful to refer a z-ratio to the table of the unit normal distribution. But now let us go back to the very same pairs of samples and calculate t-ratios for them. That is, for each pair of samples we separately calculate the ratio

$$t = \frac{M_{Xa} - M_{Xb}}{\text{est.:}\sigma_{M\text{-}M}}.$$

If we were then to take all these t-ratios and lay *them* out in the form of a frequency distribution, they too would form an orderly, symmetrical outline. It would not be a normal distribution–not in this particular case, anyway–but orderly all the same, and equally useful for making precise probability judgments. Figure shows what these two distributions of z and t would look like. The most readily visible difference between them is that the t-distribution is somewhat flatter and more spread out. The implication is that values of t that lie toward the center of the distribution are somewhat less probable than comparable values of z, while those that lie toward the extremes are somewhat more probable than comparable values of z.

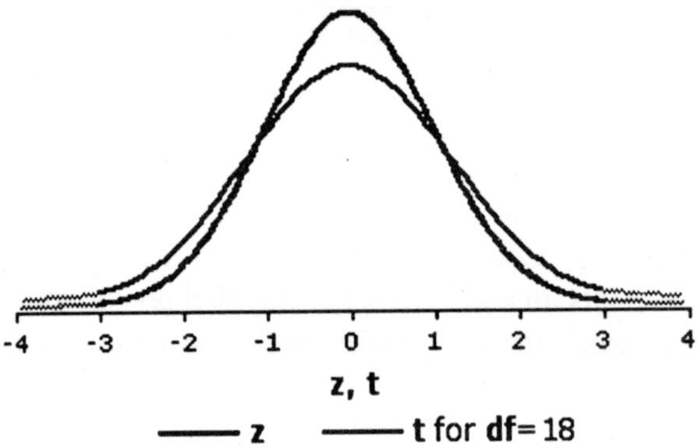

Fig. The Unit Normal Distribution Compared with the t-Distribution for df=18

Another difference between the distributions of z and t is that, while there is one and only one unit normal distribution, there is a whole range of different t-distributions. Indeed, there is a separate t-distribution for each possible value of degrees of freedom. The concept of degrees of freedom in the present context is much the same as the one introduced in connection with chi-square procedures. Essentially, it is an index of the amount of random variability that can be present in the particular situation. If you have a single sample of size *N*, the number of degrees of freedom is

$$\text{df} = N-1$$

If you have two samples of sizes N_a and N_b, it is

$$\text{df} = (N_a-1) + (N_b-1)$$

In the present case we have two samples of sizes N_a=10 and N_b=10, so the number of degrees of freedom is (10 – 1) + (10 – 1)=18.

The general pattern of resemblances and differences among the various t-distributions is illustrated in Figure, which shows the distributions of t for df=5 and df=40.

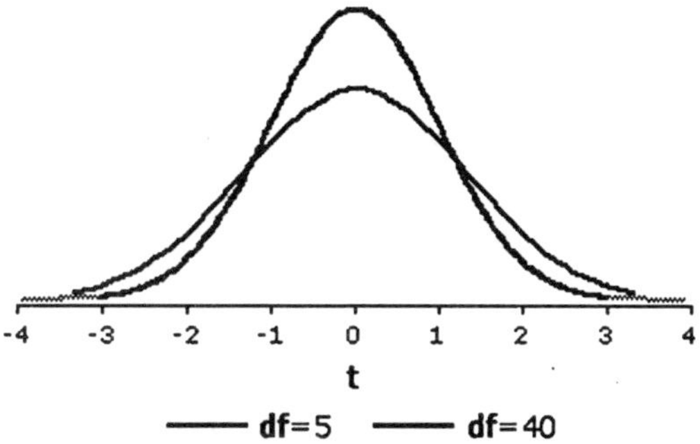

Fig. The Distributions of t for df=5 and df=40

Notice that the t-distribution for df=5 is even flatter and more spread out than the one shown in Figure for df=18, whereas the one for df=40 is markedly less flat and spread out. Indeed, to the naked eye the t-distribution for df=40 is scarcely distinguishable from the normal distribution. It is also a very close fit mathematically. In general, the smaller the number of degrees of freedom, the greater the difference between the corresponding t-distribution and the unit normal distribution. Conversely, the larger the number of degrees of freedom, the more closely the corresponding t-distribution will approximate the unit normal distribution.

The logic of probability assessment is the same with t-distributions as with the unit normal distribution. The probability of getting a result "this large or larger" is equal to the proportion of the distribution that lies to the right (+) or left (–) of the calculated value of t, depending of the direction of the hypothesis the researcher is seeking to test. For a non-directional two-way hypothesis it is the total proportion that lies both to the right of +t and to the left of —t. The full table of critical values of t can be found in Appendix C. The abbreviated version shown below will give you an idea of how it is laid out.

	Level of Significance					
	.05	.025	.01	.005	.0005	directional test
df	–	.05	.02	.01	.001	non-directional test
5	2.02	2.57	3.36	4.03	6.87	
10	1.81	2.23	2.76	3.17	4.59	
18	1.73	2.23	2.55	2.88	3.92	
20	1.72	2.09	2.53	2.85	3.85	

To the right of each value of df are the values of +t or –t that must be met or exceeded in order for a result to be significant at or beyond a particular level of significance. To illustrate, note that the first two values lying to the right of df=18 are 1.73 and 2.10.

These and the other values in that row (2.55, 2.88, 3.92) refer to specific locations within the sampling distribution of t for the case where degrees of freedom is equal to 18. As shown graphically in Figure, t=+1.73 marks the point in this distribution beyond which (to the right, away from the mean) falls 5% of the total area of the distribution. An obtained t-ratio equal to or greater than +1.73 is accordingly significant at or beyond the .05 level for a directional ("one-tailed") test.

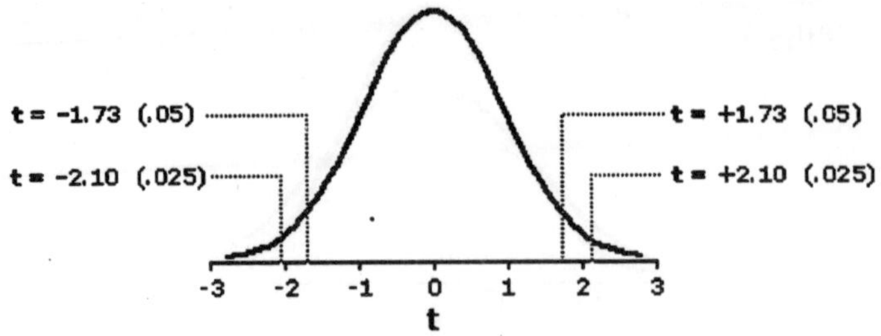

Fig. Sampling Distribution of t for df=18

Similarly, t=+2.10 marks the point beyond which falls 2.5% of the distribution, so an obtained t-ratio equal to or greater than +2.10 is significant at or beyond the .025 level for a directional test. As the distribution is symmetrical, the same logic and procedure apply to negative t-values, which lie on the left side of the distribution.

For a non-directional ("two-tailed") test, the probability associated with an obtained value of t is twice that associated with the same value of t for a directional test. Thus, for a non-directional test with df=18, an obtained value of t would have to fall at or beyond +2.10 or —2.10 in order to be significant at or beyond the .05 level. In practical terms, this means that the absolute (unsigned) value of t must be equal to or greater than 2.10.

BINARY FUNCTION

DEFINITION OF BINARY RELATION

Here we are going to define relation formally, first binary relation, then general n–ary relation. A relation in everyday life shows an association of objects of a set with objects of other sets (or the same set) such as John owns a red Mustang, Jim has a green Miata etc. The essence of relation is these associations. A collection of these individual associations is a relation, such as the ownership relation between peoples and automobiles. To represent these individual associations, a set of "related" objects, such as John and a red Mustang, can be used. However, simple sets such as {John, a red Mustang} are not sufficient here.

The order of the objects must also be taken into account, because John owns a red Mustang but the red Mustang does not own John, and simple sets do not deal with orders. Thus sets with an order on its members are needed to describe a relation. Here the concept of ordered pair and, more generally, that of ordered n–tuple are going to be defined first. A relation is then defined as a set of ordered pairs or ordered n–tuples.

Ordered pair: An ordered pair is a set of a pair of objects with an order associated with them.

If objects are represented by x and y, then we write an ordered pair as $<x, y>$ or $<y, x>$. In general $<x, y>$ is different from $<y, x>$.

Equality of ordered pairs: Two ordered pairs $<a, b>$ and $<c, d>$ are equal if and only if $a = c$ and $b = d$.

For example, if the ordered pair <*a, b*> is equal to <*1, 2*>, then $a = 1$, and $b = 2$. <*1, 2*> is not equal to the ordered pair <*2, 1*>.

Binary relation: A binary relation from a set *A* to a set *B* is a set of ordered pairs <*a, b*> where *a* is an element of *A* and *b* is an element of *B*.

When an ordered pair <*a, b*> is in a relation *R*, we write *a R b*, or <*a, b*> $\in R$. It means that element *a* is related to element *b* in relation *R*.

When $A = B$, we call a relation from *A* to *B* a (binary) relation on *A* .

Cartesian product: The set of all ordered pairs <*a, b*>, where *a* is an element of *A* and *b* is an element of *B*, is called the Cartesian product of *A* and *B* and is denoted by $A \times B$.

Thus a binary relation from *A* to *B* is a subset of Cartesian product $A \times B$.

Examples: If $A = \{1, 2, 3\}$ and $B = \{4, 5\}$, then {<*1, 4*>, <*2, 5*>, <*3, 5*>}, for example, is a binary relation from *A* to *B*.

However, {<*1, 1*>, <*1, 4*>, <*3, 5*>} is not a binary relation from *A* to *B* because *1* is not in *B*.

The parent–child relation is a binary relation on the set of people. <*John, John Jr.*>, for example, is an element of the parent–child relation if John is the father of John Jr.

DEFINITION OF N–ARY RELATION

Here we are going to formally define general *n*-ary relation using the concept of ordered n-tuple.

Ordered n-tuple: An ordered *n*-tuple is a set of *n* objects with an order associated with them. If *n* objects are represented by $x_1, x_2, ..., x_n$, then we write the ordered *n*-tuple as $<x_1, x_2, ..., x_n>$.

Cartesian product: Let $A_1, ..., A_n$ be *n* sets. Then the set of all ordered *n*-tuples $<x_1, ..., x_n>$, where $x_i \in A_i$ for all i, $1 \le i \le n$, is called the Cartesian product of $A_1, ..., A_n$, and is denoted by $A_1 \times ... \times A_n$.

Equality of n-tuples: Two ordered *n*-tuples $<x_1, ..., x_n>$ and $<y_1, ..., y_n>$ are equal if and only if $x_i = y_i$ for all i, $1 \le i \le n$.

For example the ordered *3*–tuple <*1, 2, 3*> can be equal to only <*1, 2, 3*> and nothing else. It is not equal to the ordered *n*–tuple <*2, 3, 1*> for example.

N-ary relation: An *n*–ary relation on sets $A_1, ..., A_n$ is a set of ordered *n*–tuples $<a_1, ..., a_n>$ where a_i is an element of A_i for all i, $1 \le i \le n$. Thus an *n*-ary relation on sets $A_1, ..., A_n$ is a subset of Cartesian product $A_1 \times ... \times A_n$.

Example: Let A_1 be a set of names, A_2 a set of addresses, and A_3 a set of telephone numbers. Then a set of *3*–tuples < name, address, telephone number > such as {< Amy Angels, 35 Mediterranean Ave, 224–1357 >, < Barbara Braves, 221 Atlantic Ave, 301–1734 >, < Charles Cubs, 312 Baltic Ave, 223–9876 >}, is a *3*–ary (ternary) relation over A_1, A_2 and A_3.

Example: Let A_1 be a set of names. Then a set of 1–tuples such as {< Amy >, < Barbara >, < Charles >}, is a *1*–ary (unary) relation over A_1.

A unary relation represents a property/characteristic, such as tall, rich etc., shared by the members of A_1 listed in the relation.

Special Relations

The empty set is certainly a set of ordered n–tuples. Therefore it is a relation. It is called the empty relation.

The Cartesian product $A_1 \times \ldots \times A_n$ of sets $A_1, \ldots, A_n$, is also a relation, and it is called the universal relation.

The relation we are going to study here is an abstraction of relations we see in our everyday life such as those between parent and child, between car and owner, among name, social security number, address and telephone number etc. We are going to focus our attention on one key property which all the everyday relations have in common, define everything that has that property as a relation, and study properties of those relations.

One of the places where relation in that sense is used is data base management systems. Along with hierarchical and network models of data, the *relational model* is widely used to represent data in a database. In this model the data in a database are represented as a collection of relations. Informally, each relation is like a table or a simple file. For example, consider the following table.

	Employee	
Name	*Address*	*Home Phone*
Amy Angels	35 Mediterranean Av.	224–1357
Barbara Braves	221 Atlantic Av.	301–1734
Charles Cubs	312 Baltic Av.	223–9876

Each row of this table represents a collection of data values such as name, address, and telephone number of a person. Each row is considered an instance of a relation and the table as the collection of the rows is considered a relation, which is the relation we are going to be studying in this chapter.

Operations such as inserting or deleting entries to or from a table, merging two table, finding the intersection of two tables, and searching for certain entries can be described simply and precisely as operations on relations and known mathematical results on relations can be utilized without reinventing them.

The relational model is flexible (easy to expand, easy to modify) and interface to query languages is simple. It is thus widely used today.

EQUALITY OF RELATIONS

Definition (equality of binary relation): Two binary relations $R_1 \subseteq A_1 \times A_2$ and $R_2 \subseteq B_1 \times B_2$ are equal if and only if $A_1 = B_1$, $A_2 = B_2$, and $R_1 = R_2$ as a set.

For example, let $R_1 = \{<1, 2>, <2, 2>\} \subseteq \{1, 2\} \times \{1, 2\}$, and $R_2 = \{<a, b>, <b, b>\} \subseteq \{a, b\} \times \{a, b\}$. Then $R_1 = R_2$ if and only if $a = 1$ and $b = 2$.

Definition (equality of n–ary relation): An n-ary relation $R_1 \subseteq A_1 \times ... \times A_n$ and an m-ary relation $R_2 \subseteq B_1 \times ... \times B_m$ are equal if and only if $m = n$, $A_i = B_i$ for each i, $1 \leq i \leq n$, and $R_1 = R_2$ as a set of ordered n-tuples.

RECURSIVE DEFINITION OF RELATION

Certain relations can be defined recursively. Note that a relation is a set. Therefore a recursive definition of a relation follows the same format as that of sets. Here only examples are given.

Example: Let us define recursively the relation "less than" denoted by $R_<$ on the set of natural numbers N.

Note that $R_< = \{<a, b> \mid a \in N \wedge b \in N \wedge a < b\} = \{<0, 1>, <0, 2>, ..., <1, 2>, ...\}$.

Basis Clause: $<0, 1> \in R_<$ (or $0\ R_<\ 1$), meaning 0 is less than 1.

Inductive Clause: For all x and y in N, if $<x, y> \in R_<$, then $<x, y + 1> \in R_<$, and $<x + 1, y + 1> \in R_<$.

Extremal Clause: Nothing is in $R_<$, unless it is obtained from the Basis and Inductive Clauses.

Informally one can see that this definition is correct as follows:

First, $<0, 1>$ is the "simplest" element in $R_<$.

Next by arranging the ordered pairs in $R_<$ as follows, one can see that the two operations in the Inductive Clause generate them all:

<0, 1>, <0, 2>, <0, 3>, ..., <0, n>, ...

<1, 2>, <1, 3>, <1, 4>, ..., <1, n>, ...

<2, 3>, <2, 4>, <2, 5>, ... , <2, n>, ...

....

....

In each row, the first (leftmost) ordered pair gives the "smallest" ordered pair with the first number. For example $<2, 3>$ is the smallest with 2 as the first component. It is obtained from the first ordered pair of the row immediately above it by using $<x + 1, y + 1>$ of the Inductive Clause, apply that to $<1, 2>$ to get $<2, 3>$, for example.

Within each row, the ordered pairs are obtained from the first one by using *<x, y + 1>* of the Inductive Clause successively.

Thus all the ordered pairs of $R_<$ are generated from *<0, 1>* by following the Inductive Clause.

Example: Let $R_{a+b=c}$ be the set of triples of natural numbers *<a, b, c>* which satisfy $a + b = c$. Then $R_{a+b=c}$ on the set of natural numbers N can be defined recursively as follows.

Basis Clause: *<0, 0, 0>* $\in R_{a+b=c}$.

Inductive Clause: For all x, y and z in N, if *<x, y, z>* $\in R_{a+b=c}$, then *<x + 1, y, z + 1>* and *<x, y + 1, z + 1>* $\in R_{a+b=c}$.

Extremal Clause: Nothing is in $R_{a+b=c}$ unless it is obtained from the Basis and Inductive Clauses.

PROPERTIES OF BINARY RELATION AND OPERATIONS

DIGRAPH

A digraph is short for directed graph, and it is a diagram composed of points called vertices (nodes) and arrows called arcs going from a vertex to a vertex.

For example the figure below is a digraph with 3 vertices and 4 arcs.

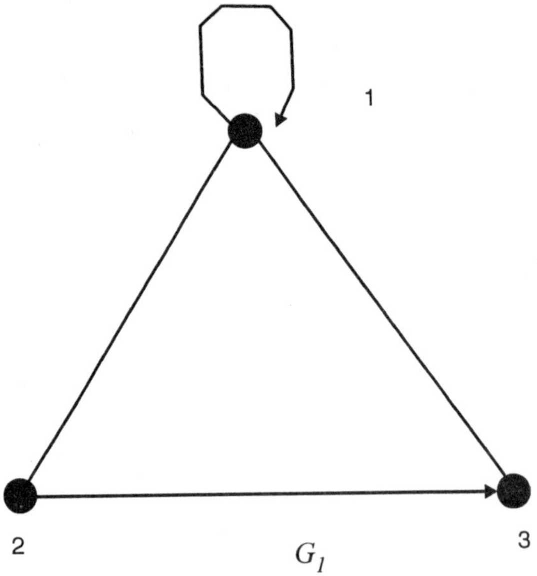

G_1

In this figure the vertices are labeled with numbers *1, 2,* and *3*.

Mathematically a digraph is defined as follows.

Digraph: A digraph is an ordered pair of sets $G = (V, A)$, where V is a set of vertices and A is a set of ordered pairs (called arcs) of vertices of V.

In the example, G_1 , given above, $V = \{1, 2, 3\}$, and A = {<1; 1>, <1, 2>, <1, 3>, <2, 3>} .

Digraph representation of binary relations: A binary relation on a set can be represented by a digraph.

Let *R* be a binary relation on a set *A*, that is *R* is a subset of $A \times A$.

Then the digraph, call it *G*, representing *R* can be constructed as follows:

1. The vertices of the digraph *G* are the elements of *A*, and
2. <*x, y*> is an arc of *G* from vertex *x* to vertex *y* if and only if <*x, y*> is in *R*.

Example: The less than relation *R* on the set of integers $A = \{1, 2, 3, 4\}$ is the set {<1, 2> , <1, 3>, <1, 4>, <2, 3> , <2, 4> , <3, 4>} and it can be represented by the following digraph.

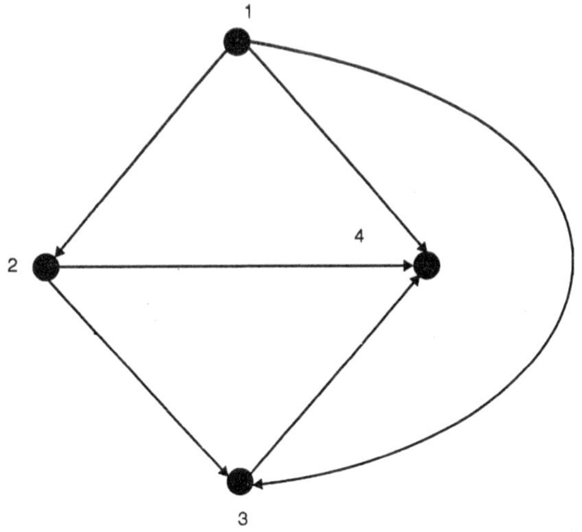

G_2

Let us now define some of the basic concepts on digraphs.

Loop: An arc from a vertex to itself such as <1, 1>, is called a loop (or self–loop) .

Degree of vertex: The in–degree of a vertex is the number of arcs coming to the vertex, and the out–degree is the number of arcs going out of the vertex.

For example, the in–degree of vertex *2* in the digraph G_2 shown above is *1*, and the out–degree is *2*.

Path: A path from a vertex x_0 to a vertex x_n in a digraph $G = (V, A)$ is a sequence of vertices $x_0, x_1, \ldots, x_n$ that satisfies the following:

for each *i*, $0 \leq i \leq n - 1$, $<x_i, x_{i+1}> \in A$, or $<x_{i+1}, x_i> \in A$, that is, between any pair of vertices there is an arc connecting them.

x_0 is the initial vertex and x_n is the terminal vertex of the path.

A path is called a directed path if $<x_i, x_{i+1}> A \in$, for every i, $0 \leq i \leq n-1$.

If the initial and the terminal vertices of a path are the same, that is, $x_0 = x_n$, then the path is called a cycle .

If no arcs appear more than once in a path, the path is called a simple path. A path is called elementary if no vertices appear more than once in it except for the initial and terminal vertices of a cycle. In a simple cycle one vertex appears twice in the sequence: once as the initial vertex and once as the terminal vertex.

Note: There are two different definitions for "simple path". Here we follow the definition of Berge[1], Liu[2], Rosen[3] and others. A "simple path" according to another group (Cormen et al[4], Stanat and McAllister[5] and others) is a path in which no vertices appear more than once.

Connected graph: A digraph is said to be connected if there is a path between every pair of its vertices.

Example: In the digraph G_3 given below,

1, 2, 5 is a simple and elementary path but not directed,

1, 2, 2, 5 is a simple path but neither directed nor elementary.

1, 2, 4, 5 is a simple elementary directed path,

1, 2, 4, 5, 2, 4, 5 is a directed path but not simple (hence not elementary),

1, 3, 5, 2, 1 is a simple elementary cycle but not directed, and

2, 4, 5, 2 is a simple elementary directed cycle.

This digraph is connected.

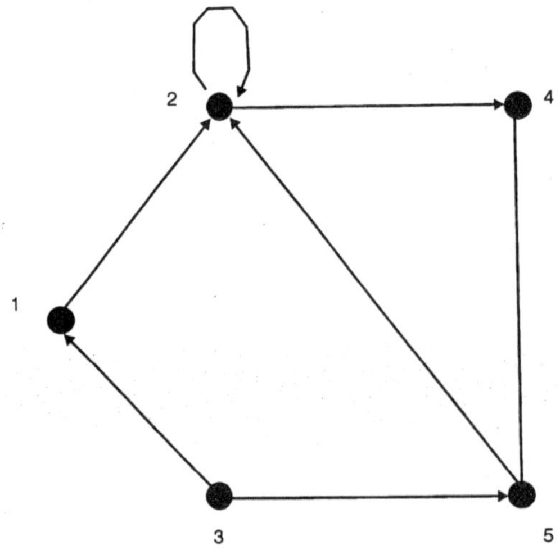

G_3

Sometimes we need to refer to part of a given digraph. A partial digraph of a digraph is a

digraph consisting of arbitrary numbers of vertices and arcs of the given digraph, while a subdigraph is a digraph consisting of an arbitrary number of vertices and all the arcs between them of the given digraph. Formally they are defined as follows:

Subdigraph, partial digraph: Let $G = (V, A)$ be a digraph. Then a digraph (V', A') is a partial digraph of G, if $V' \subseteq V$, and $A' \subseteq A \cap (V' \times V')$. It is a subdigraph of G, if $V' \subseteq V$, and $A' = A\ (V' \times V')$.

DIGRAPH REPRESENTATION OF BINARY RELATION

A binary relation on a set can be represented by a digraph. Let *R* be a binary relation on a set *A*, that is *R* is a subset of $A \times A$. Then a digraph representing *R* can be constructed as follows:

Let the elements of *A* be the vertices of the digraph *G*, and let <*x*, *y*> be an arc of *G* from vertex *x* to vertex *y* if and only if <*x*, *y*> is in *R*.

Example: The less than relation *R* on th e set of integers *A* = {*1*, *2*, *3*, *4*} is the set {<*1*, *2*> , <*1*, *3*>, <*1*, *4*>, <*2*, *3*> , <*2*, *4*> , <*3*, *4*>} and it can be represented by the following digraph.

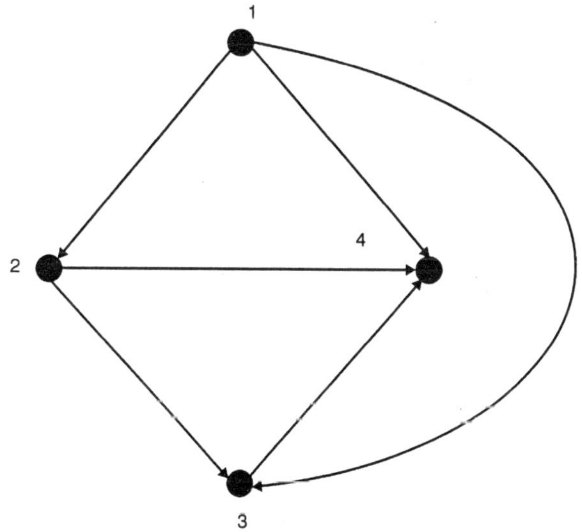

PROPERTIES OF BINARY RELATION

Certain important types of binary relation can be characterized by properties they have. Here we are going to learn some of those properties binary relations may have. The relations we are interested in here are binary relations on a set.

reflexive relation: A relation *R* on a set *A* is called reflexive if and only if < a, a > ∈*R* for every element *a* of *A*.

Example: The relation ≤ on the set of integers {*1*, *2*, *3*} is {<*1*, *1*>, <*1*, *2*>, <*1*, *3*>, <*2*, *2*>, <*2*, *3*>, <*3*, *3*>} and it is reflexive because <*1*, *1*>, <*2*, *2*>, <*3*, *3*> are in this relation. As a matter of fact ≤ on any set of numbers is also reflexive. Similarly ≥ and = on any set of numbers are reflexive. However, < (or >) on any set of numbers is not reflexive.

Example: The relation ⊆ on the set of subsets of {*1, 2*} is $\{<\emptyset, \emptyset>, <\emptyset, \{1\}>, <\emptyset, \{2\}>, <\emptyset, \{1, 2\}>, <\{1\}, \{1\}>, <\{1\}, \{1, 2\}>, <\{2\}, \{2\}>, <\{2\}, \{1, 2\}>, <\{1, 2\}, \{1, 2\}>\}$ and it is reflexive. In fact relation on any collection of sets is reflexive.

Irreflexive relation: A relation R on a set A is called irreflexive if and only if $<a, a> \notin R$ for every element a of A.

Example: The relation > (or <) on the set of integers {*1, 2, 3*} is irreflexive. In fact it is irreflexive for any set of numbers.

Example: The relation $\{<1, 1>, <1, 2>, <1, 3>, <2, 3>, <3, 3>\}$ on the set of integers {*1, 2, 3*} is neither reflexive nor irreflexive.

Symmetric relation: A relation R on a set A is called symmetric if and only if for any a, and b in A, whenever $<a, b> \in R$, $<b, a> \in R$.

Example: The relation = on the set of integers {*1, 2, 3*} is $\{<1, 1>, <2, 2> <3, 3>\}$ and it is symmetric. Similarly = on any set of numbers is symmetric. However, < (or >), ≤ (or ≥ on any set of numbers is not symmetric.

Example: The relation "being acquainted with" on a set of people is symmetric.

Antisymmetric relation: A relation R on a set A is called antisymmetric if and only if for any a, and b in A, whenever $<a, b> \in R$, and $<b, a> \in R$, $a = b$ must hold. Equivalently, R is antisymmetric if and only if whenever $<a, b> \in R$, and $a \neq b$, $<b, a> \notin R$. Thus in an antisymmetric relation no pair of elements are related to each other.

Example: The relation < (or >) on any set of numbers is antisymmetric. So is the equality relation on any set of numbers.

Transitive relation: A relation R on a set A is called transitive if and only if for any a, b, and c in A, whenever $<a, b> \in R$, and $<b, c> \in R$, $<a, c> \in R$.

Example: The relation ≤ on the set of integers {*1, 2, 3*} is transitive, because for $<1, 2>$ and $<2, 3>$ in ≤, $<1, 3>$ is also in , for $<1, 1>$ and $<1, 2>$ in ≤, $<1, 2>$ is also in , and similarly for the others. As a matter of fact ≤ on any set of numbers is also transitive. Similarly and = ≥ on any set of numbers are transitive.

The following figures show the digraph of relations with different properties.

1. Is reflexive, antisymmetric, symmetric and transitive, but not irreflexive.
2. Is neither reflexive nor irreflexive, and it is antisymmetric, symmetric and transitive.
3. Is irreflexive but has none of the other four properties.
4. Is irreflexive, and symmetric, but none of the other three.
5. Is irreflexive, antisymmetric and transitive but neither reflexive nor symmetric.

(a)

(b)

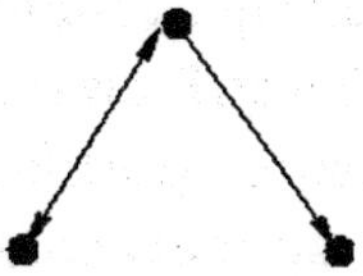

(c)

(d)

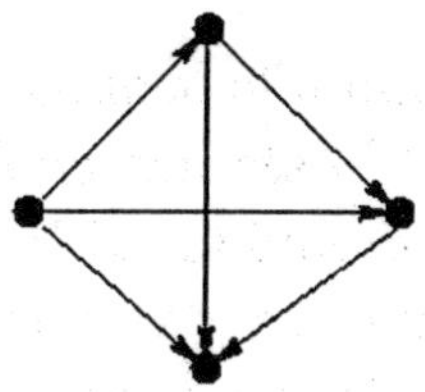

(e)

OPERATIONS ON RELATIONS

SET OPERATIONS

A relation is a set. It is a set of ordered pairs if it is a binary relation, and it is a set of ordered *n*–tuples if it is an *n*–ary relation. Thus all the set operations apply to relations such as $\cup$, $\cap$, and complementing.

For example, the union of the "less than" and "equality" relations on the set of integers is the "less than or equal to" relation on the set of integers. The intersection of the "less than" and "less than or equal to" relations on the set of integers is the "less than" relation on the same set. The complement of the "less than" relation on the set of integers is the "greater than or equal to" relation on the same set.

COMPOSITE RELATIONS

If the elements of a set *A* are related to those of a set *B*, and those of *B* are in turn related to the elements of a set *C*, then one can expect a relation between *A* and *C*. For example, if Tom is my father(parent–child relation) and Sarah is a sister of Tom (sister relation), then Sarah is my aunt (aunt–nephew/niece relation). Composite relations give that kind of relations.

· *Composite relation*: Let R_1 be a binary relation from a set *A* to a set *B*, R_2 a binary relation from *B* to a set *C*. Then the composite relation from *A* to *C* denoted by R_1R_2(also denoted by R_1 O R_2 is defined as

$$R_1R_2 = \{<a, c> \mid a \in A \wedge c\ C \wedge \exists\ b\ [b \in B < \wedge\ a, b> \in R_1 \wedge <b, c> \in R_2]\}.$$

In English, this means that an element *a* in *A* is related to an element *c* in *C* if there is an element *b* in *B* such that *a* is related to *b* by R_1 and *b* is related to *c* by R_2 . Thus R_1R_2 is a relation from *A* to *C* via *B* in a sense. If R_1 is a parent–child relation and R_2 is a sister relation, then R_1R_2 is an aunt–nephew/niece relation.

Example: Let $A = \{a_1, a_2\}$, $B = \{b_1, b_2, b_3\}$, and $C = \{c_1, c_2\}$. Also let $R_1 = \{<a_1, b_1>, <a_1, b_2>, <a_2, b_3>\}$, and $R_2 = \{<b_1, c_1>, <b_2, c_1>, <b_2, c_2>, <b_3, c_1>\}$. Then $R_1R_2 = \{<a_1, c_1>, <a_1, c_2>, <a_2, c_1>\}$.

The dashed lines in the figure of R_1R_2 indicate the ordered pairs in R_1R_2, and dotted lines show ordered pairs that produce the dashed lines. (The lines in the left figure are all supposed to be solid lines.)

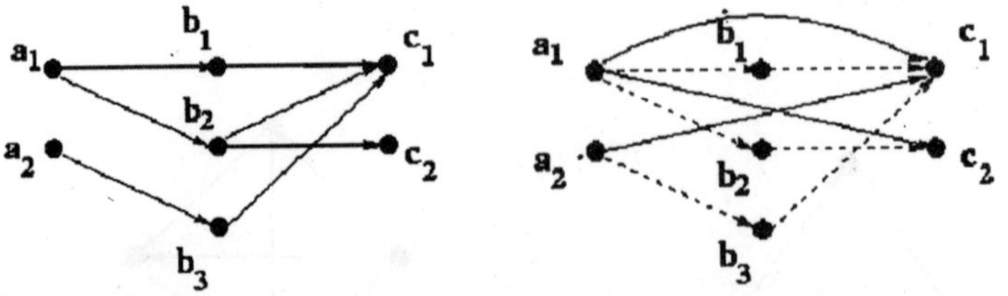

R_1 R_2 R_1R_2

Example: If R is the parent–child relation on a set of people A, then RR, also denoted by R^2, is the grandparent–grandchild relation on A.

More examples: The digraphs of R^2 for several simple relations R are shown below:

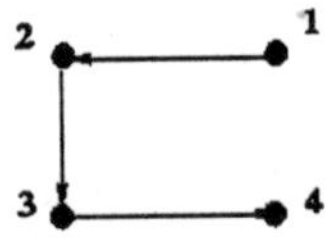

R_1

R_1^2

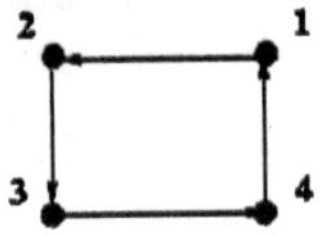

R_2

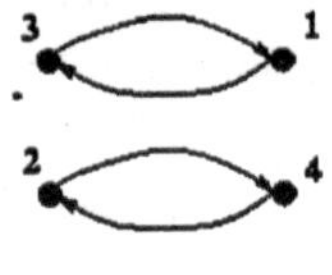

R_2^2

R_3

R_3^2

PROPERTIES OF COMPOSITE RELATIONS

Composite relations defined above have the following properties. Let R_1 be a relation from A to B, and R_2 and R_3 be relations from B to C. Then

1. $R_1(R_2R_3) = (R_1R_2)R_3$
2. $R_1(R_2 \cup R_3) = R_1R_2 \cup R_1R_3$
3. $R_1(R_2 \cap R_3) \subseteq R_1R_2 \cap R_1R_3$

Proofs for these properties are omitted.

POWERS OF RELATION

Let R be a binary relation on A. Then R^n for all positive integers n is defined recursively as follows:

Definition(power of relation):

Basis Clause: $R^0 = E$, where E is the equality relation on A.

Inductive Clause: For an arbitrary natural number n , $R^{n+1} = R^n R$.

Note that there is no need for extremal clause here.

Thus for example $R^1 = R$, $R^2 = RR$, and $R^3 = R^2R = (RR)R = R(RR) = RRR$.

The powers of binary relation R on a set A defined above have the following properties.

1. $R^{m+n} = R^m R^n$,
2. $(R^m)^n = R^{mn}$.

SPECIAL RELATIONS

CLOSURE OF BINARY RELATION

In our everyday life we often talk about parent-child relationship. This is a binary relation on the set of people in the world, dead or alive. Also we are often interested in ancestor-descendant relations. Although the latter relation can be obtained from the former, hence it is redundant in that sense, we do use ancestor-descendant relations which give us necessary information more directly. This ancestor-descendant relation relates two people if there is a sequence of parent-child relations from one to the other.

It includes the parent-child relation as a subset. The ancestor-descendant relation is an example of the closure of a relation, in particular the transitive closure of the parent-child relation. In general, the closure of a relation is the smallest extension of the relation that has a certain specific property such as the reflexivity, symmetry or transitivity. Formally they are defined as follows:

Reflexive closure: A relation R' is the reflexive closure of a relation R if and only if

1. R' is reflexive,
2. R R', and
3. For any relation R'', if $R \subseteq R''$ and R'' is reflexive, then $R' \subseteq R''$, that is, R' is the smallest relation that satisfies (1) and (2).

Example: Let R be the less-than relation on the set of integers I, that is $R = \{<a, b> \mid a \in I \wedge b \in I \wedge a < b\}$.

Then the reflexive closure $r(R)$ of R is the union of R and the equality relation on I, that is $r(R) = \{<a, b> \mid a \in I \wedge b \in I\ a \in b\}$

The digraph of the reflexive closure of a relation is obtained from the digraph of the relation by adding a self-loop at each vertex if one is already not there.

Symmetric and transitive closures can be defined similarly.

Symmetric closure: A relation R' is the symmetric closure of a relation R if and only if

1. R' is symmetric,
2. $R\ R'$, and
3. For any relation R'', if $R \subseteq R''$, and R'' is symmetric, then $R' \subseteq R''$, that is, R' is the smallest relation that satisfies (1) and (2).

Example: Let R be the less–than relation on the set of integers I. Then the symmetric closure of R, denoted by $s(R)$ is $s(R) = \{<a, b> \mid a\ I\ b\ I\ [a < b\ a > b]\}$ that is $\{<a, b> \mid a\ I\ b\ I\ a\ b\}$

The digraph of the symmetric closure of a relation is obtained from the digraph of the relation by adding for each arc the arc in the reverse direction if one is already not there.

.*Transitive closure*: A relation R' is the transitive closure of a relation R if and only if

1. R' is transitive,
2. $R \subseteq R'$, and
3. For any relation R'', if $R \subseteq R''$ and R'' is transitive, then $R' \subseteq R''$, that is, R' is the smallest relation that satisfies (1) and (2).

Examples: The transitive closure of a parent–child relation is the ancestor–descendant relation as mentioned above, and that of the less–than relation on I is the less–than relation itself.

The digraph of the transitive closure of a relation is obtained from the digraph of the relation by adding for each directed path the arc that shunts the path if one is already not there.

Two more examples of closures are given below in terms of digraphs.

The arrows with two heads represent arrows going in opposite directions.

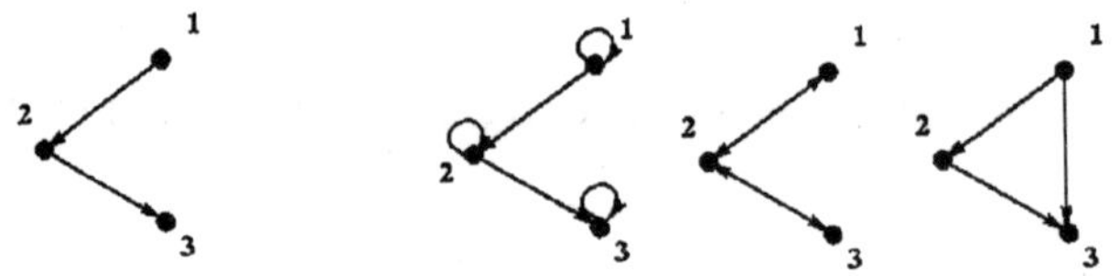

R_1 $r(R_1)$ $s(R_1)$ $t(R_1)$

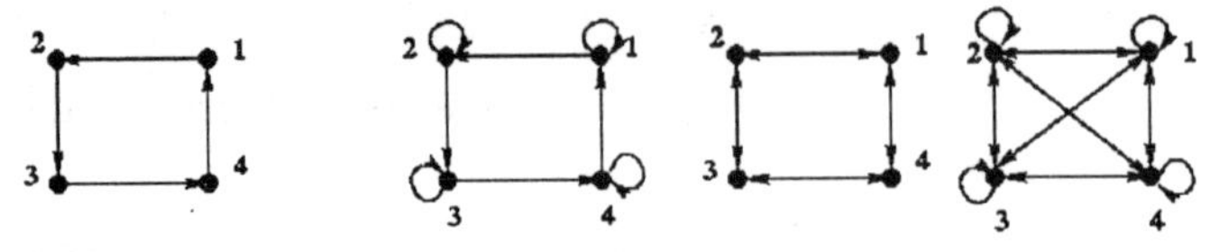

R_2 $r(R_2)$ $s(R_2)$ $t(R_2)$

PROPERTIES OF CLOSURE

The closures have the following properties. They are stated here as theorems without proof.

Theorem: Let E denote the equality relation, and R^c the inverse relation of binary relation R, all on a set A, where $R^c = \{< a, b > | < b, a \in R\}$. Then

1. $r(R) = R \cup E$
2. $s(R) = R \cup R^c$
3. $t(R) = \cup_{i=1}^{\infty} R^i = \cup_{i=1}^{n} R^i$, if $|A| = n$.
4. R is reflexive if and only if $r(R) = R$.
5. R is symmetric if and only if $s(R) = R$.
6. R is transitive if and only if $t(R) = R$.

EQUIVALENCE RELATION

On the face of most clocks, hours are represented by integers between 1 and 12. However, since a day has 24 hours after 12 hours, a clock goes back to hour 1, and starts all over again from there. Thus each pair of hours such as 1 and 13, 2 and 14, etc. share one number 1, 2, ...etc., respectively.

The same applies when we are interested in more than 24 hours. 25th hour is 1, so are 37th, 49th etc. What we are doing here essentially is that we consider the numbers in each group such as 1, 13, 25, ..., equivalent in the sense that they all are represented by one number. Being representable by one number such as we see on clocks is a binary relation on the set of natural numbers and it is an example of equivalence relation we are going to study here.

The concept of equivalence relation is characterized by three properties as follows:

Definition(equivalence relation): A binary relation R on a set A is an equivalence relation if and only if

- R is reflexive
- R is symmetric, and
- R is transitive.

Example: The equality relation (=) on a set of numbers such as $\{1, 2, 3\}$ is an equivalence relation.

Example: The congruent modulo m relation on the set of integers i.e. $\{<a, b>| a\ b \pmod{m}\}$, where m is a positive integer greater than 1, is an equivalence relation.

Note that the equivalence relation on hours on a clock is the congruent mod 12, and that when $m = 2$, i.e. the congruent mod 2, all even numbers are equivalent and all odd numbers are equivalent. Thus the set of integers are divided into two subsets: evens and odds.

Example: Taking this discrete structures course together this semester is another equivalence relation.

Equivalence relations can also be represented by a digraph since they are a binary relation on a set. For example the digraph of the equivalence relation congruent mod *3* on $\{0, 1, 2, 3, 4, 5, 6\}$ is as shown below. It consists of three connected components.

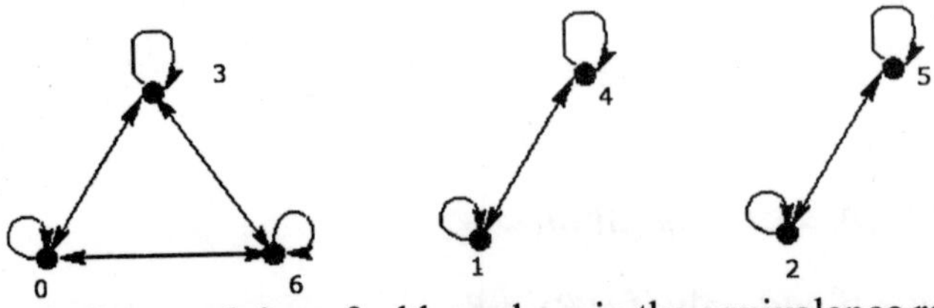

The set of even numbers and that of odd numbers in the equivalence relation of congruent mod *2*, and the set of integers equivalent to a number between 1 and 12 in the equivalence relation on hours in the clock example are called an equivalence class. Formally it is defined as follows:

Equivalence class: For an equivalence relation *R* on a set *A*, the set of the elements of *A* that are related to an element, say *a*, of *A* is called the equivalence class of element *a* and it is denoted by $[a]$.

Example: For the equivalence relation of hours on a clock, equivalence classes are

$[1] = \{1, 13, 25, \ldots\} = \{1+ 12n: n \in N\}$,

$[2] = \{2, 14, 26, \ldots\} = \{2+ 12n: n \in N\}$,

........,

where *N* is the set of natural numbers. There are altogether twelve of them.

For an equivalence relation *R* on a set *A*, every element of *A* is in an equivalence class. For if an element, say *b* , does not belong to the equivalence class of any other element in *A*, then the set consisting of the element *b* itself is an equivalence class. Thus the set *A* is in a sense *covered* by the equivalence classes. Another property of equivalence class is that equivalence classes of two elements of a set *A* are either disjoint or identical, that is either $[a] = [b]$ or $[a] \cap [b] = \emptyset$ for arbitrary elements *a* and *b* of *A*. Thus the set *A* is partitioned into equivalence classes by an equivalence relation on *A*. This is formally stated as a theorem below after the definition of partition.

Partition: Let *A* be a set and let $A_1, A_2, \ldots, A_n$ be subsets of *A*. Then $\{A_1, A_2, \ldots, A_n\}$ is a partition of *A*, if and only if

1. $\cup_{i=1}^{n} A_i = A$, and
2. $A_i \cap A_j = \emptyset$, if $A_i \neq A_j$, $1 \leq i, j \leq n$.

Example: Let $A = \{1, 2, 3, 4, 5\}$, $A_1 = \{1, 5\}$, $A_2 = \{3\}$, and $A_3 = \{2, 4\}$. Then $\{A_1, A_2, A_3\}$ is a partition of *A*. However, $B_1 = \{1, 2, 5\}$, $B_2 = \{2, 3\}$, and $B_3 = \{4\}$ do not form a partition for *A* because $B_1 \cap B_2 \neq \emptyset$, though $B_1 \neq B_2$.

Theorem: The set of equivalence classes of an equivalence relation on a set A is a partition of A.

Conversely, a partition of a set A determines an equivalence relation on A.

Theorem: Let $\{A_1, ..., A_n\}$ be a partition of a set A. Define a binary relation R on A as follows: $<a, b> \in R$ if and only if $a \in A_i$ and $b \in A_i$ for some i, $1 \leq i \leq n$. Then R is an equivalence relation.

Theorem : Let R_1 and R_2 be equivalence relations. Then $R_1 \cap R_2$ is an equivalence relation, but $R_1 \cup R_2$ is not necessarily an equivalence relation.

ORDER RELATION

Shoppers in a grocery store are served at a cashier on the first–come–first–served basis. When there are many people at cashiers, lines are formed. People in these lines are ordered for service: Those at the head of a line are served sooner than those at the end. Cars waiting for the signal to change at an intersection are also ordered similarly. Natural numbers can also be ordered in the increasing order of their magnitude. Those are just a few examples of order we encounter in our daily lives. The order relations we are going to study here are an abstraction of those relations. The properties common to orders we see in our daily lives have been extracted and are used to characterize the concepts of order. Here we are going to learn three types of order: partial order, total order, and quasi order.

Partial order: A binary relation R on a set A is a partial order if and only if it is :

1. Reflexive,
2. Antisymmetric, and
3. Transitive.

The ordered pair $<A, R>$ is called a poset (partially ordered set) when R is a partial order.

Example: The less-than-or-equal-to relation on the set of integers I is a partial order, and the set I with this relation is a poset.

Example: The subset relation on the power set of a set, say $\{1, 2\}$, is also a partial order, and the set $\{1, 2\}$ with the subset relation is a poset.

Total order: A binary relation R on a set A is a total order if and only if it is

1. A partial order, and
2. For any pair of elements a and b of A, $< a, b > \in R$ or $< b, a > \in R$.

That is, every element is related with every element one way or the other.

A total order is also called a *linear order*.

Example: The less-than-or-equal-to relation on the set of integers I is a total order.

The strictly-less-than and proper-subset relations are not partial order because they are not reflexive. They are examples of some relation called quasi order.

Quasi order: A binary relation R on a set A is a quasi order if and only if it is

1. Irreflexive, and
2. Transitive.

A quasi order is necessarily antisymmetric as one can easily verify.

Caution Like many other definitions there is another fairly widely used definition of quasi order in the literature. According to that definition a quasi order is a relation that is *reflexive* and transitive. That is something quite different from "quasi order" we have defined here.

Example: The less-than relation on the set of integers I is a quasi order.

Example: The proper subset relation on the power set of a set, say $\{1, 2\}$, is also a quasi order.

The concept of *least/greatest number* in a set of integers can be generalized for a general poset. We start with the concepts of minimal/maximal elements.

minimal/maximal element: Let $< A, \preceq >$ be a poset, where $\preceq$ represents an arbitrary partial order. Then an element $b \in A$ is a minimal element of A if there is no element $a \in A$ that satisfies $a \preceq b$. Similarly an element $b \in A$ is a maximal element of A if there is no element $a \in A$ that satisfies $b \preceq a$.

Example: The set of $\{\{1\}, \{2\}, \{1, 2\}\}$ with $\subseteq$ has two minimal elements $\{1\}$ and $\{2\}$. Note that $\{1\}$, and $\{2\}$ are not related to each other in $\subseteq$. Hence we can not say which is "smaller than" which, that is, they are not comparable.

least/greatest element: Let $< A, \preceq >$ be a poset. Then an element $b \in A$ is the least element of A if for every element $a \in A$, $b \preceq a$.

Note that the least element of a poset is unique if one exists because of the antisymmetry of $\preceq$.

Example: The poset of the set of natural numbers with the less–than–or–equal–to relation has the least element 0.

Example: The poset of the powerset of $\{1, 2\}$ with $\in$ has the least element Ø.

Well order: A total order R on a set A is a well order if every non–empty subset of A has the least element.

Example: The poset of the set of natural numbers with the less–than–or–equal–to relation is a well order, because every set of natural numbers has the least element.

The poset of the set of positive real numberes with the less–than–or–equal–to relation is not a well order, because the set itself does not have any least element (0 is not in the set).

A digraph of a binary relation on a set can be simplified if the relation is a partial order. Hasse diagrams defined as follows are such graphs.

Hasse diagram: A Hasse diagram is a graph for a poset which does not have loops and arcs implied by the transitivity. Further, it is drawn so that all arcs point upward eliminating arrowheads.

To obtain the Hassse diagram of a poset, first remove the loops, then remove arcs < a, b > if and only if there is an element c that < a, c > and < c, b > exist in the given relation.

Example: For the relation {< *a, a* >, < *a, b* >, < *a, c* >, < *b, b* >, < *b, c* >, < *c, c* >} on set {*a, b,c*}, the Hasse diagram has the arcs {< *a, b* >, < *b, c* >} as shown below.

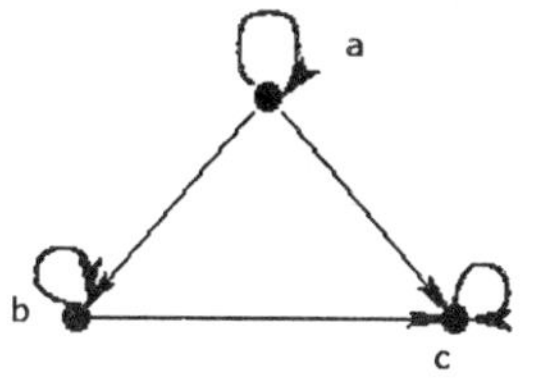

Digraph for Portial Order

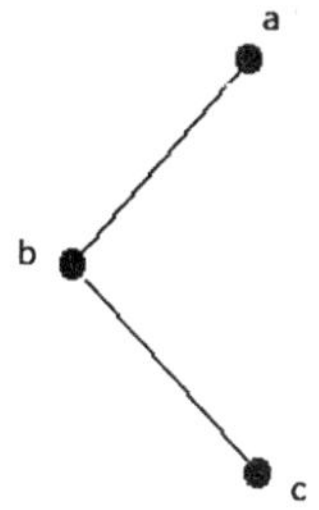

Hasse Digram

TOPOLOGICAL SORTING

The elements in a finite poset can be ordered linearly in a number of ways while preserving the partial order. For example {Ø, {1}, {2}, {1, 2}} wit h the partial order ⊆, can be ordered linearly as Ø, {1}, {2}, {1, 2}, or Ø, {2}, {1}, {1, 2}. In these orders a set appears before (to the left of) another set if it is a subset of the other. In real life, tasks for manufacturing goods in general can be partially ordered based on the prerequisite relation, that is certain tasks must be completed before certain other tasks can be started. For example the arms of a chair must be carved before the chair is assembled. Scheduling those tasks is essentially the same as arranging them with a linear order (ignoring here some possible concurrent processing for simplicity's sake).

The topological sorting is a procedure to find from a partial order on a finite set a linear order that does not violate the partial order. It is based on the fact that a finite poset has at least one minimal element. The basic idea of the topological sorting is to first remove a minimal element from the given poset, and then repeat that for the resulting set until no more elements are left. The order of removal of the minimal elements gives a linear order. The following algorithm formally describes the topological sorting.

Algorithm Topological Sor

Input: A finite poset <*A*, *R*>.

Output: A sequence of the elements of *A* preserving the order *R*.

integer *i*;

i : = 1;

while (*A*) {

pick a minimal element *b* from *A*;

$A := A - \{b\}$;
$i := i + 1$;
output b
}

Example: Let $A = \{\{1\}, \{2\}, \{3\}, \{1, 2\}, \{1, 3\}, \{2, 3\}, \{1, 2, 3\}\}$ with the partial order $\subseteq$.

This given A has three minimal elements $\{1\}$, $\{2\}$, and $\{3\}$.

Select $\{2\}$ and remove it from A. Let A denote the resultant set i.e. $A := A - \{2\}$. The new A has two minimal elements $\{1\}$, and $\{3\}$.

Select $\{1\}$ and remove it from A. Denote by A the resultant set, that is $A = \{\{3\}, \{1, 2\}, \{1, 3\}, \{2, 3\}, \{1, 2, 3\}\}$.

This new A has two minimal elements $\{3\}$ and $\{1, 2\}$.

Select $\{1, 2\}$ and remove it from A.

Proceeding in like manner, we can obtain the following linear order:

$\{\{2\}, \{1\}, \{1, 2\}, \{3\}, \{1, 3\}, \{2, 3\}, \{1, 2, 3\}\}$.

FUNCTION

A function is something that associates each element of a set with an element of another set (which may or may not be the same as the first set). The concept of function appears quite often even in nontechnical contexts. For example, a social security number uniquely identifies the person, the income tax rate varies depending on the income, the final letter grade for a course is often determined by test and exam scores, homeworks and projects, and so on.

In all these cases to each member of a set (social security number, income, tuple of test and exam scores, homeworks and projects) some member of another set (person, tax rate, letter grade, respectively) is assigned.

As you might have noticed, a function is quite like a relation. In fact, formally, we define a function as a special type of binary relation.

Definition (function): A function, denote it by *f*, from a set *A* to a set *B* is a relation from *A* to *B* that satisfies:

1. For each element *a* in *A*, there is an element *b* in *B* such that <*a, b*> is in the relation, and
2. If <*a, b*> and <*a, c*> are in the relation, then $b = c$.

The set *A* in the above definition is called the domain of the function and *B* its codomain.

Thus, *f* is a function if it *covers* the domain (maps every element of the domain) and it is *single valued*.

The relation given by f between a and b represented by the ordered pair $<a, b>$ is denoted as $f(a) = b$, and b is called the image of a under f.

The set of images of the elements of a set S under a function f is called the image of the set S under f, and is denoted by $f(S)$, that is,

$f(S) = \{f(a) \mid a S\}$, where S is a subset of the domain A of f.

The image of the domain under f is called the range of f.

Example: Let f be the function from the set of natural numbers N to N that maps each natural number x to x^2. Then the domain and codomain of this f are N, the image of, say 3, under this function is 9, and its range is the set of squares, i.e. $\{0, 1, 4, 9, 16, ...\}$.

Sum and product: Let f and g be functions from a set A to the set of real numbers R.

Then the sum and the product of f and g are defined as follows:

For all x, $(f + g)(x) = f(x) + g(x)$, and

for all x, $(f{*}g)(x) = f(x){*}g(x)$,

where $f(x){*}g(x)$ is the product of two real numbers $f(x)$ and $g(x)$.

Example: Let $f(x) = 3x + 1$ and $g(x) = x^2$. Then $(f + g)(x) = x^2 + 3x + 1$ and $(f{*}g)(x) = 3x^3 + x^2$

One-to-one: A function f is said to be one–to–one (injective), if and only if whenever $f(x) = f(y)$, $x = y$.

Example: The function $f(x) = x^2$ from the set of natural numbers N to N is a one–to–one function. Note that $f(x) = x^2$ is not one–to–one if it is from the set of integers(negative as well as non–negative) to N, because for example $f(1) = f(-1) = 1$.

Onto: A function f from a set A to a set B is said to be onto(surjective), if and only if for every element y of B, there is an element x in A such that $f(x) = y$, that is, f is onto if and only if $f(A) = B$.

Example: The function $f(x) = 2x$ from the set of natural numbers N to the set of non–negative even numbers E is an onto function. However, $f(x) = 2x$ from the set of natural numbers N to N is not onto, because, for example, nothing in N can be mapped to 3 by this function.

Bijection: A function is called a bijection, if it is onto and one–to–one.

Example: The function $f(x) = 2x$ from the set of natural numbers N to the set of non–negative even numbers E is one–to–one and onto. Thus it is a bijection.

Every bijection has a function called the inverse function.

The points on the left are in the domain and the ones on the right are in the codomain, and arrows show $< x, f(x) >$ relation.

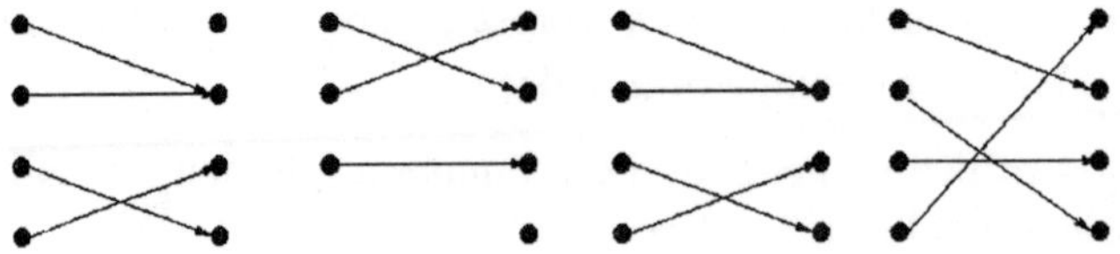

A function | A one-to-one function (Not onto) | An onto function Not one-to-one) | A bijection

Inverse: Let f be a bijection from a set A to a set B. Then the function g is called the inverse function of f, and it is denoted by f^{-1}, if for every element y of B, $g(y) = x$, where $f(x) = y$. Note that such an x is unique for each y because f is a bijection.

For example, the rightmost function in the above figure is a bijection and its inverse is obtained by reversing the direction of each arrow.

Example: The inverse function of $f(x) = 2x$ from the set of natural numbers N to the set of non–negative even numbers E is $f^{-1}(x) = 1/2\, x$ from E to N. It is also a bijection.

A function is a relation. Therefore one can also talk about composition of functions.

Composite function: Let g be a function from a set A to a set B, and let f be a function from B to a set C. Then the composition of functions f and g, denoted by fg, is the function from A to C that satisfies

$fg(x) = f(g(x))$ for all x in A.

Example: Let $f(x) = x^2$, and $g(x) = x + 1$. Then $f(g(x)) = (x + 1)^2$.

GROWTH OF FUNCTIONS

One of the important criteria in evaluating algorithms is the time it takes to complete a job. To have a meaningful comparison of algorithms, the estimate of computation time must be independent of the programming language, compiler, and computer used; must reflect on the size of the problem being solved; and must not depend on specific instances of the problem being solved. The quantities often used for the estimate are the worst case execution time, and average execution time of an algorithm, and they are represented by the number of some key operations executed to perform the required computation.

As an example for an estimate of computation time, let us consider the sequential search algorithm.

Example: Algorithm for Sequential Search

Algorithm SeqSearch(L, n, x)

L is an array with n entries indexed $1, .., n$, and x is the key to be searched for in L.

Output: if x is in L, then output its index, else output 0.

Index := *1;*

while (*index* $\leq n$ and $L[index] \neq x$)

Index := *index* + *1* ;

if (*index* > *n*) , then *index* := *0*

return *index* .

The worst case time of this algorithm, for example, can be estimated as follows: First the key operation is comparison of keys comparing $L[index]$ with x . Most search algorithms (if not all) need "comparison of keys". The largest number of execution of this comparison is n , which occurs when x is not in L or when x is at $L[n]$, and the while loop is executed n times. This quantity n thus obtained is used as an estimate of the worst case time of this sequential search algorithm.

Note that in the while loop two comparisons and one addition are performed. Thus one could use $3n$ as an estimate just as well. Also note that the very first line and the last two lines are not counted in. The reasons for those are firstly that differences in implementation details such as languages, commands, compilers and machines make differences in constant factors meaningless, and secondly that for large values of n , the highest degree term in n dominates the estimate. Since we are mostly interested in the behavior of algorithms for large values of n , lower terms can be ignored compared with the highest term. The concept that is used to address these issues is something called big–oh, and that is what we are going to study here.

BIG – OH

The following example gives the idea of one function growing more rapidly than another. We will use this example to introduce the concept the big–Oh.

Example: $f(n) = 100\ n^2$, $g(n) = n^4$, the following table and figure show that $g(n)$ grows faster than $f(n)$ when $n > 10$. We say f is big–Oh of g.

n	$f(n)$	$g(n)$
10	10,000	10,000
50	250,000	6,250,000
100	1,000,000	100,000,000
150	2,250,000	506,250,000

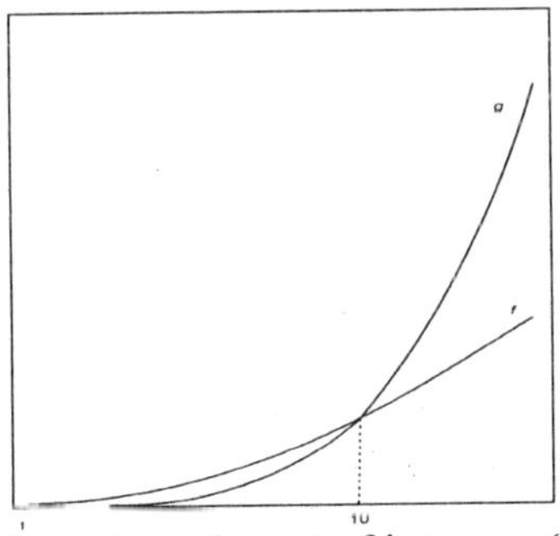

Big-oh: Let f and g be functions from the set of integers (or the set of real numbers) to the set of real numbers. Then $f(x)$ is said to be $O(g(x))$, which is read as $f(x)$ is big-oh of $g(x)$, if and only if there are constants C and n_0 such that

$$| f(x) |\ C\ | g(x) |$$

whenever $x > n_0$.

Note that big–oh is a binary relation on a set of functions (What kinds of properties does it have ? reflexive ? symmetric ? transitive ?).

For example, $5x + 10$ is big–oh of x^2, because $5x + 10 < 5x^2 + 10x^2 = 15x^2$ for $x > 1$.

Hence for $C = 15$ and $n_0 = 1$, $|5x + 10| \leq C|x^2|$.

Similarly it can be seen that $3x^2 + 2x + 4 < 9x^2$ for $x > 1$. Hence $3x^2 + 2x + 4$ is $O(x^2)$.

In general, we have the following theorem:

Theorem: $a_n x^n + \ldots + a_1 x + a_0$ is $O(x^n)$ for any real numbers $a_n, \ldots, a_0$ and any nonnegative number n.

Note: Let $f(x) = 3x^2 + 2x + 4$, $g(x) = x^2$, from the above illustration, we have that $f(x)$ is $O(g(x))$. Also, since $x^2 < 3x^2 + 2x + 4$, we can also get $g(x)$ is $O(f(x))$. In this case, we say these two functions are of the *same order*. Please refer to *big–theta* for more information.

GROWTH OF COMBINATIONS OF FUNCTIONS

Big–oh has some useful properties. Some of them are listed as theorems here.

Let use start with the definition of max function.

Definition(max function): Let $f_1(x)$ and $f_2(x)$ be functions from a set A to a set of real numbers B. Then $\max(f_1(x), f_2(x))$ is the function from A to B that takes as its value at each point x the larger of $f_1(x)$ and $f_2(x)$.

Theorem: If $f_1(x)$ is $O(g_1(x))$, and $f_2(x)$ is $O(g_2(x))$, then $(f_1 + f_2)(x)$ is $O(\max(g_1(x), g_2(x)))$.

From this theorem it follows that if $f_1(x)$ and $f_2(x)$ are $O(g(x))$, then $(f_1 + f_2)(x)$ is $O(g(x))$, and

$(f_1 + f_2)(x)$ is $O(\max(f_1(x), f_2(x)))$.

Theorem: If $f_1(x)$ is $O(g_1(x))$, and $f_2(x)$ is $O(g_2(x))$, then $(f_1 * f_2)(x)$ is $O(g_1(x) * g_2(x))$.

BIG – OMEGA AND BIG – THETA

Big–oh concerns with the "less than or equal to" relation between functions for large values of the variable. It is also possible to consider the "greater than or equal to" relation and "equal to" relation in a similar way. Big–Omega is for the former and big–theta is for the latter.

Big–omega: Let f and g be functions from the set of integers (or the set of real numbers) to the set of real numbers. Then $f(x)$ is said to be $\Omega(g(x))$, which is read as $f(x)$ is big–omega of $g(x)$, if there are constants C and n_0 such that

$$|f(x)| \geq C|g(x)|$$

whenever $x > n_0$.

Definition (big–theta): Let f and g be functions from the set of integers (or the set of real numbers) to the set of real numbers. Then $f(x)$ is said to be $(g(x))$, which is read as $f(x)$ is big–theta of $g(x)$, if $f(x)$ is $\Theta(g(x))$, and $(g(x))$. We also say that $f(x)$ is of order $g(x)$. (We have seen this at Big–Oh when introducing the concept of *same order*.)

For example, $3x^2 - 3x - 5$ is $\Omega(x^2)$, because $3x^2 - 3x - 5 \geq x^2$ for integers $x > 2$ ($C = 1$, $n_0 = 2$).

Hence by Theorem it is (x^2).

In general, we have the following theorem:

Theorem: $a_n x^n + ... + a_1 x + a_0$ is $\Theta(x^n)$ for any real numbers $a_n, ..., a_0$ and any nonnegative number n.

LITTLE – OH AND LITTLE – OMEGA

If $f(x)$ is $O(g(x))$, but not $\Theta(g(x))$, then $f(x)$ is said to be $o(g(x))$, and it is read as $f(x)$ is little–oh of $g(x)$. Similarly for little–omega (ω).

For example x is $o(x^2)$, x^2 is $o(2^x)$, 2^x is $o(x\,!)$, etc.

CALCULATION OF BIG – OH

Basic knowledge of limits and derivatives of functions from calculus is necessary here. Big–oh relationships between functions can be tested using limit of function. as follows:

Let $f(x)$ and $g(x)$ be functions from a set of real numbers to a set of real numbers.

Then

1. If , then $f(x)$ is $o(g(x))$. Note that if $f(x)$ is $o(g(x))$, then $f(x)$ is $O(g(x))$.
2. If , then $g(x)$ is $o(f(x))$.
3. If , then $f(x)$ is $(g(x))$.
4. If , then $f(x)$ is $O(g(x))$.

For example,

$(4/x + 3/x^2 + 5/x^4)/(1 - 3/x - 5/x^3 - 4/x^4) = 0$.

Hence

$(4x^3 + 3x^2 + 5)$ is $o(x^4 - 3x^3 - 5x - 4)$,

or equivalently, $(x^4 - 3x^3 - 5x - 4)$ is $(4x^3 + 3x^2 + 5)$.

Let us see why these rules hold. Here we give a proof for 4. Others can be proven similarly.

Proof: Suppose.

By the definition of limit this means that, n_0 such that whenever $x > n_0$.

Hence

Let =, then whenever $x > n_0$.

Since we are interested in non–negative functions f and g , this means that $| f(x) |$ C $| g(x)$

Hence $f(x) = O(g(x))$ . .

L'Hospital (L'Hτpital)'s Rule

. $f(x)/g(x)$ is not always easy to calculate. For example take $x^2/3^x$. Since both x^2 and 3^x go to as x goes to and there is no apparent factor common to both, the calculation of the limit is not immediate. One tool we may be able to use in such cases is L'Hospital's Rule, which is given as a theorem below.

Theorem (L'Hospital):

If $f(x) =$ and $g(x) =$, and $f(x)$ and $g(x)$ have the first derivatives, $f'(x)$ and $g'(x)$, respectively, then $f(x)/g(x) = f'(x)/g'(x)$.

This also holds when $f(x) = 0$ and $g(x) = 0$, instead of $f(x) =$ and $g(x) =$.

For example, $x/e^x = 1/e^x = 0$, because $(e^x)' = e^x$, where e is the base for the natural logarithm.

Similarly $\ln x/x = (1/x)/1 = 1/x = 0$.

Note that this rule can be applied repeatedly as long as the conditions are satisfied.

So, for example, $x^2/e^x = 2x/e^x = 2/e^x = 0$.

SUMMARY OF BIG – OH

Sometimes, it is very necessary to compare the order of some common used functions including the following:

1 $\log n$ n $n\log n$ n^2 2^n $n!$ n^n

Now, we can use what we've learned above about the concept of big–Oh and the calculation methods to calculate the order of these functions. The result shows that each function in the above list is big–oh of the functions following them. The figure below displays the graphs of these funcions, using a scale for the values of the functions that doubles for each successive marking on the graph.

RECURRENCE RELATIONS

A recurrence relation is an equation that defines a sequence recursively: each term of the sequence is defined as a function of the preceding terms.

A difference equation is a specific type of recurrence relation.

An example of a recurrence relation is the logistic map:

$$x_{n+1} = rx_n (1 - x_n).$$

Some simply defined recurrence relations can have very complex (chaotic) behaviours and are sometimes studied by physicists and mathematicians in a field of mathematics known as nonlinear analysis.

Solving a recurrence relation means obtaining a closed-form solution: a non-recursive function of n.

LINEAR HOMOGENEOUS RECURRENCE RELATIONS

The term *linear* means that each term of the sequence is defined as a linear function of the preceding terms. The *order* of a linear recurrence relation is the number of preceding terms required by the definition—so the relation $a_n = a_{n-2}$ is of order two, because there must be at least two preceding terms (whether they are both used or not).

The general form of a linear recurrence relation of order d is as follows:

$$a_n = c_1 a_{n-1} + c_2 a_{n-2} + \ldots + c_d a_{n-d} + c$$

where c and c_i (for all i) are allowed to depend on n, but a_i (for all) is not. If is independent of (for all), then the recurrence relation is said to have constant coefficients. Additionally, if then the recurrence relation is homogeneous; such a sequence is also called a linear recursive sequence or LRS.

The linear recurrence, together with *seed values* (initial conditions) for, determines the sequence uniquely.

Fibonacci Numbers

The Fibonacci numbers are defined using the linear recurrence relation

$$F_n = F_{n-1} + F_{n-2}$$

with seed values:

$$F_1 = 1$$

$$F_2 = 1.$$

The sequence of Fibonacci numbers begins:

1, 1, 2, 3, 5, 8, 13, 21, 34, 55, 89...

It can be solved via the matrix solution described below, yielding the closed form expression.

Solving via Linear Algebra

Given an LRS, one can write down its companion matrix, then put it in Jordan normal form (which is diagonal if the eigenvalues are distinct). Expressing the seed in terms of the eigenbasis, say

$$\begin{bmatrix} a_0 \\ \vdots \\ a_{d-1} \end{bmatrix} = b_1 v_1 + \ldots + b_d v_d$$

yields

$$\begin{bmatrix} a_n \\ \vdots \\ a_{n+(d-1)} \end{bmatrix} = C_n \begin{bmatrix} a_0 \\ \vdots \\ a_{d-1} \end{bmatrix} = C^n(b_1v_1 + ... + b_dv_d) = \lambda_1^n b_1v_1 + ... + \lambda_d^n b_dv_d$$

which is a closed form expression (expand on the first coordinate to obtain a closed form expression for a_n).

If the companion matrix is not diagonalizable, then the resulting expression is more complicated, but conceptually the same.

Solving Generally

Solutions to recurrence relations are found by systematic means, often by using generating functions (formal power series) or by noticing the fact that r^n is a solution for particular values of r.

Consider, for example, a recurrence relation of the form

$$a_n = Aa_{n-1} + Ba_{n-2}.$$

Suppose that it has a solution of the form $a_n = r^n$. Substituting this guess in the recurrence relation, we find:

$$r^n = Ar^{n-1} + Br^{n-2}.$$

Dividing through by r^{n-2}, we get

$$r^2 = Ar + B,$$

$$r^2 = Ar - B = 0$$

This is known as the *characteristic equation* of the recurrence relation. Solve for r to obtain the two roots λ_1, λ_2, and if these roots are distinct, we have the solution

$$a_n = C\lambda_1^n + D\lambda_2^n$$

while if they are identical (when $A^2 + 4B = 0$), we have

$$a_n = C\lambda^n + D\lambda^n$$

where constants C and D can be found from the "side conditions" that are often given as $a_0 = a$, $a_1 = b$.

Different solutions are obtained depending on the nature of the roots of the characteristic equation.

GENERAL LINEAR HOMOGENEOUS RECURRENCE RELATIONS

Many linear homogeneous recurrence relations may be solved by means of the hypergeometric series. Special cases of these lead to recurrence relations for the orthogonal polynomials, and many special functions. For example, the solution to

$$J_{n+1} = \frac{2n}{z} J_n - J_{n-1}$$

is given by the $J_n = J_n(z)$, the Bessel function, while

$$(b - n)M_{n-1} \quad + (2n - b - z)M_n - nM_{n+1} = 0$$

is solved by $M_n = M(n,b;z)$ the confluent hypergeometric series.

SOLVING NON-HOMOGENEOUS RECURRENCE RELATIONS

If the recurrence is inhomogeneous, a particular solution can be found by the method of undetermined coefficients and the solution is the sum of the solution of the homogeneous and the particular solutions. Another method to solve an inhomogeneous recurrence is the method of *symbolic differentiation*. For example, consider the following recurrence:

$$a_{n+1} = a_n + 1$$

This is an inhomogeneous recurrence. If we substitute $n \mapsto n + 1$, we obtain the recurrence

$$a_{n+2} = a_{n+1} + 1$$

Subtracting the original recurrence from this equation yields

$$a_{n+1} - a_{n+1} = a_{n+1} - a_n$$

or equivalently

$$a_{n+2} = 2a_{n+1} - a_n$$

This is a homogeneous recurrence which can be solved by the methods explained above. In general, if a linear recurrence has the form

$$a_{n+k} = \lambda_{k-1}a_{n+k-1} + \lambda_{k-2}a_{n+k-2} + \ldots + \lambda_1 a_{n+1} + \lambda_0 a_n + p(n)$$

where $\lambda_0, \lambda_1, \ldots, \lambda_{k-1}$ are constant coefficients and $p(n)$ is the inhomogeneity, then if $p(n)$ is a polynomial with degree r, then this inhomogeneous recurrence can be reduced to a homogeneous recurrence by applying the method of symbolic differentiation r times.

Example: Let's solve the following inhomogeneous recurrence relation using the method of undetermined coefficients:

$$a_{n+1} = 2a_n + 3^n + 5_n$$

The general solution for the analogous homogeneous relation

$$b_{n+1} = 2b_n$$

is

$$b_n = c_1 2^n$$

The inhomogeneous part $(3^n + 5n)$ leads us to guess the particular solution

$$a_n = c_2 a + c_3 n + c_4$$

(the guess for a^x is a^x, and the guess for a degree-n polynomial is a degree-n polynomial)

Substituting back into the recurrence relation, we get:

$$c_2 3a^{n+1} + c_3(n+1) + c_4 = 2(c_2 3^n + c_3 n + c_4) + 3^n + 5n$$

Looking only at the coefficients of 3^n:

$$3c_2 = 2c_2 + 1$$

$$c_2 = 1$$

Looking only at the coefficients of n:

$$c_3 = 2c_3 + 5$$

$$c_3 = -5$$

Looking only at the coefficients of 1 (constants):

$$c_3 + c_4 = 2c_4$$

$$c_4 = c_3 = -5$$

So the general solution is

$$a_n = c_1 2^n + 3^n - 5n - 5$$

A technique is described for using Newton's identities to compute the coefficients of a multi-step recurrence formulae for simulating the solutions of ordinary linear differential equations of high order. This approach gives the exact root-matched recurrence formula without having to determine the complex roots of the characteristic polynomials.

Let x(t) and y(t) denote continuous functions of time related according to the Nth order differential equation

$$\sum_{k-0}^{N} a_k \frac{d^k y}{dt} = \sum_{k=0}^{N} b_k \frac{d^k x}{dt^k} \qquad a_N, a_0, b_N {}^1 0$$

where the coefficients a_k and b_k are constants. Simple linear relationships of this form are widely used in numerical simulations of heat transfer, chemical reactions, fluid flows, mechanical dynamics, Markov processes, etc. (Relationships of this form are also used in signal filtering and control-system compensation, but in those applications the frequency response is usually emphasized over the exact time-domain response.) A variety of methods have been used to numerically generate the solutions y(t) of such equations for arbitrary forcing functions x(t). In the digital environment equation is often simulated by means of a multi-step linear recurrence relation of the form

$$h\,Y_n = g\,X_n$$

where Xn and Yn are the column vectors

$$Y_n = \begin{bmatrix} y((n-N)\Delta t) \\ \vdots \\ y((n-N)\Delta t) \\ y(n\Delta t) \end{bmatrix} \qquad X_n = \begin{bmatrix} y((n-N)\Delta t) \\ \vdots \\ y((n-N)\Delta t) \\ x(n\Delta t) \end{bmatrix}$$

for some constant Δt (called the *step-size* of the simulation), and g and h are constant row vectors

$$h = [h_0\ h_1 \ldots\ h_N]\ g = [g_0\ g_1 \ldots\ g_N]$$

Given the coefficients ak and bk, our objective is to compute, as quickly and efficiently as possible, the components of h and g.

It is well known that if the recurrence is required to match the exact homogeneous response for y(t) when x(t) is constantly zero, then the components of h must be proportional to the elementary symmetric functions (with alternating signs) of the quantities $e^{\alpha_i \Delta t}$, i = 1,2,..,N, where αi are the roots of the characteristic polynomial.

The usual procedure for determining this "root-matched" recurrence is to solve the characteristic polynomial for the roots ai, and then compute the elementary symmetric functions of these roots. For high order relationships the most troublesome aspect of this approach is the determination of all the complex roots of the characteristic polynomials, especially when there are multiple roots.

The most common way of simplifying the calculations is to use a substitution method, e.g., bilinear substitution, but such methods do not reproduce the exact homogeneous response. The purpose of this note is to describe how, using Newton's Identities, we can precede directly from the coefficients to the exact root-matched recurrence coefficients without solving the characteristic polynomials.

DETERMINING THE ROOT MATCHED RECURRENCE

For any integer k, let w(k) denote the sum of the kth powers of the roots $a_1, a_2, .., a_N$. Clearly, w(0) = N. Using Newton's Identities [8,9] the values of w(k) for k = 1,2,... are given by

$$a_N\, w(1) + 1a_{N-1} = 0$$

$$a_N\, w(2) + a_{N-1}\, w(1) + 2a_{N-2} = 0$$

$$a_N\, w(3) + a_{N-1}\, w(2) + a_{N-2}\, w(1) + 3a_{N-3} = 0$$

for k < N, and the recurrence

$$a_N\, w(k) + a_{N-1}\, w(k-1) + \ldots + a_0\, w(k-N) = 0$$

for all k ≥ N. Now let E(k) denote the sum of the kth powers of the values of $e^{\alpha_i \Delta t}$, i=1,2,..,N. Expanding each of the exponential functions as a Taylor's series and combining terms by powers of Δt gives the following expression for E(k):

$$E(k) = \sum_{j-0}^{\infty} \frac{(k\Delta t)^j}{j!} w(j)$$

Letting σ_k denote (-1)k times the kth elementary symmetric function of the quantities $e^{\alpha_i \Delta t}$, we can now compute σ_0 through σ_N using the identities

$E(1) + 1\sigma_1 = 0$

$E(2) + \sigma_1 E(1) + 2\sigma_2 = 0$

$E(3) + \sigma_1 E(2) + \sigma_2 E(1) + 3\sigma_3 = 0$

The precision of these computations, including the convergence, can be checked by verifying the equality

$$\sigma_N = (-1)^N e^{-a_{N-1}\Delta t / a_N}$$

The general recurrence for the homogeneous equation in y(t) can now be expressed as:

$$S_A Y_n = 0$$

where SA is the row vector

$$S_A = [\sigma_N \; \sigma_{N-1} \cdots \sigma_0]$$

Figure presents an implementation of the preceding algorithm in BASIC. The inputs to this program are the order N, the step-size Dt, and the coefficients $a_0, a_1 .., a_N$. The program computes the values of $\sigma_0, \sigma_1, .., \sigma_N$. The parameter "sumto" determines the number of terms to be evaluated in the summation.

By entering the coefficients bk in place of ak into the preceding algorithm we can determine the recurrence for the homogeneous equation in x(t)

$$S_B X_n = 0$$

We can then combine the two sides of the recurrence by applying the appropriate scale factors, which are easily deduced from steady-state conditions. The result is

$$\frac{a_0}{\sum S_A} S_A Y_n = \frac{b_0}{\sum S_B} S_B X_n$$

where ΣS_A and ΣS_B denote the sums of the components of SA and SB respectively.

EQUILATERAL RECURRENCES

If $b_N = 0$ then the characteristic polynomial of the right side has fewer than N roots. In this case the recurrence would involve fewer "past values" of x(t) than of y(t). Note that the coefficients of x_n were uniquely determined by the requirement that x(t) conform to the true homogeneous behavior when y(t) is constantly zero.

However, in practice x(t) is an arbitrary forcing function so we are not really concerned with reproducing the homogeneous "response" of x(t). Instead, we are trying to reconstruct the function x(t) from the discrete samples. Thus it is often considered more appropriate to make use of as many past values of x(t) as is computationally convenient. This typically leads to recurrences that use the same number of past values of x(t) as of y(t), i.e., *equilateral* recurrences. The usual approach in signal processing is to augment the roots of the right-hand characteristic polynomial with fictitious roots of the form ±p i, and then proceed as before.

The effect is to add roots equalling -1 to the characteristic polynomial of the recurrence, which essentially mimics bi-linear substitution.

The rationale for the "method of fictitious roots" just described is based on the frequency response characteristics. In the time domain it is often more appropriate to convert into an equilateral equation by a simple change of variables. If we define the variable z(t) such that $y=\lambda\,(x+z)$ for some constant λ, then (1) can be written

$$\sum_{k-0}^{N} a_k \frac{d^k z}{dt^k} = \sum_{k-0}^{N} \left(\frac{b_k}{\lambda} - a_k \right) \frac{d^k x}{dt^k}$$

For any choice of λ not equal to 0, (b_N/a_N), or (b_0/a_0), this equation is equilateral, so we can apply the method to give the recurrence

$$h\, Z_n = g\lambda\, X_n$$

where $g\lambda$ is based on the coefficients of the right hand side of (5). Reverting back to the y(t) variable gives

$$h\, Y_n = \lambda\,(g\lambda + h)\, X_n$$

For large values of λ the vector $\lambda\,(g\lambda + h)$ is quite insensitive to changes in λ, and it converges in the limit as $|\lambda| \to \infty$. In practice, any λ greater than about 10 times the largest b_i will give very nearly the limiting vector. This "change-of-variables method" enables us to determine equilateral recurrences for non-equilateral differential equations without resorting to fictitious roots.

THE EXPECTED VALUE METHOD

For comparison we briefly describe a method that explicitly treats x(t) as an *independent* variable and y(t) as a *dependent* variable. From this point of view it can be argued that the *expected value* of x(t) based on the N+1 most recent samples is given by the Nth degree polynomial that passes through those samples. Although the method is a straight-forward application of polynomial curve fitting, no explicit formulation of the general Nth-order recurrence formula appears in the literature, so a brief description is given below.

As in our previous recurrence formulas, we require that the homogeneous response of y(t) be matched exactly, so we use the algorithm shown in Figure to determine SA. Combining this with the particular solution based on the Nth order polynomial curve fit of x(t) leads directly to the total recurrence

$$S_A Y_n = S_A T A^{-1} B T^{-1} X_n$$

where A, B, and T are square matrices such that the elements in the jth columns of the kth rows (where j,k range from 0 to N) are given by

$$A_{kj} = \begin{cases} \dfrac{j!}{k!} a_{j-k} & \text{if } j \geq k \\ 0 & \text{if } j \geq k \end{cases} \qquad B_{kj} = \begin{cases} \dfrac{j!}{k!} b_{j-k} & \text{if } j \geq k \\ 0 & \text{if } j < k \end{cases} \qquad T_{kj} = \begin{cases} (k\Delta t)^j & \text{if } k+j > 0 \\ 0 & \text{if } k+j = 0 \end{cases}$$

This method gives, arguably, the best possible reconstruction of arbitrary forcing functions x(t), although the matrix inversions make it somewhat laborious when applied to high order equations.

Examples: To illustrate the basic algorithm for determining the root-matched recurrence relation for a homogeneous equation, consider the differential equation

$$\frac{d^5y}{dt^5}+7\frac{d^4y}{dt^4}+21\frac{d^3y}{dt^3}+33\frac{d^3y}{dt^2}+28\frac{dy}{dt}+10y=0$$

To simplify the recurrence expressions we will denote y(nΔt) by y_n. Taking a time interval of

Δt = 0.1, the algorithm described gives the following recurrence formula for y_n

$$y_n = -h_4\,y_{n-1} - h_3\,y_{n-2} - h_2\,y_{n-3} - h_1\,y_{n-4} - h_0\,y_{n-5}$$

where

h_0 = -0.4965853038

h_1 = 2.8479524294

h_2 = -6.5428490609

h_3 = 7.5263050469

h_4 = -4.3347524368

(Notice that we've scaled the coefficients to make h_5 equal to 1.) The accuracy of this recurrence can be illustrated by noting that the characteristic roots -1, -2+i, -2-i, -1+i, and -1-i, so one particular analytical solution has the simple form

$$y_n = -8e^{-n\Delta t} + (-6e^{-2n\Delta t} + 14e^{-n\Delta t})\cos(n\Delta t)$$

Taking y_0 through y_4 from as initial values, it is easily verified that the recurrence numerically reproduces the exact analytical values of y_5, y_6, y_7... etc., very precisely. After 35 iterations of the recurrence formula (carried out to 10 significant decimal digits) the value of y_{40} differs from the analytical value by only 0.0000013, this discrepancy being due entirely to accumulated round-off error. In contrast, using the recurrence relation given by bilinear substitution, the error at y_{40} is 0.0013094. For larger step-sizes the error using bilinear substitution becomes progressively worse, whereas the root-matched recurrence method is equally applicable to any step-size.

We chose a homogeneous equation with simple characteristic roots for the preceding example so that the recurrence values could be easily compared to the analytical solution. Also, the characteristic roots in the preceding example all have negative real parts, so the solutions are "stable". However, our interest is not limited to "stable" systems. For example, problems involving combustion and thermal "run-away" often lead to "unstable" systems. To illustrate the various methods discussed in this note as applied to one of these "unstable" systems, and to include "numerator dynamics", consider the differential equation

$$\frac{d^5y}{dt^5} - \frac{d^3y}{dt^3} - 2\frac{d^2y}{dt^2} + y = \frac{d^3x}{dt^3} - \frac{dx}{dt} - x$$

Taking a time interval of $\Delta t = 1$, the relation between x and y can be simulated using a linear recurrence of the form (2), where the components of h and g are as listed in the table below (normalized for $h_5 = 1$).

i	*All except bilinear substitut*	*Bilinear substitut*	*Expected value method*	*Change of variables method*	*Fictitious roots method*	*Bilinear substit*
0	−1.000000	−1.823529	−0.283164	−0.083311	−0.250012	−0.411765
1	5.060084	8.764706	−0.530963	−0.537182	0.386879	0.529412
2	−11.732310	−18.235294	2.456330	2.489417	0.365333	0.588235
3	14.904837	21.294118	−3.428658	−3.485152	−1.780006	−1.764706
4	−7.146380	−9.117647	0.407632	0.446686	−0.658437	−1.117647
5	1.000000	1.000000	0.092613	0.083311	0.250012	0.294118

All the methods except Bilinear Substitution made use of the BASIC program in Figure to determine the components of h and to assist in determining the components of g. Interestingly, the Change-of-Variables Method gives components of g that closely match those given by the Expected Value Method, even though the former makes no use of a 5th order curve fit of x(t). In contrast, the methods of fictitious roots and bilinear substitution give very different g vectors, and therefore will clearly yield significantly different values of y(t) in response to certain forcing functions x(t).

RECURRENCES FOR HARMONIC SUMS

Let h(n) denote the nth partial sum of the harmonic series, and define $H[k] = h(2^k)$. It's not hard to show that H[n] approaches a linear recurring sequence satisfying

$$H[k] - 2\,H[k-1] + H[k-2] = 0$$

as k increases. The error of this recurrence decreases by a factor of two at each step. Correcting for this error, we arrive at the third- order recurrence

$$2\,H[k] - 5\,H[k-1] + 4\,H[k-2] - H[k-3] = 0$$

Eliminating the main error term from this relation leads to a fourth- order recurrence, and so on. The coefficients of these recurrences up through order six are tabulated below (without signs):

(2)	1	2	1			
(4)	2	5	4	1		
(16)	8	22	21	8	1	
(64)	128	360	358	149	24	1
(256)	8192	23168	23272	9894	1685	88 1

The numbers in parentheses on the left indicate the factor by which the error of the respective recurrence decreases at each step. If c[i, j] denotes the coefficient in the ith row and the jth column, and r_i denotes the "error factor" of the ith row, then the rule of formation for these coefficients can be expressed as

$$c[i, j] = (r_{(i-1)})\, c[i-1, j] + c[i-1, j-1]$$

Note that the sum of the elements of each row, with alternating signs, is zero. Also, the "error factor" increases by a multiple of 4 at each level except the first,

so we have

$$r_1 = 2 \text{ and } r_i = 4^{(i-1)} \text{ for } i>1.$$

This same approach works for other geometric arguments, e.g., if we define $G[k] = h(3^k)$, we can derive a sequence of recurrences with the coefficients listed below (without signs):

(3)	1	2	1		
(9)	3	7	5	1	
(81)	27	66	52	14	1

The rule of formation is the same, except that in this case we have $r_1 = 3$ and $r_i = 9^{(i-1)}$ for i>1.

To generalize this a bit, for any fixed integer m>1 define

$$H[k] = h(m^k)$$

Since h(n) increases logrithmically, it's natural to expect that the above recurrence is a good starting point for any m, because

$$\ln(m^k) + A \ln(m^{(k-1)}) + B \ln(m^{(k-2)}) = k \ln(m) + A\,(k-1) \ln(m) + B\,(k-2) \ln(m)$$

Setting this equal to zero and dividing by ln(m) gives

$$k + A\,(k-1) + B\,(k-2) = k(1+A+B) - (A+2B) = 0$$

which implies the two conditions

$$1 + A + B = 0 \qquad A + 2B = 0$$

so we must have A=-2 and B=1. However, the third order recurrence

for $h(m^k)$ gives only two constraints on three coefficients, i.e.,

$$1 + A + B + C = 0 \qquad A + 2B + 3C = 0$$

and similarly for higher order recurrences. Thus, all recurrences for $\ln(m^k)$ of order greater than 2 are underspecified. Anyway, taking the order 2 recurrence we see that the characteristic polynomial splits into linear factors

$$x^2 - 2x + 1 = (x-1)^2.$$

It turns out that the characteristic polynomials of all the asymptotic recurrences for H[k] described previously also split into linear factors. Specifically, for m=2 we have

$$2x^3 - 5x^2 + 4x - 1 = (2x-1)(x-1)^2$$

$$8x^4 - 22x^3 + 21x^2 - 8x + 1 = (4x-1)(2x-1)(x-1)^2$$

$$128x^5 - 360x^4 + 358x^3 - 149x^2 + 24x - 1 = (16x-1)(4x-1)(2x-1)(x-1)^2$$

and so on. Similarly for m=3 we have

$$3x^3 - 7x^2 + 5x - 1 = (3x-1)(x-1)^2$$

$$27x^4 - 66x^3 + 52x^2 - 14x + 1 = (9x-1)(3x-1)(x-1)^2$$

and so on. In general, the characteristic polynomial of the asymptotic Jth-order recurrence (J>3) for H[k] with any fixed m is

$$P(x) = (mx-1)\,(x-1)^2 \prod_{t=1}^{J-3} (m^{(2t)}\,x - 1)$$

so the characteristic roots (eigen values) are

$$1,\ 1,\ 1/m,\ 1/m^2,\ 1/m^4,\ 1/m^6, \ldots,\ 1/m^{(2(J-3))}.$$

This allows us to give an explicit formula for the terms of the Jth-order linear recurring sequence of values that approximate H[k].

Let Q be a JxJ matrix with the components

$Q[i,1] = 1$	for i = 1 to J
$Q[i,2] = i$	for i = 1 to J
$Q[i,3] = 1/m^i$	for i = 1 to J
$Q[i,j] = 1/m^{2i}(j-3)$	for i = 1 to J and j = 4 to J

and let R be a column vector with the components

$R[i] = H[i-1]$	for i=1 to J

Then the asymptotic recurrence for H[k] has the following closed-form solution using the exact values H[0] through H[J-1] as initial values:

$$H[k-1] = [1\ k\ 1/m^k\ 1/m^{2k}\ 1/m^{4k} \ldots 1/m^{(2k(J-3))}]\ Q^{-1}\ R$$

Clearly, as k increases, all but the first two terms of this expression go to zero, so for large values of k we have approximately

$$H[k-1] = C[1] + C[2]*k$$

where C[1] and C[2] are the first two components of the column vector $C = Q^{-1} R$. The difference between H[k-1] and $h(2^{(k-1)})$ is therefore approximated by

$$C[1] + C[2]\,k - (k-1)\ln(2) = C[1] + \ln(2) + k\,(C[2] - \ln(2))$$

For high order recurrence this value becomes virtually independent of k, which suggests that C[2] approaches ln(2). Making this substitution,

we would expect to find that

$$\text{gamma} \sim = C[1] + C[2]$$

To illustrate, with m = 2 the closed–form solution for the 6th–order recurrence gives

$$C[1] + C[2] = 0.5772175....$$

which is approximately equal to gamma = 0.5772156...

INTERLEAVING FIBONACCI NUMBERS

The sequence of Fibonacci numbers is defined by the initial values $f_0 = 0$, $f_1 = 1$, and the recurrence $f_n = f_{n-1} + f_{n-2}$. The first several values are

n:	0	1	2	3	4	5	6	7	8	9	10	11	...
f_n:	0	1	1	2	3	5	8	12	21	34	55	89	...

Notice that the even-ordered terms satisfy the relations

$$f_2 = 1$$
$$f_4 = 1f_2 + 2$$
$$f_6 = 1f_4 + 2f_2 + 3$$
$$f_8 = 1f_8 + 2f_4 + 3f_2 + 4$$
$$f_{10} = 1f_8 + 2f_6 + 3f_4 + 4f_2 + 5$$

and so on. Similarly the odd-ordered terms satisfy the relations

$$f_1 = 1$$
$$f_3 = 1f_1 + 1$$
$$f_5 = 1f_3 + 2f_1 + 1$$
$$f_7 = 1f_5 + 2f_3 + 3f_1 + 1$$
$$f_9 = 1f_7 + 2f_5 + 3f_3 + 4f_1 + 1$$

and so on. One way of proving that this pattern continues is by induction. We know by inspection up to a certain n that consecutive Fibonacci numbers are given by

$$f_{2n} = 1f_{2(n-1)} + 2f_{2(n-2)} \quad +... +(n-1)f_2+n$$
$$f_{2n+1} = 1f_{2(n-1)} +1+ 2f_{2(n-2)+1} \quad +... +(n-1)f_3+(n)f_1 + 1$$

(Note that this requires $f_1 = f_2 = 1$.) Adding these together, term by term, it follows from the basic Fibonacci recurrence that the next even-ordered value in the sequence is

$$f_{2(n+1)} = 1f_{2n} + 2f_{2(n-1)}+... +(n-1)f_4+(n)f_2 +(n+1)$$

which is of the same form as the previous even-ordered term, but with n+1 in place of n. Likewise the next odd-ordered value in the sequence can be found by adding this expression for $f_{2(n+1)}$ to the preceding expressing for f_{2n+1}, which gives

$$f_{2(n+1)+1} = 1f_{2n} + 2f_{2(n-1)} + \dots + (n-1)f_4 + (n)f_2 + (n+1)$$

where we have used the fact that $f_1 = 1$. This is of the same form as the previous odd-ordered term, but with n+1 in place of n, so the induction is complete. There, the interleaving odd and even parts of the Fibonacci numbers satisfy the relations

$$f_{2n} = n + \sum_{k=1}^{n-1} kf_{2(n-k)} \quad f_{2n+1} = 1 + \sum_{k=1}^{n} kf_{2(n-k)+1}$$

Naturally these identities are closely related to the generating functions for the recurrence relations. We know the ratios of successive values of fn asymptotically approach successive powers of the dominant root ϕ of the characteristic equation $x^2 - x - 1 = 0$. Therefore, with n = 6, the even-ordered relation approaches

$$\phi^{12} = 1\phi^{10} + 2\phi^8 + 3\phi^6 + 4\phi^4 + 5\phi^2 + 6$$

and if we divide this by ϕ^{10} we get

$$\phi^2 = 1 + 2\phi^{-2} + 3\phi^{-4} + 3\phi^{-6} + 5\phi^{-8} + 6\phi^{-10}$$

Taking this to the limit as n goes to infinity, and substituting for the derivative of the geometric series, we get

$$\phi^2 = \frac{1}{\left(1-\phi^{-2}\right)^2}$$

which is equivalent to the characteristic equation $\phi^2 - \phi - 1 = 0$. However, this isn't the only possible expansion of the terms of the even (or odd) ordered Fibonacci sequences. For example, given the even-ordered sequence 0, 1, 3, 8, 21, 55, 144, … we could simply apply the "greedy algorithm" by expressing each term as the sum of the maximum number of the immediately preceding terms. Thus we have

$$2(1) + 1 = 3$$
$$2(3) + 1(1) + 1 = 8$$
$$2(8) + 1(3) + 1(1) + 1 = 21$$
$$2(21) + 1(8) + 1(3) + 1(1) + 1 = 5$$
$$1(55) + 1(21) + 1(8) + 1(3) + 1(1) + 1 = 144$$

From this it's easy to see that $f_{2n} = f_{2(n-1)} - f_{2(n-2)} + 2f_{2(n-1)}$, so the even-ordered Fibonacci numbers satisfy the recurrence

$$f_{2n} = 3f_{2(n-1)} - f_{2(n-2)}$$

which has the characteristic equation

$$x^4 - 3x^2 + 1 = \left(x^2 - x - 1\right)\left(x^2 + x - 1\right)$$

This shows that another way of arriving at the recurrence for the kth Fibonacci numbers is to solve for the (k-2)th degree polynomial that gives, when multiplied by the characteristic polynomial of the Fibonacci sequence, a polynomial whose only non-zero coefficients are for powers that are multiples of k. For example, to find the recurrence for every third Fibonacci number, we evaluate the product

$$\left(x^2 - x - 1\right)\left(x^4 + Ax^3 + Bx^2 + Cx + D\right)$$
$$= x^6 + (A-1)x^5 + (B-A-1)x^4 + (C-A-B)x^3 + (D-C-B)x^2 + (-C-D)x - D$$

We need the coefficients of x^5, x^4, x^2, and x to vanish, so we require A = 1, B = 2, C = –1, and D = 1. Hence the characteristic polynomial for every third Fibonacci number is

$$x^4 - 4x^3 - 1 = \left(x^2 - x - 1\right)\left(x^4 + x^3 + 2x^2 - x + 1\right)$$

with the corresponding recurrence relation

$$f_{3n} = 4f_{3(n-1)} + f_{3(n-2)}$$

Naturally the same type of relations can be derived for any linear recurring sequence. For example, consider the "tribonacci" sequence, defined by the initial values $t_0 = t_1 = t_2 = 1$, and the recurrence $t_n = t_{n-1} + t_{n-2} + t_{n-3}$. The first several values are shown below.

n:	0	1	2	3	4	5	6	7	8	9	10	11	12	13	14...
t_n:	1	1	1	3	5	9	17	31	57	105	193	355	653	1201	2209...

To find the recurrence formula for every third term of this sequence, we can apply the greedy algorithm to express the terms 1, 9, 57, 355, 2209, 13745, 85525, ... as follows

$$6(57) + 1(9) + 4(1) = 355$$
$$6(355) + 1(57) + 2(9) + 4(1) = 2209$$
$$6(2209) + 1(355) + 2(57) + 2(9) + 4(1) = 13745$$
$$6(12745) + 1(2209) + 2(355) + 2(57) + 2(9) + 4(1) = 85525$$

From this it's easy to see that $t_{3n} = t_{3(n-1)} - 5t_{3(n-2)} + t_{3(n-3)} + 6t_{3(n-1)}$, so we have the recurrence

$$t_{3n} = 7\,t_{3(n-1)} - 5t_{3(n-2)} + t_{3(n-3)}$$

corresponding to the characteristic polynomial

$$x^9 - 7x^6 + 5x^3 - 1 = \left(x^3 - x^2 - x - 1\right)\left(x^4 + x^5 + 2x^4 - 3x^3 - x + 1\right)$$

We can relate this back to the greedy recurrence by examining the asymptotic expression in terms of the dominant root f, which can be written in the form

$$\phi^3 = 6 + \phi^{-3} + 2\phi^{-6} + 2\phi^{-9} + \ldots$$
$$= 4 - \phi^{-3} + 2\left(\frac{1}{1-\phi^{-3}}\right)$$

This implies that f is a root of the characteristic polynomial, as expected. Incidentally, the "poles" of this recurrence are 0 and the cube roots of 1.

ROOT-MATCHED RECURRENCES FOR DIFF EQS

Here is a method for determining the "root-matched" recurrence corresponding to a given ordinary differential equation without having to explicitly solve for the roots of the characteristic polynomial. More generally, this method enables us to find the exact root-matched recurrence corresponding to any continuous linear system without having to solve for the eigen values of the system. The following description covers the just simple homogeneous case. Suppose you have an ordinary linear differential equation with constant coefficients:

$$a5\ y''''' + a4\ y'''' + a3\ y''' + a2\ y'' + a1\ y' + a0\ y = 0$$

where the (') denotes derivatives with respect to t. The response of y(t) can be reproduced discretely by a recurrence relation of the form

$$c5\ y(5dt) + c4\ y(4dt) + c3\ y(3dt) + c2\ y(2dt) + c1\ y(dt) + c0\ y(0) = 0$$

where "dt" is the time increment of the simulation. The question is, given the coefficients ai of the differential equation, how can you compute the coefficients ci of the corresponding recurrence relation.

The coefficients ci must be proportional to the elementary symmetric functions (with alternating signs) of the quantities $e^{(ri\ dt)}$, where ri (i=1,2,..,5) are the roots of the characteristic polynomial. Thus, a straightforward method of computing the ci is to solve the characteristic polynomial for the roots ri, and then compute the elementary symmetric functions of these roots. However, determining all the roots of a high order polynomial can be computationally tricky, particularly for embedded code, where the use of iterative processes can cause problems.

Fortunately, there is a way to compute the ci without having to solve for the characteristic roots of the equation. The technique makes use of "Newton's Identities" and the Taylor's series expansion of the exponential function.

Assume our basic ODE is of order N. For any integer k, let w(k) denote the sum of the kth powers of the roots r1,r2,...,rN. Clearly, w(0)=N. Using Newton's Identities, the values of w(k) for k=1,2,... are given by

$$w(1)a(N) + 1a(N-1) = 0$$

$$w(2)a(N) + w(1)a(N-1) + 2a(N-2) = 0$$

$$w(3)a(N) + w(2)a(N-1) + w(1)a(N-2) + 3a(N-3) = 0 \text{ etc}$$

for k=N. Now let E(k) denote the sum of the kth powers of the values $e^{(ri\ dt)}$, i=1,2,..,N.

Expanding each exponential function as a Taylor series and combining terms by powers of dt gives the following expression for E(k):

$$E(k) = \text{SUM from } j=0 \text{ to inf of } \{((k\ dt)^j)\ w(j)/\ j!\}$$

for k=1,2,..,N. The coefficients of the recurrence relation can then be computed using the identities

$$c(N) = 1$$

$$E(1) + 1\ c(N-1) = 0$$

$$E(2) + E(1)\ c(N-1) + 2\ c(N-2) = 0$$

$$E(3) + E(2)\ c(N-1) + E(1)\ c(N-2) + 3\ c(N-3) = 0$$

This gives the exact root-matched recurrence coefficients without ever having to determine the roots. It also has the advantage that no special provisions are needed to cover the case of repeated roots.

Chapter 4

t-Procedures

An archeologist has found 19 intact specimens of a certain prehistoric artifact. Although she finds the specimens individually interesting in and of themselves, her broader research interest is in what they can tell her about the entire population of such artifacts that must once have existed within this particular prehistoric setting. As a first step she undertakes, on the basis of information contained within her sample, to estimate the mean length of such artifacts within the population. Listed below are the lengths of the 19 items in the sample, each measured to the nearest tenth of a centimeter. Also shown are the mean of the sample, the sum of squared deviates, the variance, and the standard deviation.

17.3	18.9	17.7	23.8	16.0	$M_X = 18.9$
22.1	18.4	18.2	13.3	26.8	$SS = 248.5$
18.6	24.5	22.8	13.4	18.1	$s^2 = 248.5/19 = 13.1$
14.8	20.6	17.4	16.1		$s = \text{sqrt}[13.1] = \pm 3.6$

The mean of a sample can be taken as an unbiased estimate of the mean of the population from which it is randomly drawn. Our archeologist is of course well aware that her sample of N=19 was not actively "drawn" from the population; it is simply the 19 specimens that happen to have been found intact. It is, however, commonly recognized within her field that the survival and discovery of intact artifacts of this particular provenience is a matter of the merest chance, so she considers it reasonable to assume that the sample, though not actively drawn, is nonetheless random and therefore an unbiased representation of the population. Having found nothing in the texture of her data to suggest anything to the contrary, she is also prepared to assume that the source population is normally distributed. Providing that these two assumptions are correct, she can then conclude that the mean of the population must lie somewhere in the vicinity of the observed mean of the sample.

$$\begin{aligned} \text{est.}\mu_{\text{source}} &= M_X \pm \text{[definition of 'vicinity']} \\ &= 18.9 \pm \text{[definition of 'vicinity']}. \end{aligned}$$

All that now remains is to work out the details for the portion of this structure that reads "±[definition of 'vicinity']." We will first develop the underlying logic of the procedure, and then translate this logic into the nuts-and-bolts of practical application.

THE UNDERLYING LOGIC

This is another of those "if only" scenarios: If only we knew in advance the variability of the population from which the sample comes. For the sake of illustration, suppose for a moment that the measures of variability found within the sample could be taken as the true, precise measures of variance and standard deviation within the entire population. In fact, they cannot be taken this way; but suppose just for the moment that they could. This would give us

$i\sigma^2_{source} = 13.1$ — Hypothetical. For illustrative purposes only. Don't do it this way in practice.

$i\sigma_{source} = \pm 3.6$

Where we go from here must be prefaced with a reminder of three points developed :

- If the source population is normally distributed, then so, too, will be the sampling distribution of sample means.
- The mean of the sampling distribution will be equal to the mean of the source population:$_i$

$$i\mu_{source} = i\mu_M$$

So if we can figure out the value of $i\sigma_M$, that will also give us the value of $i\mu_{source}$.

- When the variability of the source population is precisely known (in the present case it actually is not, but we are pretending for the moment that it is), the standard deviation of the sampling distribution can be calculated directly as:

$$i\sigma_M = \frac{\sigma_{source}}{\text{sqrt}[N]} \quad \text{From Ch.9, Pt.1}$$

which in the present case would yield

$$i\sigma_M = \frac{\pm 3.6}{\text{sqrt}[19]} = \pm 0.83.$$

This would entail that the mean of the sample belongs to a sampling distribution in which each unit of standard deviation is equal to 0.83 units of the original scale of measurement.

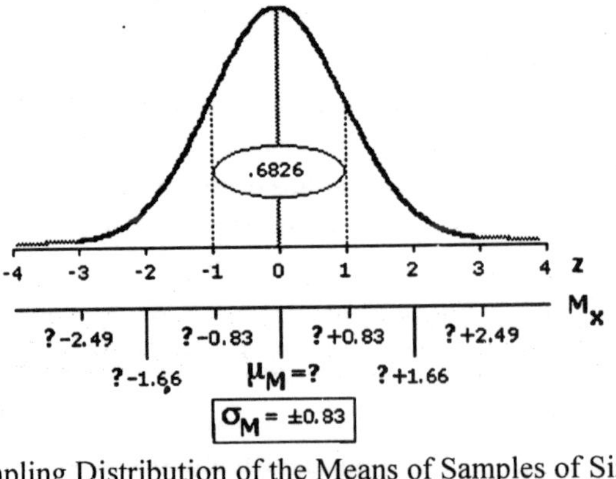

Fig. Sampling Distribution of the Means of Samples of Size N=19 Randomly Drawn from a Normally Distributed Source Population with σ_{source}=±3.6

The meaning of the red-bordered ellipse in the center of the distribution will be fairly obvious by now. As we have noted on several previous occasions, 68.26% of all the constituents of a normal distribution (in the present case, the constituents are sample means) fall within one standard deviation of the distribution's mean. We of course do not know what the mean of the distribution is—but, whatever it might be, we know that any particular randomly selected constituent has a 68.26% chance of falling within ±1 standard deviation of it.

The next step is to recognize the symmetry of this relationship. If the mean of any particular sample has a 68.26% chance of falling within ±1 standard deviation of the distribution's mean, then the distribution's mean has an identical 68.26% chance of falling within ±1 standard deviation of the mean of any particular sample. In the present case the observed mean of the sample is M_X =18.9 and the calculated value of the standard deviation of the sampling distribution is $i\sigma_M$= ±0.83. So our investigator can be 68.26% confident that the true mean of the sampling distribution, hence the true mean of the source population, lies somewhere within ±0.83 centimeters of 18.9; that is, between 18.08cm and 19.73cm, inclusive. A procedure of this general type is typically spoken of as a point and interval estimation. "Point" refers to the estimated location of the mean of the source population, and "interval" denotes the plus-or-minus range around this point that defines the concept of "somewhere in the vicinity." This latter is also spoken of as the confidence interval. In the present example the "point" is 18.9cm and the 68.26% confidence interval extends ±0.83cm on either side of that point.

Expressing it formulaically, our archeologist could therefore conclude with 68.26% confidence that

$$\begin{aligned} \text{est.m}_{\text{source}} &= M_X \pm s_M \\ &= 18.9 \pm 0.83. \end{aligned}$$

I have begun with this example of the 68.26% level of confidence only because its connection with ±1 standard deviation makes it easy to grasp intuitively. In actual scientific work the preference is usually for the more stringent 95% level, sometimes for the even stricter 99% level. The next version of Figure shows the details of the sampling distribution for this scenario that would pertain to the 95% confidence level.

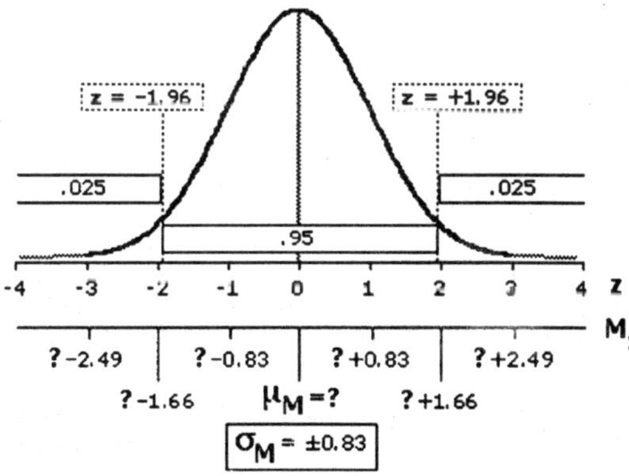

In the table of the normal distribution you will find that 2.5% of the distribution falls to the left of z=–1.96 and an equivalent 2.5% falls to the right of z=+1.96. The obvious implication is that the remaining 95% of the distribution lies *between* z=–1.96 and z=+1.96. It is the same logic as before, but with a somewhat different cast of characters. Any particular randomly selected constituent of the distribution has a 95% chance of falling within ±1.96 standard deviations of the distribution's mean. Symmetrically, there is a 95% chance that the mean of the distribution will lie within ±1.96 standard deviations of any particular randomly selected constituent. Recalling that each unit of z in the present case corresponds to 0.83cm on the original scale of measurement, our point and interval estimate for the 95% confidence level would accordingly be

$$\text{est.}\mu_{\text{source}} = M_X \pm (1.96\ x\sigma_M)$$
$$= 18.9 \pm (1.96 \times 0.83)$$
$$= 18.9 \pm 1.63.$$

The final version of Figure shows the details pertaining to the 99% confidence level.

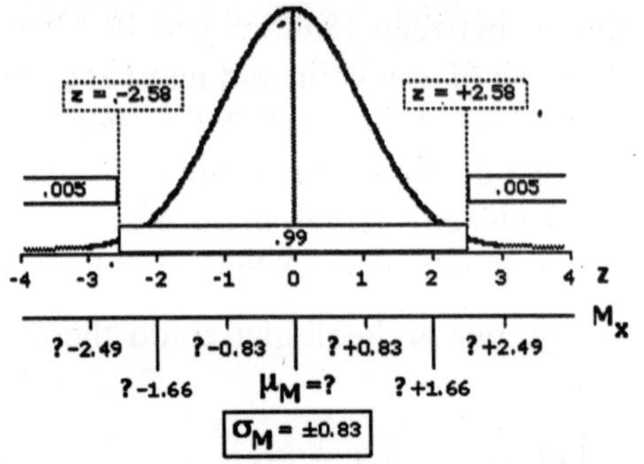

Here the critical value of z is ±2.58. To the left of z=–2.58 falls one-half of one percent of the distribution; an equivalent one-half of one percent falls to the right of z=+2.58; and the remaining 99% lies between these two markers. So our point and interval estimate for the 99% confidence level would be

$$\text{est.}\mu_{\text{source}} = M_X \pm (2.58\ x_M)$$
$$= 18.9 \pm (2.58 \times 0.83)$$
$$= 18.9 \pm 2.14$$

THE PRACTICAL APPLICATION

I began the foregoing section by asking you to suppose that the measures of variability found within the sample could be taken as the true, precise measures of variance and standard deviation within the entire source population. This assumption is of course completely untenable, for in reality it is highly unlikely that the variability of any particular sample would be identical with that of the population. The variability observed within the sample does provide a basis for rationally and systematically estimating the variability of the source population. Beyond this, the only difference between the procedure outlined above and the one we will now examine is the difference between z and t. When the variability of the source population is known only through estimation, the sampling distribution to which you ultimately refer is one or another of the distributions of t. For the present example, it is the t-distribution pertaining to df=18.

To refresh your memory of details, here again are the N=19 measures of length observed by our archeologist, along with their mean and sum of squared deviates.

17.3	18.9	17.7	23.8	16.0	$M_X = 18.9$
22.1	18.4	18.2	13.3	26.8	$SS = 248.5$
18.6	24.5	22.8	13.4	18.1	
14.8	20.6	17.4	16.1		

The variance of the source population can be estimated as

$$\{s^2\} = \frac{SS}{N-1} \quad \frac{\text{estimate of:}\sigma^2_{\text{source}}}{\text{From Ch.9, Pt.2}}$$

$$= \frac{248.5}{18} = 13.81.$$

That, in turn, allows us to estimate the standard deviation of the sampling distribution of sample means as

$$\text{est.}\sigma_M = \text{sqrt}\left[\frac{\{s^2\}}{N}\right]$$

$$= \text{sqrt}\left[\frac{13.81}{19}\right] = \pm 0.85.$$

The following abbreviated table lists the critical values of t for df=18, and Figure below it shows in graphic form the sampling distribution to which these values refer. Note in particular the two critical values listed in red.

Level of Significance						
	.05	.025	.01	.005	.0005	directional test
df	–	.05	.02	.01	.001	non-
18	1.73	2.10	2.55	2.88	3.92	directional test

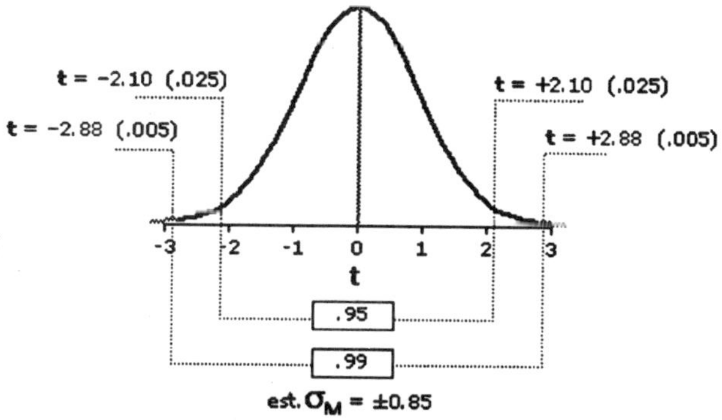

Fig. 10.2. Sampling Distribution of t for df=18

For the next step we do exactly what we did earlier with z, except now the reference is to t. In the sampling distribution of t for df=18, 95% of the distribution falls within ±2.10 units of t from the mean. Given that each unit of t in this case is equal to 0.85 units on the original scale of measurement, the estimate of the source population's mean for the 95% level of confidence would be

$$\begin{aligned} \text{est.}\mu_{\text{source}} &= M_X \pm (2.10 \times \text{est.}\sigma_M) \\ &= 18.9 \pm (2.10 \times 0.85) \\ &= 18.9 \pm 1.79. \end{aligned}$$

And similarly for the 99% level of confidence. In the sampling distribution of t for df=18, 99% of the distribution falls within ±2.88 units of t from the mean. Hence

$$\begin{aligned} \text{est.}\mu_{\text{source}} &= M_X \pm (2.88 \times \text{est.}\sigma_M) \\ &= 18.9 \pm (2.88 \times 0.85) \\ &= 18.9 \pm 2.45. \end{aligned}$$

Our archeological researcher could therefore estimate with 99% confidence that the true mean of the population from which her sample comes lies somewhere between 16.45cm and 21.35cm, inclusive. If she were willing to settle for the 95% level of confidence, she could narrow that range down to between 17.11cm and 20.69cm, inclusive.

Step-by-Step Computational Procedure for Estimating the Mean of a Population on the Basis of Information Contained within a Sample [sh]

Note that this process makes the following assumptions and can be meaningfully applied only insofar as these assumptions are met:

That the scale of measurement has the properties of an equal interval scale.

That the sample is randomly drawn from the source population.

That the source population can be reasonably supposed to have a normal distribution.

Step 1. For the sample of N values of X_i, calculate

M_X the mean of the sample

SS the sum of squared deviates

Step 2. Estimate the variance of the source population as

$$\{s^2\} = \frac{SS}{N-1}.$$

Step 3. Estimate the standard deviation of the sampling distribution of sample means (the "standard error" of the mean) as

$$\text{est.}\sigma_M = \text{sqrt}\left[\frac{\{s^2\}}{N}\right]$$

Note that Steps 2 and 3 can be combined into the more streamlined formula

$$\text{est.}\sigma_M = \text{sqrt}\left[\frac{\text{SS}/(N-1)}{N}\right].$$

Step 4. Perform the point and interval estimate as

$$\text{est.}_{\text{source}} = M_X \pm (t_{\text{critical}} \times \text{est.}\sigma_M)$$
$$\text{with df} = N\text{–}1.$$

For the 95% confidence level, t_{critical} is the value of t that is listed in the conventional table of t as significant at the .05 level for a non-directional ("two-tailed") test. For the 99% confidence level, it is the value listed as significant at the .01 level for a non-directional test. In the following abbreviated version of the table of t, the values of t_{critical} for the 95% and 99% confidence levels are listed in blue and red, respectively. The particular value of t_{critical} employed would of course depend on the relevant value of df. The full table of t can be found in Appendix C.

Level of Significance						
	.05	.025	.01	.005	.0005	directional test
df	–	.05	.02	.01	.001	non-
5	2.02	2.57	3.36	4.03	6.87	directional test
10	1.81	2.23	2.76	3.17	4.59	
18	1.73	2.10	2.55	2.88	3.92	
20	1.72	2.09	2.53	2.85	3.85	

When you perform a point and interval estimate at the 95% confidence level, the implication is that you have a 95% chance of being right and a 5% chance of being wrong. Similarly for an estimate at the 99% level: you have a 99% chance of being right and a 1% chance of being wrong. It all sounds fine in principle, but would it really work out that way in practice? I hope the following demonstration will help to convince you that it would.

Each time you click the button labeled "Sample," your computer will perform a 95% point and interval estimate based on a sample of size N=19 randomly drawn from the reference source population. As in the archeological example above, degrees of freedom is 18 and t_{critical} for the 95% confidence level is ±2.10. So the calculation for each sample will take the form

$$\text{est.}\mu_{\text{source}} = M_X \pm (2.10 \times \text{est.}\sigma_M).$$

In this particular situation we know the true mean of the population to be

$$i\mu_{\text{source}} = 18.0$$

so it will be determinable in each instance whether the estimate is right or wrong. The estimate is "right" when the true mean of the population lies within the calculated 95% confidence interval, and it is "wrong" when the true population mean falls outside that interval. Each right estimate will be marked by a blue square next to it, and each wrong one by a red square. If you click the button a sufficiently large number of times, you will find the percentage of right estimates closely approximating the theoretical value of 95%. However, even as few as twenty or thirty clicks should make it clear that all but a few of such estimates will in fact include the true mean of the population within their specified plus-or-minus range. (If red squares turn up fairly frequently at the outset, keep in mind that there is always the possibility of ending up with something improbable and extraordinary within any particular small set of samples. Keep clicking several hundred times over, and you will see the proportions of right and wrong estimates closely approximating 0.95 and 0.05, respectively.)

T-TEST FOR THE SIGNIFICANCE

t-TESTS

Because there is a distinction between the common statistical vernacular definition of *t*-tests and the more technical definition, *t* tests can be a little confusing. The common-use definition or description of *t* tests is simply comparing two means to see if they are significantly different from each other. The more technical definition or description of a *t* test is any statistical test that uses the *t*, or Student's *t*, family of distributions. In this chapter, we will briefly describe the family of distributions known as the *t* distribution. Then we will discuss the two most commonly conducted *t* tests, the independent samples *t* test and the paired or dependent samples *t* test.

t -DISTRIBUTIONS

As discussed in previous Chapters the normal distribution and how to use the normal distribution to find z scores. The probabilities that are based on the normal distribution are accurate when (a) the population standard deviation is known, and/or (b) we have a large sample (i.e., $n > 120$). If neither of these is true, then we cannot assume that we have a nicely shaped bell curve and we cannot use the probabilities that are based on this normal distribution. Instead, we have to adjust our probability estimates by taking our sample size into account. As we discussed in earlier Chapter, we are fortunate to have a set of distributions that have already been created for us that do this, and this is known as the family of *t* distributions. Which specific *t* distribution you use for a given problem depends on the size of your sample. There is a table of probabilities based on the different *t* distributions in Appendix B.

THE INDEPENDENT SAMPLES t -TEST

One of the most commonly used *t* tests is the independent samples *t* test. You use this test when you want to compare the means of two *independent* samples on a given variable. For example, if you wanted to compare the average height of 50 randomly selected men to that of 50 randomly selected women, you would conduct an independent samples *t* test. Note that the sample of men is not related to the sample of women, and there is no overlap between these two samples (i.e., one cannot be a member of both groups). Therefore, these groups are *independent,* and an independent samples *t* test is appropriate. To conduct an independent samples *t* test, you need one categorical or nominal independent variable and one continuous or intervally scaled dependent variable. A dependent variable is a variable on which the scores may differ, or *depend* on the value of the independent variable. An independent variable is the variable that may cause, or simply be used to predict, the value of the dependent variable. The independent variable in a *t* test is simply a variable with two categories (e.g., men and women, fifth graders and ninth graders, etc.). In this type of *t* test, we want to know whether the average scores on the dependent variable differ according to which group one belongs to (i.e., the level of the independent variable). For example, we may want to know if the average height of people (height is the dependent, continuous variable) *depends* on whether the person is a man or a woman (gender of the person is the independent, categorical variable).

DEPENDENT SAMPLES t -TEST

A dependent samples *t* test is also used to compare two means on a single dependent variable. Unlike the independent samples test, however, a dependent samples *t* test is used to compare the means of a single sample or of two matched or paired samples. For example, if a group of students took a math test in March and that same group of students took the same math test two months later in May, we could compare their average scores on the two test dates using a dependent samples *t* test. Or, suppose that we wanted to compare a sample of boys' Scholastic Aptitude Test (SAT) scores with their fathers' SAT scores. In this example, each boy in our study would be matched with his father. In both of these examples, each score is matched, or paired with, a second score. Because of this pairing, we say that the scores are *dependent* upon each other, and a dependent samples *t* test is warranted.

INDEPENDENT SAMPLES t -TESTS IN DEPTH

To understand how *t* tests work, it may be most helpful to first try to understand the conceptual issues and then move to the more mechanical issues involved in the formulas. Because the independent and dependent forms of the *t* tests are quite different, each of them will be discussed separately. Let's begin with the independent samples *t* test.

Conceptual Issues With the Independent Samples t Test

The most complicated conceptual issue in the independent samples *t* test involves the standard error for the test. If you think about what this *t* test does, you can see that it is designed to answer a fairly straightforward question: Do two independent samples differ from each other *significantly* in their average scores on some variable? Using an example to clarify this question, we might want to know whether a random sample of 50 men differs *significantly* from a random sample of 50 women in their average enjoyment of a new television show. Suppose that we arranged to have each sample view our new television show and than rate, on a scale from 1 to 10, how much they enjoyed the show, with higher scores indicating greater enjoyment. In addition, suppose that our sample of men gave the show an average rating of 7.5 and our sample of women gave the show an average rating of 6.5.

In looking at these two means, we can clearly see that our sample of men had a higher mean enjoyment of the television show than did our sample of women. We asked whether they differed *significantly* in their average enjoyment of the show. The word *significantly* is critical in much of statistics, so we discuss it briefly here as it applies to independent *t* tests.

When we conduct an independent samples *t* test, we generally must collect data from two samples and compare the means of these two samples. But we are interested not only in whether these two samples differ on some variable. We are also interested in whether the differences in the two sample means are large enough to suggest that there are also differences in the two *populations* that these samples represent. So, returning to our previous example, we already know that the 50 men in our sample enjoyed the television show more, on average, than did the 50 women in our sample. So what? Who really cares about these 50 men and these 50 women, other than their friends and families? What we really want to know is whether the difference between these two samples of men and women is large enough to indicate that men *in general* (i.e., the population of men that this sample represents) will like the television show more than women *in general* (i.e., the population of women that this sample represents). In other words, is this difference of 1.0 between our two samples large enough to represent a real difference between the populations of men and women? The way of asking this question in statistical shorthand is to ask "Is the difference between the means of these two samples statistically significant?" (or significant for short).

To answer this question, we must know how much difference we should *expect* to see between two samples of this size drawn from these two populations. If we were to randomly select a different sample of 50 men and a different sample of 50 women, we might get the opposite effect, where the women outscore the men. Or, we might get an even larger difference, where men outscore the women by 3 points rather than 1. So the critical question here is this: What is the *average expected difference between the means* of two samples of this size (i.e., 50 each) selected randomly from these two populations? In other words, *what is the*standard error of the difference between the means?

Understanding the concept of standard error provides the key to understanding how inferential statistics work, so take your time and reread the preceding four paragraphs to make sure you get the gist. Regarding the specific case of independent samples *t* tests, we can conclude that the question we want to answer is whether the difference between our two samples means is large or small compared to the amount of difference we would expect to see just by selecting two different samples. Phrased another way, we want to know whether our *observed* difference between our two samples means is large relative to the standard error of the difference between the means. The general formula for this question is as follows:

$$t = \frac{observed\, difference\, between\, sample\, means}{s\tan dard\, error\, of\, the\, difference\, between\, the\, means}$$

or

$$t = \frac{\overline{X}_1 - \overline{X}_2}{s_{\bar{x}1-\bar{x}2}}$$

where $\overline{X}_1$ is the mean for sample 1

is $\overline{X}_2$ the mean for sample 2

$s_{\bar{x}1-\bar{x}2}$ is the standard error of the difference between the means

The Standard Error of the Difference Between Independent Sample Means

The standard error of the difference between independent samples means is a little bit more complex than the standard error of the mean discussed. That's because instead of dealing with a single sample, now we have to find a single standard error involving two samples. Generally speaking, this involves simply combining standard errors of the two samples. In fact, when the two samples are roughly the same size, the standard error for the difference between the means really is similar to simply combining the two sample standard errors of the mean, as the formula presented in Table indicates.

When the two samples are not roughly equal in size, there is a potential problem with using the formulas in Table 8.1 to calculate the standard error. Because these formulas essentially blend the standard errors of each sample together, they also essentially give each sample equal weight and treat the two samples as one new, larger sample. But if the two samples are not of equal size, and especially if they do not have equal standard deviations, than we must adjust the formula for the standard error to take these differences into account. The only difference between this formula and the formula for the standard error when the sample sizes are equal is that the unequal sample size formula adjusts for the different samples sizes.

In practice, let us hope that you will never need to actually calculate any of these standard errors by hand. Because computer statistical programmes compute these for us these days, it may be more important to understand the concepts involved than the components of the formulas themselves. In this spirit, try to understand what the standard error of the difference between independent samples means is and why it may differ if the sample sizes are unequal. Simply put, the standard error of the difference between two independent samples means is the average expected difference between any two samples of a given size randomly selected from a population on a given variable. In our example comparing men's and women's enjoyment of the new television show, the standard error would be the average (i.e., *standard)* amount of difference (i.e., error) we would expect to find between any two samples of 50 men and 50 women selected randomly from the larger populations of men and women.

Determining the Significance of the t Value for an Independent Samples t Test

Once we calculate the standard error and plug it into our formula for calculating the *t* value, we are left with an *observed t* value. How do we know if this *t* value is statistically significant? In other words, how do we decide if this *t* value is large enough to indicate that the difference between our sample means probably represents a real difference between our population means? To answer this question, we must find the probability of getting a *t* value of that size by chance. In other words, what are the odds that the difference between our two samples means is just due to the luck of the draw when we selected these two samples at random rather than some real difference between the two populations? Fortunately, statisticians have already calculated these odds for us, and a table with such odds is included in Appendix B. Even more fortunately, statistical software programmes used on computers calculate these probabilities for us, so there will hopefully never be a need for you to use Appendix B. We provide it here so because we think the experience of calculating a t value by hand and determining whether it is statistically significant can help you understand how t tests work.

We saw how statisticians generated probabilities based on the normal distribution. With t distributions, the exact same principles are involved, except that now we have to take into account the size of the samples we are using. This is because the shape of the t distribution changes as the sample size changes, and when the shape of the distribution changes, so do the probabilities associated with it. The way that we take the sample size into account in statistics is to calculate degrees of freedom. The explanation of exactly what a degree of freedom is may be a bit more complicated than is worth discussing here (although you can read about it in most statistics textbooks if you are interested). At this point, suffice it to say that in an independent samples t test, you find the degrees of freedom by adding the two sample sizes together and subtracting 2. So the formula is $df = n^1 + n^2 - 2$. Once you have your degrees of freedom and your t value, you can look in the table of t values in Appendix B to see if the difference between your two sample means is significant.

To illustrate this, let's return to our example comparing men's and women's enjoyment of the new television programme. Let's just suppose that the standard error of the difference between the means is .40. When we plug this number into the t value formula we get the following:

$$t = \frac{7.5 - 6.5}{.40}$$

$$t = \frac{1.0}{.40} = 2.50$$

$$df = 50 + 50 - 2 = 98$$

Now that we have a t value and our degrees of freedom, we can look in Appendix B to find the probability of getting a t value of this size (t=2.50) by chance when we have 98 degrees of freedom. Our table tells us that the probability of getting a t value this large by chance (i.e., due strictly to random sampling) is between 1% and 2%. In other words, when we randomly select two samples of 50 each from two different populations, we would expect to a t value of this size less than 2% of the time *when there is no real difference between the population means* . Because this is such a small probability, we conclude that the difference between our sample of 50 men and our sample of 50 women that we observed in the average ratings of enjoyment of the television show probably represents a real difference between the larger populations of men and women rather than some fluke difference that emerged simply because of who we happened to get in our samples (i.e., *random sampling error*).

It is important to remember that although this difference between the means was *statistically* significant, that does not necessarily mean that it is *practically* significant. Just as the standard error of the mean is influenced by the size of the sample, the standard error of the *difference* between the means is also affected by sample size. The larger the samples, the smaller the standard error and the more likely it is that you will find a statistically significant result. To determine whether this difference between men and women is *practically* significant, we should consider the actual *raw score* difference.

Men in our sample scored an average of one point higher on a 10-point scale than did women. Is that a big difference? Well, that is a judgment call. We would consider that a fairly inconsequential difference because we are talking about preferences for a television show. We don't consider a one-point difference on a 10-point scale regarding television preferences to be important. But potential advertisers might consider this a meaningful difference. Those wanting to advertise female-oriented products may not select this show, which seems to appeal more to male viewers.

Another way to determine whether this difference in the means is practically significant is to calculate an effect size.

The formula for the effect size for an independent samples *t* test is presenfed in Table. To calculate the effect size, you must first calculate the denominator. Using our example where the sample size for one group is 50 and the standard error of the difference between the means is .40, we get the following:

$$\hat{s} = \sqrt{50}(.40)$$

$$\hat{s} = 7.07(.40)$$

$$\hat{s} = 2.83$$

We can then plug this into the formula for the effect size, along with the two sample means:

$$d = \frac{7.5 - 6.5}{2.83} \Rightarrow d = .35$$

So our effect size for this problem is .35, which would be considered a small- to medium-size effect.

$$d = \frac{\overline{X}_1 - \overline{X}_2}{\bar{s}}$$

$$\bar{s} = \sqrt{n_1}(s_{it-k2})$$

where $\overline{X}_1$ is the mean for the first sample

$\overline{X}_2$ is the mean for the second sample

n_1•is the sample size for one sample

is the standard error of the difference between the means

THE DIFFERENCE BETWEEN THE MEANS OF TWO INDEPENDENT SAMPLES

This is probably the most widely used statistical test of all time, and certainly the most widely known. It is simple, straightforward, easy to use, and adaptable to a broad range of situations. No statistical toolbox should ever be without it.

Its utility is occasioned by the fact that scientific research very often examines the phenomena of nature two variables at a time, with an eye toward answering the basic question: Are these two variables related? If we alter the level of one, will we thereby alter the level of the other? Or alternatively: If we examine two different levels of one variable, will we find them to be associated with different levels of the other?

Here are three examples to give you an idea of how these abstractions might find expression in concrete reality. On the left of each row of cells is a specific research question, and on the right is a brief account of a strategy that might be used to answer it.

The first two examples illustrate a very frequently employed form of experimental design that involves randomly sorting the members of a subject pool into two separate groups, treating the two groups differently with respect to a certain independent variable, and then measuring both groups on a certain dependent variable with the aim of determining whether the differential treatment produces differential effects. (Variables: Independent and Dependent.)

A quasi-experimental variation on this theme, illustrated by the third example, involves randomly selecting two groups of subjects that already differ with respect to one variable, and then measuring both groups on another variable to determine whether the different levels of the first are associated with different levels of the second.

Question	*Strategy*
Does the presence of a certain kind of mycorrhizal fungus enhance the growth of a certain kind of plant?	Begin with a "subject pool" of seeds of the type of plant in question. Randomly sort them into two groups, A and B. Plant and grow them under conditions that are identical in every respect except one: namely, that the seeds of group A (the experimental group) are grown in a soil that contains the fungus, while those of group B (the control group) are grown in a soil that does not contain the fungus. After some specified period of time, harvest the plants of both groups and take the relevant measure of their respective degrees of growth. If the presence of the fungus does enhance growth, the average measure should prove greater for group A than for group B.
Do two types of music, type-I and type-II, have different	Begin with a subject pool of college students, relatively homogeneous with respect to age, record of academic

effects upon the ability of college students to perform a series of mental tasks requiring concentration?	achievement, and other variables potentially relevant to the performance of such a task. Randomly sort the subjects into two groups, A and B. Have the members of each group perform the series of mental tasks under conditions that are identical in every respect except one: namely, that group A has music of type-I playing in the background, while group B has music of type-II. (Note that the distinction between experimental and control group does not apply in this example.) Conclude by measuring how well the subjects perform on the series of tasks under their respective conditions. Any difference between the effects of the two types of music should show up as a difference between the mean levels of performance for group A and group B.
Do two strains of mice, A and B, differ with respect to their ability to learn to avoid an aversive stimulus?	With this type of situation you are in effect starting out with two subject pools, one for strain A and one for strain B. Draw a random sample of size N_a from pool A and another of size N_b from pool B. Run the members of each group through a standard aversive-conditioning procedure, measuring for each one how well and quickly the avoidance behavior is acquired. Any difference between the avoidance-learning abilities of the two strains should manifest itself as a difference between their respective group means.

In each of these cases, the two samples are independent of each other in the obvious sense that they are separate samples containing different sets of individual subjects. The individual measures in group A are in no way linked with or related to any of the individual measures in group B, and vice versa. The version of a t-test examined in this chapter will assess the significance of the difference between the means of two such samples, providing:

- That the two samples are randomly drawn from normally distributed populations;
- Tthat the measures of which the two samples are composed are equal-interval.

To illustrate the procedures for this version of a t-test, imagine we were actually to conduct the experiment described in the second of the above examples. We begin with a fairly homogeneous subject pool of 30 college students, randomly sorting them into two groups, A and B, of sizes N_a=15 and N_b=15. (It is not essential for this procedure that the two samples be of the same size.) We then have the members of each group, one at a time, perform a series of 40 mental tasks while one or the other of the music types is playing in the background. For the members of group A it is music of type-I, while for those of group B it is music of type-II. The following table shows how many of the 40 components of the series each subject was able to complete. Also shown are the means and sums of squared deviates for the two groups.

Group A music of type-I			*Group B music of type-II*		
26	21	22	18	23	21
26	19	22	20	20	29
26	25	24	20	16	20
21	23	23	26	21	25
18	29	22	17	18	19
N_a=15			N_b=15		
M_a=23.13			M_b=20.87		
SS_a=119.73			SS_b=175.73 3		
		$M_a - M_b$=2.26			

NULL HYPOTHESIS

Whenever you perform a statistical test, what you are testing, fundamentally, is the null hypothesis. In general, the null hypothesis is the logical antithesis of whatever hypothesis it is that the investigator is seeking to examine. For the present example, the research hypothesis is that the two types of music have different effects, so the null hypothesis is that they do not have different effects. Its immediate implication is that any difference we find between the means of the two samples should not significantly differ from zero.

If the investigator specifies the direction of the difference in advance as either

task performance will be better with type-I music than with type-II	which would be supported by finding the mean of sample A to be significantly greater than the mean of sample B (M_a>M_b)
or	
task performance will be better with type-II music than with type-I	which would be supported by finding the mean of sample B to be significantly greater than the mean of sample A (M_b>M_a)

then the research hypothesis is directional and permits a one-tail test of significance. A non-directional research hypothesis would require a two-tail test, as it is the equivalent of saying "I'm expecting a difference in one direction or the other, but I can't guess which." For the sake of discussion, let us suppose we had started out with the directional hypothesis that task performance will be better with type-I music than with type-II music. Clearly our observed result, M_a–M_b=2.26, is in the hypothesized direction. All that remains is to determine how confident we can be that it comes from anything more than mere chance coincidence.

LOGIC AND PROCEDURE

The groundwork for the following points is laid down.

- The mean of a sample randomly drawn from a normally distributed source population belongs to a sampling distribution of sample means that is also normal in form. The overall mean of this sampling distribution will be identical with the mean of the source population:

$$i\mu_M = i\mu_{\text{source}} \text{ From Ch.9, Pt.1}$$

- For two samples, each randomly drawn from a normally distributed source population, the difference between the means of the two samples,

$$M_a - M_b$$

belongs to a sampling distribution that is normal in form, with an overall mean equal to the difference between the means of the two source populations

$$i\mu_{M\text{-}M} = i\mu_{\text{source }A} - \mu_{\text{source }B} \quad \text{From Ch.9, Pt.1}$$

- On the null hypothesis, $i\mu_{\text{source }A}$ and $i\mu_{\text{source }B}$ are identical, hence

$$i\mu_{M\text{-}M} = 0$$

- For the present example, the null hypothesis holds that the two types of music do not have differential effects on task performance. This is tantamount to saying that the measures of task performance in groups A and B are all drawn indifferently from the same source population of such measures. In items 3 and 4 below, the phrase "source population" is a shorthand way of saying "the population of measures that the null hypothesis assumes to have been the common source of the measures in both groups."
- If we knew the variance of the source population, we would then be able to calculate the standard deviation (aka "standard error") of the sampling distribution of sample-mean differences as

$$x_{M\text{-}M} = \text{sqrt}\left[\frac{x^2_{\text{source}}}{N_a} + \frac{x^2_{\text{source}}}{N_b}\right] \quad \text{From Ch.9, Pt.1}$$

- This, in turn, would allow us to test the null hypothesis for any particular $M_a - M_b$ difference by calculating the appropriate z-ratio

$$z = \frac{M_{Xa} - M_{Xb}}{\sigma_{M\text{-}M}} \quad \text{From Ch.9, Pt.1.}$$

- and referring the result to the unit normal distribution.

In most practical research situations, however, the variance of the source population, hence also the value of $i\sigma_{M\text{-}M}$, can be arrived at only through estimation. In these cases the test of the null hypothesis is performed not with z but with t:

$$t = \frac{M_{Xa} - M_{Xb}}{\text{est.}i\sigma_{M\text{-}M}} \quad \text{From Ch.9, Pt.2}$$

- The resulting value belongs to the particular sampling distribution of t that is defined by df= $(N_a - 1) + (N_b - 1)$.

- To help you keep track of where the particular numerical values are coming from beyond this point, here again are the summary statistics for our hypothetical experiment on the effects of two types of music:

Group A *music of type-I*	*Group B* *music of type-II*
N_a=15	N_b=15
M_a=23.13	M_b=20.86
SS_a=119.73	SS_b=175.73
M_a-M_b=2.26	

- The variance of the source population can be estimated as

$$\{s^2_p\} = \frac{SS_a + SS_b}{(N_a - 1)+(N_b - 1)}$$

- which for the present example comes out as

$$\{s^2_p\} = \frac{119.73 + 175.73}{14 + 14} = 10.55$$

- This, in turn, allows us to estimate the standard deviation of the sampling distribution of sample-mean differences as

est.$i\sigma_{M\text{-}M}$

$$= \text{sqrt}\left[\frac{\{s^2_p\}}{N_a} + \frac{\{s^2_p\}}{N_b}\right]$$

$$= \text{sqrt}\left[\frac{10.55}{15} + \frac{10.55}{15}\right] = \pm 1.19$$

- And with this estimated value of $i\sigma_{M\text{-}M}$ in hand, we are then able to calculate the appropriate t-ratio as

$$t = \frac{M_{Xa} - M_{Xb}}{\text{est.}i\sigma_{M\text{-}M}}$$

$$= \frac{23.13 - 20.87}{1.19} = +1.9$$

- with df=(15—1)+(15—1)=28

In the calculation of a two-sample t-ratio, note that the sign of t depends on the direction of the difference between M_{Xa} and M_{Xb}. $M_{Xa} > M_{Xb}$ will produce a positive value of t, while $M_{Xa} < M_{Xb}$ will produce a negative value of t.

INFERENCE

Figure shows the sampling distribution of t for df=28. Also shown is the portion of the table of critical values of t that pertains to df=28. The designation "t_{obs}" refers to our observed value of t=+1.9. We started out with the directional research hypothesis that task performance would be better for group A than for group B, and as our observed result, $M_{Xa}-M_{Xb}=2.26$, proved consistent with that hypothesis, the relevant critical values of t are those that pertain to a directional (one-tail) test of significance: 1.70 for the .05 level of significance, 2.05 for the .025 level, 2.47 for the .01 level, and so on.

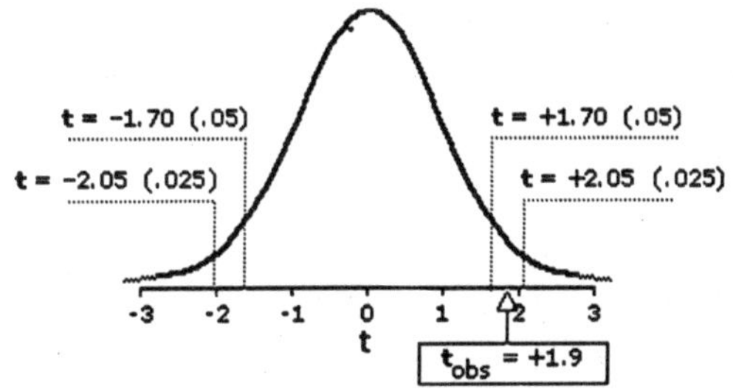

	Level of Significance for a Directional Test				
	.05	.025	.01	.005	.0005
	Level of Significance for a Non-Directional Test				
	---	.05	.02	.01	.001
df = 28	1.70	2.05	2.47	2.76	3.67

Fig. Sampling Distribution of t for df=28

If our observed value of t had ended up smaller than 1.70, the result of the experiment would be non-significant vis-ΰ-vis the conventional criterion that the mere-chance probability of a result must be equal to or less than .05. If it had come out at precisely 1.70, we would conclude that the result is significant at the .05 level. As it happens, the observed t meets and somewhat exceeds the 1.70 critical value, so we conclude that our result is significant somewhat beyond the .05 level. If the observed t had been equal to or greater than 2.05, we would have been able to regard the result as significant at or beyond the .025 level; and so on.

The same logic would have applied to the left tail of the distribution if our initial research hypothesis had been in the opposite direction, stipulating that task performance would be better with music of type-II than with music of type-I. In this case we would have expected M_{Xa} to be smaller than M_{Xb}, which would have entailed a negative sign for the resulting value of t.

If, on the other hand, we had begun with no directional hypothesis at all, we would in effect have been expecting

$$\text{either } M_{Xa} > M_{Xb} \text{ or } M_{Xa} < M_{Xb}$$

and that disjunctive expectation ("either the one or the other") would have required a non-directional, two-tailed test. Note that for a non-directional test our observed value of t=+1.9 (actually, for a two-tailed test it would have to be regarded as t=±1.9) would not be significant at the minimal .05 level.

In this particular case, however, we did begin with a directional hypothesis, and the obtained result as assessed by a directional test is significant beyond the .05 level. The practical, bottom-line meaning of this conclusion is that the likelihood of our experimental result having come about through mere random variability–mere chance coincidence, "sampling error," the luck of the scientific draw—is a somewhat less that 5%; hence, we can have about 95% confidence that the observed result reflects something more than mere random variability. For the present example, this "something more" would presumably be a genuine difference between the effects of the two types of music on the performance of this particular type of task.

Step-by-Step Computational Procedure

Note that this test makes the following assumptions and can be meaningfully applied only insofar as these assumptions are met:

That the two samples are independently and randomly drawn from the source population(*s*).

That the scale of measurement for both samples has the properties of an equal interval scale.

That the source population(*s*) can be reasonably supposed to have a normal distribution.

Step 1. For the two samples, A and B, of sizes of N_a and N_b respectively, calculate

$$M_{Xa} \text{ and } SS_a$$

the mean and sum of squared deviates of sample A.

$$M_{Xb} \text{ and } SS_a$$

the mean and sum of squared deviates of sample B.

Step 2. Estimate the variance of the source population as

$$\{s^2{}_p\} = \frac{SS_a + SS_b}{(N_a - 1) + (N_b - 1)}.$$

Recall that "source population" in this context means "the population of measures that the null hypothesis assumes to have been the common source of the measures in both groups."

Step 3. Estimate the standard deviation of the sampling distribution of sample-mean differences (the "standard error" of M_{Xa}–M_{Xb}) as

$$\text{est.}i\sigma_{M\text{-}M} = \text{sqrt}\left[\frac{\{s^2{}_p\}}{N_a} + \frac{\{s^2{}_p\}}{N_b}\right].$$

Step 4. Calculate t as

$$t = \frac{M_{Xa} - M_{Xb}}{\text{est.}i\sigma_{M\text{-}M}}$$

Step 5. Refer the calculated value of t to the table of critical values of t, with df=$(N_a-1)+(N_b-1)$. Keep in mind that a one-tailed directional test can be applied only if a specific directional hypothesis has been stipulated in advance; otherwise it must be a non-directional two-tailed test.

THE MANN-WHITNEY TEST

Twenty-one persons seeking treatment for claustrophobia are independently and randomly sorted into two groups, the first of size n_a=11 and the second of size n_b=10. The members of the first group each individually receive Treatment A over a period of 15 weeks, while those of the second group receive Treatment B. The investigators' directional hypothesis is that Treatment A will prove the more effective.

At the end of the experimental treatment period, the subjects are individually placed in a series of claustrophobia test situations, knowing that their reactions to these situations are being recorded on videotape. Subsequently three clinical experts, uninvolved in the experimental treatment and not knowing which subject received which treatment, independently view the videotapes and rate each subject according to the degree of claustrophobic tendency shown in the test situations. Each judge's rating takes place along a 10-point scale, with 1="very low" and 10="very high"; and the final measure for each subject is the simple average of the ratings of the three judges for that subject. The following table shows the average ratings for each subject in each of the two groups.

	Group A			*Group B*	
4.6	4.7	4.9	5.2	5.3	5.4
5.1	5.2	5.5	5.6	6.2	6.3
5.8	6.1	6.5	6.8	7.7	8.0
6.5	7.2		8.1		
mean	5.6			6.5	

The investigators expected Treatment A to prove the more effective, and sure enough it is group A that appears to show the lower mean level of claustrophobic tendency in the test situations. You might suppose that all they need do now is plug their data into an independent-samples t-test to see whether the observed mean difference is significant.

If you ever find yourself with a set of data of this sort, and are seized by the impulse to plug the numbers into a t-test, please lie down until the impulse passes. For the fact is that a t-test could not legitimately be used in this particular situation, nor in any other where the assumptions of the test are so patently violated. The t-test for independent samples can be meaningfully applied only insofar as

1. The two samples are independently and randomly drawn from the source population(*s*);$_{\text{T}}$
2. The scale of measurement for both samples has the properties of an equal interval scale; and$_{\text{T}}$
3. The source population(*s*) can be reasonably supposed to have a normal distribution.

While our initial sorting of the subjects into groups A and B satisfies the first requirement in the list, the second and third requirements are clearly not met.

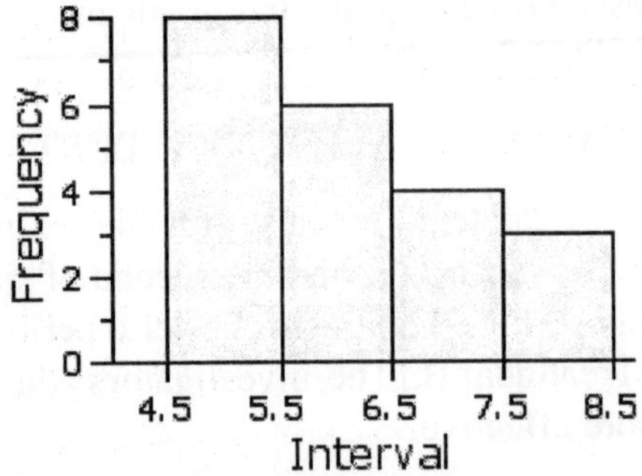

A rating scale cannot be assumed to possess the properties of an equal interval scale. Moreover, as you can see from the adjacent graph, there is nothing in the data to suggest an underlying normal distribution. The indication, indeed, is that the distribution of the underlying population might have a rather pronounced positive skew. (The graph shows the frequency distribution for the 21 subjects of both groups combined, sorted into the intervals 4.5 to 5.4, 5.5 to 6.4, 6.5 to 7.4, and 7.5 to 8.4.)

The admonition to refrain from using a t-test in situations of this sort is not simply a matter of good manners or good taste. The t-test makes these particular assumptions about the data you are feeding into it. Feed it the type of information it assumes it is getting, and the result it gives you in return will provide a firm basis for drawing rational conclusions. Feed it information that violates any one or several of its assumptions, and it will still patiently crunch the numbers and crank out a result. But that result will be nonsense, and so too will be any conclusion you might draw from it. The "significant" result that would come from applying a t-test to this set of data would in fact be no result at all. It would be merely the semblance of a result.

Assumption 2, an equal-interval scale of measurement; assumption 3, an underlying normal distribution. In cases where data from two independent samples fail to meet either or both of these requirements, one is well advised to forego the t-test and turn instead to its non-parametric alternative, the Mann-Whitney Test. The only assumptions of the Mann-Whitney test are

1. That the two samples are randomly and independently drawn;$_T$
2. That the dependent variable (e.g., claustrophobic tendency) is intrinsically continuous, capable in principle, if not in practice, of producing measures carried out to the nth decimal place; and$_T$
3. That the measures within the two samples have the properties of at least an ordinal scale of measurement, so that it is meaningful to speak of "greater than," "less than," and "equal to."

We will begin with the sheer mechanics of the Mann-Whitney procedure, and then come back to consider its underlying logic. Actually, there are two separate sequences in the logic of the Mann-Whitney test. The first, described here as Method I, is the more widely applicable. The second, Method II, is needed only when one or another of the samples is fairly tiny.

MECHANICS

It begins by assembling the measures from samples A and B into a single set of size $N=n_a+n_b$. These measures are then rank-ordered from lowest (rank#1) to highest (rank#N), with tied ranks included where appropriate. For the present set of data, this process would yield the following result.

Raw Measure	*Rank*	*from Sample*
4.6	1	A
4.7	2	A
4.9	3	A
5.1	4	A
5.2	5.5	A
5.2	5.5	B
5.3	7	B
5.4	8	B
5.5	9	A
5.6	10	B
5.8	11	A
6.1	12	A
6.2	13	B
6.3	14	B
6.5	15.5	A
6.5	15.5	A
6.8	17	B
7.2	18	A
7.7	19	B
8.0	20	B
8.1	21	B

$$N = n_a + n_b$$
$$= 11 + 10 = 21.$$

Tied Ranks. Note that two of the entries in the raw-measures column each have the value of 5.2. As these two entries fall in the sequence where ranks #5 and #6 would be, in the absence of such a tie, they are each given the average of these two ranks, which is 5.5. Similarly for the two raw-measure entries whose value is 6.5; each is accorded the average of ranks #15 and #16, which is 15.5. If there were three raw-measure entries tied for ranks 8, 9, and 10, each would receive the average of those ranks, which is 9. And so on.

Once they have been sorted out in this fashion, the rankings are then returned to the sample, A or B, to which they belong and substituted for the raw measures that gave rise to them. Thus, the raw measures that appear in the following table on the left are replaced by their respective ranks, as shown in the table on the right.

Raw Measures		*Ranked Measures*		
Group A	*Group B*	*Group A*	*Group B*	
4.6	5.2	1	5.5	
4.7	5.3	2	7	
4.9	5.4	3	8	
5.1	5.6	4	10	
5.2	6.2	5.5	13	
5.5	6.3	9	14	
5.8	6.8	11	17	
6.1	7.7	12	19	
6.5	8.0	15.5	20	
6.5	8.1	15.5	21	A & B
7.2		18		Combined
sum of ranks	96.5	134.5	231	
average of ranks	8.8	13.5	11	

Our final bit of mechanics is just a dab of symbolic notation.

T_A = The sum of the n_a ranks in group A

T_B = The sum of the n_b ranks in group B

T_{AB} = The sum of the N ranks in groups A and B combined

For the present example:

$T_A = 96.5$ [with n_a=11]

$T_B = 134.5$ [with n_b=10]

$T_{AB} = 231$ [with N=21]

Logic & Procedure: Method I [sh]

The effect of replacing raw measures with ranks is two-fold. The first is that it brings us to focus only on the ordinal relationships among the raw measures–"greater than," "less than," and "equal to"—with no illusion or pretense that these raw measures (4.6, 4.7, 4.9, etc.) derive from an equal-interval scale. The second is that it transforms the data array into a kind of closed system, many of whose properties can then be known by dint of sheer logic.

For example: If you had a total of N=4 items ranked in this fashion, the sum of those ranks would inescapably be

$$T_{AB} = 1+2+3+4 = 10$$

This would also be true if some of the ranks were tied; for example:

$$T_{AB} = 1+2.5+2.5+4 = 10$$

Similarly for N=5 ranked items:

$$T_{AB} = 1+2+3+4+5 = 15$$

And for N=21 ranked items, as in the present example:

$$T_{AB} = 1+2+3+4+\cdots+18+19+20 = 231$$

In general, the sum of any set of N ranks will be equal to

$$T_{AB} = \frac{N(N+1)}{2}$$

for N=4: $(4\times5)/2=10$

for N=5: $(5\times6)/2=15$

for N=21: $(21\times22)/2=231$

and the average of the N ranks will of course be that sum divided by N. In any particular case you could calculate the value of T_{AB} in the manner shown above and then divide it by N, though a rather more direct way is to calculate it simply as $(N+1)/2$, which is what the mean of the N ranks boils down to, once you perform the requisite algebraic juggling:

$$\text{mean rank}_{AB} = \frac{N(N+1)}{2} \times \frac{1}{N} = \frac{N+1}{2}$$

for N=4: $5/2=2.5$

for N=5: $6/2=3$

for N=21: $22/2=11$

Either way, the null hypothesis in our example is that treatments A and B do not differ with respect to their effectiveness. If this were true, then the raw measures within samples A and B would be about the same, on balance, and the rankings that derive from them would be evenly mixed within samples A and B, like cards in a well shuffled deck.

So if the null hypothesis were true, we would expect the separate averages of the A ranks and the B ranks each to approximate this same overall mean value of $(N+1)/2=11$. This entails that the rank-sums of the two groups, T_A and T_B, would approximate the values

$$T_A = n_a(N+1)/2 = 11(21+1)/2 = 121$$

and

$$T_B = n_b(N+1)/2 = 10(21+1)/2 = 110$$

Thus we know at the outset that

- the observed value of T_A=96.5 belongs to a sampling distribution whose mean is equal to 121_T

  and$_T$
- the observed value of T_B=134.5 belongs to a sampling distribution whose mean is equal to 110

Now and again in writings on subjects more or less mathematical, you will find an author saying "It can be shown that such-and-such is so." This typically occurs at junctures where the point to be made is not an obvious one; where it would be a distraction to stop and make it obvious; and where you must therefore ask the reader to take it on faith. This is one of those junctures. For any particular situation of the sort now under consideration, it can be shown that the sampling distributions for T_A and T_B both have the same variance and the same standard deviation. Here I will show only the structure for the standard deviation, which in the general case is equal to

$$-\sigma_T = \text{sqrt}\left[\frac{n_a n_b (N+1)}{12}\right].$$

Thus, for the present example, the sampling distributions of T_A and T_B each have a standard deviation equal to

$$-\sigma_T = \text{sqrt}\left[\frac{(11)(10)(21+1)}{12}\right] = \pm 14.2.$$

Another assertion of the "it can be shown" variety is this. In situations where N ranks are distributed within two samples of sizes n_a and n_b, it can be shown that the sampling distributions of T_A and T_B tend to approximate the form of a normal distribution, providing that n_a and n_b are both equal to or greater than 5. The source of this tendency is similar when considering the tendency of binomial sampling distributions to approximate the normal.

If the preceding two steps have not been obvious, the next one surely will be. Given the mean and standard deviation of a normally distributed sampling distribution, you can then fold the observed value of either T_A or T_B into an appropriate version of a z-ratio and refer the result to the unit normal distribution. In this particular context the z-ratio must include a "±.5" correction for continuity, just as with the binomial z-ratio, to accommodate the fact that the sampling distributions of T are intrinsically discrete. (T_A and T_B can assume decimal values such as 96.5 and 30.5 only as an artifact of the process of assigning tied ranks. Intrinsically, the ranks–1, 2, 3, 4, etc.–on which they are based are all integers.)

Designating

T_{obs} as the observed value of either T_A or T_B;

$_{-}\mu_T$ as the the mean of the corresponding sampling distribution of T; and

$_{=}\sigma_T$ as the standard deviation of that sampling distribution,

the general structure of the ratio is

$$z = \frac{(T_{obs} - \mu_T) \pm .5}{\sigma_T} \qquad \begin{array}{l}\text{correction for continuity:}\\ -.5 \text{ when } T_{obs} > \mu T\\ +.5 \text{ when } T_{obs} < \mu T\end{array}$$

Shown below are the details of calculation for both of our observed T values, T_A=96.5 and T_B=134.5. Keep in mind that the means of the sampling distributions for T_A and T_B are 121 and 110, respectively, and that the standard deviation for both distributions is ±14.2.

$$Z_A = \frac{(96.5-121)+.5}{14.2} \qquad Z_B = \frac{(134.5-110)-.5}{14.2}$$

$$= \frac{-24}{14.2} = -1.69 \qquad\qquad = \frac{+24}{14.2} = +1.69\,.$$

The remarkable symmetry of z_A and z_B is no accident. In all instances, z_A and z_B will have the same absolute value and opposite signs, one positive and the other negative. It therefore makes no difference whether the test of statistical significance is performed with z_A or z_B. The only requirement is that we keep a clear idea of what a positive or negative sign for z means in any particular case.

In our example, the investigators begin with the directional hypothesis that Treatment A will prove the more effective. This entails that sample A will tend to have the smaller values among the raw measures (indicating lower levels of claustrophobic tendency), hence the smaller ranks, hence an observed value of T_A smaller than its null-hypothesis value of 121, hence a value of z_A with a negative sign. With the opposite directional hypothesis–that Treatment B will prove the more effective–the expectation would be for z_B to end up with the negative sign. With a non-directional hypothesis there would of course be no specific expectation one way or the other.

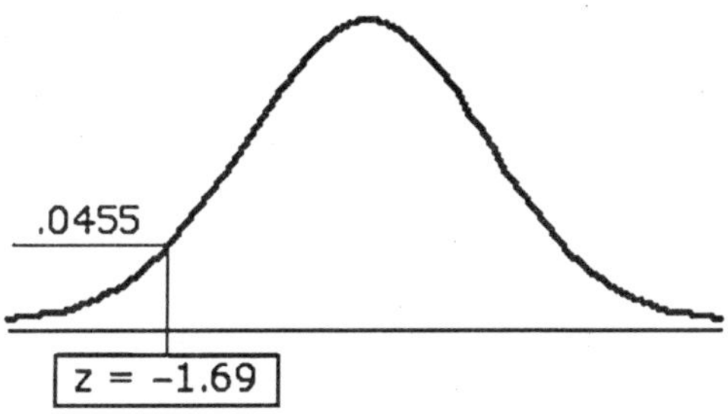

Here, at any rate, is the bottom line. If the null hypothesis were true, the mere-chance probability of ending up with a value of z_A falling at or beyond (to the left of) z=–1.69 within the normal distribution would be a scant P=.0455. For a directional test, our observed value of z_A= -1.69 is therefore significant a bit beyond the basic .05 level. (For a non-directional test it would of course be non-significant, since the mere-chance probability in this case would have to be figured as P=.0455+.0455=.091.)

Critical Values of ±z Level of Significance for a Directional Test				
.05	025	01	005	.0005
		Non-Directional Test		
–	05	02	01	.001
		$z_{critical}$		
1.645	1.960	2.326	2.576	3.291

Logic & Procedure: Method II

The limitation of the simple procedure described above is that it is valid only when n_a and n_b are both equal to or greater than 5. This is probably not much of a limitation in actual practice, for one would rarely choose to rely on such tiny samples, except as a last resort. At any rate, if either sample is of a size smaller than 5, it is a different path that must be taken. Up to a point, it is also possible to take this second path even when the sample sizes are both equal to or greater than 5, as in our present example. In this event, the probability assessments arrived at via the two methods are essentially equivalent.

Here as well we begin with some things that can be known by dint of sheer logic. Continuing with our claustrophobia example, recall that T_A=96.5 is the sum of the n_a=11 ranks in sample *A* and that T_B=134.5 is the sum of the n_b=10 ranks in sample *B*.

This first point will be fairly obvious. The maximum possible value of T_A in our example would be the sum of the highest n_a=11 of the ranks:

$$11+12+13+14+15+16+17+18+19+20+21=176$$

Similarly, the maximum possible value of T_B would be the sum of the highest n_b=10 of the ranks:

$$12+13+14+15+16+17+18+19+20+21=165$$

For any particular combination of n_a and n_b, these maximum possible values can be reached through the formulas

$$\mathrm{T}_{A[\max]} = n_a n_b + \frac{n_a(n_a+1)}{2}$$

which for the present example comes out as

$$\mathrm{T}_{A[\max]} = (11)(10) + \frac{11(12)}{2} = 176$$

and

$$T_{B[\max]} = n_a n_b + \frac{n_b(n_b+1)}{2}$$

which comes out as

$$T_{B[\max]} = (11)(10) + \frac{10(11)}{2} = 165.$$

Method II of the Mann-Whitney test turns upon the calculation of a measure known as *U*, which for either sample is equal to the difference between the maximum possible value of *T* for the sample versus the actually observed value of *T*. Thus for sample *A* it would be

$$U_A = T_{A[\max]} - T_A$$

$$U_A = n_a n_b + \frac{n_a(n_a+1)}{2} - T_A$$

and for sample B

$$U_B = T_{B[\max]} - T_B$$

$$U_B = n_a n_b + \frac{n_b(n_b+1)}{2} - T_B.$$

For the present example we have already obtained the maximum-possible and observed values for T_A and T_B:

	maximum possible value	*observed value*
T_A	176	96.5
T_B	165	134.5

so we can easily calculate the value of U for this example as either

$$U_A = 176–96.5 = 79.5$$

or

$$U_B = 165–134.5 = 30.5.$$

It does not matter which of these values you use, so long as you are consistent. The reason it does not matter is that U_A and U_B are mirror images of each other. For any given values of n_a and n_b, the sum of U_A and U_B will always be equal to the product of n_a times n_b. Hence the following identities:

$$U_A + U_B = n_a n_b$$
$$U_A = n_a n_b - U_B$$
$$U_B = n_a n_b - U_A.$$

The upshot is that once you know one of the values of U, you also implicitly know the other. Note as well that U_A and U_B are inversely related. The larger the value of one, the smaller must be the value of the other.

We determined earlier that the values of T_A and T_B that would be expected on the null hypothesis would be T_A=121 (versus the observed T_A=96.5) and T_B=110 (versus the observed T_B=134.5). From this you can infer that the values of U to be expected under the null hypothesis are

$$U_A = 176–121 = 55 \qquad \text{[vs observed } U_A = 79.5]$$

and

$$U_B = 165–110 = 55 \qquad \text{[vs observed } U_B = 30.5]$$

More generally, the null-hypothesis values of U are given by the identity

$$U_A = U_B = (n_a n_b)/2$$

which for our example would work out as

$$U_A = U_B = (11)(10)/2 = 55$$

On the null hypothesis we would therefore expect the values of U_A and U_B both to approximate $(n_a n_b)/2=55$. So if the null hypothesis is true, if there really is no difference in effectiveness between Treatment A and Treatment B, how likely is it that we could end up, by mere chance, with an observed value of U_A as large as 79.5; or alternatively, how likely that we could end up with an observed value of U_B as small as 30.5?

The principle underlying the answer to this question is but one more verse of a by-now familiar song. First you figure out the total number of possible ways in which $N=21$ ranks can be combined within two groups of sizes $n_a=11$ and $n_b=10$; and then you figure out the proportion of such combinations that would produce a value of U_A as large as 79.5, or alternatively a value of U_B as small as 30.5.

When I tell you that the total number of possible combinations in this example is

$$\frac{N!}{na!nb!} = \frac{21!}{11!10!} = 352{,}716$$

you will see in a flash that here is yet another instance of "easier said than done." Fortunately, it is also another instance where the bedeviling details have already been worked out. In most hard-copy statistics textbooks you will find a multi-page table of critical values of U, applicable to cases where n_a and n_b are both equal to or less than 20.

I cannot show you the full scope of this table, because it is under copyright and I would have to pay a fee to reproduce it. Fortunately the underlying principles cannot be held in copyright, so what I have done instead is program the following table to calculate some of the critical values of U directly. Enter any particular values of n_a and n_b into the designated cells, click the "Calculate" button, and the corresponding critical values of U will appear in the table. The only restriction is that n_a and n_b must both be between 5 and 20, inclusive. (For cases where the size of either sample is smaller than 5, you will need to consult a table in the back of some hard-copy textbook.) If the two sample-size values are different, you should designate the larger as n_a and the smaller as n_b.

Before you start plugging any new numbers, however, please be sure to read the explanatory text that follows the table.

Critical Values of U for n_a= 11 , n_b= 10

	Level of Significance for a Directional Test		
	.05	.025	.01
Non-Directional Test	.10	.05	.02
lower limit	31	26	21
upper limit	79	84	89

As you can see, I have already inserted our current values of $n_a=11$ and $n_b=10$, along with the corresponding critical values of U. As illustrated in the next table, each of the pairs of numbers designated "lower limit" and "upper limit" mark the boundaries between which falls a certain proportion of all possible mere-chance combinations of U_A and U_B in the case where $n_a=11$ and $n_b=10$.

lower limit:	31	26	21	
upper limit:	79	84	89	
	–.90	.95	.98	proportion between lower and upper limits
	.10	.05	.02	proportion outside middle range; both "tails" combined
	.05	.025	.01	proportion outside middle range at either end; each "tail" considered separately.

For the present example our focus can be chiefly on the first pair, 31 and 79. If the null hypothesis is true, there is a 90% chance that U_A and U_B will both be larger than 31 and both smaller than 79. The remaining possibilities fall outside that range, 5% below it and 5% above. At the one end, there is a 5% chance that U_A might end up as large as 79 and U_B as small as 31. At the other, there is a 5% chance that U_B might end up as large as 79 and U_A as small as 31. In our example, the investigators begin with the directional hypothesis that Treatment A will prove the more effective. Specifically, what they expect is

- That the raw-measure ratings of claustrophobic tendency will be lower in sample A than in sample B; and$_T$
- That sample A will accordingly tend to have the lower ranks; hence $T_B > T_A$ and $U_A > U_B$

And that is exactly what they do find: U_A=79.5, U_B=30.5. As these observed values of U fall just outside the range bounded by 31 and 79, we can say that the result is significant just beyond the .05 level for a directional test.

For a non-directional test the mere-chance probability of the obtained result would be only a shade below 10%, hence non-significant.

Similarly, the second set of numbers in the table, 26 and 84, mark the boundaries between which fall 95% of all possible mere-chance outcomes. Observed values of U falling at or beyond these boundary points would therefore be significant at or beyond the .05 level for a non-directional test and at or beyond the .025 level for a directional test. And so on for the third set of numbers in the table, marking the lower and upper boundaries for the directional .01 and the non-directional .02 levels of significance.

THE DIFFERENCE BETWEEN THE MEANS OF TWO CORRELATED SAMPLES

A certain researcher developed the inspired hypothesis that people are taller when they are wearing shoes than when they are not wearing shoes. To test this hypothesis, he took a random sample of 15 adults, measuring the height of each individual subject first with shoes on, and then again with shoes off. The result was two samples of height measures, A and B, of sizes N_a=15 and N_b=15. The following table shows the shoes-on and shoes-off measures of height, in inches, for each subject.

Subject	Sample A shoes on	Sample B shoes off
1	64.8	63.5
2	70.5	68.8
3	69.3	67.6
4	55.5	54.1
5	61.4	59.9
6	69.7	68.6
7	68.8	66.7
8	64.6	63.0
9	63.8	61.8
10	61.9	59.4
11	69.4	68.4
12	63.0	61.1
13	75.5	73.9
14	69.4	68.2
15	59.1	58.1
mean	65.8	64.2- $M_A–M_B = 1.6$
SS	378.4	384.1
variance	25.2	25.6
standard deviation	±5.0	±5.1

Aha! says the investigator. The null hypothesis here is that the distance from the top of a person's head to the horizontal surface on which he or she erectly stands is unrelated to whether the person is wearing shoes. I, on the other hand, have begun with the directional hypothesis that people are in fact taller with shoes on than with shoes off–and the outcome of my experiment is clearly consistent with that hypothesis. On average, my subjects were 1.6 inches taller when they had their shoes on than when they took them off. Moreover, each individual subject, without exception, was taller with shoes on than with shoes off.

Well, you're right, of course. The word to describe our investigator's hypothesis is not "inspired," but "banal." People obviously are taller with their shoes on than with their shoes off, and one hardly needs the labors of science to prove that commonplace. But please bear with me a moment, for the general structure of this example is illustrative of a wide range of real-life research situations where the questions are by no means trivial and the answers are not at all obvious in advance. The point I want to make with it is that, if you were to plug the data from the above table into the independent-samples t-test, you would find the 1.6 inch difference between the means of the two sets of measures to be non-significant by a substantial margin (t=+0.84, df=28; for significance at the basic .05 level for a directional test, the observed value of t would have to be at least 1.70). Clearly people are taller when shod than when unshod–but the t-test for independent samples is unable to detect that simple fact. It would have us conclude that the mean difference between the shoes-on and shoes-off conditions could easily have occurred through mere chance coincidence.

The reason for this oddity can be approached through the relationship shown in Figure. The blue circles in the top row show the measures of height for individual subjects with shoes on, and those in the bottom row show the measures with shoes off. Notice in particular that the difference between the means of these two sets of measures is rather tiny in comparison with the variability (the spread of the circles in each row) that exists inside the two sets.

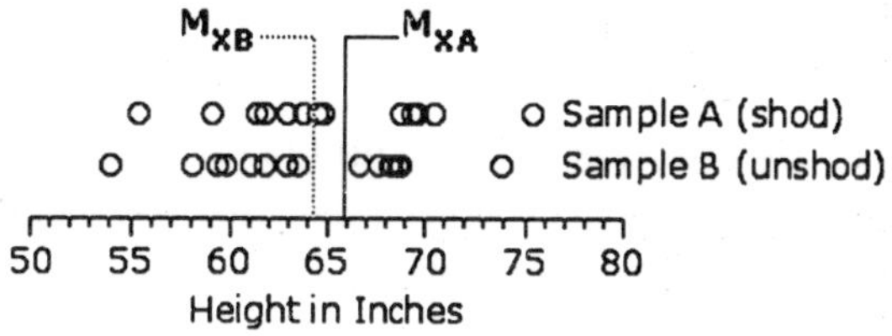

Fig. Distributions of Samples A and B

The import of this relationship—small mean difference versus large internal variability—can be seen by looking closely at what is going on inside the independent-samples t-test.

$$t = \frac{M_{Xa} - M_{Xb}}{\text{est.}i\sigma_{M\text{-}M}}$$ Formula for independent-samples t-test, from Ch. 11.

The denominator in this formula derives ultimately from SS_A and SS_B, the raw measures of variability that exists inside the two samples. Hence, the greater the amount of variability there is within the two samples, the larger will be the denominator; and the smaller, accordingly, will be the value of t. This is no mere game of numbers. The variability that exists inside samples A and B in this situation reflects nothing other than the fact that there are substantial individual differences among people with respect to the variable of height. A person who is relatively tall without shoes will also be relatively tall with shoes. A person who is relatively short without shoes will also be relatively short with shoes. The only difference between the respective variabilities of the two sets of measures will be occasioned by the differing amounts by which their shoes raise them off the floor. If all subjects had worn shoes of identical height, the respective variabilities of the two samples would be identical.

The point here is that these pre-existing individual differences with respect to height are entirely extraneous to the question of whether people on average are taller with shoes than without. The t-test for independent samples treats this extraneous variability as though it were not extraneous, and in consequence it overestimates the standard deviation of the relevant sampling distribution. That, in turn, results in an underestimate of the significance of the observed mean difference. The procedure to be introduced in the present chapter avoids this pitfall by disregarding the extraneous variability and looking only at what is relevant to the question at hand.

This procedure is spoken of as the t-test for correlated samples, for the simple reason that the two sets of measures in such a situation are arranged in pairs and are thus potentially correlated. You will also find this procedure spoken of as the repeated-measures or within-subjects t-test, because it typically involves situations in which each subject is measured twice, once in condition A, and then again in condition B. However, it is not essential that the measures in conditions A and B come from the same subjects. You could equally well start out with matched pairs of subjects (to be illustrated in a moment). The only requirement of the correlated-samples design, vis-ΰ-vis the structure of the data, is that each individual item in sample A is intrinsically linked with a corresponding item in sample B. The height of subject 1 while wearing shoes is linked with the height of subject 1 while not wearing shoes; and so, too, for subjects 2 through 15.

Indeed, insofar as we are concerned only with the difference between the shoes-on and shoes-off conditions, there is really only one sample here. The variable in this single sample can be denoted as D (for "difference") and defined for each linked pair as

$$D_i = X_{Ai} - X_{Bi}$$

where X_{Ai} is the height measure for subject i (i = 1, 2, 3, etc.) in the shoes-on condition and X_{Bi} is the height measure for the same subject in the shoes-off condition. The next table shows the same data you saw earlier, but now with the calculation of D for each subject. Notice that while the mean of all these D-values, M_D=1.6, is precisely the same as the difference we noted above between M_A and M_B, we now have much smaller measures of variability.

Subject	*Sample A shoes on*	*Sample B shoes off*	D_i
1	64.8	63.5	+1.3
2	70.5	68.8	+1.7
3	69.3	67.6	+1.7
4	55.5	54.1	+1.4
5	61.4	59.9	+1.5
6	69.7	68.6	+1.1
7	68.8	66.7	+2.1
8	64.6	63.0	+1.6
9	63.8	61.8	+2.0
10	61.9	59.4	+2.5
11	69.4	68.4	+1.0
12	63.0	61.1	+1.9
13	75.5	73.9	+1.6
14	69.4	68.2	+1.2
15	59.1	58.1	+1.0
Recall that$_x$ D_i=X_{Ai}– X_{Bi}		M_D	1.6
		SS_D	2.59
		variance standard	0.17
		deviation	±0.42

From this point on, the logic of the situation will be familiar. If there were no tendency for people to be taller with shoes than without, then we would expect the mean of the D-values in such a sample to approximate zero. The question, therefore, is whether our observed value of M_D=1.6 is significantly different from zero. Once again, it is a task of determining where the observed fact falls within the appropriate sampling distribution.

The t-test for correlated samples is especially useful in research involving human or animal subjects precisely because it is so very effective in removing the extraneous effects of pre-existing individual differences. This is not to suggest that individual differences are "extraneous" in every context. In some cases they might be the very essence of the phenomena that are of interest. But there are also situations where the facts that are of interest are merely obscured by the variability of individual differences. Suppose we were interested in determining whether two types of music, A and B, differ with respect to their effects on sensory-motor coordination. One way to approach the question would be to assemble two separate, independent samples of human subjects, measuring the members of one group on a task of sensory-motor coordination performed in the presence of type-A music, and the members of the other group on the same task in the presence of type-B music.

However, if your life, fame, fortune, or tenure depend on ending up with a significant result, this would probably not be a good strategy. For even if the two types of music do, in reality, have different effects, the difference would probably be eclipsed by the pre-existing individual differences among your subjects. Much of this variability would stem from inherent differences in sensory-motor coordination itself, although one can readily imagine other kinds of individual differences as well, such as motivation for this particular task, anxiety in test situations, ability to work under pressure, prior adaptation to one or the other type of music, and so on. In any event, it would be completely extraneous to the simple question of whether the two types of music have different effects on sensory-motor coordination.

With a research design involving two correlated samples on the other hand, we can hold the obscuring effects of such individual differences to a bare minimum. In the design involving independent samples, we test some subjects in the presence of type-A music and other subjects in the presence of type-B music.

With the design for correlated samples we test all subjects in both conditions and focus on the difference between the two measures for each subject. To obviate the potential effects of practice and test sequence in this case, we would also want to arrange that half the subjects are tested first in the type-A condition, then later in the type-B condition, and vice versa for the other half.

Suppose we were actually to conduct an experiment of this sort with a sample of 15 subjects, measuring how well each subject performs the task of sensory-motor coordination in each of the two conditions. We begin with the expectation that the two types of music might have different effects on sensory-motor coordination, though we have no particular hunches about the likely direction of the difference. Our research hypothesis is therefore non-directional.

The following table shows the measures we end up with, along with the relevant summary statistics. We will stipulate that the measures of task performance in columns A and B derive from an equal-interval scale; the measures of D are therefore also on an equal-interval scale.

Subject	Music Type A	B	D_i
1	10.2	13.2	.–3.0
2	8.4	7.4	+1.0
3	17.8	16.6	+1.2
4	25.2	27.0	.–1.8
5	23.8	27.5	.–3.7
6	25.7	26.6	.–0.9
7	16.2	18.0	.–1.8
8	21.5	21.2	+0.3
9	21.1	23.4	.–2.3
10	16.9	21.1	.–4.2
11	24.6	23.8	+0.8
12	20.4	20.2	+0.2
13	25.8	29.1	.–3.3
14	17.1	17.7	.–0.6
15	14.4	19.2	.–4.8
Recall that$_x$ $D_i=X_{Ai}-X_{Bi}$		M_D	.–1.53
		SS_D	55.45
		variance	3.70
		standard deviation	±1.92

In the column of D-values, a positive sign indicates that the subject's performance on the task was better in condition A than in condition B, while a negative sign indicates the opposite. As you can see, the negative signs preponderate, suggesting at first glance that sensory-motor coordination is on average better with music of type-B than with music of type-A. This, of course, is also what is suggested by the observed value of M_D=–1.53. All that remains is to determine whether this observed value significantly differs from the zero that would have been expected on the basis of the null hypothesis.

Logic and Procedure

The groundwork for each of the following points.

- According to the null hypothesis, the values of D_i in the sample derive from a source population whose mean is

$$i\mu_{\text{source}} = 0$$

- If we knew the variance of the source population, we would then be able to calculate the standard deviation ("standard error") of the sampling distribution of M_D as

$$x\sigma_{MD} = \text{sqrt}\left[\frac{x\ \sigma^2\text{source}}{N}\right].$$

- This, in turn, would allow us to test the null hypothesis for any particular instance of M_D by calculating the appropriate z-ratio

$$z = \frac{M_D}{\sigma_{MD}}.$$

- Referring the result to the unit normal distribution.

In actual practice, however, the variance of the source population of D-values, hence also the value of $i\sigma_{MD}$, can be arrived at only through estimation. In these cases the test of the null hypothesis is performed not with z but with t:

$$t = \frac{M_D}{\text{est.}i\sigma_{MD}}.$$

- The resulting value belongs to the particular sampling distribution of t that is defined by df=N–1, where N is the number of D-values.
- For this next point, recall that the relevant numerical values for the present example are N=15, $M_D = -1.53$, and SS_D=55.45.

The variance of the source population can be estimated as

$$\{s^2\} = \frac{SS_D}{N-1}$$

- Which for the present example comes out as

$$\{s^2\} = \frac{55.45}{14} = 3.96.$$

- This, in turn, allows us to estimate the standard deviation of the sampling distribution of M_D as

$$\text{est.}i\sigma_{MD} = \text{sqrt}\left[\frac{\{s^2\}}{N}\right] \quad \text{From Ch.9,Pt.2}$$

$$= \text{sqrt}\left[\frac{3.96}{15}\right] = \pm 0.51.$$

- The estimated value of $i\sigma_{MD}$ then permits the calculation of t as

$$t = \frac{M_D}{\text{est.}i\sigma_{MD}}$$

$$= \frac{-1.53}{0.51} = -3.0 \text{ with df=14}$$

INFERENCE

The outline and numerical details the sampling distribution of t for df=14. We started out with a non-directional research hypothesis, so the relevant critical values of t are those that pertain to a non-directional, two-tailed test of significance: 2.15 for the .05 level of significance, 2.62 for the .02 level, 2.98 for the .01 level, and so on. For purposes of a two-tailed test, the observed value of t=–3.0 must be conceived of as t=±3.0.

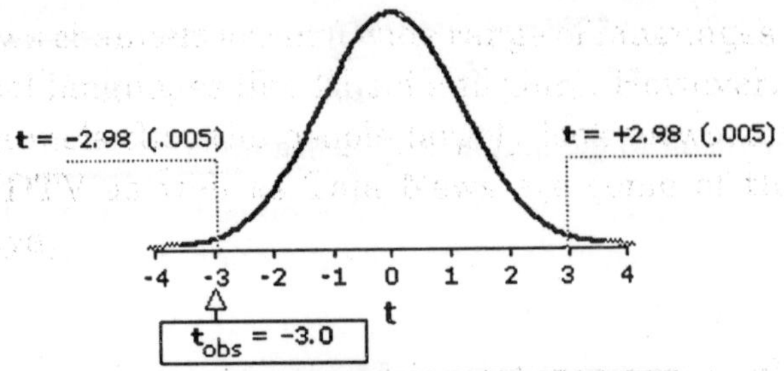

	Level of Significance for a Directional Test				
	.05	.025	.01	.005	.0005
	Level of Significance for a Non-Directional Test				
	---	.05	.02	.01	.001
df = 14	1.76	2.15	2.62	2.98	4.14

Fig. Sampling Distribution of t for df=14

Our observed t meets and slightly exceeds the critical value for the .01 level, hence can be regarded as significant slightly beyond the .01 level. Here again, the practical, bottom-line meaning of such a conclusion is that the likelihood of our experimental result having come about through mere chance coincidence is a bit less that 1%. So we can be quite confident, at the level of about 99%, that it reflects something more than mere random variability. If we had started out with the directional hypothesis that performance would be better under condition B, we would have performed a one-tailed test and found the result to be significant beyond the .005 level.

Step-by-Step Computational Procedure

Note that this test makes the following assumptions and can be meaningfully applied only insofar as these assumptions are met:

That the scale of measurement for X_A and X_B has the properties of an equal-interval scale.

That the values of D_i have been randomly drawn from the source population.

That the source population from which the values of D_i have been drawn can be reasonably supposed to have a normal distribution.

Step 1. For the sample of N values of D_i, where each instance of D_i is equal to X_{Ai}–X_{Bi}, calculate the mean of the sample as

$$M_D = \frac{i\Sigma D_i^{\,i}}{N_i}.$$

and the sum of squared deviates as

$$SS_D = iD^2{}_i - \frac{(i\Sigma D_i)^2}{{}_iN_i}$$

Step 2. Estimate the variance of the source population as

$$\{s^2\} = \frac{SS_D}{N-1}.$$

Step 3. Estimate the standard deviation of the sampling distribution of M_D as

$$\text{est.}\sigma_{MD} = {}_i\text{sqrt}_i \left[\frac{\{s^2\}}{{}_iN_i}\right].$$

Note that Steps 2 and 3 can be combined into the more streamlined formula

$$\text{est.}\sigma_{MD} = {}_i\text{sqrt}_i \left[\frac{SS_D/(N-1)}{{}_iN_i}\right].$$

Step 4. Calculate t as

$${}_it =_i \frac{M_D}{\text{est.}i\sigma_{MD}}.$$

Step 5. Refer the calculated value of t to the table of critical values of t, with df=N—1. Keep in mind that a one-tailed directional test can be applied only if a specific directional hypothesis has been stipulated in advance; otherwise it must be a non-directional two-tailed test.

So as not to leave you lying awake at night wondering about it, I'll conclude by noting that if you were to apply the correlated-samples t-test to the data of the shoes-on versus shoes-off example, you would find the observed value of M_D=+1.6 to be very significant indeed (t=+14.17, df=14). If the null hypothesis were true–if it makes no difference in height whether a person is shod or unshod–the one-tailed probability of ending up with a value of M_D this large or larger by mere chance coincidence would be a miniscule 0.0000000005.

THE WILCOXON SIGNED-RANK TEST

In Subchapter 11a we examined a non-parametric alternative to the t-test for independent samples. We now turn to consider a somewhat analogous alternative to the t-test for correlated samples. The correlated-samples t-test makes certain assumptions and can be meaningfully applied only insofar as these assumptions are met. Namely,

1. That the scale of measurement for X_A and X_B has the properties of an equal-interval scale;$_{\text{T}}$
2. That the differences between the paired values of X_A and X_B have been randomly drawn from the source population; and$_{\text{T}}$
3. That the source population from which these differences have been drawn can be reasonably supposed to have a normal distribution.

Here again, it is not simply a question of good manners or good taste. If there is one or more of these assumptions that we cannot reasonably suppose to be satisfied, then the t-test for correlated samples cannot be legitimately applied.

Of all the correlated-samples situations that run afoul of these assumptions, I expect the most common are those in which the scale of measurement for X_A and X_B cannot be assumed to have the properties of an equal-interval scale. The most obvious example would be the case in which the measures for X_A and X_B derive from some sort of rating scale. In any event, when the data within two correlated samples fail to meet one or another of the assumptions of the t-test, an appropriate non-parametric alternative can often be found in the Wilcoxon Signed-Rank Test.

To illustrate, suppose that 16 students in an introductory statistics course are presented with a number of questions concerning basic probabilities. In each instance, the question takes the form "What is the probability of such-and-such?" However, the students are not allowed to perform calculations. Their answers must be immediate, based only on their raw intuitions. They are instructed to frame each answer in terms of a zero to 100 percent rating scale, with 0% corresponding to P=0.0, 27% corresponding to P=.27, and so forth. They are also told that they can give non-integer answers if they wish to make really fine-grained distinctions; for example, 49.0635...%.

The instructor of the course is particularly interested in student's responses to two of the questions, which we will designate as question A and question B. He reasons that if students have developed a good, solid understanding of the basic concepts, they will tend to give higher probability ratings for question A than for question B; whereas, if they were sleeping through that portion of the course, their answers will be mere shots in the dark and there will be no overall tendency one way or the other. The instructor's hypothesis is of course directional: he expects his students have mastered the concepts well enough to sense, if only intuitively, that the event described in question A has the higher probability. The following table shows the probability ratings of the 16 subjects for each of the two questions.

Subj.	X_A	X_B	X_A-X_B
1	78	78	0
2	24	24	0
3	64	62	+2
4	45	48	--3
5	64	68	–4
6	52	56	–4
7	30	25	+5
8	50	44	+6
9	64	56	+8
10	50	40	+10
11	78	68	+10
12	22	36	–14
13	84	68	–16
14	40	20	+20
15	90	58	+32
16	72	32	+40
mean difference = +7.75			

Voilŏ! The observed results are consistent with the hypothesis. The probability ratings do on average end up higher for question A than for question B. Now to determine whether the degree of the observed difference reflects anything more than some lucky guessing.

MECHANICS

The Wilcoxon test begins by transforming each instance of X_A–X_B into its absolute value, which is accomplished simply by removing all the positive and negative signs. Thus the entries in column 4 of the table below become those of column 5. In most applications of the Wilcoxon procedure, the cases in which there is zero difference between X_A and X_B are at this point eliminated from consideration, since they provide no useful information, and the remaining absolute differences are then ranked from lowest to highest, with tied ranks included where appropriate.

The guidelines for assigning tied ranks are described in Subchapter 11a in connection with the Mann-Whitney test.

The result of this step is shown in column 6. The entries in column 7 will then give you the clue to why the Wilcoxon procedure is known as the signed-rank test. Here you see the same entries as in column 6, except now we have re-attached to each rank the positive or negative sign that was removed from the X_A–X_B difference in the transition from column 4 to column 5.

1 *Subj.*	*2* X_A	*3* X_B	*4* *original* X_A–X_B	*5* *absolute* X_A–X_B	*6* *rank of absolute* X_A–X_B	*7* *signed rank*
1	78	78	0	0	–	–
2	24	24	0	0	–	–
3	64	62	+2	2	1	+1
4	45	48	–3	3	2	–2
5	64	68	–4	4	3.5	–3.5
6	52	56	–4	4	3.5	–3.5
7	30	25	+5	5	5	+5
8	50	44	+6	6	6	+6
9	64	56	+8	8	7	+7
10	50	40	+10	10	8.5	+8.5
11	78	68	+10	10	8.5	+8.5
12	22	36	–14	14	10	–10
13	90	68	+16	16	11	+11
14	90	20	+20	20	12	+12
15	90	58	+32	32	13	+13
16	72	32	+40	40	14	+14
						W = 67.0 ^{T}N = 14

The sum of the signed ranks in column 7 is a quantity symbolized as W, which for the present example is equal to 67. Two of the original 16 subjects were removed from consideration because of the zero difference they produced in columns 4 and 5, so our observed value of W is based on a sample of size N=14.

LOGIC & PROCEDURE

Here again, as with the Mann-Whitney test, the effect of replacing the original measures with ranks is two-fold. The first is that it brings us to focus only on the ordinal relationships among the measures–"greater than," "less than," and "equal to"–with no illusion that these measures have the properties of an equal-interval scale. And the second is that it transforms the data array into a kind of closed system whose properties can then be known by dint of sheer logic. For openers, we know that the sum of the N unsigned ranks in column 6 will be equal to

$$\text{sum} = \frac{N(N+1)}{2} \quad \text{From Subchapter 11a}$$

$$= \frac{14(14+1)}{2} = 105.$$

Thus the maximum possible positive value of W (in the case where all signs are positive) is W=+105, and the maximum possible negative value (in the case where all signs are negative) is W=–105. For the present example, a preponderance of positive signs among the signed ranks would suggest that subjects tend to rate the probability higher for question A than for question B. A preponderance of negative signs would suggest the opposite. The null hypothesis is that there is no tendency in either direction, hence that the numbers of positive and negative signs will be approximately equal. In that event, we would expect the value of W to approximate zero, within the limits of random variability.

For fairly small values of N, the properties of the sampling distribution of W can be figured out through simple (if tedious) enumeration of all the possibilities. Suppose, for example, that we had only N=3 subjects, whose absolute (unsigned) $X_A - X_B$ differences produced the untied ranks 1, 2, and 3. The following table shows the possible combinations of plus and minus signs that could be distributed among these ranks, along with the value of W that each combination would produce.

Ranks			
1	2	3	W
+	+	+	+6
–	+	+	+4
+	–	+	+2
+	+	–	0
–	–	+	0
–	+	–	–2
+	–	–	–4
–	–	–	–6

There is a total of 8 equally probable mere-chance combinations, of which exactly one would yield a positive value of W as large as +6, exactly two would yield a positive value as large as +4, and so on. And similarly at the other end of the distribution: exactly one combination yields a negative value of W as large as –6, exactly two yield negative values of W as large as –4, and so on. Hence the probability of ending up with a positive value of W as large as +4 is 2/8=.25; the probability of obtaining a negative value of W as large as –4 is 2/8=.25; and the "two-tailed" probability of finding a value of ±W as large as ±4 (in either direction) is (2/8)+(2/8) = .5.

The first of the following graphs shows the sampling distribution of this N=3 situation in pictorial form, and the other two show the corresponding distributions for the situations where N=4 and N=5. Note that for any such situation, the number of possible combinations of plus and minus signs is equal to 2^N. Thus for N=3, 2^3=8; for N=4, 2^4=16; for N=5, 2^5=32, and so on.

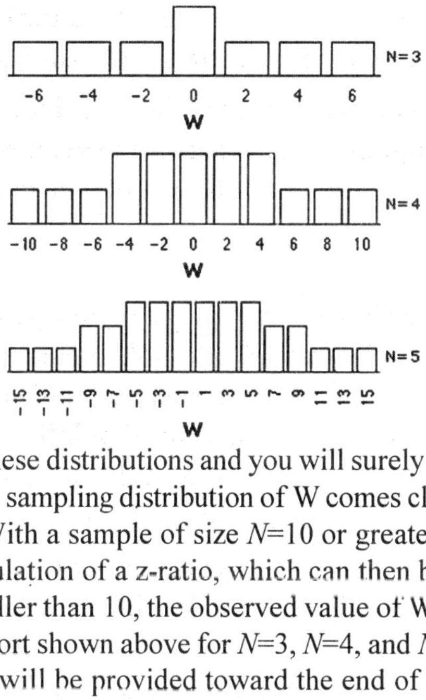

Examine the shapes of these distributions and you will surely see where things are heading. As the size of N increases, the sampling distribution of W comes closer and closer to the outlines of the normal distribution. With a sample of size N=10 or greater, the approximation is close enough to allow for the calculation of a z-ratio, which can then be referred to the unit normal distribution. (When N is smaller than 10, the observed value of W must be referred to an exact sampling distribution of the sort shown above for N=3, N=4, and N=5. A table of critical values of W for small sample sizes will be provided toward the end of this subchapter.)

We noted earlier that on the null hypothesis we would expect the value of W to approximate zero, within the limits of random variability. This is tantamount to saying that any particular observed value of W belongs to a sampling distribution whose mean is equal to zero. Hence

$$-\mu_W = 0$$

Considerably less obvious is the standard deviation of the distribution. As it would be a distraction to try to make it obvious, I will resort to another of those "it can be shown" assertions and say simply: For any particular value of N, it can be shown that the standard deviation of the sampling distribution of W is equal to

$$\sigma_W = \text{sqrt}\left[\frac{N(N+1)(2N+1)}{6}\right]$$

which for the present example, with N=14, works out as

$$\sigma_W = \text{sqrt}\left[\frac{14(14+1)(28+1)}{6}\right] = \pm 31.86.$$

When considering the Mann-Whitney test in Subchapter 11a we noted that the z-ratio must include a "±.5" correction for continuity. The same is true for the Wilcoxon test, and for the same sort of reason. The measure designated as W can assume decimal values only as an artifact of the process of assigning tied ranks. Intrinsically, the absolute ranks–1, 2, 3, 4, etc.– on which W is based are all integers. Thus, the structure of the z-ratio for the Wilcoxon test is

$$z = \frac{(W - \mu_W) \pm .5}{\sigma_W}.$$

The correction for continuity is "–.5" when W is greater than μ_W and "+.5" when W is less than μ_W. Since μ_W is in all instances equal to zero, the simpler computational formula is

$$z = \frac{W - .5}{\sigma_W}.$$

For the present example, with N=14, W=67, and σ_W=±31.86, the result is

$$z = \frac{67 - .5}{31.86} = +2.09.$$

From the following table of critical values of z, you can see that the observed value of z=+2.09 is significant just a shade beyond the .025 level for a directional test, which is the form of test called for by our investigator's directional hypothesis. For a two-tailed non-directional test, it would be significant just beyond the .05 level.

Critical Values of ±z

Level of Significance for a Directional Test				
.05	.025	.01	.005	.0005
Non-Directional Test				
–	.05	.02	.01	.001
$z_{critical}$				
1.645	1.960	2.326	2.576	3.291

When N is smaller than 10, the observed value of W must be referred to an exact sampling distribution of the sort described earlier. The following table shows the critical values of W for N=5 through N=9. For sample sizes smaller than N=5 there are no possible values of W that would be significant at or beyond the baseline .05 level.

Critical Values of ±W for Small Samples:

	Level of Significance for a Directional Test			
	05	.025	.01	.005
	Non-Directional Test			
N	–	.05	.02	.01
5	15	–	–	–
6	17	21	–	–
7	22	24	28	–
8	26	30	34	36
9	29	35	39	43

The assumptions of the Wilcoxon test are:

- That the paired values of X_A and X_B are randomly and independently drawn (i.e., each pair is drawn independently of all other pairs);$_T$
- That the dependent variable (e.g., a subject's probability estimate) is intrinsically continuous, capable in principle, if not in practice, of producing measures carried out to the nth decimal place; and$_T$
- That the measures of X_A and X_B have the properties of at least an ordinal scale of measurement, so that it is meaningful to speak of "greater than," "less than," and "equal to."

Chapter 5

Analysis of Variance

The independent-samples t-test with the example of an experiment aimed at determining whether two types of music have different effects on the performance of a mental task. Suppose that we were instead interested in assessing the relative effects of three types of music. In this case, the experimental procedure is the same in every detail, except that now we carry it out with three groups, one for each of the three types of music. As shown abstractly in the following table, what we end up with are three independent samples of measures, A, B, and C. If the three types of music have different effects on task performance, we would expect this fact to be reflected in significant differences among the means of the three samples.

A	B	C
X_{a1}	X_{b1}	X_{c1}
X_{a2}	X_{b2}	X_{c2}
X_{a3}	X_{b3}	X_{c3}
etc.	etc.	etc.
M_a	M_b	M_{c3}

At first glance, you might suppose you could determine whether the three group means significantly differ form one another by performing a separate independent-samples t-test for each possible pair of means: that is,

- One test for M_a versus M_b
- Another for M_a versus M_c
- Yet another for M_b versus M_c

A moment's reflection, however, will show why this simple strategy would not be advisable.

Essentially it is an exercise in disjunctive probabilities, along with a reminder of what it means to say than that some particular result is "significant." If an observed result is found to be significant at the basic .05 level, what this means is that there is only a 5% chance of its having occurred through mere chance. But 5% is still 5%! For any particular one of the three pair-wise t-test comparisons listed above, there would be a 5% probability by mere chance, even if the null hypothesis were true, of ending up with a difference that is "significant" at the .05 level. And three times 5% is 15%! If you were to perform all three of the pair-wise comparisons listed above, the disjunctive probability that one or another of them might end up "significant" at the .05 level by mere chance, absent any genuine differences among the effects of the three types of music, would be on the order of .05+.05+.05=.15. I say "on the order of" because the true disjunctive probability here would not be precisely .15, owing to some rather complex conditional probabilities that creep into situations of this sort. The main point, however, is independent of any particular numbers. If you are performing t-tests on multiple pairs of sample means, the probability that one or another of the comparisons might end up "significant" at the .05 level, by the merest chance, is substantially greater than .05.

The analysis of variance, commonly referred to by the acronym ANOVA, was first developed as a strategy for dealing with this sort of complication. At its lowest level it is essentially an extension of the logic of t-tests to those situations where we wish to compare the means of three or more samples concurrently.

As its name suggests, the analysis of variance focuses on variability. It involves the calculation of several measures of variability, all of which come down to one or another version of the basic measure of variability, the sum of squared deviates. Before we get going, here is a brief reminder of how this "raw" measure of variability is obtained.

For any set of N values of X_i that derive from an equal-interval scale of measurement, a deviate is the difference between an individual value of X_i and the mean of the set:

$$\text{deviate} = X_i - M_X$$

a squared deviate is the square of that quantity:

$$\text{squared deviate} = (X_i - M_X)^2$$

and the sum of squared deviates is the sum of all the squared deviates in the set:

$$SS = (X_i - M_X)^2$$

For practical computational purposes, it is often convenient to calculate the sum of squared deviates via the algebraically equivalent formula

$$SS = \Sigma X^2_i - \frac{(\Sigma X_i)^2}{N_i}.$$

THE LOGIC OF ANOVA [H]

Here again is our scenario for three independent samples of measures, A, B, and C, except now we plug some specific numbers into the cells. Listed below the values of X_i in each of the three groups are five relevant summary statistics. I have also added a fourth column to include summary measures for the total array of data (all three groups combined). In each case, these latter summary measures are subscripted "T" for "Total." The sizes of the three groups separately are N_a=5, N_b=5, and N_c=5; hence N_T=15. The mean of all 15 of these combined values of X_i is M_T=18.2, and the sum of squared deviates of all 15 combined is SS_T=70.4.

A	B	C	Total Array		
16	20	18	16	20	18
15	19	19	15	19	19
17	21	18	17	21	18
15	16	23	15	16	23
20	18	18	20	18	18
N_a=5	N_b=5	N_c=5	N_T=15		
ΣX_{ai}=83	ΣX_{bi}=94	ΣX_{ci}=96	ΣX_{Ti}=273		
ΣX^2_{ai}=1395	ΣX^2_{bi}=1782	ΣX^2_{ci}=1862	ΣX^2_{Ti}=5039		
M_a=16.6	M_b=18.8	M_c=19.2	M_T=18.2		
SS_a=17.2	SS_b=14.8	SS_c=18.8	SS_T=70.4		

The Measure of Aggregate Differences among Sample Means [sh]

The central question here is: Do the means of the three samples significantly differ from one another?

$$M_a=16.6 \quad M_b=18.8 \quad M_c=19.2$$

So the first part of our task is to figure out a way of measuring the degree of their differences. If there were only two samples, the task would be quite easy: simply subtract the mean of one from the mean of the other. But that, of course, will not work when there are more than two samples. What you need in this more complex case is a measure of the aggregate degree to which the three (or more) group means differ. As it happens, there is a form of measurement you have already encountered that will perform this task quite handily. It is none other than that elemental "raw" measure of variability, the sum of squared deviates. The basic concept is that, whenever you have three or more numerical values, the measure of their variability is equivalent to the measure of their aggregate differences. That, indeed, is precisely what "variability" means: aggregate differences.

Here is how you could construct a sum of squared deviates measure for the three group means of the present example. For any particular group mean (the subscript "g" means "any particular group"), the deviate would be the difference between the group mean and M_T, the mean of the total array of data:

$$M_g - M_T$$

and the squared deviate would of course be the square of that quantity:

$$(M_g - M_T)^2$$

The following table shows the numerical details of this procedure for each of the three groups.

M_a=16.6	M_b=18.8	M_c=19.2	M_T=18.2
$(16.6–18.2)^2$	$(18.8–18.2)^2$	$(19.2–18.2)^2$	
^{2}=2.56	^{2}=0.36	^{2}=1.0	

These measures give you the squared deviates for each of the three group means, but they do not yet give you a sum of squared deviates. Actually, what we will now end up with is not so much a sum as a weighting. That is, we will be weighting (adding weight) to the squared deviate of each of the group means in accordance with the number of individual values of X_i on which the group mean is based: N_a=5 for group A, N_b=5 for group B, and N_c=5 for group C. Thus, for any particular group mean, the squared deviate is

$$(M_g–M_T)^2$$

and the corresponding "sum" of squared deviates is the squared deviate multiplied by the appropriate value of N

$$N_g(M_g–M_T)^2$$

Here again are the numerical details.

N_a=5	N_b=5	N_c=5	
M_a=16.6	M_b=18.8	M_c=19.2	M_T=18.2
$5(16.6–18.2)^2$	$5(18.8—18.2)^2$	$5(19.2–18.2)^2$	
^{2}=12.8	^{2}=1.8	^{2}=5.0 --	

The sum of these three resulting values, 12.8+1.8+5.0=19.6, will give you a quantity spoken of within the context of the analysis of variance as the sum of squared deviates between-groups, rendered symbolically as SS_{bg}. And that is our aggregate measure of the degree to which the three sample means differ from one another: SS_{bg}=19.6.

The Measure of Background Random Variability [sh]

Once we have this measure, all that remains is to figure a way to determine whether it differs significantly from the zero that would be specified by the null hypothesis. As a first step, return for a moment to the formula for the independent-samples t-test:

$$t = \frac{M_{Xa} - M_{Xb}}{\text{est.}i\sigma_{M\text{-}M}}.$$

Clearly our measure of SS_{bg} is analogous to the numerator of this formula: $M_{Xa}–M_{Xb}$ is the difference between two means; SS_{bg} is the difference among three or more means. What we now need to find is an appropriate analogy for the denominator. The way to find it is to look behind the abstract notation, $i\sigma_{M–M'}$, to see the more general concept that it embodies.

What it refers to in the first instance, of course, is that elaborate mouthful, "the standard deviation of the sampling distribution of sample-mean differences." More generally, you can say it refers to "the standard deviation of the sampling distribution that happens to be appropriate in this particular situation." And even more generally, you can say it refers to "the measure of sheer, cussed random variability that happens to pertain to this particular situation." Whenever you perform a test of statistical significance, what you are essentially doing is comparing an observed fact to a measure of the random variability in which it is embedded; the aim of the comparison is to determine whether the fact–in this case, the aggregate difference among the three sample means–goes beyond anything that mere random variability might have produced.

In the independent-samples t-test, the ultimate source of your measure of random variability is the variability that appears inside each of the two samples, A and B, as measured by SS_a and SS_b. The same is true when you have more than two samples. In our original table of data we observed the following values of SS for our three samples, A, B, and C:

$$SS_a=17.2 \quad SS_b=14.8 \quad SS_c=18.8$$

Taken together, they constitute a quantity known as sum of squared deviates within-groups, symbolized as SS_{wg}. Thus,

$$\begin{aligned} SS_{wg} &= SS_a + SS_b + SS_c \\ &= 17.2+14.8+18.8 \\ &= 50.8. \end{aligned}$$

So here is where we stand up to this point: $SS_{bg}=19.6$ is the aggregate raw measure of the degree to which our three sample means differ, and $SS_{wg}=50.8$ is the raw measure of the random variability in which these sample-mean differences are embedded. I say "raw" in both cases, because each of these measures will have to be refined somewhat before we can actually use them. But more of that in a moment. First I want to call your attention to a remarkable connectedness among the various numbers that you have been scrolling through in this chapter. At the very top came $SS_T=70.4$, the measure of variability within the entire array of data for all three groups combined. Then came $SS_{bg}=19.6$, then $SS_{wg}=50.8$. Add the latter two together and the result is the same 70.4 that was found for SS_T. The correspondence is no accident. When SS_T, SS_{bg}, and SS_{wg} are calculated in the manner just described, it is always and necessarily the case that SS_{bg} and SS_{wg} add up to SS_T (within the limits of rounding error). It is hardly surprising. SS_T is the total variability of the array, and there are only two places where portions of that total can be distributed: either within the groups or between the groups. The relationships among these three measures of variability are described by the following identities:

- $SS_T = SS_{wg} + SS_{bg}$
- $SS_{bg} = SS_T - SS_{wg}$
- $SS_{wg} = SS_T - SS_{bg}$
- The Refinement of Between-Groups and Within-Groups Measures

In Chapters 9 through 12 you encountered several versions of the basic concept that the variance of a source population can be estimated as

$$\text{est.}\sigma^2_{\text{source}} = \frac{\text{sum of squared deviates}}{\text{degrees of freedom}}.$$

You have also seen in previous chapters that the basic concept of degrees of freedom in this context is "*N*–1," where "*N*" refers to the number of items on which the measure of sum of squared deviates is based. Using these same concepts, we now proceed to form two separate estimates of source population variance, one on the basis of SS_{bg} and the other on the basis of SS_{wg}. I will first go through the mechanics of the process, and then come back to try to explain just what it is that these variance estimates are aiming to estimate.

Within the context of the analysis of variance, an estimate of a source population variance is spoken of as a mean square (shorthand for "mean of the squared deviates") and conventionally symbolized as MS. The value of the between-groups *SS* in this example is based on the means of three groups, so the number of degrees of freedom associated with SS_{bg} is $df_{bg}=3-1=2$, and the variance estimate is

$$MS_{bg} = SS_{bg} / df_{bg}$$

$$MS_{bg} = 19.6 / 2 = 9.8$$

The value of the within-groups *SS* is the sum of the separate *SS* measures for each of the three samples: SS_a, SS_b, and SS_c. Each of these separate within-groups measures of *SS* is associated with a certain number of degrees of freedom: N_a-1, N_b-1, and N_c-1, respectively. So the number of degrees of freedom associated with the composite within-groups measure, SS_{wg}, is

$$(N_a-1) + (N_b-1) + (N_c-1)$$

which in the present case comes out to $df_{wg}=12$. So here the variance estimate is

$$MS_{wg} = SS_{wg} / df_{wg}$$

$$MS_{wg} = 50.8 / 12 = 4.23$$

Now back to the question of just what it is that these two variance estimates, $MS_{bg}=9.8$ and $MS_{wg}=4.23$, are aiming to estimate. The simplest answer is that they are both estimates of the same thing. In the general case where you have three (or more) independent samples of measures arrayed in the fashion shown abstractly at the beginning of this chapter,

A	*B*	*C*
X_{a1}	X_{b1}	X_{c1}
X_{a2}	X_{b2}	X_{c2}
X_{a3}	X_{b3}	X_{c3}
etc.	etc.	etc.
M_a	M_b	M_c

the null hypothesis is that the values of X_i in the three samples have all been drawn indifferently from the same underlying source population. Our two values of MS, for between-groups and within-groups, are both estimates of the variance of the population assumed by the null hypothesis to be the common source of all three samples. The relationship between these two estimates is rather complicated, and we will not try to go into it in fine-grained detail. Suffice it to say that;

- When the null hypothesis is true, MS_{bg} will tend to be equal to or less than MS_{wg};
- When the null hypothesis is not true, MS_{bg} will tend to be greater than MS_{wg}.

The F-Ratio

The relationship between two values of *MS* is conventionally described by a ratio known as F, which is defined for the general case as

$$F = \frac{MS_{\text{effect}}}{MS_{\text{error}}}$$

where MS_{effect} is a variance estimate pertaining to the particular fact whose significance you wish to assess (e.g., the differences among the means of several independent samples), and MS_{error} is a variance estimate reflecting the amount of sheer, cussed random variability that is present in the situation. For the present example, MS_{effect} would be the same as MS_{bg} and MS_{error} would be the same as MS_{wg}. When the null hypothesis is true, the F-ratio will tend to be equal to or less than 1.0, within the limits of random variability; and when the null hypothesis is false, the F-ratio will tend to be significantly greater than 1.0. In the present example, the ratio comes out as

$$F = \frac{MS_{bg}}{MS_{wg}} = \frac{9.8}{4.23} = 2.32.$$

You will certainly be able to anticipate what comes next. Granted that $F=2.32$ is greater than the null hypothesis stipulation of $F<1.0$: How likely is it that a difference this large or larger might have occurred through mere chance coincidence? Same song, new verse. As with z, t, and chi-square, the destiny of a calculated F-ratio is to be referred to its appropriate sampling distribution.

Before we get into the theoretical details of this point, here is an exercise that will allow you to simulate the sampling distribution of F that applies to this particular example. A normally distributed reference population that has a mean of 18 and a standard deviation of ±3. Each time you click the button below, labeled "Samples," your computer will draw three random samples of X_i from this common source population, each sample of size $N=5$. As the null hypothesis in this case is patently true for each set of samples, you would expect most of the resulting F-ratios to be equal to or less than 1.0; though of course some will be greater than 1.0, perhaps even substantially greater, through sheer, cussed random variability.

Click the button repeatedly and try to get a sense of how often these mere-chance F-ratios come out equal to or greater than our calculated value of 2.32. If the proportion of such cases over the long run is 5% or less, then $F=2.32$ can be regarded as significant at or beyond the basic .05 level. If it is more than 5%, then $F=2.32$ is non-significant. The cell labeled "PCT" will perform a running calculation of the percentage of cases in which the mere-chance F-ratio is equal to or greater than 2.32. If you continue clicking the button long enough (at least 50 times, preferably several hundred), you will see this percentage beginning to settle in at about 14%, which is of course quite a lot larger than the 5% or less that you are looking for. So our calculated value of 2.32 is, in a word, non-significant. In a situation of this particular type–three independent samples, each of size *N*=5–an F-ratio this large or larger could readily occur by mere chance.

THE SAMPLING DISTRIBUTIONS OF F

You have already seen that for values of z there is only one sampling distribution, whereas for t and chi-square there is in principle a different sampling distribution for each possible value of df, degrees of freedom. With F the situation becomes even more complex, for here there are two distinct values of df to be taken into account: one pertaining to the numerator of the ratio and the other pertaining to the denominator. Recall that when you calculate an F-ratio of the general form

$$F = \frac{MS_{bg}}{MS_{wg}}$$

what you are actually calculating is

$$F = \frac{MS_{bg}}{MS_{wg}} = \frac{SS_{bg}/\mathrm{df}_{bg}}{SS_{wg}/\mathrm{df}_{wg}}.$$

Hence the numerator of the F-ratio is associated with df_{bg}, the denominator with df_{wg}. For the present example, df_{bg}=2 and df_{wg}=12, so the F-ratio in this case is associated with 2 and 12 degrees of freedom for the numerator and denominator, respectively. The conventional notation for a pairing of numerator/denominator df values of this type is "df=2,12."

There is a separate sampling distribution of F for each possible pair of such numerator/denominator df values. Thus, there is one sampling distribution for df=2,12, another for df=2,15, another for df=12,160, and so on. The shapes of these various sampling distributions lie within the range of the two extreme forms shown below in Figure. With quite large values of df, the highest point of the curve will tend to be reached at F=1.0, while with smaller values of df it will fall somewhere to the left of F=1.0. In all cases, however, the portion of the curve that falls to the right of F=1.0 has much the same shape: the curve drops steeply at first, and then more gradually.

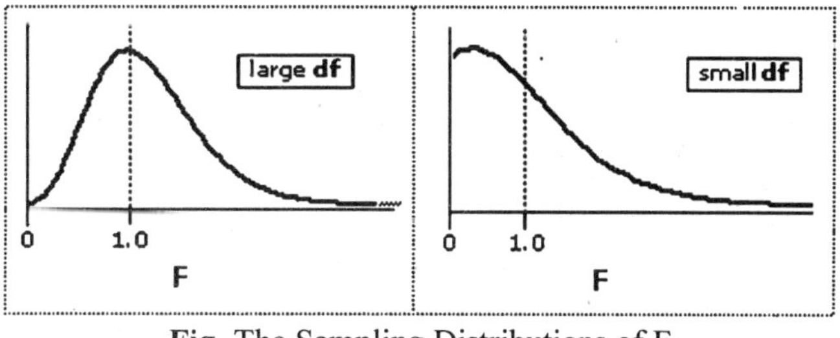

Fig. The Sampling Distributions of F

The next figure shows the sampling distribution of F that applies to the example described above, where the df values for numerator and denominator are 2 and 12. For the sake of simplicity, the horizontal axis in this graph begins at F=1.0. The stippled blue patch is just a reminder that a portion of the distribution lies invisibly to the left of 1.0.

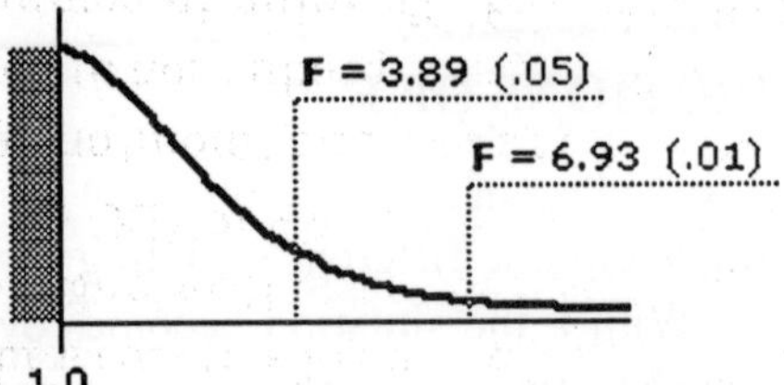

Fig. Sampling Distribution of F for df=2,12

As indicated by the accompanying notations, F=3.89 and F=6.93 mark the points beyond which fall 5% and 1%, respectively, of all possible mere-chance outcomes, assuming the null hypothesis to be true. An observed F-ratio equal to or greater than 3.89 would therefore be significant at or beyond the .05 level, while one equal to or greater than 6.93 would be significant at or beyond the .01 level.

Because of the very large number of separate sampling distributions, tables of critical values for F typically list only the values for the .05 level and the .01 level. The following abridged table will give you an idea of how it is all laid out. Across the top are values of df for the numerator of the F-ratio, and down the side on the left are df values for the denominator. For each combination of $df_{numerator}$ and $df_{denominator}$, the critical value for the .05 level is listed in plain type, while the critical value for the .01 level is given in bold-face type. Thus, for df=2,12 the listings appear as "3.89" and "6.93." For df=4,10, it is "3.48" and "5.99." And so on. A more complete table of critical values of F is found.

df denominator	*df numerator*			
	1	2	3	4
10	4.96	4.10	3.71	3.48
	10.04	7.56	6.55	5.99
11	4.84	3.98	3.59	3.36
	9.65	7.21	6.22	5.67
12	4.75	3.89	3.49	3.26
	9.33	6.93	5.95	5.41
13	4.67	3.81	3.41	3.18
	9.07	6.70	5.74	5.20

THE RELATIONSHIP BETWEEN F AND T

Although it is not normally done this way, the analysis of variance can also be applied to situations where you have only two groups of measures–in particular, to situations of the sort described in Chapters 11 and 12 in connection with t-tests for independent samples and correlated samples. And when you do apply it to two-sample situations of this type, the results are equivalent to what would be found in the corresponding t-test. Within the limits of rounding errors, the F-ratio obtained in such an analysis will be equal to the square of the corresponding t-ratio. The only difference is that a t-test can be either directional or non-directional, whereas the analysis of variance is (like chi-square) intrinsically non-directional.

APPLICATIONS OF F

Of course, if you had only two groups you would not really want to bother with the apparatus of the F-ratio, because you could arrive at the same place more easily with a simple t-test. The advantage of the F-ratio is that its logic and procedure, once developed, can then be extended to those numerous cases where the researcher might wish to examine three or more samples concurrently.

The simplest application is for the case where you have three or more independent samples. The procedure for three or more correlated samples is a bit more complex, although its obvious analogy with the correlated-samples t-test will smooth the path considerably. Both of these relatively simple applications fall under the heading of one-way analysis of variance, so named because they can consider only one independent variable at a time–type of music (A, B, C), loudness of a particular type of music (low, medium, high), type of drug (*X*, *Y*, *Z*), dosage of a particular type of drug (0mg, 5mg, 10mg), and so on.

Considerably more complex, though at the same time far more useful, is the two-way analysis of variance. As its name suggests, this is a procedure that allows you to examine the effects of two independent variables concurrently; for example, the effects of two different drugs, A and B, each at three different dosage levels, 0mg, 5mg, and 10mg. Here the levels of drug A would constitute one independent variable, the levels of drug B would constitute another. This two-way version of the analysis of variance would not only allow you to test the two drugs concurrently; it would also potentially be able to tell you whether the two drugs interact. I introduce this concept with the example of two drugs only because that is a form of interaction you have no doubt already heard about or read about, or perhaps even experienced at first hand, for better or worse. The possibility of two variables interacting, however, is by no means limited to the realm of drugs. There are many precincts of science where the disentangling of interaction effects is one of the main challenges of the enterprise. This is especially so within the domains of the biological and behavioral sciences.

Whether simple or complex, however, all analysis of variance procedures end up with the calculation of one or more F-ratios of the general form described above:

$$F = \frac{MS_{\text{effect}}}{MS_{\text{error}}}$$

where MS_{effect} is a measure pertaining to the particular fact whose significance you wish to assess, and MS_{error} is a variance estimate reflecting the amount of sheer, cussed random variability that is present in the situation. Once again: same song, new verse. It is all cut from the same pattern you have seen building up.

Chapter 6

One-way Analysis for Independent Samples

As most of the logic and procedure for this simplest version of the analysis of variance was developed, the main portion of the present chapter can be fairly brief and to the point. Part 1 covers the essentials of one-way ANOVA for independent samples.

This version of ANOVA applies to the case where you have one independent variable and three or more independent samples of subjects, each sample measured at a different level of the variable. To avoid having to repeat the cumbersome phrase "three or more," we will henceforth refer to the number of independent samples (which is the same as the number of levels of the independent variable) as k. Thus, with three groups of subjects and three levels of the independent variable, k=3; with four groups and four levels, k=4; and so on. We will illustrate the procedure with an example involving k=4.

One of the many complications in the lives of persons who have Alzheimer's disease is that they tend to suffer from frequent periods of intense agitation. Some investigators have theorized that this agitation stems from old aversive conditionings–the conditioned fears and anxieties that a person accumulates over the course of his or her life–with which the Alzheimer's patient is no longer able to cope, on account of severely diminished cognitive capacities.

Against this background, a team of investigators has developed an experimental medication that they believe will substantially decrease the effects of old aversive conditionings. As a preliminary test of the medication, they trained 20 laboratory rats, by standard aversive conditioning procedures, to flee from a certain visual stimulus. These 20 subjects were then randomly and independently sorted into k=4 groups–A, B, C, and D–of 5 subjects each. In the subsequent experimental procedure, the members of each group received via sub-cutaneous injection one or another of four dosage levels of the medication. The members of group A, serving as a control group, received only an inert placebo containing zero units of the medication, while the members of groups B, C, and D received 1 unit, 2 units, and 3 units of the medication, respectively. Five minutes after receiving its injection, each subject was then presented with the aversively conditioned stimulus, and a measure was taken of how hard the subject pulled against a restraining harness in trying to move away from the stimulus. The smaller the pull, the smaller the degree of agitation presumed to be occasioned by the aversive stimulus.

The following table shows the measure of "pull" for each of the 5 subjects in each of the four groups. Also shown are the means of the groups along with M_T, the mean of the total array. This table is only our first pass at the data by way of an overview. A fuller listing of summary statistics will be given in a moment.

A *0 units*	*B* *1 unit*	*C* *2 units*	*D* *3 units*	*Total* *Array*
27.0	22.8	21.9	23.5	
26.2	23.1	23.4	19.6	
28.8	27.7	20.1	23.7	All groups
33.5	27.6	27.8	20.8	combined.
28.8	24.0	19.3	23.9	
M_a=28.86	M_b=25.04	M_c=22.50	M_d=22.30	M_T=24.68

Figure provides the same overview in graphical form. Whichever way you look at it, there clearly are differences among the means of the four groups, and these are consistent with what the investigators would have expected if the medication has the effect they suppose it to have. The greatest mean pull, hence presumably the greatest level of agitation, was found with the group that received only the placebo. For the group that received 1 unit of the medication, the mean pull was smaller; and for the groups that received 2 units and 3 units, it was smaller still.

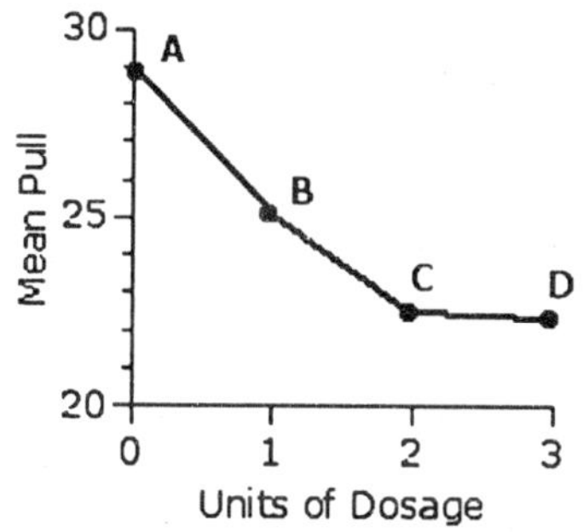

Fig. Dosage Level and Mean Pull

But of course, here as elsewhere, there is always the possibility that the observed "effect" results from nothing more than mere random variability. And until that possibility is rationally assessed, no conclusions can be drawn, one way or the other. The one-way analysis of variance for independent samples performs that assessment by taking the ratio of two quantities

$$F = \frac{MS_{bg}}{MS_{wg}} = \frac{\text{a measure of the aggregate differences among the means of the } k \text{ groups}}{\text{a measure of the amount of random variability that exists inside the } k \text{ groups}}$$

which is then referred to the appropriate sampling distribution of F, as defined by df_{bg} and df_{wg}. For purposes of practical computation, the first step is to calculate the values of X_i and X^2_i for each of the k groups and for all k groups combined. These, in conjunction with the relevant values of N (N_a, N_b, etc.), will then permit the calculation of all other quantities required for the analysis. The following table shows the full array of the preliminary summary statistics.

	Units of Dosage			
0	*1*	*2*	*3*	*Total Array*
27.0	22.8	21.9	23.5	
26.2	23.1	23.4	19.6	
28.8	27.7	20.1	23.7	All groups
33.5	27.6	27.8	20.8	combined.
28.8	24.0	19.3	23.9	
N_a=5	N_b=5	N_c=5	N_d=5	N_T=20
ΣX_{ai}=144.30	ΣX_{bi}=125.20	ΣX_{ci}=112.50	ΣX_{di}=111.50	ΣX_{Ti}=493.50
ΣX^2_{ai}= 4196.57	ΣX^2_{bi}= 3158.50	ΣX^2_{ci}=2576.51	ΣX^2_{di}=2501.95	ΣX^2_{Ti}=12433.53
M_a=28.86	M_b=25.04	M_c=22.50	M_d=22.30	M_T=24.68
SS_a=32.07	SS_b=23.49	SS_c=45.26	SS_d=15.50	SS_T=256.42

As indicated in the table, the raw measure of variability within the entire array of data, with all k groups combined, is SS_T=256.42. This, in turn, is composed of two complementary components: SS_{bg}, which is the raw measure of the aggregate differences among the means of the k groups; and SS_{wg}, which is the raw measure of the variability that exists inside the k groups. With k=4, the latter measure comes out as

$$\begin{aligned} SS_{wg} &= SS_a + SS_b + SS_c + SS_d \\ &= 32.07 + 23.49 + 45.26 + 15.50 \\ &= 116.32. \end{aligned}$$

Assuming that all number-crunching up to this point has been performed without error, you could then calculate the remaining component simply as

$$\begin{aligned} SS_{bg} &= SS_T - SS_{wg} \\ &= 256.42 — 116.32 \\ &= 140.10. \end{aligned}$$

However, it is always good practice to check the accuracy of one's calculations by calculating SS_{bg} also from scratch. The weighted squared deviate for each of the k group means can be calculated as

$$N_g(M_g - M_T)^2$$

where M_g is the mean of a particular group and N_g is the number of values of X_i on which that mean is based. Thus, for each of the four groups

A: $5(28.86–24.68)^2 = 87.36$

B: $5(25.04–24.68)^2 = 0.65$

C: $5(22.50–24.68)^2 = 23.76$

D: $5(22.30–24.68)^2 = 28.32$

$SS_{bg} = 140.09$

140.09 versus 140.10.

Close enough. The slight difference between the two derives from rounding errors in the present calculation. (Each mean value starts out rounded to two decimal places.)

An alternative method for calculating SS_{bg} from scratch would be by way of the following computational formula. Although this rather cumbersome-looking device will not provide as clear an idea of the structure of SS_{bg}, it requires fewer computational steps and is also less susceptible to rounding errors.

$$SS_{bg} = \frac{(\Sigma X_{ai})^2}{N_a} + \frac{(\Sigma X_{bi})^2}{N_b} + \frac{(\Sigma X_{ci})^2}{N_c} + \frac{(\Sigma X_{di})^2}{N_d} - \frac{(\Sigma X_T)^2}{N_T}$$

$$= \frac{(144.3)^2}{5} + \frac{(125.2)^2}{5} + \frac{(112.5)^2}{5} + \frac{(111.5)^2}{5} - \frac{(493.5)^2}{20}$$

$$= 140.10.$$

Here then, in summary, are our two component values of raw variability:

$$SS_{bg} = 140.10 \text{ and } SS_{wg} = 116.32.$$

The next step is to refine them into measures of MS through dividing each by its corresponding number of degrees of freedom. For the between-groups measure, degrees of freedom , except now we replace the phrase "number of groups" with the simple designation *k*.

$$df_{bg} = k-1$$
$$= 4-1 = 3.$$

Similarly for the within-groups measure, except now the groups are not A|B|C but A|B|C|D.

$$df_{wg} = (N_a-1) + (N_b-1) + (N_c-1) + (N_d-1)$$
$$= (5-1) + (5-1) + (5-1) + (5-1) = 16$$

However, once you clearly understand the structure of df_{wg}, its numerical value can be reached more simply through the algebraically equivalent formula

$$df_{wg} = N_T - k$$
$$= 20 - 4 = 16$$

Note that df_T, the number of degrees of freedom for the entire array of data, is

$$df_T = N_T - 1$$
$$df^T = 20 - 1 = 19$$

and that $df_{bg} + df_{wg} = df_T$.

With the values of *SS* and df for between-groups and within-groups, we can then calculate

$$MS_{bg} \frac{SS_{bg}}{df_{bg}} = \frac{140.10}{3} = 46.70$$

and

$$MS_{bg} \frac{SS_{wg}}{df_{wg}} = \frac{116.32}{16} = 7.27.$$

And this, in turn, permits the calculation of the F-ratio as

$$F = \frac{MS_{bg}}{MS_{wg}} = \frac{46.70}{7.27} = 6.42 \text{ with df=3,16.}$$

Figure shows the sampling distribution of F for df=3,16. As indicated, F=3.24 and F=5.29 mark the points beyond which fall 5% and 1%, respectively, of all possible mere-chance outcomes, assuming the null hypothesis to be true. These are the same numbers that appear in the table of critical values of F, the relevant portion of which is shown adjacent to the graph.

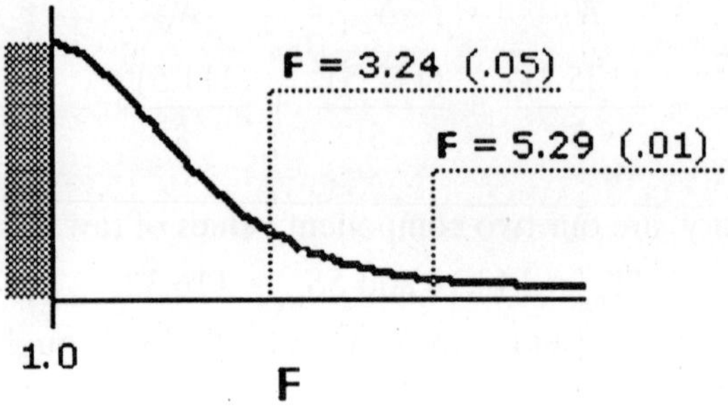

Fig. Sampling Distribution of F for df=3,16

df denominator		df numerator	
	1	2	3
16	4.49	6.23	3.24
8.53	3.63	5.29	

As the observed value of F=6.42 falls to the right of F=5.29, our investigators can regard the aggregate differences among the means of their four samples as significant beyond the .01 level. Recall once again that this term "significant" always has an If/Then logical structure embedded within it, and that the center-point of the structure is always the null hypothesis. For the present example the structure is this: If the null hypothesis were true—if the differences among the means of the four samples were occasioned by nothing more than random variability—then the likelihood of ending up with an F-ratio this large or larger would be less than 1% (P<.01). The investigators can accordingly reject the null hypothesis with a level of confidence somewhat greater than 99%, provisionally concluding that their experimental medication does have the effect they supposed it to have.

It is possible to calculate the probability associated with an F-ratio in more fine-grained detail than is provided by the .05 and .01 critical values conventionally listed in tables of F. Although the details of this calculation are more complex than most users of statistical methods would care to go into, it can be easily performed with the built-in functions of certain computer spreadsheet applications such as Microsoft Excel. For the present example, with F=6.42 and df=3,16, the more fine-grained probability value is P=.005.

ANOVA Summary Tables

When reporting the results of an analysis of variance, it is good practice to present a summary table such as the following. Clearly identifying each component of *SS*, df, and MS, it allows the reader to take in the main details of the analysis at a single glance.

Source	*SS*	*df*	*MS*	*F*	*P*
between groups ("effect")	140.10	3	46.70	6.42	<.01*
within groups ("error")	116.32	16	7.27		
TOTAL	256.42	19			

*As indicated above, the probability value for this analysis could also be reported as P=.005.

ASSUMPTIONS OF THE ONE-WAY ANOVA FOR INDEPENDENT SAMPLES [H]

This particular version of the analysis of variance makes the following assumptions about the data that are being fed into it:

1. That the scale on which the dependent variable is measured has the properties of an equal interval scale;
2. That the *k* samples are independently and randomly drawn from the source population(*s*);
3. That the source population(*s*) can be reasonably supposed to have a normal distribution; and
4. That the *k* samples have approximately equal variances.

You have already encountered the first three of these assumptions in connection with the t-test for independent samples. The fourth assumption will require a bit of explaining. Listed below are sums of squared deviates within each of our 4 samples. Dividing each by N_g—1 (recall that the subscript "g" means "any particular group") will give an estimate of the variance of the population from which the sample comes. These four variance estimates are all approximately equal. As a practical rule of thumb, you can take the phrase "all approximately equal" to entail that the ratio of the largest sample variance to the smallest should not exceed 1.5.

Group	*SS*	N_g-1	*Variance Estimate*
A	32.07	4	8.02
B	23.49	4	5.87
C	45.26	4	11.32
D	15.50	4	3.88

As you can see, this equal-variance assumption is potentially a bit of bad news. The largest of our sample variances (11.32) is nearly three times as great as the smallest (3.88) and nearly twice as great as the next smallest (5.87). This bad news does not apply only to the present example. It very often happens in real-life applications of ANOVA that sample variances do not satisfy the assumption of being "approximately equal."

But now for the good news. The analysis of variance is a very robust test, in the sense that it is relatively unperturbed when the equal-variance assumption is not met. This is especially so when the k samples are all of the same size, as in the present example. Hence, for this or any other version of ANOVA, it is always a good idea to ensure that all samples are of the same size. (When the several samples are of different sizes, the rule of thumb mentioned above remains in force: the ratio of the largest sample variance to the smallest should not exceed 1.5.)

When the samples are of the same size, the analysis of variance is also robust with respect to the assumption that the source populations are normally distributed. So, in brief, the one-way ANOVA for independent samples can be applied to virtually any set of data that will fit into it, providing that all k of the samples are of equal size and that the first two of the assumptions are met:

1. That the scale on which the dependent variable is measured has the properties of an equal interval scale; and
2. That the k samples are independently and randomly drawn from the source population(*s*).

Step-by-Step Computational Procedure

I will show the procedures for the case where the number of groups is k=4. The modifications required for different values of K will be fairly obvious. The steps listed below assume that you have already done the basic number-crunching to get X_i and $X^2{}_i$ for each of the k groups and for all k groups combined.

Step 1. Combining all k groups together, calculate

$$SS_T = \Sigma X^2{}_i \frac{(\Sigma X_i)^2}{N_T}.$$

Step 2. For each of the k groups separately, calculate the sum of squared deviates within the group ("g") as

$$SS_g = \Sigma X^2{}_{gi} \frac{(\Sigma X_{gi})^2}{N_g}.$$

Step 3. Take the sum of the SS_g values across all k groups to get

$$SS_{wg} = SS_a + SS_b + SS_c + SS_d$$

Step 4. Calculate SS_{bg} as

$$SS_{bg} = SS_T - SS_{wg}$$

Step 4a. Check your calculations up to this point by calculating SS_{bg} separately as

$$SS_{bg} = \frac{(\Sigma X_{ai})^2}{N_a} + \frac{(\Sigma X_{bi})^2}{N_b} + \frac{(\Sigma X_{ci})^2}{N_c} + \frac{(\Sigma X_{di})^2}{N_d} - \frac{(\Sigma X_T)^2}{N_T}.$$

Step 5. Calculate the relevant degrees of freedom as

$$\text{df}_T = N_T - 1$$
$$\text{df}_{bg} = k - 1$$
$$\text{df}_{wg} = N_T - k.$$

Step 6. Calculate the relevant mean-square values as

$$MS_{bg} = \frac{SS_{bg}}{\text{df}_{bg}}$$

and

$$MS_{wg} = \frac{SS_{wg}}{\text{df}_{wg}}$$

Step 7. Calculate F as

$$F = \frac{MS_{bg}}{MS_{wg}}.$$

Step 8. Refer the calculated value of F to the table of critical values of F, with the appropriate pair of numerator/denominator degrees of freedom.

M_a=28.86	M_b=25.04	M_c=22.50	M_d=22.30		
Source	*SS*	*df*	*MS*	*F*	*P*
between groups ("effect")	140.10	3	46.70	6.42	<.01
within groups ("error")	116.32	16	7.27		
TOTAL	256.42	19			

POST-ANOVA COMPARISONS: THE TUKEY HSD TEST

A significant F-ratio tells you only that the aggregate difference among the means of the several samples is significantly greater than zero. It does not tell you whether any particular sample mean significantly differs from any particular other. For some research purposes this might be entirely sufficient. Since the investigators in the present example regard their experiment with laboratory rats as only a first step in testing the medication, we can imagine they might be content simply with the global conclusion suggested by the graph of their data: namely, that the curve of "pull" (presumably a reflection of the effect of the medication) slopes downward from A to B to C, then levels off between C and D.

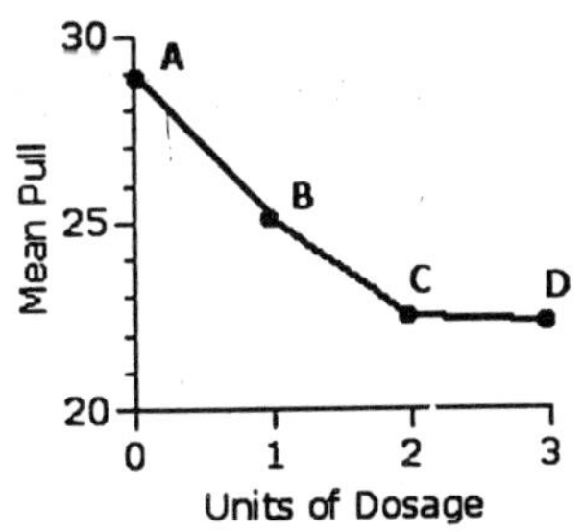

There are, however, many situations in which the investigator might wish to determine specifically whether M_a significantly differs from M_b, or M_b from M_c, and so on. This comparison of sample means two at a time cannot be done by way of simple t-tests, because it potentially involves 3 or more comparisons, depending on the number of samples, *k*, involved in the original analysis. With *k*=3, there would be 3 potential comparisons:

A·B, A·C, B·C

With *k*=4, as in the present case, there would be 6 potential comparisons:

A·B, A·C, A·D, B·C, B·D, C·D

With *k*=5, there would be 10:

A·B, A·C, A·D, A·E, B·C, B·D, B·E, C·D, C·E, D·E

and so forth. The performance of any one or several of these pair-wise comparisons requires a procedure that takes the full range of potential comparisons into account.

The subject of post-ANOVA comparisons is a rather complex one, and most of it lies beyond the scope of an introductory presentation. I will describe here only one of the available procedures, which I think will serve the beginning student well enough for most practical purposes. It goes under the name of the Tukey HSD test, the "HSD" being an acronym for the forthright phrase "honestly significant difference."

The Tukey test revolves around a measure known as the Studentized range statistic, which we will abbreviate as Q. For any particular pair of means among the *k* groups, let us designate the larger and smaller as M_L and M_S, respectively. The Studentized range statistic can then be calculated for any particular pair as

$$Q = \text{sqrt}\frac{M_L - M_S}{[MS_{wg} / N_{\text{p/s}}]}$$

where MS_{wg} is the within-groups MS obtained in the original analysis and $N_{\text{p/s}}$ is the number of values of X_i per sample ("*p/s*"=per sample). For the present example, MS_{wg}=7.27 and $N_{\text{p/s}}$=5.

If the *k* samples are of different sizes, the value of $N_{\text{p/s}}$ can be set as equal to the harmonic mean of the sample sizes. For *k*=3 this would be

$$N_{p/s} = \frac{3}{(1/N_a) + (1/N_b) + (1/N_c)}$$

For *k*=4,

$$N_{p/s} = \frac{3}{(1/N_a) + (1/N_b) + (1/N_c) + (1/N_d)}$$

And so on for *k*=5, *k*=6, etc.

If the *k* samples are of different sizes, the value of $N_{p/s}$ can be set as equal to the harmonic mean of the sample sizes. For *k*=3 this would be

As it happens, you do not really need to worry about calculating Q, because there is a simpler way of applying the Tukey test. However, I will pause to calculate one instance of it, just to give you an idea of what it looks like. For the present example, M_a=28.86, M_b=25.04, MS_{wg}=7.27, and $N_{p/s}$=5. Thus, for the comparison between M_a and M_b the Studentized range statistic would be

$$Q = \frac{28.86 - 25.04}{\text{sqrt}[7.27 / 5]} = 3.16.$$

And similarly for any of the other pair-wise comparisons one might wish to make among the means of this particular set of 4 groups.

In any particular case, this Studentized range statistic belongs to a sampling distribution defined by two parameters: the first is *k*, the number of samples in the original analysis; and the second is df_{wg}, the number of degrees of freedom associated with the denominator of the F-ratio in the original analysis. Within any particular one of these sampling distributions you can define the value of Q required for significance at any particular level. The following calculator will fetch the critical values of Q at the .05 and .01 levels of significance for any value of *K* between 3 and 10, inclusive, and for various values of df_{wg}. To proceed, enter *K* and df_{wg} in the designated cells, then click «Calculate». For the present example, with *k*=4 and df_{wg}=16, you will end up with $Q_{.05}$=4.05 and $Q_{.01}$=5.2.

The Tukey HSD test then uses these critical values of Q to determine how large the difference between the means of any two particular groups must be in order to be regarded as significant. The other participants in this determination, MS_{wg} and $N_{p/s}$, are the same items you saw in the earlier formula for Q. The following two "HSD" formulas are simply algebraic jugglings of the original formula, in which the value of Q is set to one or the other of the two critical values, $Q_{.05}$ and $Q_{.01}$.

For the .05 level:

$$\begin{aligned} \text{HSD}_{.05} &= Q_{.05} \times \text{sqrt}\left[\frac{MS_{wg}}{N_{p/s}}\right] \\ &= 4.05 \times \text{sqrt}\left[\frac{7.27}{5}\right] \\ &= 4.88. \end{aligned}$$

That is: In order to be considered significant at or beyond the .05 level, the difference between any two particular group means (larger—smaller) must be equal to or greater than 4.88.

And for the .01 level:

$$\text{HSD}_{.01} = Q_{.01} \times \text{sqrt}\left[\frac{MS_{wg}}{N_{p/s}}\right]$$

$$= 5.2 \times \text{sqrt}\left[\frac{7.27}{5}\right]$$

$$= 6.27.$$

That is: In order to be considered significant at or beyond the .01 level, the difference between any two particular group means (larger—smaller) must be equal to or greater than 6.27.

The blue entries in the following table show the differences between each pair of group means in our example. As you can see, two of the comparisons (A·C and A·D) are significant beyond the .01 level, while all the others fail to achieve significance even at the basic .05 level.

A·B	$M_a = 28.86$ $M_b = 25.04$	3.82	$HSD_{.05} = 4.88$ $HSD_{.01} = 6.27$
A·C	$M_a = 28.86$ $M_c = 22.50$	6.36	
A·D	$M_a = 28.86$ $M_d = 22.30$	6.56	
B·C	$M_b = 25.04$ $M_c = 22.50$	2.54	
B·D	$M_b = 25.04$ $M_d = 22.30$	2.74	
C·D	$M_c = 22.50$ $M_d = 22.30$	0.20	

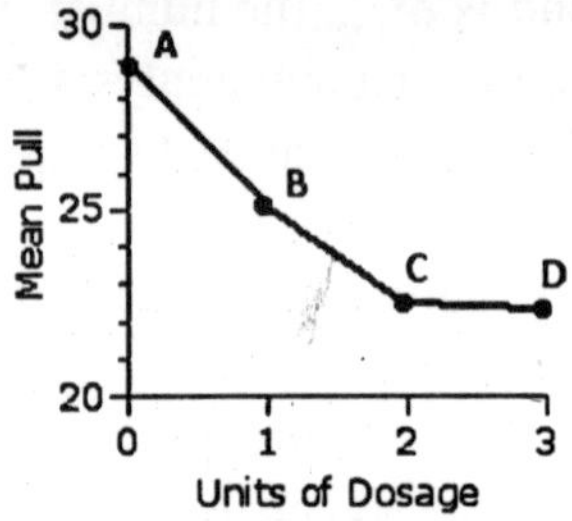

Our investigators would therefore be able to conclude that 2 units and 3 units of the experimental medication each produced significantly lower mean levels of "pull" than was found in the zero-unit control group. They would not be able to conclude that the effect of 2 units or 3 units was significantly greater than the effect of 1 unit, nor that the mean "pull" of the 1-unit group was significantly smaller that that of the zero-unit control group. Please note carefully, however, that failing to find a significant difference between M_a and M_b would not entail that 1 unit of the medication has no effect at all. It merely means that the Tukey HSD test does not detect a significant difference between the two in this particular situation. If the investigators had found approximately the same array of group means with samples of twice the size (10 per group, rather than 5), they would very likely have found all of the pair-wise comparisons to be significant, except for the one between M_c and M_d.

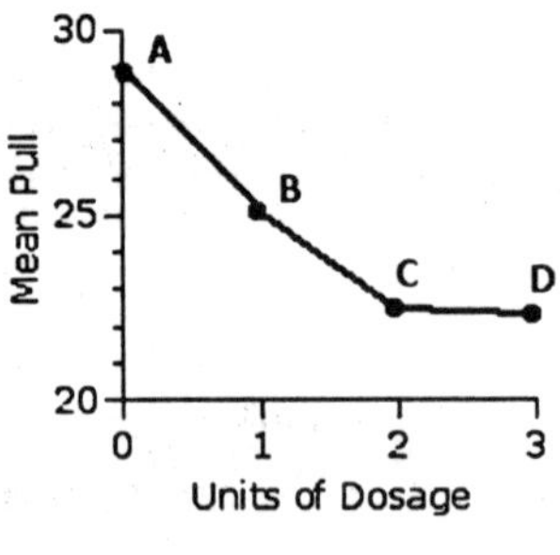

ONE-WAY ANOVA AND CORRELATION [H]

Here yet again is Figure, which you have now seen several times over. It will be fairly obvious to the naked eye that the two variables, dosage and pull, are correlated in the sense that variations in the one are associated with variations in the other. It will be equally obvious that the relationship is not of the rectilinear (straight-line) sort. It is better described by a curved line, hence "curvilinear." Within the context of a one-way analysis of variance for independent samples, a useful measure of the strength of a curvilinear relationship between the independent and dependent variable is given by a quantity known as as eta-square ("eta" to rhyme with "beta"), which is simply the ratio of SS_{bg} to SS_T. For the medication experiment it comes out as

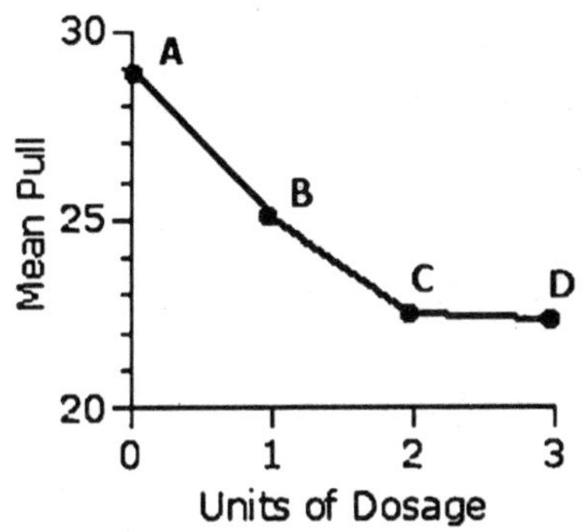

$$\text{eta}^2 = \frac{SS_{bg}}{SS_T} = \frac{140.10}{256.42} = 0.55.$$

The essential meaning of "eta^2=0.55" is this: Of all the variability that exists within the dependent variable "pull," 55% is associated with variability in the independent variable "dosage level." A moment's reflection of what we observed will remind you that this is also the essential meaning of the coefficient of determination, r^2. The only intrinsic difference between the two is that r^2 can measure the strength of a correlation only insofar as it is linear (can be described by a straight line), while eta^2 provides a measure of the strength of correlation irrespective of whether it is linear or curvilinear. If the relationship is linear–fully describable by a straight line–then the values of r^2 and eta^2 will be the same. In the degree that a curved line describes the relationship better than a straight line, then eta^2 will be greater than r^2.

This point is illustrated by the two panels of Figure, which show the data for the N_T=20 individual subjects of the experiment laid out in the form of a scatter plot. Applying the procedures of linear correlation to this set of bivariate data will yield the straight regression line shown in the panel on the left, along with r^2=0.48. The panel on the right shows the same data with a curvilinear line of best fit, corresponding to our calculated value of eta^2=0.55.

Please note, however, that it is meaningful to speak of eta^2 as analogous to r^2 only when the levels of the independent variable are quantitative and linear, as in the present example where zero units, 1 unit, 2 units, and 3 units of the medication represent points along an equal-interval scale. If the levels of the independent variable are only categorical (several different types of medication, several different types of music, etc.), the meaning of eta^2 reverts back to a version of the more general statement given above: Of all the variability that exists within the dependent variable, such-and-such percent is associated with the differences among the levels of the independent variable.

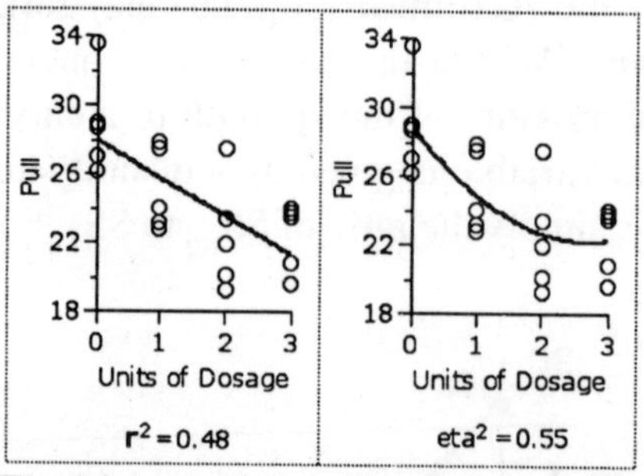

Fig. Linear and Curvilinear Correlation

THE KRUSKAL-WALLIS TEST [MH]

As a reminder, the assumptions of the one-way ANOVA for independent samples are

1. That the scale on which the dependent variable is measured has the properties of an equal interval scale;$_T$
2. That the k samples are independently and randomly drawn from the source population(s);$_T$
3. That the source population(s) can be reasonably supposed to have a normal distribution; and$_T$
4. That the k samples have approximately equal variances.

We need not worry very much about the first, third, and fourth of these assumptions when the samples are all the same size. For in that case the analysis of variance is quite robust, by which we mean relatively unperturbed by the violation of its assumptions. But of course, the other side of the coin is that when the samples are not all the same size, we do need to worry. In this case, should one or more of assumptions 1, 3, and 4 fail to be met, an appropriate non-parametric alternative to the one-way independent-samples ANOVA can be found in the Kruskal-Wallis Test.

To assess the effects of expectation on the perception of aesthetic quality, an investigator randomly sorts 24 amateur wine aficionados into three groups, A, B, and C, of 8 subjects each. Each subject is scheduled for an individual interview. Unfortunately, one of the subjects of group B and two of group C fail to show up for their interviews, so the investigator must make do with samples of unequal size: n_a=8, n_b=7, and n_c=6, for a total of N=21. The subjects who do show up for their interviews are each asked to rate the overall quality of each of three wines on a 10-point scale, with "1" standing at the bottom of the scale and "10" at the top.

As it happens, the three wines are the same for all subjects. The only difference is in the texture of the interview, which is designed to induce a relatively high expectation of quality in the members of group A; a relatively low expectation in the members of group C; and a merely neutral state, tending in neither the one direction nor the other, for the members of group B. At the end of the study, each subject's ratings are averaged across all three wines, and this average is then taken as the raw measure for that particular subject. The following table shows these measures for each subject in each of the three groups.

	Group		
	A	*B*	*C*
	6.4	2.5	1.3
	6.8	3.7	4.1
	7.2	4.9	4.9
	8.3	5.4	5.2
	8.4	5.9	5.5
	9.1	8.1	8.2
	9.4	8.2	
	9.7		
mean	8.2	5.5	4.9

MECHANICS [H]

The preliminaries of the Kruskal-Wallis test are much the same as those of the Mann-Whitney test described in Subchapter 11a. We begin by assembling the measures from all k samples into a single set of size N. These assembled measures are rank-ordered from lowest (rank#1) to highest (rank#N), with tied ranks included where appropriate; and the resulting ranks are then returned to the sample, A, B, or C, to which they belong and substituted for the raw measures that gave rise to them. Thus, the raw measures that appear in the following table on the left are replaced by their respective ranks, as shown in the table on the right.

Raw Measures			*Ranked Measures*			
A	*B*	*C*	*A*	*B*	*C*	
6.4	2.5	1.3	11	2	1	
6.8	3.7	4.1	12	3	4	
7.2	4.9	4.9	13	5.5	5.5	
8.3	5.4	5.2	17	8	7	
8.4	5.9	5.5	18	10	9	
9.1	8.1	8.2	19	14		
9.4	8.2		20	15.5		
9.7			21	15.5		A, B, C Combined
sum of ranks			131	58	42	231
average of ranks			16.4	8.3	7.0	11

With the Kruskal-Wallis test, however, we take account not only of the sums of the ranks within each group, but also of the averages. Thus the following items of symbolic notation:

T_A = the sum of the n_a ranks in group A

M_A = the mean of the n_a ranks in group A

T_B = the sum of the n_b ranks in group B

M_B = the mean of the n_b ranks in group B

T_C = the sum of the n_c ranks in group C

M_C = the mean of the n_c ranks in group C

T_{all} = the sum of the N ranks in all groups combined

M_{all} = the mean of the N ranks in all groups combined

LOGIC AND PROCEDURE

The Measure of Aggregate Group Differences

You will sometimes find the Kruskal-Wallis test described as an "analysis of variance by ranks." Although it is not really an analysis of variance at all, it does bear a certain resemblance to ANOVA up to a point. In both procedures, the first part of the task is to find a measure of the aggregate degree to which the group means differ. With ANOVA that measure is found in the quantity known as SS_{bg}, which is the between-groups sum of squared deviates. The same is true with the Kruskal-Wallis test, except that here the group means are based on ranks rather than on the raw measures. As a reminder that we are now dealing with ranks, we will symbolize this new version of the between-groups sum of squared deviates as $SS_{bg(R)}$. The following table summarizes the mean ranks for the present example. Also included are the sums and the counts (n_a, n_b, n_c, and N) on which these means are based.

	A	B	C	All
counts	8	7	6	21
sums	131	58	42	231
means	16.4	8.3	7.0	11.0

The squared deviate for any particular group mean is equal to the squared difference between that group mean and the mean of the overall array of data, multiplied by the number of observations on which the group mean is based. Thus, for each of our current three groups

$$\text{A: } -8(16.4-11.0)^2 = 233.3$$
$$\text{B: } -7(8.3-11.0)^2 = 051.0$$
$$\text{C: } -6(7.0-11.0)^2 = 096.0$$
$$SS_{bg(R)} = 380.3.$$

On analogy with the formulaic structures for SS_{bg} developed in Chapters 13 and 14, we can write the conceptual formula for $SS_{bg(R)}$ as

$$SS_{bg(R)} = ([n_g(M_g - M_{all})^2]$$

Here as well, the subscript "g" means "any particular group."

and the computational formula as

$$SS_{bg(R)} = \sum \frac{(T_g)^2}{n_g} - \frac{(T_{\text{all}})^2}{N_a}.$$

With k=3 samples, this latter structure would be equivalent to

$$SS_{bg(R)} = \frac{(T_A)^2}{n_a} + \frac{(T_B)^2}{n_b} + \frac{(T_C)^2}{n_c} - \frac{(T_{all})^2}{N_a}.$$

For k=4 it would be

$$SS_{bg(R)} = \frac{(T_A)^2}{n_a} + \frac{(T_B)^2}{n_b} + \frac{(T_C)^2}{n_c} + \frac{(T_D)^2}{n_d} - \frac{(T_{all})^2}{N_a}.$$

And so forth for other values of k.

Here, in any event, is how it would work out for the present example. The discrepancy between what we get now and what we got a moment ago (380.3) is due to rounding error in the earlier calculation. As usual, it is the computational formula that is the less susceptible to rounding error, hence the more reliable.

$$SS_{bg(R)} = \frac{(131)^2}{8} + \frac{(58)^2}{7} + \frac{(42)^2}{6} - \frac{(231)^2}{21} = 378.7.$$

The Null-Hypothesis Value of $SS_{bg(R)}$

The null hypothesis in this or any comparable situation involving several independent samples of ranked data is that the mean ranks of the k groups will not substantially differ. On this account, you might suppose that the null-hypothesis value of $SS_{bg(R)}$, the aggregate measure of group differences, would be simply zero. A moment's reflection, however, will show why this cannot be so.

A	*B*	*C*
x	x	x
x	x	x

Consider the very simple case where there are 3 groups, each containing 2 observations. By way of analogy, imagine you had six small cards representing the ranks "1," "2," "3," "4," "5," and "6." If you were to sort these cards into every possible combination of two ranks per group, you would find the total number of possible combinations to be

$$\frac{N!}{n_a!\, n_b!\, n_c!} = \frac{6!}{2!\, 2!\, 2!} = 90.$$

And the values of $SS_{bg(R)}$ produced by these 90 combinations would constitute the sampling distribution of $SS_{bg(R)}$ for this particular case. Of these 90 possible combinations, a few (6) would yield values of $SS_{bg(R)}$ equal to exactly zero. All the rest would produce values greater than zero. (It is mathematically impossible to have a sum of squared deviates less than zero.) Accordingly, the mean of this sampling distribution—the value that observed instances of

$SS_{bg(R)}$ will tend to approximate if the null hypothesis is true—is not zero, but something greater than zero.

In any particular case of this sort, the mean of the sampling distribution of $SS_{bg(R)}$ is given by the formula

$$(k-1) \times \frac{N(N+1)}{12}$$

which for the simple case just examined works out as

$$(3-1) \times \frac{6(6+1)}{12} = 7.0.$$

For our main example, we therefore know that the observed value of $SS_{bg(R)}$=378.7 belongs to a sampling distribution whose mean is equal to

$$(3-1)\times\frac{21(21+1)}{12} = 77.0.$$

All that now remains is to figure out how to turn this fact into a rigorous assessment of probability.

The Kruskal-Wallis Statistic: H

In case you have been girding yourself for some heavy slogging of the sort encountered with the Mann-Whitney test, you can now relax, for the rest of the journey is quite an easy one. The Kruskal-Wallis procedure concludes by defining a ratio symbolized by the letter H, whose numerator is the observed value of $SS_{bg(R)}$ and whose denominator includes a portion of the above formula for the mean of the sampling distribution of $SS_{bg(R)}$. Note that most textbooks give a very different-looking formula for the calculation of H – a rather impenetrable structure to which we will return in a moment. This first version affords a much clearer sense of the underlying concepts.

$$H = \frac{SS_{bg(R)}}{N(N+1)/12}.$$

And now for the denouement. When each of the k samples includes at least 5 observations (that is, when n_a, n_b, n_c, etc., are all equal to or greater than 5), the sampling distribution of H is a very close approximation of the chi-square distribution for df=k–1. It is actually a fairly close approximation even when one or more of the samples includes as few as 3 observations.

For our present example, we can therefore calculate the value of H as

$$H = \frac{SS_{bg(R)}}{N(N+1)/12} = \frac{378.7}{21(21+1)/12} = 9.84.$$

And then, treating this result as though it were a value of chi-square, we can refer it to the sampling distribution of chi-square with df=3–1=2. The following graph, will remind you of the outlines of this particular chi-square distribution. In brief: by the Kruskal-Wallis test, the observed aggregate difference among the three samples is significant a bit beyond the .01 level.

Theoretical Sampling Distribution of Chi-Square for df=2

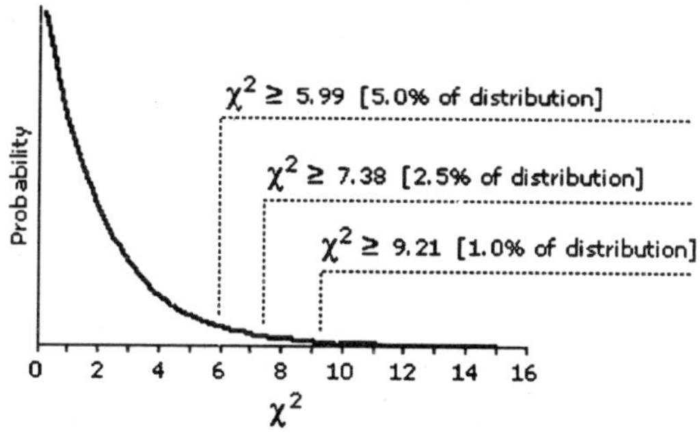

An Alternative Formula for the Calculation of H

I noted a moment ago that textbook accounts of the Kruskal-Wallis test usually give a different version of the formula for H. If you are a beginning student calculating H by hand, I would recommend using the version given above, as it gives you a clearer idea of just what H is measuring. Once you get the hang of things, however, you might find this alternative computational formula a bit more convenient.

$$H = \frac{12}{N(N+1)}\left(\sum \frac{(T_g)^2}{n_g}\right) - 3(N+1).$$

In any event, as you can see below, this version yields exactly the same result as the other.

$$H = \frac{12}{21(21+1)}\left(\frac{(131)^2}{8} + \frac{(58)^2}{7} + \frac{(42)^2}{6}\right) - 3(21+1)$$
$$= 9.84.$$

Chapter 7

One-way Analysis for Corelated Samples

You can think of this version of the analysis of variance as an extension of the correlated-samples t-test. The most conspicuous similarity between the two is in the way the data are arrayed. In the correlated-samples t-test we typically have a certain number of subjects, each measured under two conditions, A and B. Or alternatively, we have a certain number of matched pairs of subjects, with one member of the pair measured under condition A and the other measured under condition B.

It is the same structure with the correlated-samples ANOVA, except that now the number of conditions is three or more: A|B|C, A|B|C|D, and so forth. When the analysis involves each subject being measured under each of the *k* conditions, it is sometimes spoken of as a repeated measures or within subjects design. When it involves subjects matched in sets of three for *k*=3, four for *k*=4, and so on, with the subjects in each matched set randomly assigned to one or another of the *k* conditions, it is described as a randomized blocks design. Thus, for *k*=3:

Repeated Measures

Subject	*A*	*B*	*C*	
1	subj1 under condition A	subj1 under condition B	subj1 under condition C	Each row represents one subject measured under each of *k* conditions.
2	subj2 under condition A	subj1 under condition B	subj2 under condition C	
3	subj3 under condition A	subj3 under condition B	subj3 under condition C	
And so on.				

Randomized Blocks

Block	*A*	*B*	*C*	
1	subj1a under condition A	subj1b under condition B	subj1c under condition C	Each row includes *k* matched
2	subj2a under	subj2b under	subj2c under	subjects, each measured

	condition A	condition B	condition C	under one or another
3	subj3a under	subj3b under	subj3c under	of the *k* conditions
	condition A	condition B	condition C-	

And so on.

In both versions, repeated measures and randomized blocks, the utility of the correlated-samples ANOVA is the same as for the correlated-samples t-test: it is highly effective in removing the extraneous variability that derives from pre-existing individual differences. In some cases individual differences might be the very essence of the phenomena that are of interest. But there are also many situations where they are merely irrelevant clutter.

Up to a point, the logic and procedure of the correlated-samples ANOVA are the same as for the independent-samples version described in Chapters 13 and 14. In the independent-samples ANOVA, SS_T is analyzed into two complementary components, SS_{bg} and SS_{wg}. Each of the latter, divided by its respective value of df, then yields a value of MS; and these, in turn, yield the F-ratio. The numerator of the ratio, MS_{bg}, reflects the aggregate differences among the means of the *k* groups of measures; and the denominator, MS_{wg}, reflects the random variability, commonly described as "error," that exists inside the *k* groups.

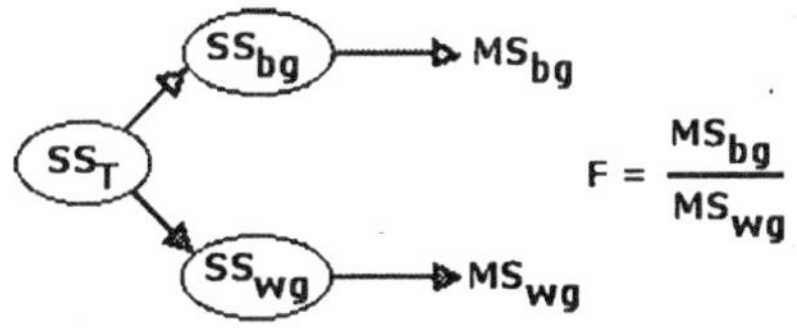

In most real-life situations, a certain amount of the variability that exists inside the *k* groups will reflect pre-existing individual differences among the subjects. It is possible that the prior aversive conditioning had been more effective in some subjects than in others; or that some subjects were more prone to agitation than others; or simply that some were stronger than others and could therefore pull harder. Each of these sources of variability would be entirely extraneous to the question the investigators were aiming to answer. Of course, one way to avoid the extraneous clutter would be to ensure at the outset that all subjects are equally well conditioned, equally prone to agitation, equally strong, and so forth. But that would surely be more easily said than done. In the correlated-samples ANOVA, the extraneous clutter is not avoided. It is faced head-on, identified, and removed. Whether by repeated measures or randomized blocks, the correlated-samples design allows us to identify the portion of SS_{wg} that is attributable to pre-existing individual differences. This portion, designated as SS_{subjects}, is dropped from the analysis; and the portion that remains, SS_{error}, is then used as the measure of sheer, cussed random variability.

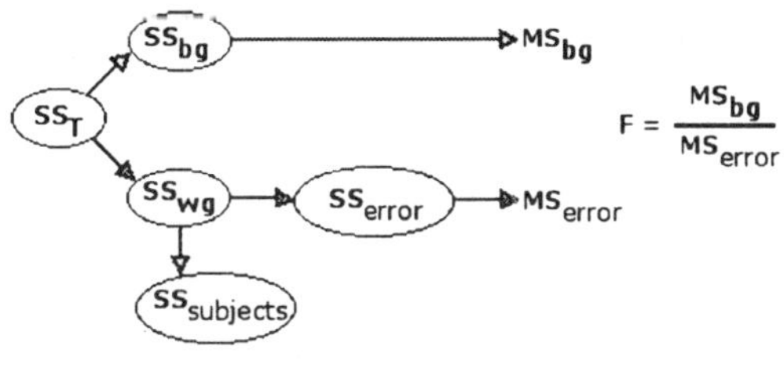

To illustrate the procedures we will take an another example involving sensory-motor coordination. The following picture and diagram represent a device known as a rotary pursuit apparatus, often used in psychological research as an instrument for measuring sensory-motor coordination. On its top is a turntable that can be set to rotate at pre-selected speeds. Toward the edge of the turntable is a metal disk (a) that revolves around the center of the turntable as the latter rotates. Connected to the apparatus is a hand-held stylus (B) with a metal tip. The subject's task is to pursue the revolving disk with the stylus, keeping the stylus in contact with the disk as much as possible. When the two are in contact an electrical circuit is closed, thus providing a record of how well the subject does on the task.

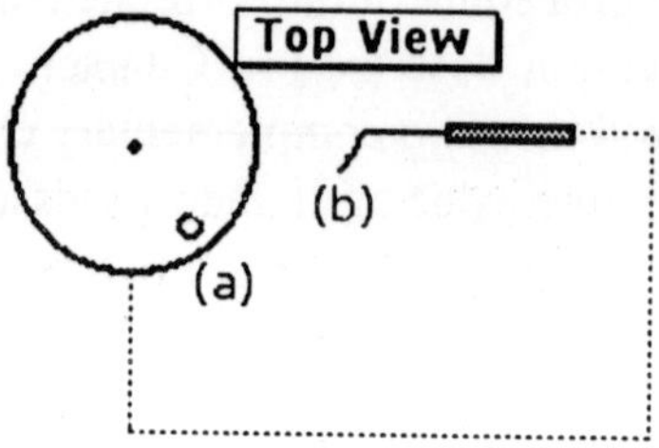

An investigator is interested in assessing the effects of rhythmic auditory stimulation on performance of the rotary pursuit task. To this end, she has each of 18 randomly selected human subjects perform the task under each of three conditions. In all conditions, the turntable rotates counter-clockwise at a constant rate of one rotation per second. In condition A there is no auditory stimulation other than the sound normally made by the apparatus. In the other two conditions the subject hears a periodic clicking sound. In condition B, the click occurs twice per second; and in condition C, six times per second. To obviate the possibility of sequence effects, three of the subjects perform the task in the sequence ABC; three perform it in the sequence ACB; and so on for all the other of the six possible sequences: BAC, BCA, CAB, and CBA.

The following table shows the measures of how well each of the 18 subjects performed under each of the k=3 conditions. Also shown are the means of the three groups of measures and the mean performance of each individual subject across the three conditions. To emphasize the substantial individual differences that occur among the subjects, they are listed in the order of their respective individual mean levels of performance, from highest to lowest. Immediately below the table is Figure, showing the pattern of group means, A|B|C, in graphic form.

	Conditions ("cps"=clicks per second)			
Subjects	*A [0cps]*	*B [2cps]*	*C [6cps]*	*Subject Means*
1	35	39	32	35.3
2	32	35	31	32.7
3	33	32	28	31.0
4	32	32	29	31.0
5	31	33	26	30.0

6	29	30	29	29.3
7	29	31	27	29.0
8	27	29	27	27.7
9	27	31	24	27.3
10	28	27	24	26.3
11	27	27	23	25.7
12	27	26	23	25.3
13	24	29	19	24.0
14	24	25	19	22.7
15	17	16	18	17.0
16	17	15	17	16.3
17	14	15	12	13.7
18	13	13	13	13.0
Group Means	25.9	26.9	23.4	

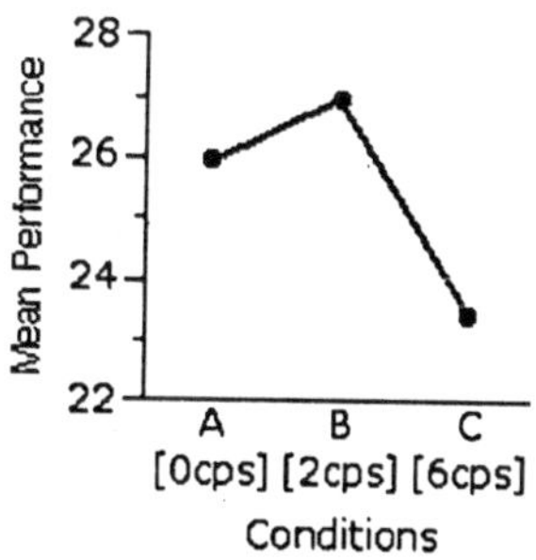

Fig. 15.1. Mean Performance in Conditions A, B, and C

Please keep in mind that the data in this example are completely imaginary. I do not know whether an actual experiment of this sort would produce a pattern of group means resembling the one shown here..

So here we are again: on the one hand, ostensible differences among the group means; and on the other, the ever-present possibility that the observed "effect" results from nothing more than mere random variability.

PARALLELS WITH THE INDEPENDENT-SAMPLES ANOVA

Basic preliminary number-crunching on the data in columns A, B, and C of the above table yields the following values of ΣX_i and ΣX^2_i for the three groups and for total array of data.

A	*B*	*C*	*All groups*
[0cps]	[2cps]	[6cps]	combined
N_a=18	N_b=18	N_c=18	N_T=54
ΣX_{ai}=466	ΣX_{bi}=485	ΣX_{ci}=421	ΣX_{Ti}=1372
ΣX^2_{ai}= 2800	ΣX^2_{bi}=14021	ΣX^2_{ci}=10443	ΣX^2_{Ti}=37264

These in turn allow for the calculation of the following sums of squared deviates.

A	*B*	*C*	*All groups*
[0cps]	[2cps]	[6cps]	combined
SS_a=735.8	SS_b=952.9	SS_c=596.3	SS_T=2405.0

As in the independent-samples ANOVA, the within-groups *SS* is the sum of the *k* separate values of SS_g (recall that the subscript "g" means "any particular group"). Thus

$$SS_{wg} = SS_a + SS_b + SS_c$$
$$= 735.8 + 952.9 + 596.3$$
$$= 2285.0.$$

Similarly, the between-groups *SS* can then be calculated as

$$SS_{bg} = SS_T - SS_{wg}$$
$$= 2405.0 - 2285.0$$
$$= 120.0.$$

Here again it is a good idea to check the accuracy of one's calculations up to this point by also fetching SS_{bg} through the computational formula

$$SS_{bg} = \frac{(\Sigma X_{ai})^2}{N_a} + \frac{(\Sigma X_{bi})^2}{N_b} + \frac{(\Sigma X_{ci})^2}{N_c} - \frac{(\Sigma X_{Ti})^2}{N_T}$$
$$= \frac{(466)^2}{18} + \frac{(485)^2}{18} + \frac{(421)^2}{18} - \frac{(1372)^2}{54}$$
$$= 120.0.$$

The aggregate differences (SS_{bg}) among the means of the 3 groups are rather tiny in comparison with the variability (SS_{wg}) that occurs inside the groups. If you were performing the analysis as an independent-samples ANOVA, this would not be great news unless you were hoping to find a non-significant result. For that huge value of SS_{wg} would also give you a large value of MS_{wg}, hence a rather puny value of F.

SS_T =2405.0

SS_{bg}=120.0

SS_{wg}=2285.0

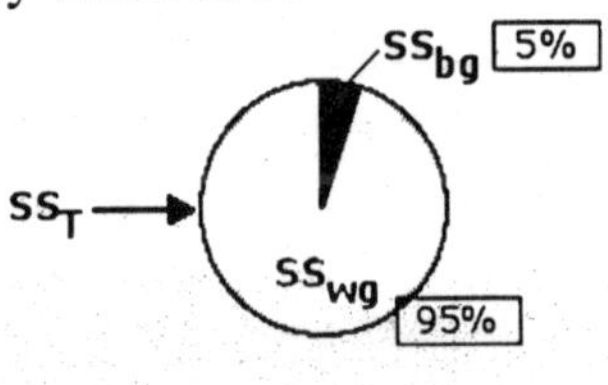

As already mentioned, the virtue of the correlated-samples procedure is that it takes the analysis a step further. It has arranged things in advance so that it can identify the portion of SS_{wg} that derives from pre-existing individual differences. And once identified, this portion—in the present case, a very large portion—can be removed.

From the example:

	A [0cps]	*B* [2cps]	*C* [6cps]	*All groups combined*
	N_a=18	N_b=18	N_c=18	N_T=54
	ΣX_{ai}=466	ΣX_{bi}=485	ΣX_{ci}=421	ΣX_{Ti}=1372
	ΣX^2_{ai}= 12800	ΣX^2_{bi}=14021	ΣX^2_{ci}=10443	ΣX^2_{Ti}=37264
	M_a=25.9	M_b=26.9	M_c=23.4	M_T=25.4
	SS_a=735.8	SS_b=952.9	SS_c=596.3	SS_T=2405.0

SS_T = 2405.0
$SS_{wg} = {}^T SS_a + SS_b + SS_c$ = 735.8+952.9+596.3 = 2285.0
$SS_{bg} = {}^T SS_T - SS_{wg}$ = 2405.0—2285.0 = 120.0

IDENTIFICATION AND REMOVAL OF SS_{SUBJ}

The first five columns of the following table are a somewhat streamlined version of the data table. The means for the individual subjects are now highlighted in blue, and next to each subject's mean, in a new column, is a calculation whose general form you will surely recognize. The difference between each subject's mean and the mean of the total array of data is taken as a deviate:

$$\text{deviate} = M_{\text{subj*}} - M_T$$

We will use the subscript "subj*" to mean "any particular subject." The first subject would be subj1; the second, subj2; and so on.

The square of that difference is accordingly a squared deviate:

$$\text{squared deviate} = (M_{\text{subj*}} - M_T)^2$$

And that squared deviate is then weighted by k=3, which is the number of measures (X_{ai}, X_{bi}, and X_{ci}) on which the subject's mean is based.

$$\text{weighted squared deviate} = k(M_{\text{subj*}} - M_T)^2$$

It is the same general structure to calculate the value of SS_{bg} from scratch, to measure the aggregate differences among the k groups. Now what we come out with is SS_{subj}, which is a measure of the aggregate differences among the 18 individual subjects.

Subjects	*A*	*B*	*C*	*Subject Means*	
^{2}1	^{2}35	^{2}39	^{2}32	^{2}35.3	$3(35.3–25.4)^2$=294
^{2}2	^{2}32	^{2}35	^{2}31	^{2}32.7	$3(32.7–25.4)^2$=159.9
^{2}3	^{2}33	^{2}32	^{2}28	^{2}31.0	$3(31.0–25.4)^2$=94.1
^{2}4	^{2}32	^{2}32	^{2}29	^{2}31.0	$3(31.0–25.4)^2$=94.1
^{2}5	^{2}31	^{2}33	^{2}26	^{2}30.0	$3(30.0–25.4)^2$=63.5
^{2}6	^{2}29	^{2}30	^{2}29	^{2}29.3	$3(29.3–25.4)^2$=45.6
^{2}7	^{2}29	^{2}31	^{2}27	^{2}29.0	$3(29.0–25.4)^2$=38.9
^{2}8	^{2}27	^{2}29	^{2}27	^{2}27.7	$3(27.7–25.4)^2$=15.9
^{2}9	^{2}27	^{2}31	^{2}24	^{2}27.3	$3(27.3–25.4)^2$=10.8

^{2}10	^{2}28	^{2}27	^{2}24	^{2}26.3	$3(26.3–25.4)^2=2.4$
^{2}11	^{2}27	^{2}27	^{2}23	^{2}25.7	$3(25.7–25.4)^2=0.3$
^{2}12	^{2}27	^{2}26	^{2}23	^{2}25.3	$3(25.3–25.4)^2=0$
^{2}13	^{2}24	^{2}29	^{2}19	^{2}24.0	$3(24.0–25.4)^2=5.9$
^{2}14	^{2}24	^{2}25	^{2}19	^{2}22.7	$3(22.7–25.4)^2=21.9$
^{2}15	^{2}17	^{2}16	^{2}18	^{2}17.0	$3(17.0–25.4)^2=211.7$
^{2}16	^{2}17	^{2}15	^{2}17	^{2}16.3	$3(16.3–25.4)^2=248.4$
^{2}17	^{2}14	^{2}15	^{2}12	^{2}13.7	$3(13.7–25.4)^2=410.7$
^{2}18	^{2}13	^{2}13	^{2}13	^{2}13.0	$3(13.0–25.4)^2=461.3$
				$M_T=25.4$	$SS_{subj}=2179.4$

I show you this particular method for calculating SS_{subj} only because it gives a clear idea of the conceptual structure of the measure. It would not be the method to use for practical computational purposes, as it is likely to end up with a considerable accumulation of rounding error. Also, if you are doing it by hand, it can be rather laborious. The following method is somewhat less laborious and will in any event minimize the risk of rounding errors. You will recognize it as analogous to the computational formula for SS_{bg} that we worked through in Part 1; namely

$$SS_{bg} = \frac{(\Sigma X_{ai})^2}{N_a} + \frac{(\Sigma X_{bi})^2}{N_b} + \frac{(\Sigma X_{ci})^2}{N_c} - \frac{(\Sigma X_{Ti})^2}{N_T}.$$

Except now the string of items to the left of the minus sign is replaced by another string that looks like this:

$$\frac{(\Sigma X_{subj1})^2}{k} + \frac{(\Sigma X_{subj2})^2}{k} + \frac{(\Sigma X_{subj3})^2}{k} + \ldots$$ and so on, for as many subjects as you have in the analysis.

That is: For each subject in the analysis, take the sum of that subject's scores across all k conditions, square that sum, and then divide the result by k, which is the number of scores on which the sum is based.

Since k is the same for each subject, this portion of the formula to the left of the minus sign can be rewritten as

$$\frac{(\Sigma X_{subj*})^2}{k}$$

That is: For each subject in the analysis, take the sum of that subject's scores across all k conditions, square that sum, add up these squared sums across all subjects, and then divide the total of squared sums by k.

and the entire formula can then be written more compactly as

$$SS_{subj} = \frac{\Sigma(X_{subj*})^2}{k} - \frac{\Sigma(X_{Ti})^2}{N_T}.$$

The next version of our data table shows the sum for each subject, along with the calculation of the square of each subject's sum.

Subjects	*A*	*B*	*C*	*Subject Sums*	$(\Sigma X_{subj*})^2$
^{2}1	^{2}35	^{2}39	^{2}32	^{2}106	106^2=11236
^{2}2	^{2}32	^{2}35	^{2}31	^{2}97	97^2=9604
^{2}3	^{2}33	^{2}32	^{2}28	^{2}93	93^2=8649
^{2}4	^{2}32	^{2}32	^{2}29	^{2}93	93^2=8649
^{2}5	^{2}31	^{2}33	^{2}26	^{2}90	90^2=8100
^{2}6	^{2}29	^{2}30	^{2}29	^{2}88	88^2=7744
^{2}7	^{2}29	^{2}31	^{2}27	^{2}87	87^2=7569
^{2}8	^{2}27	^{2}29	^{2}27	^{2}83	83^2=6889
^{2}9	^{2}27	^{2}31	^{2}24	^{2}82	82^2=6724
^{2}10	^{2}28	^{2}27	^{2}24	^{2}79	79^2=6241
^{2}11	^{2}27	^{2}27	^{2}23	^{2}77	77^2=5929
^{2}12	^{2}27	^{2}26	^{2}23	^{2}76	76^2=5776
^{2}13	^{2}24	^{2}29	^{2}19	^{2}72	72^2=5184
^{2}14	^{2}24	^{2}25	^{2}19	^{2}68	68^2=4624
^{2}15	^{2}17	^{2}16	^{2}18	^{2}51	51^2=2601
^{2}16	^{2}17	^{2}15	^{2}17	^{2}49	49^2=2401
^{2}17	^{2}14	^{2}15	^{2}12	^{2}41	41^2=1681
^{2}18	^{2}13	^{2}13	^{2}13	^{2}39	39^2=1521
					$(\Sigma X_{subj*})^2$=111122

Recalling that X_{Ti}=1372 and N_T=54, our calculation of SS_{subj} is accordingly

$$SS_{subi} = \frac{\Sigma(X_{subj*})^2}{k} - \frac{(\Sigma X_{Ti})^2}{N_T}$$

$$= \frac{111122}{3} - \frac{(1372)^2}{54}$$

$$= 2181.7.$$

As always, one must be careful not to get so lost in the calculations as to lose sight of the concepts. What we have just calculated with SS_{subj} is the amount of variability within the total array of data that derives from individual differences among the subjects. And now that this amount has been identified, it can be removed. The process of removal is one of simple subtraction. The total amount of within-groups variability is SS_{wg}=2285.0. Of this amount, SS_{subj}=2181.7 is attributal not to sheer, cussed random variability, but to extraneous pre-existing individual differences among the 18 subjects. Remove the latter from the former, and what remains is

$$SS_{error} = SS_{wg} - SS_{subj}$$

$$= 2285.0 - 2181.7$$

$$= 103.3.$$

The specific numerical values have been included for the several values of SS. The bottom line is that a huge chunk of SS_{wg} is attributable to individual differences among the subjects, and that chunk is now removed. From this point on, it is all a smooth ride across familiar terrain.

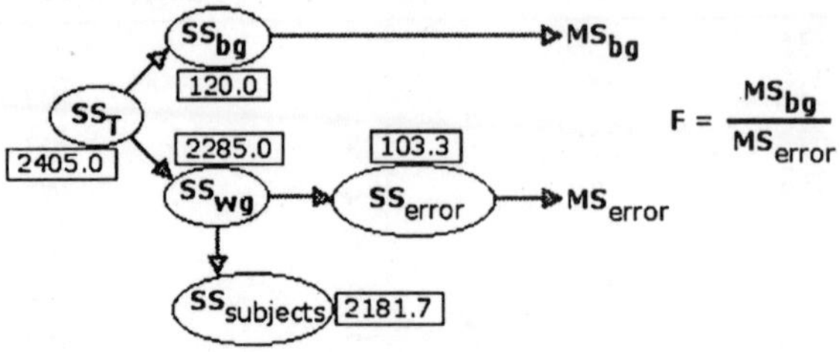

CALCULATION OF MS_{BG}, MS_{ERROR}, AND F

We begin by sorting out the several components of degrees of freedom. The first three are the same as we would have with a one-way independent-samples ANOVA:

$$df_T = N_T - 1 = 54 - 1 = 53$$

$$df_{bg} = k - 1 = 3 - 1 = 2$$

and

Note that: $df_T = df_{bg} + df_{wg}$

$$df_{wg} = N_T - k = 54 - 3 = 51$$

The only difference is that now we break df_{wg} into two further components, corresponding respectively to the two components of within-groups variability, SS_{subj} and SS_{error}. For df_{subj} the basic concept is once again "the number of items minus one." In this case the number of items is the number of subjects, which we will designate as N_{subj}. Thus

$$df_{subj} = N_{subj} - 1 = 18 - 1 = 17$$

Remove this component from df_{wg}, and what remains is

$$df_{error} = df_{wg} - df_{subj} = 51 - 17 = 34$$

With these, we can then calculate the two relevant MS values as

$$MS_{bg} = \frac{SS_{bg}}{df_{bg}} = \frac{120.0}{2} = 60.0$$

and

$$MS_{error} = \frac{SS_{error}}{df_{error}} = \frac{103.3}{34} = 3.0$$

which in turn give us

$$F = \frac{MS_{bg}}{MS_{error}} = \frac{60.0}{3.0} = 20.0 \quad \text{with df=2,34.}$$

Figure shows the sampling distribution of F for df=2,34, and the adjacent table shows the corresponding portion. As indicated, F=3.28 and F=5.29 mark the points in this distribution beyond which fall 5% and 1%, respectively, of all possible mere-chance outcomes, assuming the null hypothesis to be true.

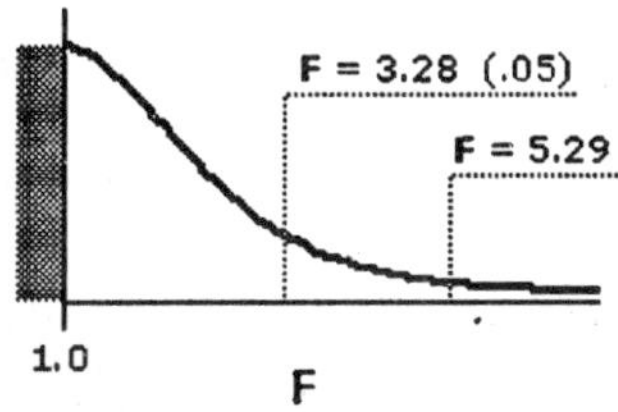

Fig. Sampling Distribution of F for df=2,34

df denominator	*df numerator* 1	2	3
34	4.13	3.28	2.88
	7.44	5.29	4.42

As the observed value of F=20.0 falls far to the right of F=5.29, the aggregate differences among the means of the three groups of measures, A|B|C, can be adjudged significant well beyond the .01 level. Our investigator can accordingly reject the null hypothesis, concluding with a high degree of confidence that the aggregate mean differences among the three groups of measures stem from something more than mere chance coincidence.

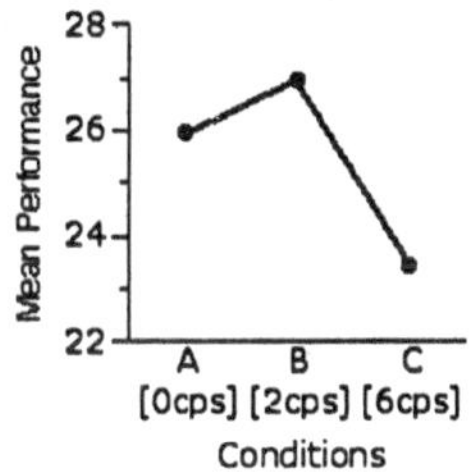

ANOVA

Source	*SS*	*df*	*MS*	*F*	*P*
between groups ("effect")	120	2	60	20	<.01
within groups	2285	51			
•error	103.3	34	3.0		
•subjects	2182	17			
Total	2405	53			

The assumptions and step-by-step computational procedures for the one-way correlated-samples ANOVA will be outlined in this chapter. We will see how the Tukey HSD test can be adapted to perform pairwise comparisons among the individual group means in a correlated-samples ANOVA.

ASSUMPTIONS OF THE ONE-WAY ANOVA FOR CORRELATED SAMPLES

The first four of these assumptions are the same as for the independent-samples ANOVA:

1. That the scale on which the dependent variable is measured has the properties of an equal interval scale;
2. That the measures within each of the *k* groups are independent of each other;
3. That the source population(*s*) from which the *k* samples of measures are drawn can be reasonably supposed to have a normal distribution; and
4. That the *k* groups of measures have approximately equal variances.

The fifth assumption is unique to the correlated-samples version, entailed by the fact that the *k* groups in such an analysis are, after all, potentially intercorrelated. Actually, it is a cluster of assumptions that go by such names as "compound symmetry," "homogeneity of covariance," "circularity," and "sphericity," all of which are more complex than can be dealt with at the introductory level. For practical purposes you can think of it this way. Suppose we were to calculate all possible correlation coefficients (*r*) among the *k* groups of measures. The homogeneity of covariance assumption requires that all of these correlation coefficients be positive and of approximately the same magnitude. Essentially, it is a requirement that the differential effects of the *k* conditions are consistent among the subjects in the repeated measures design, or among the matched sets of subjects in the randomized blocks design.

Figure shows these intercorrelations for the *k*=3 example considered in the present chapter. As you can see, they do all come out positive and of approximately the same magnitude.

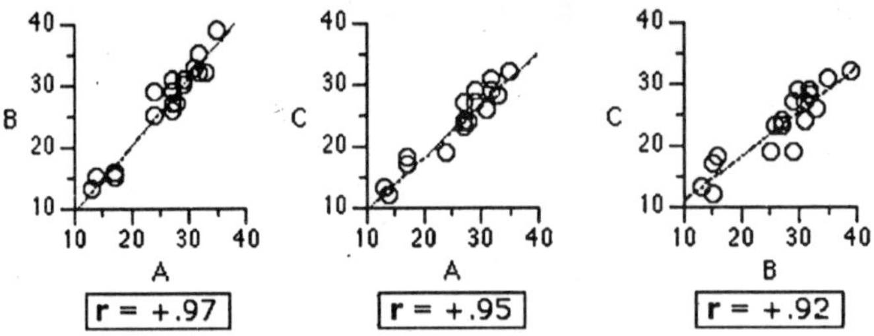

Fig. Intercorrelations among Groups A, B, and C

Note incidentally that the degree of intercorrelation among the *k* groups of measures is directly related to SS_{subj}, which is the aggregate measure of individual differences among the subjects in a repeated measures design, or among the matched sets of subjects in a randomized blocks design. The greater the degree of intercorrelation, the greater will be the size of SS_{subj}.

POST-ANOVA COMPARISONS: THE TUKEY HSD TEST

In the independent-samples version of ANOVA, the "error term" of the F-ratio (the quantity that appears in the denominator) is MS_{wg}. In the correlated-samples version, it is MS_{error}. The Tukey HSD test ican be extended to the correlated-samples version through the simple strategy of using MS_{error} in place of MS_{wg} and df_{error} in place of df_{wg}. Since the number of measures per group is equal to the number of subjects in a repeated measures design, or to the number of matched sets of subjects in a randomized blocks design, we will also be substituting N_{subj} for $N_{p/s}$.

MS_{error}=3.0, df_{error}= 34, N_{subj}=18, K=3

Here again is the calculator for critical values of Q for the .05 and .01 levels of significance. It is exactly the same calculator that appeared, except now the entry for df is subscripted as df_{error}. Recall that values of K must fall between 3 and 10, inclusive.

Entering k=3 and df_{error}=34, the critical values come out as $Q_{.05}$=3.47 and $Q_{.01}$=4.42. Our calculations for HSD are therefore for the .05 level:

That is: In order to be considered significant at or beyond the .05 level, the difference between any two particular group means (larger–smaller) must be equal to or greater than 1.42.

$$\text{HSD.05} = \left[\frac{MS_{error}}{N_{subi}}\right]$$

$$= 3.47 \times \text{sqrt}\left[\frac{3.0}{18}\right]$$

$$= 1.42.$$

and for the .01 level:

$$\text{HSD}_{.01} = \text{Q.01} \times \text{sqrt}\left[\frac{MS_{error}}{N_{subj}}\right]$$

$$= 4.42 \times \text{sqrt}\left[\frac{3.0}{18}\right]$$

$$= 18.$$

That is: In order to be considered significant at or beyond the .01 level, the difference between any two particular group means (larger–smaller) must be equal to or greater than 1.8.

The blue entries in the following table show the differences (larger–smaller) between each pair of group means in the example. As you can see, the comparisons for A·C and B·C are significant beyond the .01 level, while the one for A·B fails to achieve significance even at the basic .05 level.

A·B	M_a=25.9 M_b=26.9	1.0	$\text{HSD}_{.05}$ = 1.42 $\text{HSD}_{.01}$ = 1.8
A·C	M_a=25.9 M_c=23.4	2.5	
B·C	M_b=26.9 M_c=23.4	3.5	

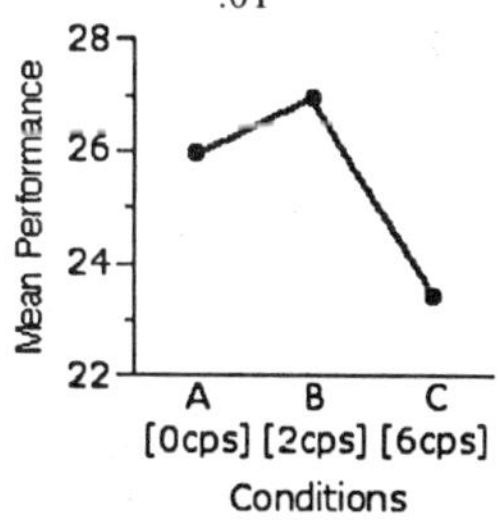

Our investigator would therefore be able to conclude that performance on the rotary pursuit task is better under the 0cps and 2cps condition than it is under the 6cps condition. Notwithstanding the fact that M_b is greater than M_a, she would not be able to conclude that performance is better under the 2cps condition than under the 0cps condition.

STEP-BY-STEP COMPUTATIONAL PROCEDURE

Here again I will show the procedures for one particular value of *k*, namely *k*=3. The modifications required for different values will be fairly obvious. The steps listed below assume that you have already done the basic number-crunching to get X_i and $X^2{}_i$ for each of the *k* groups of measures and for all *k* groups combined. They also require the prior calculation of X_{subj*} for each subject in a repeated measures design, or for each set of matched subjects in a randomized blocks design.

Step 1. Combining all *k* groups of measures together, calculate

$$SS_T = \Sigma X^2 i - \frac{(\Sigma X_i)^2}{N_T}.$$

Step 2. For each of the *k* groups separately, calculate the sum of squared deviates within the group ("g") as

$$SS_g = \Sigma X^2 gi - \frac{(\Sigma X_{gi})^2}{N_g}.$$

Step 3. Take the sum of the SS_g values across all *k* groups to get

$$SS_{wg} = SS_a + SS_b + SS_c$$

Step 4. Calculate SS_{bg} as

$$SS_{bg} = SS_T - SS_{wg}$$

Step 4a. Check your calculations up to this point by calculating SS_{bg} separately as

$$SS_{bg} = \frac{(\Sigma X_{ai})^2}{N_a} + \frac{(\Sigma X_{bi})^2}{N_b} + \frac{(\Sigma X_{bi})^2}{N_b} - \frac{(\Sigma X_T)^2}{N_T}.$$

Step 5. Calculate SS_{subj} as

$$SS_{subj} = \frac{\Sigma(\Sigma X_{subj}{}^*)^2}{k} - \frac{(\Sigma X_{subj}{}^*)^2}{N_T}.$$

Step 6. Calculate SS_{error} as

$$SS_{error} = SS_{wg} - SS_{subj}$$

Step 7. Calculate the relevant degrees of freedom as

$$\text{df}_T = N_T - 1 \quad \text{df}_{bg} = k - 1 \quad \text{df}_{wg} = N_T - k$$

$$.\ \text{df}_{subj} = N_{subj} - 1 \quad \text{df}_{error} = \text{df}_{wg} - \text{df}_{subj}$$

Step 8. Calculate the relevant mean-square values as

$$MS_{bg} = \frac{SS_{bg}}{df_{bg}}$$

and

$$MS_{error} = \frac{SS_{error}}{\text{df}_{error}}.$$

Step 9. Calculate F as

$$F = \frac{MS_{bg}}{MS_{error}}.$$

Step 10. Refer the calculated value of F to the table of critical values of F, with the appropriate pair of numerator/denominator degrees of freedom.

THE FRIEDMAN TEST FOR 3 OR MORE CORRELATED SAMPLES

Assumptions of one-way ANOVA for correlated samples:

- Equal-interval scale of
- Measurement
- Independence of measures
- Within each group
- Normal distribution of source
- Population(*s*)
- Equal variances among groups
- Homogeneity of covariance

We have noted several times that the analysis of variance is quite robust with respect to the violation of its assumptions, providing that the *k* groups are all of the same size. In the correlated-samples ANOVA this provision is always satisfied, since the number of measures in each of the groups is necessarily equal to the number of subjects in the repeated-measures design, or to the number of matched sets in the randomized blocks design.

Still, there are are certain kinds of correlated-samples situations where the violation of one or more assumptions might be so thorough-going as to cast doubt on any result produced by an analysis of variance. In cases of this sort, a useful non-parametric alternative can be found in a rank-based procedure known as the Friedman Test.

There are two kinds of correlated-samples situations where the advisability of the non-parametric alternative would be fairly obvious. The first would be the case where the k measures for each subject start out as mere rank-orderings.

E.g.: To assess the likely results of an upcomming election, the 30 members of a presumably representative "focus group" of eligible voters are each asked to rank the 3 candidates, A, B, and C, in the order of their preference (1=most preferred, 3=least preferred).

And the second would be the case where these measures start out as mere ratings.

E.g.: The members of the "focus group" are instead asked to rate candidates on a 10-point scale (1=lowest rating, 10=highest).

In both of these situations the assumption of an equal-interval scale of measurement is clearly not met. There's a good chance that the assumption of a normal distribution of the source population(*s*) would also not be met. Other cases where the equal-interval assumption will be thoroughly violated include those in which the scale of measurement is intrinsically non-linear: for example, the decibel scale of sound intensity, the Richter scale of earthquake intensity, or any logarithmic scale.

The Friedman test with a rating-scale example that is close to my amateur violinist's heart. The venerable auction house of Snootly & Snobs will soon be putting three fine 17th-and 18th-century violins, A, B, and C, up for bidding. A certain musical arts foundation, wishing to determine which of these instruments to add to its collection, arranges to have them played by each of 10 concert violinists. The players are blindfolded, so that they cannot tell which violin is which; and each plays the violins in a randomly determined sequence (BCA, ACB, etc.).

		Violin	
Subjects	*A*	*B*	*C*
1	9.0	7.0	6.0
2	9.5	6.5	8.0
3	5.0	7.0	4.0
4	7.5	7.5	6.0
5	9.5	5.0	7.0
6	7.5	8.0	6.5
7	8.0	6.0	6.0
8	7.0	6.5	4.0
9	8.5	7.0	6.5
10	6.0	7.0	3.0

They are not informed that the instruments are classic masterworks; all they know is that they are playing three different violins. After each violin is played, the player rates the instrument on a 10-point scale of overall excellence (1=lowest, 10=highest). The players are told that they can also give fractional ratings, such as 6.2 or 4.5, if they wish. The results are shown in the adjacent table. For the sake of consistency, the N=10 players are listed as "subjects."

LOGIC AND PROCEDURE

The Friedman test begins by rank-ordering the measures for each subject. For the present example we will assign the rank of "3" to the largest of a subject's three measures, "2" to the intermediate of the three, and "1" to the smallest. Thus for subject 1, the largest measure is in column A, the next largest in column B, and the smallest in column C; so the sequence of ranks across the row for subject 1 is 3,2,1. For subject 2 it is 3,1,2. And so forth. (The guidelines for assigning tied ranks are described in Subchapter 11a in connection with the Mann-Whitney test.)

	Original Measures			*Ranked Measures*		
Subjects	*A*	*B*	*C*	*A*	*B*	*C*
1	9.0	7.0	6.0	3	2	1
2	9.5	6.5	8.0	3	1	2
3	5.0	7.0	4.0	2	3	1
4	7.5	7.5	6.0	2.5	2.5	1
5	9.5	5.0	7.0	3	1	2
6	7.5	8.0	6.5	2	3	1
7	8.0	6.0	6.0	3	1.5	1.5
8	7.0	6.5	4.0	3	2	1
9	8.5	7.0	6.5	3	2	1
10	6.0	7.0	3.0	2	3	1

The null hypothesis in this scenario is that the three violins do not differ with respect to whatever it is that blindfolded expert players would judge to be the overall excellence of an instrument.

	Ranked Measures		
Subjects	*A*	*B*	*C*
1	3	2	1
2	3	1	2
3	2	3	1
4	2.5	2.5	1
5	3	1	2
6	2	3	1
7	3	1.5	1.5
8	3	2	1
9	3	2	1
10	2	3	1
sums	26.5	21	12.5
means	2.65	2.1	1.25

This would entail that each of the six possible sequences of A,B,C ranks 1,2,3/1,3,2/2,1,3/2,3,1/3,1,2/3,2,1 is equally likely, hence that the three columns will tend to include a random jumble of 1's, 2's, and 3's, in approximately equal proportions. In this case, the sums and the means of the columns would also tend to come out approximately the same.

In most respects you will find the logic of the Friedman test quite similar to that of the Kruskal-Wallis test examined in Subchapter 14a. For any particular value of k (the number of measures per subject), the mean of the ranks for any particular one of the N subjects is $(k+1)$ 2.

Thus for k=3, as in the present example, it is 4/2=2; for k=4, it would be 5/2=2.5; and so on. On the null hypothesis, this would also be the expected value of the mean for each of the k columns. Similarly, the expected value for each of the column sums would be this amount multiplied by the number of subjects: $N(k+1)/2$. For the present example, with N=10, it would be (10)(4)/2=20.

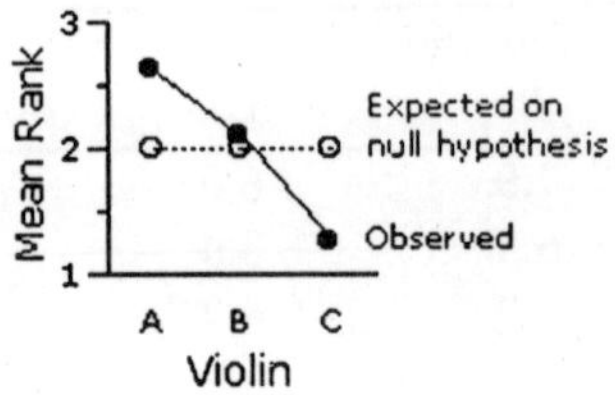

The following items of symbolic notation are the same ones we used in connection with the Kruskal-Wallis test:

T_A	= the sum of the N ranks in column A
M_A	= the mean of the N ranks in column A
T_B	= the sum of the N ranks in column B
M_B	= the mean of the N ranks in column B
T_C	= the sum of the N ranks in column C
M_C	= the mean of the N ranks in column C
T_{all}	= the sum of the nk ranks in all columns combined [In all cases equal to nk$(k+1)/2]_T$
M_{all}	= the mean of the nk ranks in all columns combined [In all cases equal to $(k+1)/2$]

The Measure of Aggregate Group Differences

Also the same as in the Kruskal-Wallis test is the measure of the aggregate degree to which the k group means differ. It is the between-groups sum of squared deviates defined in Subchapter 14a as $SS_{bg(R)}$, the "(R)" serving as a reminder that this particular version of SS_{bg} is based on ranks. The following table summarizes the values needed for the calculation of $SS_{bg(R)}$.

	A	*B*	*C*	*All*	
counts	10	10	10	30	n=10 [subjects]T
sums	26.5	21	12.5	60	k=3 [measures per subject]T
means	2.65	2.1	1.25	2.0	nk=30

As in all other cases heretofore examined, the squared deviate for any particular group mean is equal to the squared difference between that group mean and the mean of the overall array of data, multiplied by the number of observations on which the group mean is based. Thus, for each of our current three groups

$$\begin{aligned} \text{A: } 10(2.65-2.0)^2 &= 4.225 \\ \text{B: } 10(2.10-2.0)^2 &= 0.100 \\ \text{C: } 10(1.25-2.0)^2 &= 5.625 \\ SS_{bg(R)} &= 9.950. \end{aligned}$$

Once again we can write the conceptual formula for $SS_{bg(R)}$ as

$SS_{bg(R)} = (\Sigma[n_g(M_g - M_{all})^2]$ As usual, the subscript "g" means "any particular group."

Except now, since each group is necessarily of the same size, it can be reduced to the simpler form

$SS_{bg(R)} = (n\Sigma(M_g - M_{all})^2 \; n =$ number of subjects

For the same reason, the computational formula (less susceptible to rounding errors) can take the simpler form

$$SS_{bg(R)} = \frac{\Sigma(T_g)^2}{n_a} - \frac{(\text{Tall})^2}{nk_a} \quad \begin{aligned} n &= \text{number of subjects } T \\ k &= \text{measures per subject} \end{aligned}$$

For the present example, with n=10, k=3, and values of T_g and T_{all} as indicated above, this would come out as

$$\begin{aligned} SS_{bg(R)} &= \frac{(26.5)^2 + (21.0)^2 + (12.5)^2}{10} - \frac{(60)^2}{30} \\ &= 9.95. \end{aligned}$$

·The Sampling Distribution of $SS_{bg(R)}$

When we examined the Kruskal-Wallis test in Subchapter 14a, we saw that $SS_{bg(R)}$ can be converted into the measure designated as H, which can then be referred to the sampling distribution of chi-square for df=k–1. The same is true of the Friedman test; the only difference is in the details of the conversion. For the Friedman test, the resulting measure is spoken of simply as a value of chi-square and takes the form

$$\times\chi^2 = \frac{SS_{bg(R)}}{k(k+1)/12}$$

which for the present example comes out as

$$\times\chi^2 = \frac{9.95}{3(3+1)/12} = \frac{9.95}{1} = 9.95.$$

When k is equal to 3, the application of this "conversion" formula is merely pro forma; - for in this case the denominator of the ratio will always come down to 3(4)/12=1, so the resulting value of chi-square will always be equal to $SS_{bg(R)}$. This, however, will not be so when k is something other than 3. With k=4 the denominator will be 4(5)/12=1.67; with k=5 it will be 5(6)/12=2.5; and so on.

As you can see, the observed value of =9.95, when referred to the appropriate sampling distribution of chi-square, is significant beyond the .01 level.

Theoretical Sampling Distribution of Chi-Square for df=2

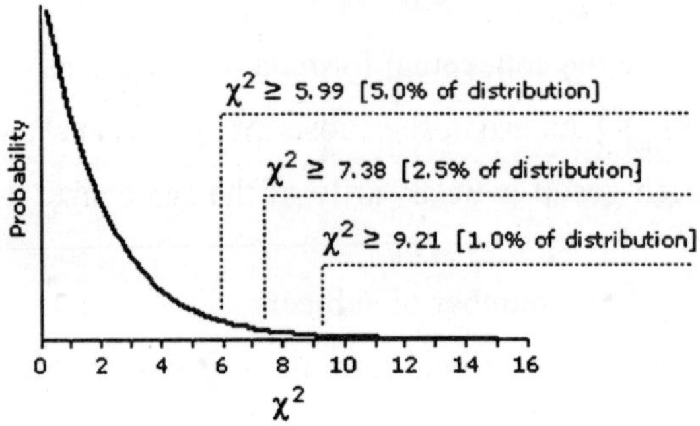

Our musical arts foundation can therefore conclude with considerable confidence that the observed differences among the mean rankings for the three violins reflect something more than mere random variability, something more than mere chance coincidence among the judgments of the expert players.

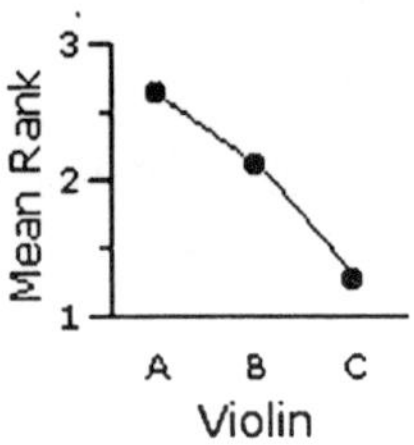

An Alternative Computational Formula

Textbook accounts of the Friedman test usually give a different computational formula for chi-square. Its advantage is that it can be (slightly) more convenient to use. Its disadvantage is that it does not give you the faintest idea of just what the measure is measuring. But here it is anyway, just in case you ever need to recognize it.

$$_{x}\chi^2 = \frac{12}{nk(k+1)}\Sigma(T_g)^2 - 3n(k+1).$$

As you can see in connection with our present example, the result comes out quite the same either way.

$$_{x}\chi^2 = \frac{12}{(10)(3)(4)}\,[(26.5)2+(21.0)2+(12.5)^2]_x - (3)(10)(4)$$

$$= (0.1 \times 1299.5) - 120 = 9.95.$$

Chapter 8

Two-way Analysis for Independent Samples

The two-way ANOVA is a procedure that examines the effects of two independent variables concurrently. This not only provides a kind of economy in allowing you to look at two things for the price of one. It also, and often much more importantly, allows you to determine whether the two independent variables interact with respect to their effect on the dependent variable.

I will introduce this critically important concept with the familiar example of drug interactions. Suppose we were interested in the effects of two drugs, A and B, on the blood level of a certain hormone. To this end, we randomly and independently sort the members of our subject pool (healthy adult human males between the ages of 30 and 40) into four equal-sized groups. As shown in the following table, group r1c1 ("*r*"=row, "c"=column) serves as a kind of control group, receiving only an inert placebo containing zero units of A and zero units of B. Group r1c2 receives zero units of A and 1 unit of B; r2c1 receives 1 unit of A and zero units of B; and r2c2 receives 1 unit each of A and B.

Note that the data structure in the two-variable case takes the form of a rows-by-columns matrix. For economy of expression, it is conventional to speak of the "row variable" and the "column variable" in accordance with how the two independent variables are arrayed in the matrix. In the present example, drug A is the row variable and drug B is the column variable.

		Drug B	
		0 units	*1 unit*
Drug A	0 units	r1c1 0 units of A 0 units of B	r1c2 0 units of A 1 unit of B
	1 unit	r2c1 1 unit of A 0 units of B	r2c2 1 unit of A 1 unit of B

The following tables and graphs show four possible scenarios. The numbers inside the blue cells in the table represent the respective mean blood levels of the hormone for each of the four groups following adminstration of the experimental treatment; the numbers to the right of each row and at the bottom of each column represent the mean for that particular row or column; and the number in the bottom right corner is the mean of the entire array. The adjacent graph in each scenario is a plot of the group mean values appearing in the blue cells.

SCENARIO 1

		Drug B		
		0 units	*1 unit*	
Drug A	0 units	5	5	5
	1 unit	5	5	5
		5	5	5

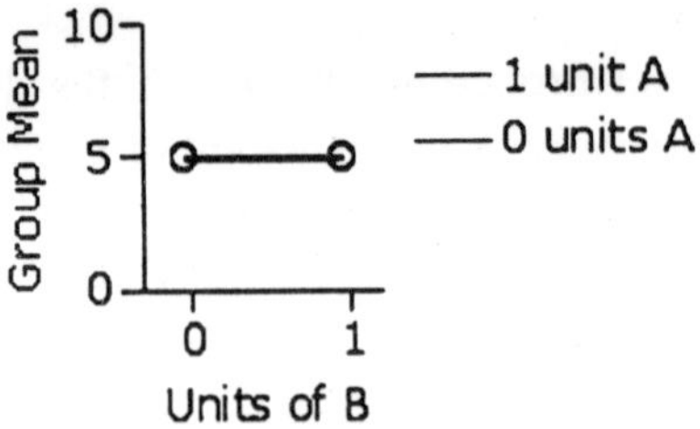

One need not know much about statistical inference to surmise that the two drugs in this scenario show no effect at all, either separately or in combination. This is what would be expected on the basis of the null hypothesis.

SCENARIO 2

		Drug B		
		0 units	*1 unit*	
Drug A	0 units	5	10	7.5
	1 unit	10	15	7.5
		7.5	12.5	10

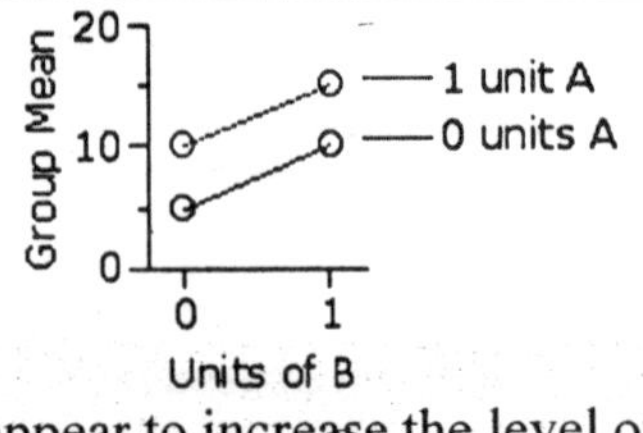

In this scenario both drugs appear to increase the level of the hormone, but note that their combined effect is one of simple addition. With zero units of A and B, the mean blood level of the hormone is 5. Add 1 unit of B and it increases to 10. Start again from zero, add 1 unit of A, and it also increases from 5 to 10. When presented alone, 1 unit of either A or B increases the level of the hormone by 5. When presented together, they increase it by 5+5 to 15. The effect of the two drugs in combination is merely the sum of the effects they have separately.

In considering the effects of the two independent variables separately, what the two-way ANOVA actually looks at are the differences among the means of the row variable and the differences among the means of the column variable. By convention, these two sets of differences are spoken of as the "main effects" of the analysis, as distinguished from the "interaction effect," which is something above and beyond the two main effects. Thus, for the present scenario, the main effect for the row variable, drug A, is the difference between M_{row1}=7.5 and M_{row2}=12.5. The main effect for the column variable, drug B, is the difference between M_{col1}=7.5 and M_{col2}=12.5. (In real life you will not normally get such a symmetrical assortment of row and column means.) In Scenario 1 there are no effects at all. In Scenario 2 there are main effects for A and B, but no interaction effect.

SCENARIO 3

		Drug B		
		0 units	*1 unit*	
Drug A	0 units	5	10	7.5
	1 unit	10	20	15
		7.5	15	11.25

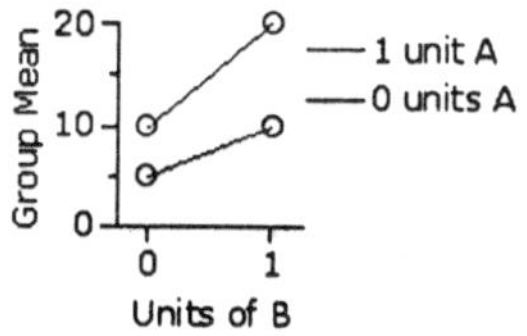

Here the drugs when presented separately are having the same effects as in Scenario 2, but now they are also interacting. The effect of the two when presented in combination is greater than the sum of their separate effects.

Note again that the main effects would consist of the difference between M_{row1}=7.5 and M_{row2}=15, and of the difference between M_{col1}=7.5 and M_{col2}=15. In this scenario there are main effects for rows and columns as well as an interaction effect.

SCENARIO 4

		Drug B		
		0 units	*1 unit*	
Drug A	0 units	5	10	7.5
	1 unit	10	5	7.5
		7.5	15	7.5

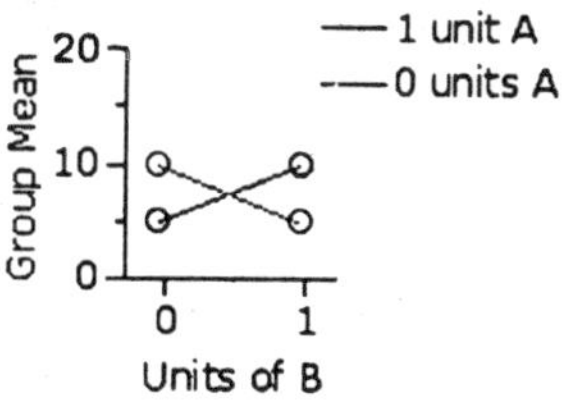

Here, too, is an interaction effect, but in the opposite direction. In Scenario 3 combining the drugs enhances their separate effects. Now the combined effect is one of mutual cancellation.

But note that in this scenario there would be no main effects for either rows or columns ($M_{row1}=M_{row2}=7.5$ and $M_{col1}=M_{col2}=7.5$), notwithstanding that drugs A and B are obviously having an effect when presented separately. The moral of this story is that the interpretation of the presence or absence of main effects in a two-way ANOVA is not always simple and straightforward. More of this later.

These are by no means the only sorts of interaction effects that can be found, though I think they will give you at least a preliminary idea of what the concept is pointing to. Whenever two variables in combination produce effects that are different from the simple sum of their separate effects, you have the makings of an interaction. When represented graphically, the absence of an interaction will appear as lines that are approximately parallel, as in Scenario 2, while the presence of an interaction will appear as lines that diverge or converge or cross over each other.

PROCEDURE

Up to a point, the two-way ANOVA for independent samples proceeds exactly like the corresponding one-way ANOVA. Here as well, SS_{bg} is the measure of the aggregate differences among the several groups, and SS_{wg} is the measure of random variability inside the groups. The latter, SS_{wg}, is treated the same way as in the one-way analysis: it ends up as the MS_{error} that appears in the denominator of the F-ratio. But now, with two independent variables and the data arrayed in the form of a rows-by-columns matrix, SS_{bg} can be further divided, like Caesar's Gaul and Freud's psychical topography, into three parts. As illustrated in the following diagram, one of these parts measures the differences among the means of the two or more rows, another measures the mean differences among the two or more columns, and the third is a measure of the degree to which the two independent variables interact. Each of these three components is then converted into a corresponding value of MS, with the result that there are now three separate F-ratios to calculate and three separate tests of significance to perform: one for the row variable, one for the column variable, and one for the interaction between the two variables.

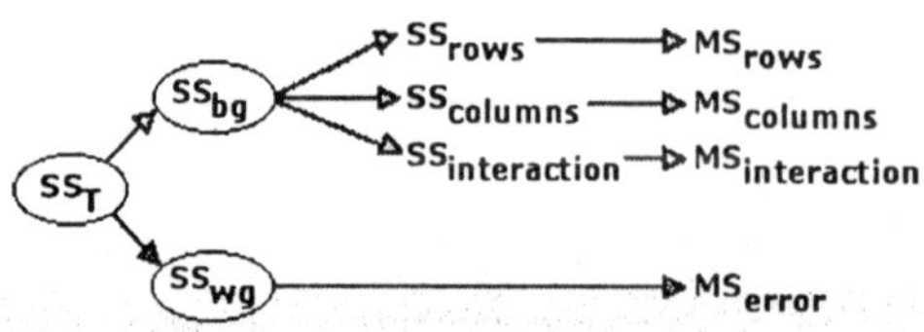

The two-way ANOVA is sufficiently important to warrant illustration with several examples. The first two are variations on the simple drug A versus drug B scenarios we have just examined. The third will be a bit more complex.

Example

In order to test the separate and mutual effects of two drugs, A and B, on physiological arousal, researchers randomly and independently sorted 40 laboratory rats into four groups of 10 subjects each. Each group received a certain dosage of drug A (zero units or 1 unit) and a certain dosage of drug B (zero units or 1 unit). The dependent variable was a standard measure of physiological arousal. As in the earlier two-drug illustration, one of the groups served as a control, receiving only an inert placebo containing zero units of A and zero units of B. The following table shows the consequent measures of physiological arousal for each subject in each of the four groups.

raw data		B: *0 units*		B: *1 unit*	
	0	20.4	17.4	20.5	26.3
	units	20.0	18.4	26.6	19.8
		24.5	21.0	25.4	28.2
		19.7	22.3	22.6	23.7
		17.3	23.3	22.5	22.6
	1	22.4	19.1	34.1	21.9
	unit	22.4	25.4	32.6	28.5
		26.2	25.1	29.0	25.8
A		28.8	21.8	29.0	27.1
		26.3	25.2	25.7	24.4

Performing the requisite number-crunching on this array yields the summary values shown in the next table. The values pertaining to the four groups of measures in the original table of raw data are subscripted as $g1$, $g2$, $g3$, and $g4$; those pertaining to the rows are subscripted as r1 and r2; and those pertaining to the columns are subscripted as c1 and c2. The values deriving from the entire array of data (all groups combined) are as usual subscripted "T." It is worth a moment of your time to look back and forth between the table above and the one below to make sure you have a clear sense of where the summary values are coming from.

summary data		B: *0 units*	B: *1 unit*	*rows*
A	0 units	$N_{g1}=10$ $\Sigma X_{g1}=204.3$ $\Sigma X^2_{g1}=4226.3$	$N_{g2}=10$ $\Sigma X_{g2}=238.2$ $\Sigma X^2_{g2}=5741.4$	$N_{r1}=20$ $\Sigma X_{r1}=442.5$
	1 unit	$N_{g3}=10$ $\Sigma X_{g3}=242.7$ $\Sigma X^2_{g3}=5961.34$	$N_{g4}=10$ $\Sigma X_{g4}=278.1$ $\Sigma X^2_{g4}=7855.3$	$N_{r2}=20$ $\Sigma X_{r2}=520.8$
columns		$N_{c1}=20$ $\Sigma X_{c1}=447.0$	$N_{c2}=20$ $\Sigma X_{c2}=516.3$	$N_T=40$ $\Sigma X_T=963.3$ $\Sigma X^2_T=23784.4$

Up to this point it is all a tangle of numbers with no immediately discernible structure. It is only when you calculate the various means—for groups, rows, columns, and total array of data—that the pattern becomes visible. These means are given in the next table, along with a plot of the group means analogous to the ones shown in the earlier illustrations.

means		*B* *0 units*	*1 unit*	*rows*
A	0 units	M_{g1}=20.43	M_{g2}=23.82	M_{r1}=22.13
	1 units	M_{g3}=24.27	M_{g4}=27.81	M_{r2}=26.04
columns		M_{c1}=22.35	M_{c2}=25.82	M_T=24.08

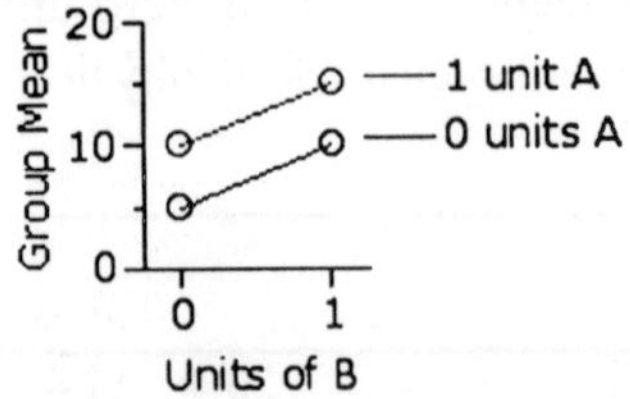

The resemblance between the above graph and the plot of our earlier Scenario 2 is no accident. When doing actual research, you must of course take your data as they come. The nice thing about generating illustrative data for a textbook is that you can shape them in any way you want. For this example I have deliberately arranged things so that there will be main effects for the row and column variables, but no interaction effect. The rationale is that understanding the presence of an interaction effect is best arrived at by first understanding its absence.

The next table shows the sums of squared deviates within each of the four groups, as well as for all four groups combined. These values are calculated exactly as they are in the corresponding one-way ANOVA, according to the general computational formula

$$SS = \Sigma X^2_i - \frac{(\Sigma X_i)^2}{N}$$

preliminary SS values		*B* *0 units*	*1 unit*	
A	0 unit	SS_{g1}=52.44	SS_{g2}=67.48	
	1 unit	SS_{g3}=71.02	SS_{g4}=121.37	
				SS_T=585.70

Also as in the one-way ANOVA, SS_{wg} is given by the sum of the *SS* measures within the several groups. In the present case there are four groups; hence

$$\begin{aligned} SS_{wg} &= SS_{g1} + SS_{g2} + SS_{g3} + SS_{g4} \\ &= 52.44 + 67.48 + 71.02 + 121.37 \\ &= 312.31. \end{aligned}$$

The similarity continues for one more step, and then we shift gears. Once you have SS_T and SS_{wg}, the measure of between-groups *SS* can be reached through simple subtraction:

$$\begin{aligned} SS_{bg} &= SS_T - SS_{wg} \\ &= 585.70 - 312.31 \\ &= 273.39. \end{aligned}$$

Here again I recommend performing a computational check by calculating SS_{bg} from scratch. The structure is the same as before. The only difference is in the new subscriptions: $g1$, $g2$, etc. Please pay particularly close attention to this structure, for we will soon be applying it twice again with different casts of characters.

$$SS_{bg} = \frac{(\Sigma X_{g1})^2}{N_{g1}} + \frac{(\Sigma X_{g2})^2}{N_{g2}} + \frac{(\Sigma X_{g3})^2}{N_{g3}} + \frac{(\Sigma X_{g4})^2}{N_{g4}} - \frac{(\Sigma X_T)^2}{N_T}$$

$$= \frac{(204.3)^2}{10} + \frac{(238.2)^2}{10} + \frac{(242.7)^2}{10} + \frac{(278.1)^2}{10} - \frac{(963.3)^2}{40}$$

$$= 273.39.$$

Breaking SS_{bg} into its Component Parts

As indicated earlier, SS_{bg} has three complementary components: one, SS_{rows}, measures the differences among the means of the two or more rows; another, SS_{cols}, measures the mean differences among the two or more columns; and the third, $SS_{interaction}$, is a measure of the degree to which the row and column variables interact. To save space, we will henceforth refer to the interaction component as SS_{rxc}, the subscription "rxc" being an abbreviation of the conventional shorthand expression "rows by columns."

For practical computational purposes, the simplest way to proceed is to calculate SS_{rows} and SS_{cols}, and then subtract those two from SS_{bg} to get the third component, SS_{rxc}. Conceptually, the calculation of SS_{rows} and SS_{cols} is exactly like the calculation of SS_{bg} when you are doing it from scratch.

SS_{rows}/conceptual.

Consider the means of the two rows in our example: M_{r1}=22.13 and M_{r2}=26.04. The null hypothesis is expecting these means to be the same, and that can happen only if they are both equal to M_T=24.08. The following conceptual procedure will be a familiar sight by now, so can be offerred without commentary.

	row 1	*row* 2
observed row mean	22.13	26.04
expected row mean	24.08	24.08
deviate	– 1.95	+1.96
squared deviate	3.80	3.84
squared deviate weighted by number in row (20)	76.0	76.8

Take the sum of these two weighted squared deviates and you have

SS_{rows} = 76.0 + 76.8 = 152.8 [tentative]

SS_{cols}/conceptual.

It is the same logic and procedure for SS_{cols}. The null hypothesis expects the means of the columns to be the same, and that can happen only if they are both equal to M_T=24.08.

	col 1	*col* 2
observed column mean	22.35	25.82
expected column mean	24.08	24.08
deviate	—1.73	+1.74
squared deviate	2.99	3.03
squared deviate weighted by number in column (20)	59.8	60.6

Hence

$SS_{cols} = 59.8 + 60.6 = 120.4$.

I have marked both of the above values as "tentative" because calculations that start out with rounded numbers are at risk of accumulating substantial rounding errors. For practical purposes it is better to use the following computational formulas. Both follow the same basic pattern as when you are calculating SS_{bg} from scratch

$$SS_{bg} = \frac{(\Sigma X_{g1})^2}{N_{g1}} + \frac{(\Sigma X_{g2})^2}{N_{g2}} + \frac{(\Sigma X_{g3})^2}{N_{g3}} + \frac{(\Sigma X_{g4})^2}{N_{g4}} - \frac{(\Sigma X_T)^2}{N_T}$$

only now the items to the left of the minus sign pertain not to the individual groups of measures, but to the rows or the columns.

SS_{rows}/computational.

$$SS_{rows} = \frac{(\Sigma X_{r1})^2}{N_{r1}} + \frac{(\Sigma X_{r2})^2}{N_{r2}} - \frac{(\Sigma X_T)^2}{N_T}$$

$$= \frac{(442.5)^2}{20} + \frac{(520.8)^2}{20} - \frac{(963.3)^2}{40}$$

$$= 153.27.$$

SS_{cols}/computational.

$$SS_{cols} = \frac{(\Sigma X_{c1})^2}{N_{c1}} + \frac{(\Sigma X_{c2})^2}{N_{c2}} - \frac{(\Sigma X_T)^2}{N_T}$$

$$= \frac{(447.0)^2}{20} + \frac{(516.3)^2}{20} - \frac{(963.3)^2}{40}$$

$$= 120.06.$$

SS_{rxc}/computational.

Once you have these two components of SS_{bg}, the SS measure of interaction can then be reached through simple subtraction:

$$SS_{rxc} = SS_{bg} - SS_{rows} - SS_{cols}$$
$$= 273.39 - 153.27 - 120.06$$
$$= 0.06.$$

As promised, the interaction effect in this example is essentially zero. The advantage of the simple subtractive procedure by which we have arrived at this conclusion is that it is quick and easy. The disadvantage is that does not give the slightest clue of the underlying logic of the process. The following procedure is more cumbersome and more prone to rounding errors, though at the same time more revealing of the inner workings of SS_{rxc}.

SS_{rxc}/conceptual.

It begins once again with the concept of the null hypothesis. To streamline things a bit, we will first lay out some items of symbolic notation:$_T$

M_{g*} = The mean of any particular one of the the individual groups of measures

M_{r*} = The mean of the row to which that group belongs

M_{c*} = The mean of the column to which that group belongs

If there is zero interaction between the row and column variables, then the mean of any particular one of the the individual groups, M_{g*}, should be a simple additive combination of M_{r*} and M_{c*}. The specific form of the combination is

$$[\text{null}]M_{g*} = M_{r*} + M_{c*} - M_T$$ Click here for a brief account of the logic of this formula.

Thus, for group 1, which falls in row 1 and column 1:

$$[\text{null}]M_{g1} = M_{r1} + M_{c1} - M_T$$
$$= 22.13 + 22.35 - 24.08$$
$$= 20.40.$$

For group 2, which falls in row 1 and column 2:

$$[\text{null}]M_{g2} = M_{r1} + M_{c2} - M_T$$
$$= 22.13 + 25.82 - 24.08$$
$$= 23.87.$$

And so forth.

Here is the same table of means you saw earlier, except now I also include (in red) the results of the calculation of [null]M_{g*} for each of the four groups. The observed means of the groups (20.43, 23.82, etc.) appear in black.

	means	*B* 0 units	1 unit	*rows*
A	0 units	20.43 20.4	23.82 23.87	M_{r1}=22.13
	1 unit	24.27 24.31	27.81 27.78	M_{r2}=26.04
columns		M_{c1}=22.35	M_{c2}=25.82	M_T=24.08

As you can see, there is only the tiniest bit of difference between the observed group means and the means that would be expected if there were no rows-by-columns interaction.

Here again is that familiar conceptual structure by which you can convert the differences between observed and expected mean values into a meaningful measure of *SS*:

	g1	g2	g3	g4
observed group mean	20.43	23.82	24.27	27.81
expected group mean	20.40	23.87	24.31	27.78
deviate	+0.03	—0.05	—0.04	+0.03
squared deviate	0.0009	0.0025	0.0016	0.0009
squared deviate weighted by number in group (10)	0.009	0.025	0.016	0.009

The sum of these weighted squared deviates comes out to the same SS_{rxc}=0.06 calculated earlier with the simple subtractive procedure.

For those whose memories, like mine, fall short of being photographic, here is a summary of the several *SS* values we have now calculated for this example:

Total:	SS_T	= 585.70
within groups:	SS_{wg}	= 312.31
between groups:	SS_{bg}	= 273.39
rows:	SS_{rows}	= 153.27
columns:	SS_{cols}	= 120.06
interaction:	SS_{rxc}	= 0.06
df		

The following table lists the respective degrees of freedom that are associated with these values of *SS*. Note that df_{rxc}, the degrees of freedom for rows-by-columns interaction, is calculated in the same way as for a two-dimensional (rows-by-columns) chi-square test; namely,

$$df_{rxc} = (r-1)(c-1) \quad \begin{matrix} r = \text{number of rows} \\ c = \text{number of columns} \end{matrix} \cdot$$

All the other df structures are much as you would expect on the basis of previously examined versions of ANOVA. Note that the number of individual groups, or cells, in a rows-by-columns matrix is always equal to the product of *r* multiplied by *c*, rendered here as "rc." Thus, for the present example, rc=2x2=4.

	in general	*for the present example*	
Total	$df_T = N_T-1$	40–1=39	
within-groups (error)	$df_{wg} = N_T-\text{rc}$	40–(2)(2)=36	Note that$_T$ $df_T = df_{wg} + df_{bg}$
between-groups	$df_{bg} = rc-1$	(2)(2)–1=3	
rows	$df_{rows} = r-1$	2–1=1	
columns	$df_{cols} = c-1$	2–1=1	Note that$_T$ $df_{bg} = df_{rows} + df_{cols} +$
df_{rxc} interaction	$df_{rxc} = (r-1)(c-1)$	(2–1)(2–1)=1	

As in previous versions of ANOVA, the relevant values of *MS* are in each case given by the ratio *SS*/df. Thus, for rows, columns, and interaction:

$$MS_{rows} = \frac{SS_{rows}}{df_{rows}} \quad MS_{cols} = \frac{SS_{cols}}{df_{cols}} \quad MS_{rxc} = \frac{SS_{rxc}}{df_{rxc}}$$

$$= \frac{153.27}{1} = \frac{120.06}{1} = \frac{0.06}{1}$$

$$= 153.27 = 120.06 = 0.06.$$

These are the values of *MS* that will appear in the numerators of the three F-ratios that will complete the analysis. The denominator in each case will be the error term,

$$MS_{error} = \frac{SS_{wg}}{df_{wg}}$$

$$= \frac{312.31}{36} = 8.68$$

F And [T]here are the three bottom lines of the analysis:

$$F_{rows} = \frac{MS_{rows}}{MS_{error}} \quad F_{cols} = \frac{MS_{cols}}{MS_{error}} \quad F_{rxc} = \frac{MS_{rxc}}{MS_{error}}$$

$$= \frac{153.27}{8.68} = \frac{120.06}{8.68} = \frac{0.06}{8.68}$$

$$= 17.67 = 13.84 = 0.01.$$

with df=1,36 with df=1,36 with df=1,36.

Figure shows the sampling distribution of F for df=1,36, and the adjacent table shows the corresponding portion of Appendix D. As indicated, F=4.11 and F=7.40 mark the points in this distribution beyond which fall 5% and 1%, respectively, of all possible mere-chance outcomes, assuming the null hypothesis to be true.

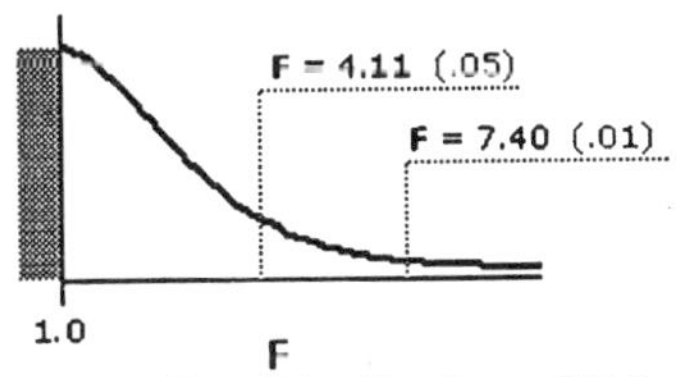

Fig. 16.1. Sampling Distribution of F for df=1,36

df denominator	*df numerator* 1	2	3
	4.11	3.26	2.87
36	7.4	5.25	4.38

Clearly our miniscule value of F_{rxc}=0.06 for rows-by-columns interaction falls nowhere near what would be needed for significance even at the basic .05 level. The values for the two main effects, however (F_{rows}=17.67 and F_{cols}=13.84), are both significant well beyond the .01 level.

The fundamental meaning of the significant row and column effects is that the difference between the two row means (22.13 vs 26.04) and the difference between the two column means (22.35 vs 25.82) each reflect something more than mere random variability. In the present example, where there is essentially zero interaction between the row and column variables, the interpretation of these two main effects would be entirely straightforward: 1 unit of A produces greater arousal than zero units of A; and 1 unit of B produces greater arousal than zero units of B.

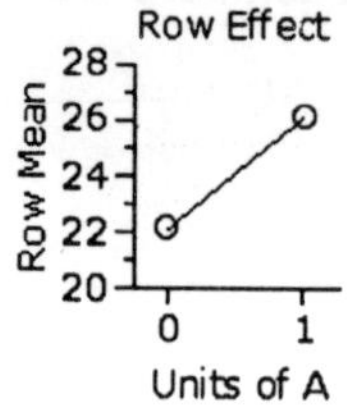

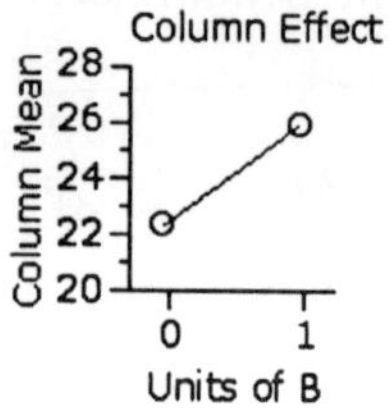

	col 1[B=0]	*col 2[B=1]*	*rowmeans*
row 1[A=0]			22.13
row 2[A=1]			26.04
columnmeans	22.35	25.82	

However, do keep in mind that the row means for the two levels of drug A are measured across the two levels of drug B, and that the column means for the two levels of drug B are measured across the two levels of drug A. This rows-by-columns complexity can make the interpretation of the main effects considerably less obvious when the two independent variables are interacting.

But for the present example it is all plain and simple. Each of the two drugs appears to increase arousal, and there is no indication that they interact with each other. When presented in combination, their effects are merely additive.

ANOVA Table

Source	*SS*	*df*	*MS*	*F*	*P*
between groups	273.4	1			
rows	153.3	1	153.3	17.67	<.01
columns	120.1	1	120.1	13.84	<.01
interaction	0.06	1	0.06	0.01	ns
within groups (error)	312.3	36	8.68		
TOTAL	585.7	39			

"ns" = "non-significant"

Our second example uses the same research setting as in Example and, as you will see in a moment, almost the same data.

First a reminder of the setting: To test the separate and mutual effects of two drugs, A and B, on physiological arousal, researchers randomly and independently sorted 40 laboratory rats into four groups of 10 subjects each. Each group received a certain dosage of drug A (zero units or 1 unit) and a certain dosage of drug B (zero units or 1 unit). The dependent variable was a standard measure of physiological arousal. One of the groups served as a control, receiving only an inert placebo containing zero units of A and zero units of B.

The following table shows the consequent measures of physiological arousal for each subject in each of the four groups. If you have a sharp eye and a keen memory, you will see that all of the values listed for groups 1, 2, and 3 are the same as in Example. The only difference is in the listing for group 4, whose members receive 1 unit of A and 1 unit of B.

raw data		*B* *0 units*		*1 unit*	
A	0 units	20.4	17.4	20.5	26.3
		20.0	18.4	26.6	19.8
		24.5	21.0	25.4	28.2
		19.7	22.3	22.6	23.7
		17.3	23.3	22.5	22.6
	1 unit	22.4	19.1	17.5	13.6
		22.4	25.4	16.9	12.4
		26.2	25.1	16.4	18.3
		28.8	21.8	13.6	19.1
		26.3	25.2	16.1	20.5

As the procedure follows the same format we worked through in Example, we will zip through it with only a minimum of commentary.

VALUES FROM PRELIMINARY NUMBER-CRUNCHING

summary data		*B* *0 units*	*1 unit*	*rows*
A	0 units	N_{g1}=10 ΣX_{g1}=204.3 ΣX^2_{g1}=4226.3	N_{g2}=10 X_{g2}=238.2 X^2_{g2}=5741.4	N_{r1}=20 ΣX_{r1}=442.5
	1 unit	N_{g3}=10 ΣX_{g3}=242.7 ΣX^2_{g3}=5961.34	N_{g4}=10 ΣX_{g4}=164.4 ΣX^2_{g4}=2764.6	N_{r2}=20 ΣX_{r2}=407.1
	columns	N_{c1}=20 ΣX_{c1}=447.0	N_{c2}=20 ΣX_{c2}=402.6	N_T=40 ΣX_T=849.6 ΣX^2_T=18693.7

MEANS AND GRAPH OF GROUP MEANS

		B		
		0 units	*1 unit*	*rows*
A	0 units	M_{g1}=20.43	M_{g2}=23.82	M_{r1}=22.13
	1 units	M_{g3}=24.27	M_{g4}=16.44	M_{r2}=20.36
	coluMns	M_{c1}=22.35	M_{c2}=20.13	M_T=21.24

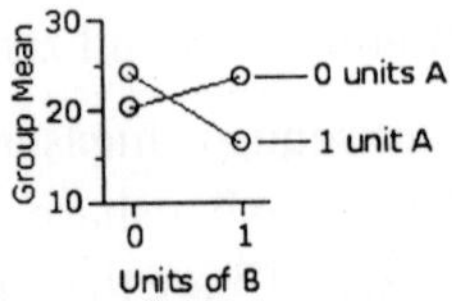

PRELIMINARY SS VALUES

		B	
		0 units	*1 unit*
A	0 units	SS_{g1}=52.44	SS_{g2}=67.48
	1 unit	SS_{g3}=71.02	SS_{g4}=61.40
			SS_T=647.53

$$SS_{wg} = SS_{g1} + SS_{g2} + SS_{g3} + SS_{g4}$$
$$= 52.44 + 67.48 + 71.02 + 61.40$$
$$= 252.34.$$

$$SS_{bg} = SS_T - SS_{wg}$$
$$= 647.53 - 252.34$$
$$= 395.19.$$

$$SS_{\text{rows}} = \frac{(\Sigma X_{r1})^2}{N_{r1}} + \frac{(\Sigma X_{r2})^2}{N_{r2}} - \frac{(\Sigma X_T)^2}{N_T}$$
$$= \frac{(442.5)^2}{20} + \frac{(407.1)^2}{20} - \frac{(849.6)^2}{40}$$
$$= 31.33.$$

$$SS_{\text{cols}} = \frac{(\Sigma X_{c1})^2}{N_{c1}} + \frac{(\Sigma X_{c2})^2}{N_{c2}} - \frac{(\Sigma X_T)^2}{N_T}$$
$$= \frac{(447.0)^2}{20} + \frac{(402.6)^2}{20} - \frac{(849.6)^2}{40}$$
$$= 49.28$$

$$SS_{\text{rxc}} = SS_{bg} - SS_{\text{rows}} - SS_{\text{cols}}$$
$$= 395.19 - 31.33 - 49.28$$
$$= 314.58.$$

The following table shows (in red) the values of [null]M_{g*} for each of the four groups, as calculated by the method described in connection with Example. As before, the observed means of the groups (20.43, 23.82, etc.) appear in black. The graphs below the table show the observed group means in comparison with the pattern that would be expected if there were zero interaction between the row and column variables. As you can see from both the table and the graphs, there is a substantial difference between the observed and the expected.

		B		
		0 units	*1 unit*	*rows*
A	0 units	20.43 23.24	23.82 21.02	M_{r1}=22.13
	1 unit	24.27 21.47	16.44 19.25	M_{r2}=20.36
	columns	M_{c1}=22.35	M_{c2}=20.13	M_T=21.24

observed expected.

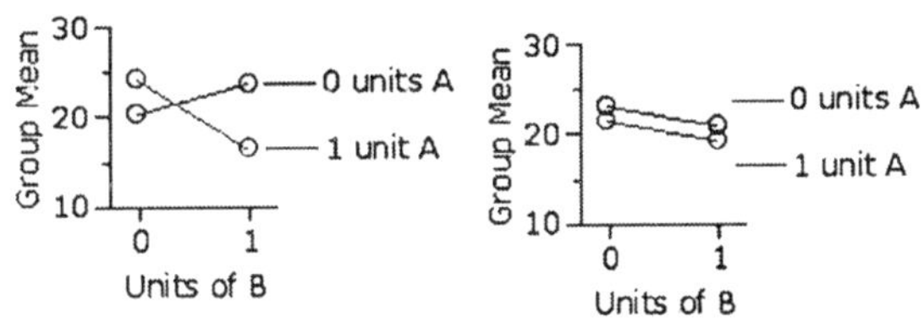

DEGREES OF FREEDOM

degrees of freedom	*in general*	*for the present example*
Total	$df_T = N_T - 1$	40–1=39
within-groups (error)	$df_{wg} = N_T - rc$	40–(2)(2)=36
between-groups	$df_{bg} = rc - 1$	(2)(2)–1=3
rows	$df_{rows} = r - 1$	2–1=1
columns	$df_{cols} = c - 1$	2–1=1
interaction	$df_{rxc} = (r-1)(c-1)$	(2–1)(2–1)=1

MS VALUES

$$MS_{rows} = \frac{SS_{rows}}{df_{rows}} \quad MS_{cols} \frac{SS_{cols}}{df_{cols}} \quad MS_{rxc} \frac{SSrxc}{df_{rxc}}$$

$$= \frac{31.33}{1} = \frac{49.28}{1} = \frac{314.58}{1}$$

$$= 31.33 = 49.28 = 314.58.$$

$$MS_{error} = \frac{SSwg}{df_{wg}}$$

$$= \frac{252.34}{36} = 7.01.$$

F-RATIOS[H]

$$F_{rows} = \frac{MS_{rows}}{MS_{error}} \quad F_{cols} = \frac{MS_{cols}}{MS_{error}} \quad F_{rxc} = \frac{MSrxc}{MS_{error}}$$

$$= \frac{31.33}{7.01} = \frac{49.28}{7.01} = \frac{314.58}{7.01}$$

$$= 4.47 = 7.03 = 44.88.$$

with df=1,36 with df=1,36 with df=1,36.

Here again is the sampling distribution of F for df=1,36, along with the corresponding tabular portion. The critical values for the .05 and .01 levels of significance are F=4.11 and F=7.40, respectively.

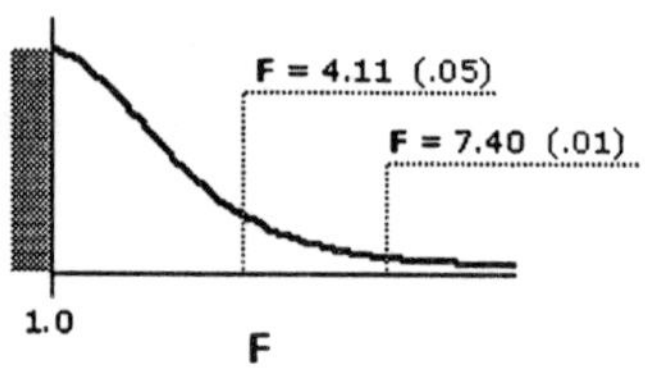

Fig. Sampling Distribution of F for df=1,36

df/*denominator*	df numerator		
	1	2	3
	4.11	3.26	2.87
36	7.4	5.25	4.38

In brief, all three of the effects are significant: the main effects for rows and columns (F_{rows}=4.47, F_{cols}=7.03) both beyond the .05 level, and the large interaction effect (F_{rxc}=44.88) far beyond the .01 level. This is the easy part of the analysis. Now for the trickier part, which is to figure out what it means.

Once again, the fundamental meaning of the significant row and column effects is that the difference between the two row means (22.13 vs 20.36) and the difference between the two column means (22.35 vs 20.13) each reflect something more than mere random variability. But note what happens when you simply take these two differences at face value, without reference to the very substantial interaction effect. As indicated by the following plots of the row and column means, the "obvious" inference is that drugs A and B both decrease arousal.

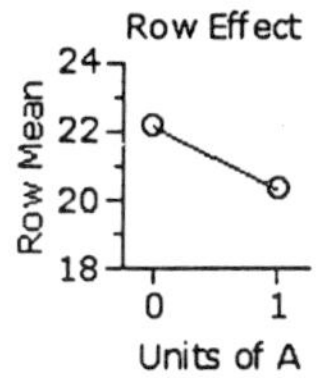

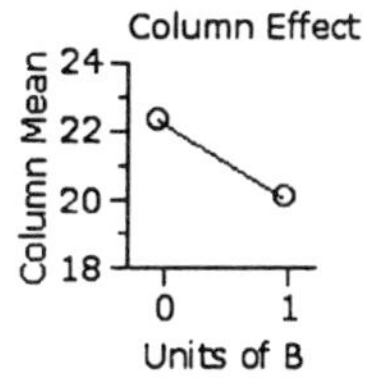

Examine the data in closer detail, however, and you will see that this conclusion is quite the opposite of what is really happening. For group 1, which receives only the inert placebo, the mean level of arousal is 20.43. For group 2, which receives 1 unit of B and none of A, it is higher: 23.82. For group 3, which receives 1 unit of A and none of B, it is also higher: 24.27. The complication comes with group 4, which receives 1 unit each of A and B; for here the effects of the two drugs in combination are not merely additive..

		B	
		0 units	*1 unit*
A	0 units	M_{g1}=20.43	M_{g2}=23.82
	1 unit	M_{g3}=24.27	M_{g4}=16.44

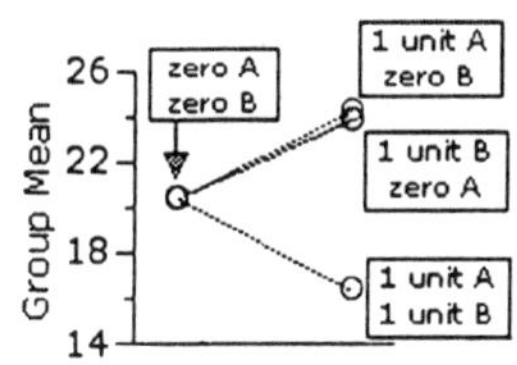

Indeed, they are the very opposite of additive: A and B in combination produce a mean arousal level of 16.44, substantially lower than the 20.43 observed with the inert placebo group. Presented separately, each of the two drugs increases arousal, while in combination they interact to produce a marked decrease in arousal.

Although the above graph captures the texture of the interaction effect in this particular example, the more generic format is the one shown earlier (now adjacent), in which the individual group means are plotted across their respective rows. The fundamental generic meaning of a significant interaction effect is that the difference between -

Interaction Effect

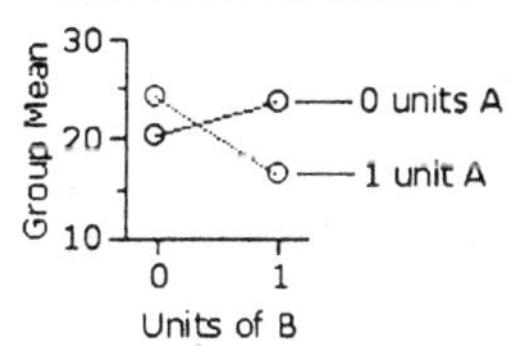

- The observed pattern of group means;
- The pattern that would be expected if the combined effects of the row and column variables were merely additive reflects something more than mere random variability.

ANOVA Table:

Source	*SS*	df	MS	F	P
between groups	395.19	3			
rows	31.33	1	31.33	4.47	<.05
columns	49.28	1	49.28	7.03	<.05
interaction	314.58	1	314.58	44.88	<.01
within groups (error)	252.34	36	7.01		
TOTAL	647.53	39			

The first difference between the present example and the first two is simply structural. The second difference, of greater conceptual importance, is that we will now be analyzing data that derive not from an equal-interval scale of measurement, but from a merely ordinal rating scale. The analysis of variance is quite robust—relatively unperturbed when its assumptions are not fully met—providing that the several groups of measures are all of the same size. These assumptions are the same for the two-way independent-samples ANOVA as for the one-way version:

1. That the scale on which the dependent variable is measured has the properties of an equal interval scale;$_{\text{T}}$
2. That the several samples are independently and randomly drawn from the source population(*s*);$_{\text{T}}$
3. That the source population(*s*) can be reasonably supposed to have a normal distribution; and$_{\text{T}}$
4. That the several samples have approximately equal variances.

Although the flexibility afforded by this robustness applies most directly to assumptions 3 and 4, there is also a case to be made for its being extended to assumption. The details of that case are more complex than we would want to get into just now. Suffice it to say that it is less compelling than the case that can be made for flexibility with respect to assumptions 3 and 4. Compelling or not, however, you will often find the analysis of variance applied to rating-scale data, and for that reason you should have a sense of what such an analysis looks like.

The question at issue in the project was whether an individual's level of self-esteem is predictive of how he or she will perceive other persons. As a first step, 72 subjects (college students) were pre-sorted according to their relative self-esteem scores (high, medium, or low) as measured by the Coopersmith Self-Esteem Inventory. Within each of these three groupings by self-esteem level, subjects were then randomly assigned to one or the other of two experimental conditions. For each condition, the subject's task was to read a description of a "target" person. In one of the conditions the description was worded so as to suggest a low level of self-esteem on the part of the target, while in the other it was worded to suggest a high level of self-esteem. After reading the description, subjects were then asked to rate the target according to three questions:

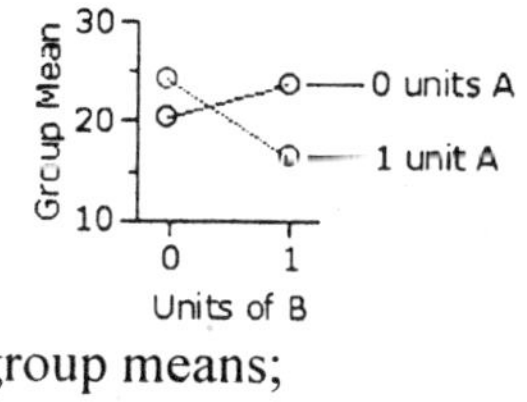

- how happy is this person?
- how similar is this person to you? and
- how likely is this person to experience divorce at some point in life?

For each question the rating was performed on a 5-point scale, with "1" representing the lowest rating and "5" the highest.

The following table shows the ratings that resulted from the second question: how similar is this person to you? Note that there is a total of rc=6 independent groups, with N_g=12 observations in each group.

		Subject's Self-Esteem		
raw data		*Low*	*Medium*	*High*
		4 4	3 3	3 1
		3 5	3 4	3 3
	Low	4 4	4 2	3 5
		5 4	4 4	3 2
		2 4	1 2	3 3
Target's Self-Esteem		4 2	2 3	3 3
		2 2	4 3	3 2
		4 2	1 2	3 2
	High	2 3	1 3	3 4
		2 4	2 4	3 4
		2 2	3 1	4 3
		2 3	1 4	3 4

Once again, we can zip through the computational details of the analysis with a minimum of commentary.

VALUES FROM PRELIMINARY NUMBER-CRUNCHING [H]

summary data		*Subject's Self-Esteem* *Low*	*Medium*	*High*	*rows*
Target's Self-Esteem	Low	N_{g1}=12 ΣX_{g1}=45 ΣX^2_{g1}=179	N_{g2}=12 ΣX_{g2}=35 ΣX^2_{g2}=113	N_{g3}=12 X_{g3}=35 X^2_{g3}=111	N_{r1}=36 ΣX_{r1}=115
	High	N_{g4}=12 ΣX_{g4}=30 X^2_{g4}=82	N_{g5}=12 X_{g5}=29 X^2_{g5}=87	N_{g6}=12 X_{g6}=38 X^2_{g6}=126	N_{r2}=36 ΣX_{r2}=97
	columns	N_{c1}=24 ΣX_{c1}=75	N_{c2}=24 ΣX_{c2}=64	N_{c3}=24 ΣX_{c3}=73	N_T=72 ΣX_T=212 ΣX^2_T=698

MEANS AND GRAPH OF GROUP MEANS

		Subject's Self-Esteem			
		Low	*Medium*	*High*	*rows*
Target's	Low	M_{g1}=3.8	M_{g2}=2.9	M_{g3}=2.9	M_{r1}=3.2
Self- EsteeM	High	M_{g4}=2.5	M_{g5}=2.4	M_{g6}=3.2	M_{r2}=2.7
	coluMns	M_{c1}=3.1	M_{c2}=2.7	M_{c3}=3.0	M_T=2.9

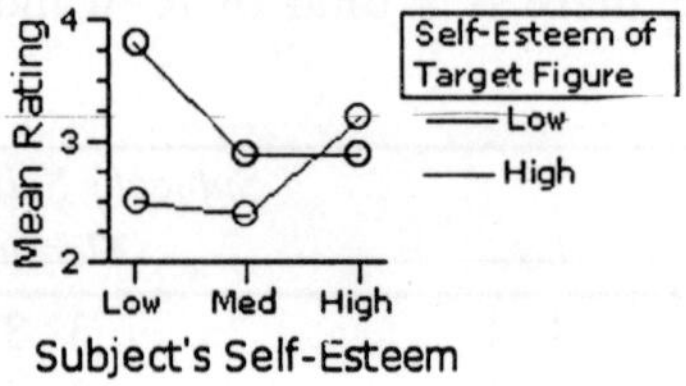

PRELIMINARY SS VALUES

		Columns			
		col1	*col2*	*col3*	
Rows	row1	SS_{g1}=10.25	SS_{g2}=10.92	SS_{g3}=8.92	
	row2	SS_{g4}=7.00	SS_{g5}=16.92	SS_{g6}=5.67	
					SS_T=73.78

$$SS_{wg} = SS_{g1} + SS_{g2} + SS_{g3} + SS_{g4} + SS_{g5} + SS_{g6}$$

$$= 10.25 + 10.92 + 8.92 + 7.00 + 16.92 + 5.67$$

$$= 59.67$$

$$SS_{bg} = SS_T - SS_{wg}$$

$$= 73.78 - 59.67$$

$$= 14.11.$$

$$SS_{rows} = \frac{(\Sigma X_{r1})^2}{N_{r1}} + \frac{(\Sigma X_{r2})^2}{N_{r2}} - \frac{(\Sigma X_T)^2}{N_T}$$

$$= \frac{(115)^2}{36} + \frac{(97)^2}{36} - \frac{(212)^2}{72}$$

$$= 4.50.$$

$$SS_{cols} = \frac{(\Sigma X_{c1})^2}{N_{c1}} + \frac{(\Sigma X_{c2})^2}{N_{c2}} + \frac{(\Sigma X_{c3})^2}{N_{c3}} - \frac{(\Sigma X_T)^2}{N_T}$$

$$= \frac{(75)^2}{24} + \frac{(64)^2}{24} + \frac{(73)^2}{24} - \frac{(212)^2}{72}$$

$$= 2.86$$

$$SS_{rxc} = SS_{bg} - SS_{rows} - SS_{cols}$$
$$= 14.11 - 4.50 - 2.86$$
$$= 6.75.$$

The following table shows (in red) the values of [null]M_{g*} for each of the six groups, as calculated by the method described in connection with Example. As before, the observed means of the groups (3.8, 2.9, etc.) appear in black. The graphs below the table show the observed group means in comparison with the pattern that would be expected if there were zero interaction between the row and column variables. Here again you can see the makings of an interaction effect.

means			*Columns*		
		col1	*col2*	*col3*	*rows*
rows	row1	3.8	2.9	2.9	M_{r1}=3.2
		3.4	3	3.3	
	row2	2.5	2.4	3.2	M_{r2}=2.7
	2.9	2.5	2.8		
	columns	M_{c1}=3.1	M_{c2}=2.7	M_{c3}=3.0	M_T=2.9

observed

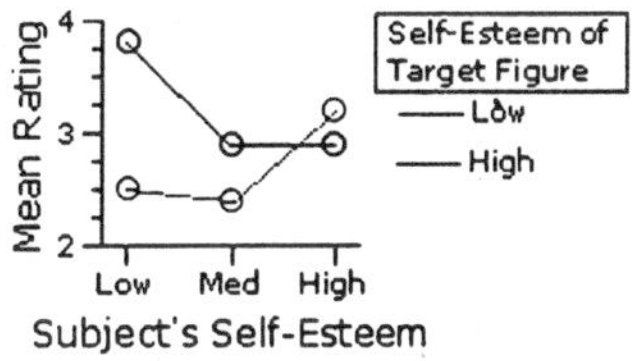

expected

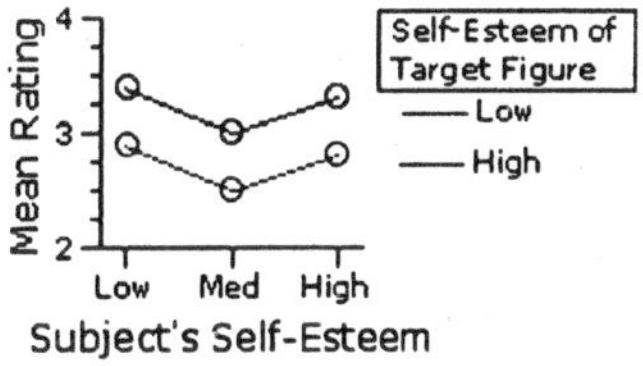

DEGREES OF FREEDOM

degrees of freedom	*in general*	*for the present example*
Total	$df_T = N_T - 1$	72–1=71
within-groups (error)	$df_{wg} = N_T - rc$	72–(2)(3)=66
between-groups	$df_{bg} = rc - 1$	(2)(3)–1=5
rows	$df_{rows} = r - 1$	2–1=1
columns	$df_{cols} = c - 1$	3–1=2
interaction	$df_{rxc} = (r-1)(c-1)$	(2–1)(3–1)=2

MS VALUES

$$MS_{rows} = \frac{SS_{rows}}{df_{rows}} \quad MS_{cols} = \frac{SS_{cols}}{df_{cols}} \quad MS_{rxc} = \frac{SS_{rxc}}{df_{rxc}}$$

$$= \frac{4.50}{1} \quad = \frac{2.86}{2} \quad = \frac{6.75}{2}$$

$$= 4.50 = 1.43 = 3.37.$$

$$MS_{\text{error}} = \frac{SS_{wg}}{\text{df}_{wg}}$$

$$= \frac{59.97}{66} = 0.90.$$

F-RATIOS

$$F_{\text{rows}} = \frac{MS_{\text{rows}}}{MS_{\text{error}}} \quad F_{\text{cols}} = \frac{MS_{\text{cols}}}{MS_{\text{error}}} \quad F_{rxc} = \frac{MS_{rxc}}{MS_{\text{error}}}$$

$$= \frac{4.50}{0.90} = \frac{1.43}{0.90} = \frac{3.37}{0.90}$$

$$= 4.98 = 1.58 = 3.73.$$

with df=1,66/with df=2,66/with df=2,66.

The critical values of F for df=1,66 and df=2,66. As you can see, F_{rows} and F_{rxc} are both significant beyond the .05 level, while F_{cols} is non-significant. Thus there is a significant main effect for the row variable and a significant rows-by-columns interaction effect, but no significant main effect for the column variable.

df denominator	*df numerator*			
	1	2	3	F_{rows}=4.98 df=1,66
66	3.99	3.14	2.74	F_{cols}=1.58 df=2,66^{T}
	7.04	4.94	4.09	F_{rxc}=3.73 df=2,66^{T}

The following table shows a plot for each of these dimensions of the analysis, along with a little commentary. Recall that the dependent variable in this research is a subject's rating of the target figure in response to the question: how similar is this person to you?

Rows

Columns

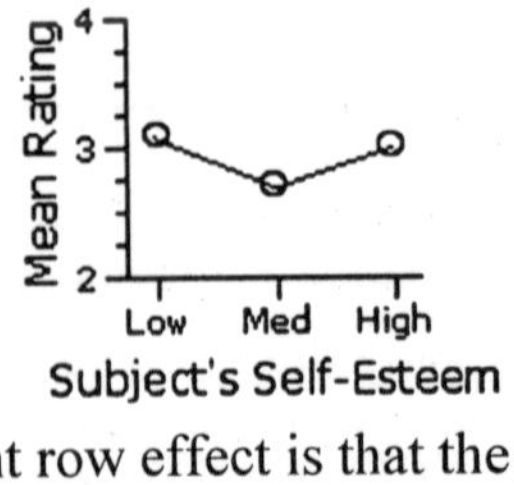

The fundamental meaning of the significant row effect is that the difference between the two row means is greater than what could be expected on the basis of mere random variability. In the present instance it indicates that subjects rating the low self-esteem targets saw the targets as more similar to themselves than did subjects rating the high self-esteem targets. Recall that the row means in this example are calculated across all three levels (low, medium, and high) of subjects' self-esteem.

The fundamental meaning of the non-significant column effect is that the aggregate differences among the three column means are no greater than what might be expected on the basis of random variability. Hence the observed results provide no evidence that similarity ratings of the target figure differ in accordance with the self-esteem level of the subjects doing the ratings. Recall that column means in the example are calculated across both levels (low and high) of the target figure's imputed self-esteem.

Interaction Observed

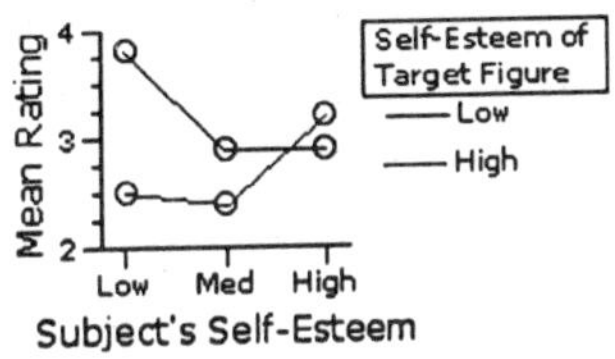

Expected Zero Interaction

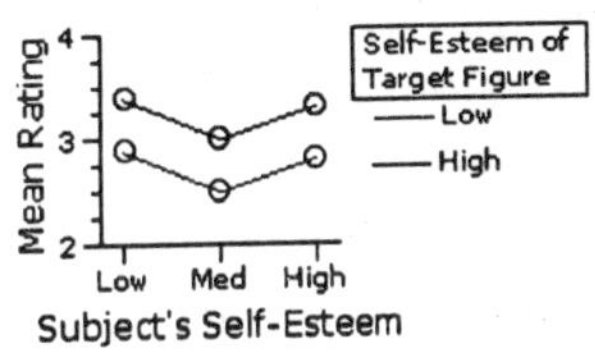

As in the earlier examples, the interaction effect refers to the difference between

- The observed pattern of group means and
- The pattern that would be expected if the combined effects of the row and column variables were merely additive.

A significant interaction indicates that this difference reflects something more than mere random variability. For the present example, the short of it is that low and high self-esteem target figures are rated by subjects as more or less similar to themselves in accordance with the self-esteem level of the subjects. As evident from the adjacent graph, this is particularly salient when subjects in the lowest self-esteem category are rating a low self-esteem target figure, and when subjects in the highest self-esteem category are rating a high self-esteem target figure. (The fact that low self-esteem target figures, overall, are rated as more similar than high self-esteem targets is a reflection of the significant main effect for the row variable.)

THE QUESTION OF ROBUSTNESS WITH ORDINALSCALE DATA [H]

When a human subject rates something on a 5-point scale, we may be reasonably confident that the scale has ordinal properties, such that "5" represents something greater than "4"; "4" represents something greater than "3"; and so on. But we can have no confidence at all that the points on the scale are separated by equal intervals:$_T$

1 2 3 4 5$_T$

For all we know, the scale could look like this:$_T$

1 2 3 4 5$_T$

or this:$_T$

1 2 3 4 5$_T$

or any one of a multitude of other non-equal-interval possibilities. We can also have no confidence that the scale intervals underlying a subject's rating on one item are the same as might underlie his or her rating on another item. By the same token, we can have no confidence that any multiplicity of subjects rating the same item are doing so on the basis of the same scale intervals.

But what does it matter if the scale of measurement is merely ordinal? A "5" is still greater than a "4," and surely it makes a kind of sense to say that the mean rating of a group of subjects on a certain item is 2.6, 3.4, or whatever it might be. Indeed, it does make a kind of sense; but please keep clearly in mind that the sense it makes is not nearly so strong as when you say, for example, that the mean weight of a sample of seeds of a certain plant species is 1.3 grams; or that the mean number of "Yes" responses to a 20-item questionnaire is 14.2; and so on for any other example where the scale of measurement clearly is equal-interval. Basic mathematical operations involving addition, subtraction, multiplication, and division assume that the numbers being fed into them derive from an equal-interval scale of measurement, and they can sometimes get indigestion when that assumption is not met.

Still; it does make a kind of sense to speak of mean ratings, and in that degree it potentially makes sense to speak of a significant difference between or among the mean ratings of two or more groups of subjects. The question is: If you plug merely ordinal rating-scale data into an analysis of variance and end up with "significant" effects, are those effects really significant in the technical statistical meaning of that term. The following table will remind you of what that technical meaning is:$_T$

Conclusion	Technical meaning
"significant at the .05 level"-	If the null hypothesis were true, the observed effect would have had only a 5% chance of occurring through mere random variability.
"significant beyond the .05 level"	If the null hypothesis were true, the observed effect would have had less than a 5% chance of occurring through mere random variability.
"significant at the .01 level"	If the null hypothesis were true, the observed effect would have had only a 1% chance of occurring through mere random variability.
"significant beyond the .01 level"	If the null hypothesis were true, the observed effect would have had less than a 1% chance of occurring through mere random variability.

and so on for any other level of significance.

With respect to the use of ordinal-scale data in the analysis of variance, one way of approaching the question of robustness is this: If you were to plug such data into the analysis, and if the null hypothesis were clearly true (such that there are really no effects at all within the population[*s*] of measures from which the data are drawn), would there still be only a 5% chance of ending up with an effect "significant" at the .05 level?; would there still be only a 1% chance of ending up with an effect "significant" at the .01 level?; and so on.

The following exercise simulates this situation through a somewhat elaborate shaping of random numbers. The procedure begins by drawing 5 random integers falling within the range of zero to 1000, with each possible integer having an equal chance of being drawn. For example:$_T$

650 806 104 792 446$_T$

These are then rank-ordered from lowest to highest, and each is associated with the corresponding value on a 5-point scale, as indicated in blue:$_T$

1 104
2 446
3 650
4 792
5 806$_T$

Then another random integer between zero and 1000 is drawn, and it is assigned the ordinal-scale value (1, 2, 3, 4, or 5) of the one that it is closest to in the previous set of 5 random integers. For example: if the sixth random integer were 43, it would be closest to 104 in the above list, hence assigned the ordinal-scale value of 1; if it were 634, it would be closest to 650, hence assigned the ordinal-scale value of 3; and so on.

This final ordinal-scale assignment is then treated as though it were the rating (1, 2, 3, 4, or 5) of one particular subject in one particular group. On analogy with the structure of Example, we perform this random drawing 72 times, so as to end up with the ratings of 72 "subjects," 12 in each of six groups arranged in a matrix of two rows by three columns.

In the first of the two tables following this paragraph there is one button labeled "1 Sample" and another labeled "10 Samples." Clicking the first will generate one sample of 72 random "ratings"; clicking the second will generate 10 samples. For each sample the F-ratios for the row, column, and interaction effects will be displayed, along with an indication ("Yes!") if the F-ratio is "significant" at or beyond the basic .05 level. You will also see an indication of the cumulative percentages of F-ratios that turn out significant at or beyond the .05 level. As you click one or another of the buttons, keep in mind that the null hypothesis in this situation is true. Any mean differences that appear among the groups therefore result from nothing more than random variability. If the analysis of variance is robust in its treatment of these merely ordinal-scale data, the numbers of "significant" F-ratios for rows, columns, and interaction should each come out at just about 5% over the long haul. To give it a really fair test you will need to accumulate about 10,000 samples, which I recognize might exceed your time or inclination. At one point in developing the programming for this exercise, I set it to crank out 10,000 samples while I tended to something else. Lo and behold! In that particular long haul, the numbers of "significant" F-ratios actually did come out at just about 5% each.

Step-by-Step Computational Procedure

I will show the procedures for the case of 2 rows and 3 columns, hence rc=6. The modifications required for different values of *r* and c will be fairly obvious. The steps listed below assume that you have already done the basic number-crunching to get ΣX_i and ΣX^2_i for each of the groups separately and for all groups combined.

Step 1. Combining all rc groups together, calculate

$$SS_T = \Sigma X^2_i \quad \frac{(\Sigma X_i)^2}{N_T}$$

Step 2. For each of the rc groups separately, calculate the sum of squared deviates within the group ("g") as

$$SS_g = \Sigma X^2_{gi} \quad \frac{(\Sigma X_{gi})^2}{N_g}$$

Step 3. Take the sum of the SS_g values acroSS all rc groups to get

$$SS_{wg} = SS_a + SS_b + SS_c + SS_d + SS_e + SS_f$$

Step 4. Calculate SS_{bg} as

$$SS_{bg} = SS_T - SS_{wg}$$

Step 5. Calculate SS_{rows} as

$$SS_{\text{rows}} = \frac{(\Sigma X_{r1})^2}{N_{r1}} + \frac{(\Sigma X_{r2})^2}{N_{r2}} - \frac{(\Sigma X_T)^2}{N_T}$$

Step 6. Calculate SS_{cols} as

$$SS_{\text{cols}} = \frac{(\Sigma X_{c1})^2}{N_{c1}} + \frac{(\Sigma X_{c2})^2}{N_{c2}} + \frac{(\Sigma X_{c3})^2}{N_{c3}} - \frac{(\Sigma X_T)^2}{N_T}$$

Step 7. Calculate SS_{rxc} as

$$SS_{rxc} = SS_{bg} - SS_{\text{rows}} - SS_{\text{cols}}$$

Step 8. Calculate the relevant degrees of freedom as

$$\text{df}_T = N_T - 1$$

$$\text{df}_{wg} = N_T - rc$$

$$\text{df}_{bg} = rc - 1$$

$$\text{df}_{\text{rows}} = r - 1$$

$$\text{df}_{\text{cols}} = c - 1$$

$$\text{df}_{rxc} = (r-1)(c-1)$$

Step 9. Calculate the relevant mean-square values as

$$MS_{\text{rows}} = \frac{SS_{\text{rows}}}{\text{df}_{\text{rows}}} \quad MS_{\text{cols}} = \frac{SS_{\text{cols}}}{\text{df}_{\text{cols}}} \quad MS_{rxc} = \frac{SS_{rxc}}{\text{df}_{rxc}} \quad MS_{\text{error}} = \frac{SS_{wg}}{\text{df}_{wg}}$$

Step 10. Calculate F as

$$F_{\text{rows}} = \frac{MS_{\text{rows}}}{MS_{\text{error}}} \quad F_{\text{cols}} = \frac{MS_{\text{cols}}}{MS_{\text{error}}} \quad F_{rxc} = \frac{MS_{rxc}}{MS_{\text{error}}}$$

Step 11. Refer the calculated values of F to the table of critical values of F, with the appropriate pair of numerator/denominator degrees of freedom.

ONE-WAY ANALYSIS OF COVARIANCE FOR INDEPENDENT SAMPLES[H]

The analysis of covariance, often referred to by its acronym ANCOVA, is the result of a felicitous marriage between

1. The analysis of variance, and
2. The concepts and procedures of linear correlation and regression.

It is, in fact, a veritable powerhouse of a statistical procedure, and I am surprised it is not used more often. Like the correlated-samples designs for t-tests and the analysis of variance, it is capable of removing the obscuring effects of pre-existing individual differences among subjects. It does this, however, without resorting to the repeated-measures strategy of testing each subject under each of the several experimental conditions. In consequence, it is able to avoid the potential complications of having subjects tested first in one condition, then in another, and then perhaps in yet another.

In the correlated-samples design, it is typically assumed that these potential sequencing effects can be obviated by systematic counter-balancing. With two experimental conditions, A and B, you would test half your subjects in the sequence A·B and the other half in the sequence B·A. With three conditions, you would test equal numbers of your subjects in the six possible sequences: A·B·C, A·C·B, B·A·C, B·C·A, C·A·B, C·B·A. And so on.

But there are certain kinds of situations where a repeated-measures design might not be feasible; and there are others where, even if it is feasible, it might not be desirable. If you were interested in determining which of two methods of elementary mathematics instruction is more effective, it would clearly make no sense to have half your subjects taught first by Method A, then by Method B, and the other half taught first by Method B, then by Method A. Analogously, suppose you were interested in determining which of two methods of hypnotic induction is the more effective. Here it would certainly be possible to test each of your subjects under each of the two conditions–half in the sequence A·B and the other half in the sequence B·A–though I suspect you would end up with a substantial sequencing effect, even with the counter-balancing. By the time the A·B subjects get to Method B, they will have already experienced Method A; and that prior experience is likely to influence their response to Method B. Similarly for the B·A subjects: their responses to Method A are likely to be substantially influenced by their prior experience with Method B. In these two cases and in many others, the analysis of covariance provides the best of both worlds: for it allows you to examine the several experimental conditions independently, in isolation from each other, while at the same time removing from the situation the obscuring effects of pre-existing individual differences.

This, however, is only the first part of its strength. An even greater advantage of the analysis of covariance is that it allows you to compensate for systematic biases among your samples. To return for the moment to the example of two methods of elementary mathematics instruction, suppose you were to begin with a subject pool of 20 third-graders, randomly sorting them into two independent groups of 10 subjects each, with the design of teaching one group by Method A and the other by Method B. The aim of the initial random sorting is to ensure that the two groups are starting out approximately equal, on average, with respect to all factors that might be pertinent to how well they are likely to respond to elementary mathematics instruction. Nonetheless it could happen through sampling error—another name for sheer, cussed random variability—that the two groups do not start out on an equal footing with respect to one or another of these factors. To take the obvious examples, it could happen through mere chance coincidence in the sorting process that one of the groups starts out with a higher average level of intelligence or motivation, either one of which would surely complicate the situation. The analysis of covariance provides a way of measuring and removing the effects of such initial systematic differences between the samples.

The procedures of ANCOVA achieve these useful results through an application of the concept of covariation, the whole apparatus of linear correlation and regression. In any particular research situation, the individual differences that appear within your measures on the variable of interest are potentially correlated with something else. If the variable of interest is a measure of how well subjects learn elementary mathematics under one or the other of two methods of instruction, the potential correlates are likely to include such items as prior relevant learning, motivation, attention, self-discipline, and sheer intelligence. We will focus for the moment on intelligence, which is surely not the least of these correlates. Suppose we had a prior measure of intelligence for each subject in each of two randomly sorted independent groups. One group receives instruction by Method A, the other by Method B; and we then follow up by measuring each subject with respect to how well he or she has learned the designated subject matter. The result is two sets of bivariate measures, one set for each group. In keeping with the language of correlation and regression, we will designate these measures as X and Y. Thus

X = The measure of intelligence

Y = The measure of learning

Suppose, now, that we were actually to perform this experiment and end up with the results listed in the following table. For the sake of simplicity, I am showing only 4 subjects per group; in real-life research you would rarely want to have such tiny samples. I am also color-coding the values of X and Y to help you keep this cast of characters straight.

	Method A			*Method B*	
Subject	X_a	Y_a	*Subject*	X_b	Y_b
*a*1	88	66	b1	90	62
*a*2	98	85	b2	100	87
*a*3	100	90	b3	110	91
*a*4	110	97	b4	120	98
Means	99	84.5		105	84.5

I have designed this set of data to make two things obvious at a glance: First, that there is a considerable range of individual differences within the values of *X* and *Y* for each of the two groups; and second, that the values of *X* and *Y* within each group have a high positive correlation. In brief, and not surprisingly: the higher the measure of intelligence (*X*), the higher the measure of learning (*Y*). Not to put too fine a point on it: the smarter a subject is at the outset, the more likely it is that he or she will learn more rather than less, irrespective of the method of instruction. So a substantial portion of the variability that occurs within each of the sets of *Y* measures is actually covariance with the corresponding set of *X* measures. Remove that covariance from *Y* and you thereby remove a substantial portion of the extraneous variability of individual differences.

The second thing I have designed into this set of data is an outcome in which the means of the two sets of *Y* measures (learning) are precisely the same: M_{Ya}=M_{Yb}=84.5. While this precise identity of two group means would not often occur in real-life settings, it is a useful contrivance to make a point. If you were to look at M_{Ya} and M_{Yb} in isolation from the rest of the data, they would seem to be suggesting that Method A and Method B do not differ. Examine them, however, in conjunction with the information provided by the *X* measures (intelligence), and you will see another possibility.

	Method A		*Method B*	
	X_a	Y_a	X_b	Y_b
Means	99.0	84.5	105.0	84.5

By mere chance coincidence, the initial random sorting gave us two groups with different mean levels of intelligence: M_{Xa}=99.0 versus M_{Xb}=105.0. If there truly were no difference between Method A and Method B, we could reasonably expect that the higher mean intelligence of group B would have produced a higher mean level of learning. As it did not produce a higher mean level of learning, we can reasonably entertain the possibility that Method A would have proved the more effective, if only the two groups of subjects had started out on an equal footing. The analysis of covariance permits you to examine a possibility of this sort by way of a kind of what-if scenario. What would have happened if the two groups had started out with equivalent mean levels of intelligence? Or more generally: What would have happened if the several (two or more) groups had started out with equivalent mean levels of the covariate (intelligence, prior learning, motivation, or whatever else the relevant covariate might be).

Comparative Effects of Two Methods of Hypnotic Induction

This example is a follow-up on an item briefly mentioned above, structurally similar to the elementary-mathematics illustration, though with a somewhat more exotic setting. A team of researchers is interested in determining whether two methods of hypnotic induction, A and B, differ with respect to their effectiveness. They begin by randomly sorting 20 volunteer subjects into two independent groups of 10 subjects each, with the aim of administering Method A to one group and Method B to the other. But then, before either of the induction methods is administered, each subject is pre-measured on a standard index of "primary suggestibility," which is a variable known to be correlated with receptivity to hypnotic induction. The dependent variable, measured during the administration of Method A or Method B, is the subject's score on a standard index of hypnotic induction. Thus

X = the score on the index of primary suggestibility

Y = the score on the index of hypnotic induction

The following table shows the scores on both variables for each subject in each of the two groups. Also shown are the means of the four sets of measures.

	Method A				*Method B*
Subject	X_a	Y_a	*Subject*	X_b	Y_b
*a*1	5	20	*b*1	7	19
*a*2	10	23	*b*2	12	26
*a*3	12	30	*b*3	27	33
*a*4	9	25	*b*4	24	35
*a*5	23	34	*b*5	18	30
*a*6	21	40	*b*6	22	31
*a*7	14	27	*b*7	26	34
*a*8	18	38	*b*8	21	28
*a*9	6	24	*b*9	14	23
*a*10	13	31	*b*10	9	22
Means	13.1	29.2		18.0	28.1

X = suggestibility score

Y = induction score

As promised, here is a bit more on the Big Picture before we get into the computational details. In particular, I want to try to convey a sense of why one might bother with the analysis of covariance in a situation of this sort, rather than just performing a simple t-test or analysis of variance. This will allow you to look back and forth at the data without having to scroll up and down within the present frame.

I'll ask you to begin by focusing on the two sets of induction scores, Y, the variable in which we are chiefly interested. That with two independent samples the procedures for ANOVA and t are logically equivalent, with $F = t^2$. If you were to plug these values of Y into either a t-test or an analysis of variance, the result in each case would fall far short of significance at the basic .05 level: t=+.40 [df=18]; F=.16 [df=1,18]. These non-significant values of t and F reflect the fact, illustrated in Figure, that the mean difference between the two groups of induction scores (29.2–28.1=1.1) is quite small in comparison with the large amount of variability that occurs within the groups.

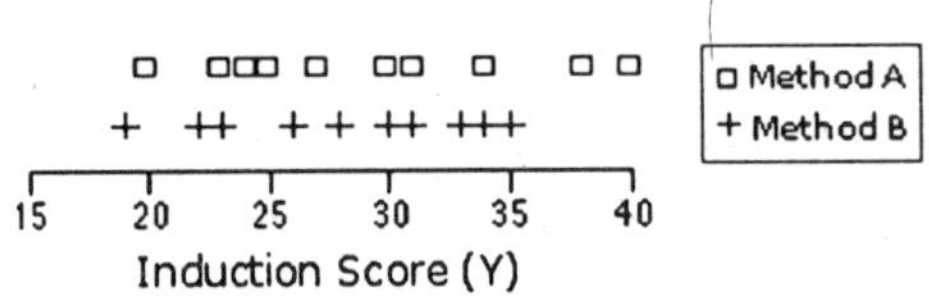

$M_{Ya} = 29.0$

$M_{Yb} = 28.1$

Fig. Variability of Induction Scores within Groups A and B

The bottom line here is that the induction scores, considered in and of themselves, would suggest that there is no particular difference between Method A and Method B, one way or the other.

But now see what happens when we look concurrently at the covariate X, "primary suggestibility." A substantial portion of the variability among the induction scores (both groups combined) is associated with pre-existing individual differences in suggestibility. In brief: the greater the level of pre-existing receptivity to hypnotic induction, the greater the response to the induction method.

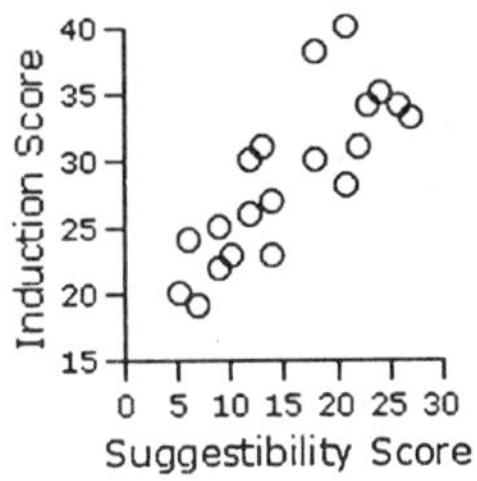

Fig. Correlation between X and Y: Both Groups Combined$_T$

The upshot of the correlation between X and Y is simply this:

Given $r^2 = .645{:}_T$

Of all the^{T}variability that exists among the induction scores (Y), 65.4% can be traced to pre-existing individual differences in primary suggestibility (X).

As you will see a bit later, this is the precise amount by which the analysis of covariance will be reducing the total variability of the induction scores (both groups combined). Following the same logic, though with different numbers, there will also be a reduction of within-groups variability and a corresponding increase in the relative weight of the mean difference between the two groups. Overall, the analysis of covariance removes from Y the portion of its variability that is attributable to covariation with X, and focuses only on what is left over (the "residual variance") following this removal.

And then there is also this to fold into the mix. The aim of the initial random sorting was to ensure that the two groups were starting out approximately equal, on average, with respect to all factors that might be relevant to how well they were likely to respond to any particular method of hypnotic induction. But random variability being the tricky thing that it is, they in fact end up with substantially different mean levels of the key covariate "primary suggestibility." The following abbreviated version of the data table will give you an overview of this portion of th picture.

	Method A		*Method B*	
	X_a	Y_a	X_b	Y_b
Means	13.1	29.2	18.0	28.1

Group A started out with the lower of the two mean levels of suggestibility (13.1 versus 18.0), yet ended up with the higher of the two mean levels of hypnotic induction (29.2

versus 28.1). Our analysis of covariance on this set of data will allow us to answer a what-if question of the type introduced in Part 1: What would have happened if the two groups had started out on an equal footing with respect to primary suggestibility?

So this, in brief, is why you would want to bother with the procedures of ANCOVA in a situation of this sort:

- First, to remove from *Y* the extraneous variability that derives from pre-existing individual differences, insofar as those differences are reflected in the covariate *X*;
- And second, to adjust the means of the *Y* measures to compensate for the fact the groups of subjects started out with different mean levels of the covariate *X*.

The calculations required to achieve these ends are given in Part 2, along with a few more observations on the Big Picture.

Comparative Effects of Two Methods of Hypnotic Induction

X = The score on the index of primary suggestibility

Y = The score on the index of hypnotic induction

	Method A			*Method B*	
Subject	*Xa*	*Ya*	*Subject*	*Xb*	*Yb*
*a*1	5	20	*b*1	7	19
*a*2	10	23	*b*2	12	26
*a*3	12	30	*b*3	27	33
*a*4	9	25	*b*4	24	35
*a*5	23	34	*b*5	18	30
*a*6	21	40	*b*6	22	31
*a*7	14	27	*b*7	26	34
*a*8	18	38	*b*8	21	28
*a*9	6	24	*b*9	14	23
*a*10	13	31	*b*10	9	22
Means	13.1	29.2		18.0	28.1

A basic one-way analysis of covariance requires four sets of calculations. In the first set you will clearly recognize the analysis-of-variance aspect of ANCOVA. The two middle sets are aimed at the covariance aspect, and the final set ties the two aspects together. As in earlier chapters, *SS* refers to the sum of squared deviates. The designation SC in the third set refers to the sum of co-deviates, the raw measure of covariance. In both cases, the subscripts "T," "wg," and "bg" refer to "Total," "within-groups," and "between-groups," respectively.

1. *SS* values for *Y*, the dependent variable in which one is chiefly interested.$_T$

   Items to be calculated:

   $SS_{T(Y)}$

   $SS_{wg(Y)}$

   $SS_{bg(Y)}$

2. *SS* values for *X*, the covariate whose effects upon *Y* one wishes to bring under statistical control.

   Items to be calculated:

   $SS_{T(X)}$

   $SS_{wg(X)}$

3. SC measures for the covariance of *X* and *Y*;

   Items to be calculated:

   SC_T

   SC_{wg}

4. And then a final set of calculations, which begin by removing from the *Y* variable the portion of its variability that is attributable to its covariance with *X*.

Calculations for the Dependent Variable *Y*

The following table shows the values of *Y* along with the several summary statistics required for the calculation of $SS_{T(Y)}$, $SS_{wg(Y)}$, and $SS_{bg(Y)}$:

	Y_a	Y_b	
	20	19	
	23	26	
	30	33	for total
	25	35	array of
	34	30	data
	40	31	
	27	34	
	38	28	
	24	23	
	31	22	
N	10	10	20
ΣY_i	292	281	573
ΣY_i^2	8920	8165	17085
SS	393.6	268.9	
Mean	29.2	28.1	28.7

Calculations for the Covariate *X*

The next table has the same structure as the one above, but now what we show are the values of *X* along with the several summary statistics required for the calculation of $SS_{T(X)}$ and $SS_{wg(X)}$. (We need not bother with $SS_{bg(X)}$, since it does not enter into any subsequent calculations.)

X_a	X_b
5	7
10	12

	12	27		
	9	24		
	23	18		
	21	22		
	14	26		
	18	21		
	6	14		
	13	9		
N	10	10	20	
$.\Sigma X_i$	131	180	311	$SS_{T(X)} = 908.9$
$.\Sigma X_i^2$	2045	3700	5745	$SS_{wg(X)} = 788.9$
SS	328.9	460		
Mean	13.1	18	15.6	

Calculations for the Covariance of *X* and *Y*

The raw measure of the covariance between two variables, *X* and *Y*, is a quantity known as the sum of co-deviates:

$$SC = \Sigma(X_i - M_X)(Y_i - M_Y)$$

$(X_i - M_X)$ = deviate$_X$

$(Y_i - M_Y)$ = deviate$_Y$

$(X_i - M_X)(Y_i - M_Y)$ = co-deviate$_{XY}$

which for practical computational purposes can be expressed as

$$SC = \Sigma(X_iY_i) - \frac{(\Sigma X_i)(\Sigma Y_i)}{N}.$$

We will be calculating two separate values of SC: one for the covariance of *X* and *Y* within the total array of data, and another for the covariance within the two groups.

The next table shows the cross-products of X_i and Y_i for each subject in each of the two groups. (E.g., the first subject in group A had X_i=5 and Y_i=20; hence X_iY_i=100.) Also shown are the separate sums of X_i and Y_i within each group and for the total array of data, as these values will also be needed for the calculations that follow.

Groups A X_aY_a	Groups B X_bY_b	
100	133	
230	312	For group A:
360	891	$\text{o}\Sigma X_{ai} = 131$
225	840	$\text{o}\Sigma Y_{ai} = 292$
782	540	For group B:
840	682	$\text{o}\Sigma X_{bi} = 180$

	378	884		$o\Sigma Y_{bi}=281$
	684	588	for total	For total array:
	144	322	array of	$o\Sigma X_{Ti}=311$
	403	198	data	$o\Sigma Y_{Ti}=573$
Sums	4146	5390	9536	
	$.\Sigma(X_{ai}Y_{ai})$	$.\Sigma(X_{bi}Y_{bi})$	$.\Sigma(X_{Ti}Y_{Ti})$	

On the basis of the summary values presented in this table you can then easily calculate SC for the total array of data as

$$SC_T = \Sigma(X_{Ti}Y_{Ti}) - \frac{(\Sigma X_{Ti})(\Sigma Y_{Ti})}{N_T}$$

$$SC_T = 9536 - \frac{(311)(573)}{20} = 625.9.$$

The values of *SC* within each of the two groups can be calculated similarly as

For Group A:

$$SC_{wg(a)} = \Sigma(X_{ai}Y_{ai}) - \frac{(\Sigma X_{ai})(\Sigma Y_{ai})}{N_a}$$

$$SC_{wg(a)} = 4146 - \frac{(131)(292)}{10} = 320.8.$$

For Group B:

$$SC_{wg(b)} = \Sigma(X_{bi}Y_{bi}) - \frac{(\Sigma X_{bi})(\Sigma Y_{bi})}{N_b}$$

$$SC_{wg(b)} = 5390 - \frac{(180)(281)}{10} = 332.0.$$

the sum of which will then yield the within-groups covariance measure of

$$SC_{wg} = SC_{wg(a)} + SC_{wg(b)}$$
$$= 320.8 + 332.0 = 652.8$$

4. The Final Set of Calculations

We begin with a summary of the values of *SS* and SC obtained so far, as these will be needed in the calculations that follow. Recall that *Y* is the variable in which we are chiefly interested, and *X* is the covariate whose effects we are seeking to remove.

X	*Y*	*Covariance*
$SS_{T(X)} = 908.9$	$SS_{T(Y)} = 668.5$	$SC_T = 625.9$
$SS_{wg(X)} = 788.9$	$SS_{wg(Y)} = 662.5$	$SC_{wg} = 652.8$
	$SS_{bg(Y)} = 6.0$	

4a. Adjustment of $SS_{T(Y)}$

The overall correlation between X and Y (both groups combined) can be calculated as

$$r_T = \frac{SC_T}{\text{sqrt}[SST(X) \times SS_{T(Y)}]}$$

$$= \frac{625.9}{\text{sqrt}\,[908.9 \times 668.5]}$$

$$= +.803.$$

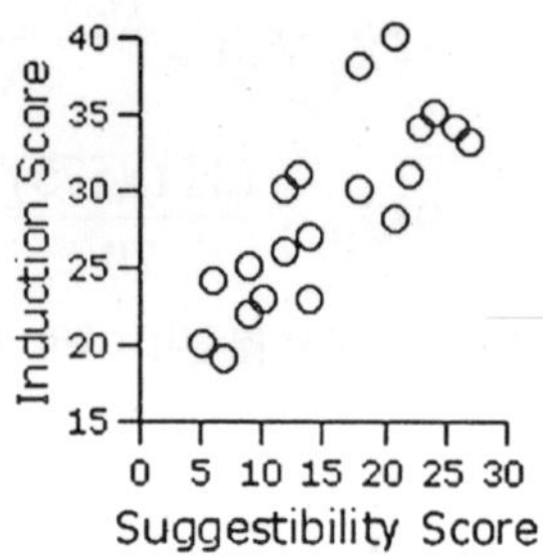

The proportion of the total variability of Y attributable to its covariance with X is accordingly

$$(r_T)^2 = (+.803)^2 = .645$$

In the first step of this series, we adjust $SS_{T(Y)}$ by removing from it this proportion of covariance. Given $SS_{T(Y)}$=668.5, the amount to be removed is

$668.5 \times .645 = 431.2$

and the adjusted value of $SS_{T(Y)}$ is therefore

$668.5 - 431.2 = 237.3$ [tentative]

I have marked the above result as tentative because a calculation based on r^2 is likely to produce rounding errors [in the present example, the 0.645 value used for $(r_T)^2$ is rounded from 0.64475...]. For practical purposes, one is better advised to use the following algebraically equivalent computational formula:

$$[\text{adj}]SS_{T(Y)} = SS_{T(Y)} - \frac{(SC_T)^2}{SS_{T(X)}} \quad \text{“[adj]” = “adjusted”}$$

$$= 668.5 - \frac{(625.9)^2}{908.9}$$

$$= 237.5.$$

4b. Adjustment of $SS_{wg(Y)}$

By analogy, the aggregate correlation between X and Y within the two groups can be calculated as

$$r_{wg} = \frac{\text{SCwg}}{\text{sqrt}[SS_{wg(X)} \times SS_{wg(Y)}]}$$

$$= \frac{652.8}{\text{sqrt}\,[788.9 \times 662.5]}$$

$$= +.903.$$

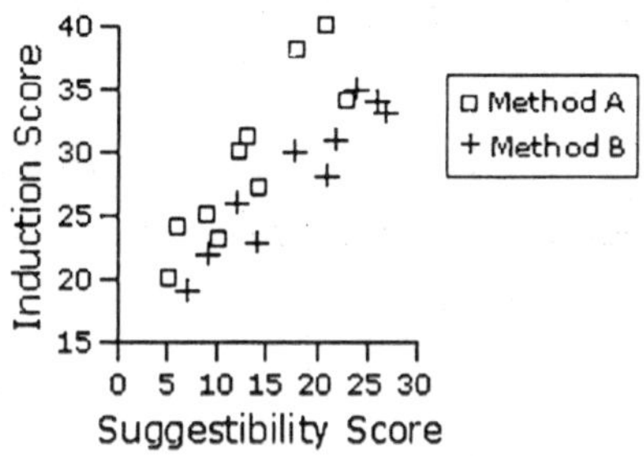

The proportion of the within-groups variability of Y attributable to covariance with X is therefore

$$(r_{wg})^2 = (+.903)^2 = .815$$

The amount of $SS_{wg(Y)}$(=662.5) to be removed is accordingly

$$662.5 \times .815 = 539.9$$

and the adjusted value of $SS_{wg(Y)}$ is therefore

$$662.5 - 539.9 = 122.6 \text{ [tentative]}$$

Here again the result is marked tentative on account of the possibility of rounding errors. The more precise result is given by the following computational formula:

$$[\text{adj}]SS_{wg(Y)} = SS_{wg(Y)} - \frac{(SC_{wg})^2}{SS_{wg(X)}}$$

$$= 662.5 - \frac{(652.8)^2}{788.9}$$

$$= 122.3.$$

4c. Adjustment of $SS_{bg(Y)}$

The adjusted value of $SS_{bg(Y)}$ can then be obtained through simple subtraction as

$$[\text{adj}]SS_{bg(Y)} = [\text{adj}]SS_{T(Y)} - [\text{adj}]SS_{wg(Y)}$$
$$= 237.5 - 122.3 = 115.2.$$

4d. Adjustment of the Means of Y for Groups A and B

A moment ago, while adjusting the value of $SS_{wg(Y)}$, we spoke of the aggregate correlation between X and Y within the two groups. Although this particular correlation is rather more abstract than the correlations one normally encounters, it is nonetheless a genuine correlation; and, as such, it can be described by a regression line defined by its slope and intercept. Our immediate interest is in its slope, which on analogy with the concept of slope can be calculated as

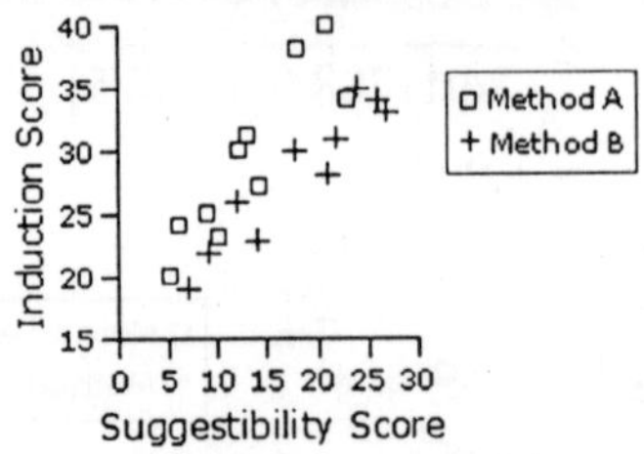

$$b_{wg} = \frac{SC_{wg}}{SS_{wg(X)}}$$

$$= \frac{652.8}{788.9} = +.83.$$

Recall that the slope of the regression line is the measure of the average amount by which Y increases or decreases as a function of X. In the present example, an increase by 1 unit of X is associated with an average increase of .83 units of Y; an increase by 2 units of X is associated with an average increase of 2x.83=1.66 units of Y; and so forth. Similarly, a decrease by 1 unit of X is associated with an average decrease of .83 units of Y; a decrease by 2 units of X is associated with an average decrease of 2x.83=1.66 units of Y; and so on.

	M_X	M_Y
group A	13.1	29.2
group B	18.0	28.1
combined	15.55	28.65

And now for our what-if scenario. Suppose that both groups had started out with the same mean level of suggestibility: namely, 15.55, which is the mean of X for both groups combined. In this case, group A would have been starting out with a mean suggestibility level 2.45 units higher than it actually started with, while group B would have been starting out with a mean suggestibility level 2.45 units lower. Given the observed dependance of Y on X, the respective means of the two groups would therefore presumably have been on the order of

$$[\text{adj}]M_{Ya} = 29.2 + (2.45 \times .83) = 31.23$$

and

$$[\text{adj}]M_{Yb} = 28.1 - (2.45 \times .83) = 26.07.$$

This is the conceptual structure for the adjustment of the means of Y for the groups. Once you have the concept, the mechanics of the process can be accomplished somewhat more expeditiously via the following computational formula:

For Group A:

$$[\text{adj}]M_{Ya} = M_{Ya} - b_{wg}(M_{Xa} - M_{XT})$$
$$= 29.2 - .83(13.1 - 15.55)$$
$$= 31.23.$$

For Group B:

$$[\text{adj}]M_{Yb} = M_{Yb} - b_{wg}(M_{Xb} - M_{XT})$$
$$= 28.1 - .83(18.0 - 15.55)$$
$$= 26.07.$$

4e. Analysis of Covariance Using Adjusted Values of SS

As with the corresponding one-way ANOVA, the final step in a one-way analysis of covariance involves the calculation of an F-ratio of the general form

$$F = \frac{MS_{\text{effec}}}{MS_{\text{error}}} = \frac{MS_{bg}}{MS_{wg}} = SS_{bg}/\text{df}_{bg}$$

$$SS_{wg}/\text{df}_{wg}$$

The only difference is that now we are using the adjusted values of $SS_{bg(Y)}$ and $SS_{wg(Y)}$, along with one adjusted value of df. In the basic analysis of variance, the degrees of freedom for the within-groups variance is N_T-k, where k is the number of groups and N_T is the total number of subjects. In the analysis of covariance the number of within-groups degrees of freedom is reduced by one to accommodate the fact that the covariance portion of within-groups variability has been removed from the analysis. Hence the adjusted value of $\text{df}_{wg(Y)}$ is

$$[\text{adj}]\ \text{df}_{wg(Y)} = N_T - k - 1 \quad \text{for the present example: } 20 - 2 - 1 = 17$$

The degrees of freedom for the between-groups condition remains the same as for the one-way *ANOVA:*

$$\text{df}_{bg(Y)} = k - 1 \quad \text{for the present example: } 2 - 1 = 1$$

Given $[\text{adj}]SS_{bg(Y)}$=115.2 and $[\text{adj}]SS_{wg(Y)}$=122.3, our F-ratio for the analysis of covariance is accordingly

$$F = \frac{[\text{adj}]SSbg(Y)/\text{df}bg_{(Y)}}{[\text{adj}]SS_{wg(Y)}/[\text{adj}]\text{df}_{wg(Y)}}$$
$$= \frac{115.2/1}{122.3/17}$$
$$= 16.01 \quad [\text{with df=1,17}]$$

df denominator	*df numerator* 1	2	3
17	4.45	3.59	3.2
	8.4	6.11	5.19

M_{Ya}= 29.2 versus M_{Yb} = 28.1.

Calculated value of F=16.01 is significant well beyond the .01 level for df=1,17. Please note carefully, however, that the interpretation of a significant F-ratio in an analysis of covariance is a bit trickier than it would be if this were just a plain, garden-variety ANOVA. It does not entail that the difference between the originally observed means of the two samples, is significant in and of itself. The claim it is making is a somewhat more complex one, tied together by several logical constructions of the if/then variety:

If the correlation between *X* and *Y* within the general population is approximately the same as we have observed within the samples; and$_T$

- If we remove from *Y* the covariance that it has with *X*, so as to remove from the analysis the pre-existing individual differences that are measured by the covariate *X*; and$_T$
- If we adjust the group means of *Y* in accordance with the observed correlation between *X* and *Y*;$_T$
- Then we can conclude that the two adjusted means

  [adj]M_{Ya}=31.23 versus [adj]M_{Yb}=26.07

  significantly differ in the degree indicated, namely, P<.01, and thus that Method A is more effective than Method B.

This, however, does not distinguish the analysis of covariance fundamentally from ANOVA or any other inferential statistical procedure, for they are all wrapped up in a chain of if/then logical constructions. It is simply that the chain for the analysis of covariance is a few links longer.

Assumptions of ANCOVA

The analysis of covariance has the same underlying assumptions as its parent, the analysis of variance. It also has the same robustness with respect to the non-satisfaction of these assumptions, providing that all groups have the same number of subjects. There is, however, one assumption that the analysis of covariance has in addition, by virtue of its co-descent from correlation and regression–namely, that the slopes of the regression lines for each of the groups considered separately are all approximately the same.

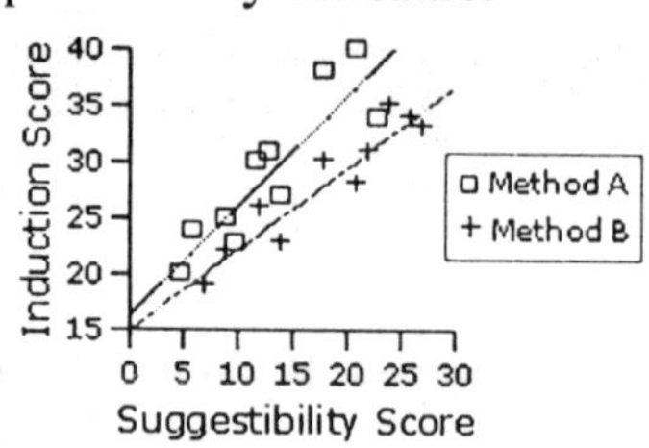

The operative word here is "approximately." Because of random variability, it would rarely happen that two or more samples of bivariate *XY* values would all end up with precisely the same slope, even though the samples might be drawn from the very same population. And so it is for the slopes of the separate regression lines for our two present samples. They are clearly not precisely the same. The question is, are they close enough to be regarded as reflecting the same underlying relationship between *X* and *Y*? In the calculations of step 4d, we found the slope of the line for the overall within-groups regression to be b_{wg}=+.83, and that was the value used in adjusting the means of group A and group B. The analysis of covariance is assuming that the slopes of the separate regression lines for the two samples do not significantly differ from +.83. We will examine this assumption more thoroughly in Part 3, after working through our second computational example.

Three Methods of Instruction for Elementary Computer Programming

To assess the relative merits of three methods of instruction for elementary computer programming, a curriculum researcher randomly selected 12 fifth graders from each of three elementary schools in a certain school district. Each group, within the setting of its home school, then received a six-week course of instruction in one or another of the three methods. The following table shows the measure of how well each of the 36 subjects, 12 per group, learned the prescribed elements of the subject matter.

	Method A	*Method B*	*Method C*
	29 24 14	15 28 13	32 27 15
	27 27 28	36 29 27	23 26 17
	27 32 13	31 33 32	25 14 29
	35 32 17	15 30 26	22 30 25
means	25.4	26.3	23.8

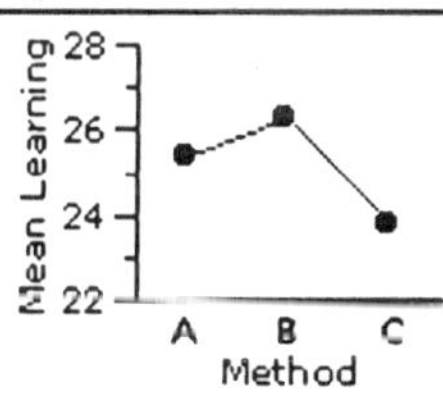

Given the differences among the means of the three groups, you might think at first glance that Method B has the edge over Method A, and that Methods B and A are both superior to Method C. As it happens, however, these differences, considered in and of themselves, are well within the range of mere random variability. A simple one-way ANOVA performed on this set of data would yield a miniscule F=0.40 [df=2,33], which falls far short of significance at the basic .05 level.

The reason for this shortfall is of course the degree of variability within the groups, which is quite large in comparison with the mean differences that appear between the groups. Well aware of the broad range of pre-existing individual differences that are likely to be found in situations of this sort, our curriculum researcher took the precaution of measuring her subjects beforehand with respect to their pre-existing levels of basic computer familiarity. The rationale is fairly obvious: the more familiar a subject is with basic computer procedures, the more of a head start he or she will have in learning the elements of computer programming; remove the effects of this covariate, and you thereby remove a substantial portion of the extraneous individual differences.

Our researcher was also well aware that her three groups, drawn from three different schools, might be starting out with substantially different levels of basic computer familiarity, in consequence of the average socio-economic differences that she knows to exist among the schools. The groups instructed by Methods A and B both reside in fairly affluent neighborhoods, while the group instructed by Method C comes from a less privileged part of town.

The following table shows the measures on both variables laid out in a form suitable for an analysis of covariance, with

X = The prior measure of basic computer familiarity

[The covariate whose effects the investigator wishes to remove from the analysis.]

and

Y = The measure of how well the subject has learned the elementary programming material

[The dependent variable in which the investigator is chiefly interested.]

	Method A			*Method B*			*Method C*	
Subject	X_a	Y_a	*Subject*	X_b	Y_b	*Subject*	X_c	Y_c
*a*1	14	29	*b*1	6	15	*c*1	15	32
*a*2	10	24	*b*2	16	28	*c*2	9	27
*a*3	7	14	*b*3	9	13	*c*3	7	15
*a*4	18	27	*b*4	19	36	*c*4	12	23
*a*5	14	27	*b*5	13	29	*c*5	12	26
*a*6	16	28	*b*6	14	27	*c*6	9	17
*a*7	13	27	*b*7	15	31	*c*7	12	25
*a*8	15	32	*b*8	18	33	*c*8	3	14
*a*9	5	13	*b*9	17	32	*c*9	13	29
*a*10	18	35	*b*10	8	15	*c*10	10	22
*a*11	16	32	*b*11	15	30	*c*11	11	30
*a*12	10	17	*b*12	16	26	*c*12	8	25
Means	13	25.4		13.8	26.3		10.1	23.8

As you can see from the means in the X columns, the investigator was right in her expectations concerning different group levels of basic computer familiarity: 13.0 and 13.8 for groups A and B, versus 10.1 for group C. So once again we encounter the what-if question: What would have happened if the three groups had all started out on the same footing?

Calculations for the Dependent Variable Y

Values of Y along with the several summary statistics required for the calculation of $SS_{T(Y)}$, and $SS_{bg(Y)}$:

	Y_a	Y_b	Y_c		
	29	15	32		
	24	28	27		
	14	13	15		
	27	36	23		
	27	29	26		
	28	27	17		
	27	31	25		
	32	33	14		
	13	32	29		
	35	15	22	for total	
	32	30	30	array of	
	17	26	25	data	
N	12	12	36		$SS_{T(Y)} = 1626.3$
$\Sigma.Y_i$	305	315	285	905	
$\Sigma.Y_i^2$	8315	8919	7143	24377	$SS_{wg(Y)} = 1587.5$
SS	562.9	650.3	374.3		
Mean	25.4	26.3	23.8	25.1	$SS_{bg(Y)} = 38.9$

Calculations for the Covariate X

Values of X along with the several summary statistics required for the calculation of $SS_{T(X)}$ and $SS_{wg(X)}$.

	X_a	X_b	X_c		
	14	6	15		
	10	16	9		
	7	9	7		
	18	19	12		
	14	13	12		
	16	14	9		
	13	15	12		
	15	18	3		
	5	17	13	for	
	18	8	10	total	
	16	15	11	array of	
	10	16	8	data	
N	12	12	12	36	
$\Sigma.X_i$	156	166	121	443	
$\Sigma.X_i^2$	2220	2482	1331	6033	$SS_{T(X)} = 581.6$
SS	192	185.7	110.9		$SS_{wg(X)} = 488.6$
Mean	13	13.8	10.1	12.3	

Calculations for the Covariance of X and Y

Cross-products of X_i and Y_i for each subject in each of the three groups, along with other summary data required for the calculation of SC_T and SC_{wg}:

A X_aY_a	B X_bY_b	C X_cY_c	
(14)(29) = 406	(6)(15) = 90	(15)(32) = 480	
(10)(24) = 240	(16)(28) = 448	(9)(27) = 243	
(7)(14) = 98	(9)(13) = 117	(7)(15) = 105	
(18)(27) = 486	(19)(36) = 684	(12)(23) = 276	
(14)(27) = 378	(13)(29) = 377	(12)(26) = 312	
(16)(28) = 448	(14)(27) = 378	(9)(17) = 153	
(13)(27) = 351	(15)(31) = 465	(12)(25) = 300	
(15)(32) = 480	(18)(33) = 594	(3)(14) = 42	
(5)(13) = 65	(17)(32) = 544	(13)(29) = 377	
(18)(35) = 630	(8)(15) = 120	(10)(22) = 220	for total
(16)(32) = 512	(15)(30) = 450	(11)(30) = 330	array of
(10)(17) = 170	(16)(26) = 416	(8)(25) = 200	data
.$\Sigma(X_{ai}Y_{ai})$=4264	.$\Sigma(X_{bi}Y_{bi})$=4683	.$\Sigma(X_{ci}Y_{ci})$=3038	.$\Sigma(X_{Ti}Y_{Ti})$=11985
oΣX_{ai}=156	oΣX_{bi}=166	oΣX_{ci}=121	oΣX_{Ti}=443
oΣY_{ai}=305	oΣY_{bi}=315	oΣY_{ci}=285	oΣY_{Ti}=905

Calculation of *SC* for the total array of data:

$$SC_T = \Sigma(X_{Ti}Y_{Ti}) - \frac{(\Sigma X_{Ti})(\Sigma Y_{Ti})}{N_T}$$

$$SC_T = 11985 - \frac{(443)(905)}{36} = 848.5.$$

Calculation of SC_{wg}:

The components for each group ("g") are calculated as:

$$SC_{wg(g)} = (\Sigma X_{gi}Y_{gi}) - \frac{(\Sigma X_{gi})(\Sigma Y_{gi})}{N_g}$$

Thus:

$$SC_{wg(a)} = 4264 - \frac{(156)(305)}{12} = 299.0.$$

$$SC_{wg(b)} = 4683 - \frac{(166)(315)}{12} = 325.5$$

$$SC_{wg(c)} = 3036 - \frac{(121)(285)}{12} = 164.3$$

$$SC_{wg} = SC_{wg(a)} + SC_{wg(b)} + SC_{wg(c)}$$
$$= 299.0 + 325.5 + 164.3 = 788.8.$$

The Final Set of Calculations

Here again is a summary of the values of *SS* and SC obtained so far. Recall that *Y* is the variable in which we are chiefly interested, and *X* is the covariate whose effects we are seeking to remove.

X	*Y*	*Covariance*
$SS_{T(X)} = 581.6$	$SS_{T(Y)} = 1626.3$	$SC_T = 848.5$
$SS_{wg(X)} = 488.6$	$SS_{wg(Y)} = 1587.4$	$SC_{wg} = 788.8$
	$SS_{bg(Y)} = 38.9$	

Adjustment of $SS_{T(Y)}$

The overall correlation between *X* and *Y* (all three groups combined) can be calculated as

$$r_T = \frac{SC_T}{\text{sqrt}[SS_{T(X)} \times SS_{T(Y)}]}$$

$$= \frac{848.5}{\text{sqrt}[581.6 \times 1626.31]}$$

$$= +.872.$$

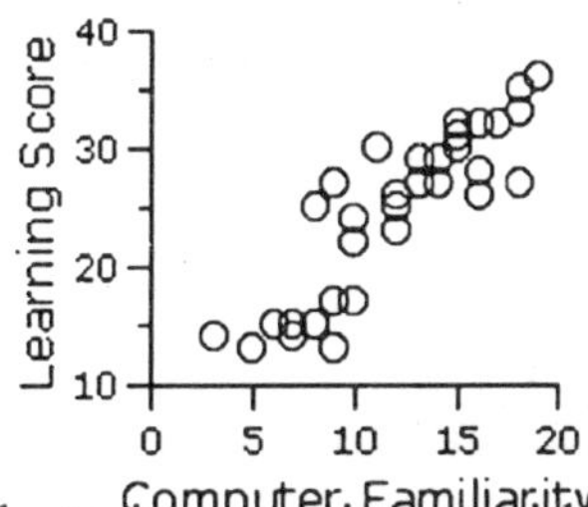

The proportion of the total variability of *Y* attributable to its covariance with *X* is accordingly

$$(r_T)^2 = (+.872)^2 = .760$$

As previously noted, the removal of this portion is best accomplished (minimizing the risk of rounding errors) through the computational formula

$$[\text{adj}]SS_{T(Y)} = SS_{T(Y)} - \frac{(SC_T)^2}{SS_{T(X)}}$$

$$= 1626.3 - \frac{(848.5)^2}{581.6}$$

$$= 388.4.$$

Adjustment of $SS_{wg(Y)}$

Similarly, the aggregate correlation between *X* and *Y* within the three groups can be calculated as

$$r_{wg} = \frac{SC_{wg}}{\text{sqrt}[SS_{wg(X)} \times SS_{wg(Y)}]}$$

$$= \frac{788.8}{\text{sqrt}[488.6 \times 1587.4]}$$

$$= +.896.$$

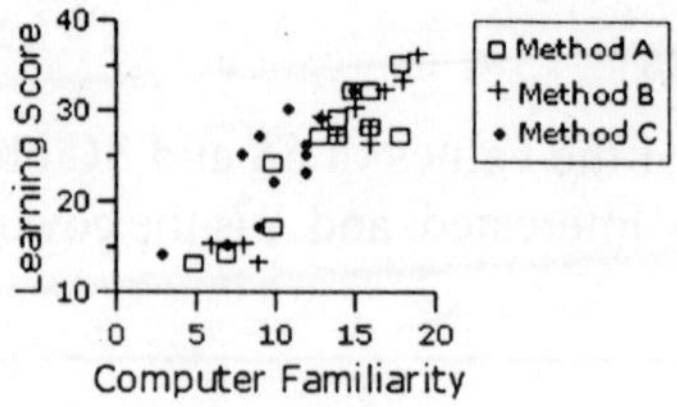

So the proportion of the within-groups variability of Y attributable to covariance with X is

$$(r_{wg})^2 = (+.896)^2 = .803.$$

Here again, the removal of this portion is best accomplished through the computational formula

$$[\text{adj}]SS_{wg(Y)} = SS_{wg(Y)} - \frac{(SC_{wg})^2}{SS_{wg(X)}}$$

$$= 1587.4 - \frac{(788.8)^2}{488.6}$$

$$= 314.0.$$

Adjustment of $SS_{bg(Y)}$

The adjusted value of $SS_{bg(Y)}$ can then again be obtained through simple subtraction as

$$[\text{adj}]\text{SS}_{bg(Y)} = [\text{adj}]\text{SS}_{T(Y)} - [\text{adj}]\text{SS}_{wg(Y)}$$

$$= 388.4 - 314.0 = 74.4$$

4d. Adjustment of the Means of Y for Groups A, B, and B

The average relationship between X and Y within the groups is given by the slope of the regression line for the within-groups correlation, which is

$$b_{wg} = \frac{SC_{wg}}{SS_{wg(X)}}$$

$$= \frac{788.8}{488.6} = +1.61.$$

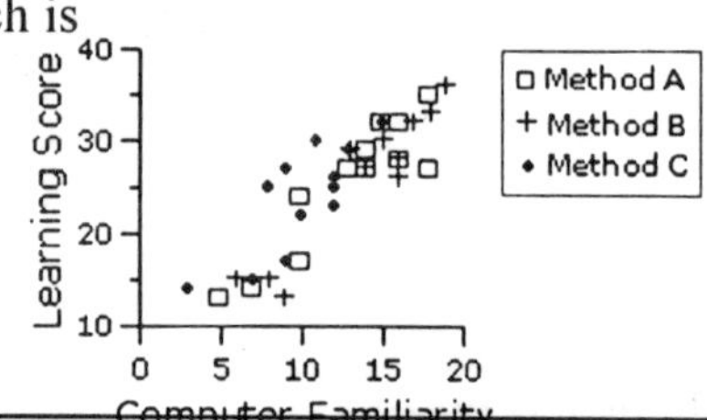

	M_X	*observed* M_Y	*adjusted* M_Y
group A	13	25.4	24.3
group B	13.8	26.3	23.9
group C	10.1	23.8	27.3
combined	12.3	25.1	

The three groups started out with different average levels of basic computer familiarity. Suppose they had instead all started out with the same level: namely, 12.3, which is the mean of X for all three groups combined. In this case, group A would have been starting out with a mean familiarity level 0.7 units lower than it actually started with; group B would have been starting out 1.5 units lower; and group C would have been starting out 2.2 units higher. Given the observed dependance of *Y* on *X*, the respective means of *Y* for the three groups would presumably therefore have been on the order of

$$[adj]M_{Ya} = M_{Ya} - b_{wg}(M_{Xa} - M_{XT})-$$
$$= 25.4 - 1.61(13.0-12.3)-$$
$$= 24.3.$$
$$[adj]M_{Yb} = M_{Yb} - b_{wg}(M_{Xb} - M_{XT})-$$
$$= 26.3 - 1.61(13.8-12.3)-$$
$$= 23.9.$$
$$[adj]M_{Yc} = M_{Yc} - b_{wg}(M_{Xc} - M_{XT})$$
$$= 23.8 - 1.61\ (10.1 - 12.3)$$
$$= 27.3.$$

Analysis of Covariance Using Adjusted Values of SS [sh]

The simple computational format for this step is the same as for Example. All we need do here is lay out its results in the form of an ANCOVA summary table:

Source	*SS*	*df*	*MS*	*F*	*P*
adjusted means [between-groups effect]	74.4	2	37.2	3.8	<.05
adjusted error [within-groups]	314	32	9.8		
adjusted total	388.4	34			

[Recall that $df_{bg(Y)}=k-1$ and $[adj]df_{wg(Y)}=N_T-k-1$]

df denominator	*df numerator* 1	2	3	
	32	4.15	3.29	2.9
	7.5	5.34	4.46	

The calculated value of F=3.8 is significant beyond the .05 level for df=2,32. Once again, let me note that a significant ANCOVA result of this sort does not refer to the observed means of the samples, but to the adjusted means. It is the same chain of if/then constructions laid out in Example :

- If the correlation between *X* and *Y* within the general population is approximately the same as we have observed within the samples; and$_T$
- If we remove from *Y* the covariance that it has with *X*, so as to remove from the analysis the pre-existing individual differences that are measured by the covariate *X*; and$_T$

- If we adjust the group means of Y in accordance with the observed correlation between X and Y;$_T$
- Then we can conclude that the three adjusted means significantly differ in the degree indicated, namely, P<.05.

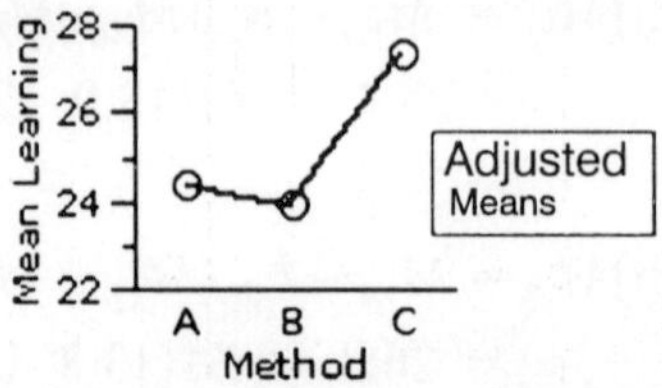

In the present case it is fairly obvious to the naked eye that the bulk of this effect is contributed by Method C, which our curriculum researcher could reasonably conclude to be more effective than either of the other two methods, given this particular population of subjects.

Test for Homogeneity of Regression [sh]

As mentioned toward the end of Part 2, the analysis of covariance assumes that the slopes of the regression lines for each of the groups considered separately do not significantly differ from the slope of the overall within-groups regression, which for the present example is b_{wg}=+1.61. If they do significantly differ, then the analysis of covariance is invalid and any positive conclusion drawn from it is potentially false and misleading.

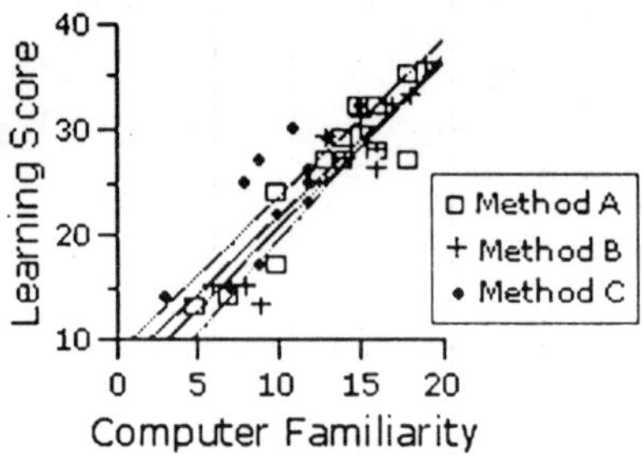

In the adjacent graph you can see that the three separate group regression lines, in red, do appear to be in close parallel with the blue line for the overall within-groups regression. However, here as elsewhere in the domain of statistical inference, a merely intuitive impression of this sort is not sufficient. For the naked eye is simply not capable of making the fine-grained distinctions that need to be made with complex sets of numerical data.

The procedure for rigorously determining whether this "homogeneity of regression" assumption is satisfied involves yet another variation on the theme of analysis of variance. As in all versions of ANOVA, the end result is an F-ratio of the general form

$$F = \frac{MS_{\text{effect}}}{MS_{\text{error}}} = \frac{SS_{\text{effect}}/\text{df}_{\text{effect}}}{SS_{\text{error}}/\text{df}_{\text{error}}}.$$

Numerator SS: In the present version, the sum of squared deviates (SS_{effect}) in the numerator of the ratio is a measure of the aggregate degree to which the separate regression-line slopes of the several groups differ from the slope (b_{wg}) of the line for the overall within-groups regression. We will henceforth designate this quantity as $SS_{b\text{-}reg}$ ("b-reg" = "between regressions"). As the rationale for this $SS_{b\text{-}reg}$ measure is a bit more intricate than we need to get into just now, I will present it simply as a matter of computational mechanics.

1. For each group ("g"), calculate

$$\frac{(SC_g)^2}{SS_{Xg}} \quad \text{Note that } SC_g/SS_{Xg} \text{ would give you the slope for group } g.$$

2. Sum up the results of step 1 and subtract from that sum the quantity

$$\frac{(SC_{wg})^2}{SS_{wg(X)}} \quad \text{Note that } SC_{wg}/SS_{wg(X)} = b_{wg} \text{ the slope for the overall within-groups regression.}$$

Hence:

$$SS_{b\text{-}reg} = \sum \frac{(SC_g)^2}{SS_{Xg}} - \frac{(SC_{wg})^2}{SS_{wg(X)}}$$

which for the present example, using previously calculated values, comes out to

$$SS_{b\text{-}reg} = \frac{(SC_a)^2}{SS_{Xa}} + \frac{(SC_b)^2}{SS_{Xb}} + \frac{(SC_c)^2}{SS_{Xc}} - \frac{(SC_{wg})^2}{SS_{wg(X)}}$$

$$SS_{b\text{-}reg} = \frac{(299.0)^2}{192.0} + \frac{(325.5)^2}{185.7} + \frac{(164.3)^2}{110.9} - \frac{(788.8)^2}{488.6}$$

$$SS_{b\text{-}reg} = 6.1.$$

Denominator SS: The sum of squared deviates (SS_{error}) in the denominator of the F-ratio is reached much more easily. We will designate this quantity as $SS_{Y(remainder)}$, because it is simply what is left over from $[adj]SS_{wg(Y)}$, the "adjusted error" of the original ANCOVA, after $SS_{b\text{-}reg}$ has been removed from it. Thus

$$\begin{aligned} SS_{Y(remainder)} &= [adj]SS_{wg(Y)} - SS_{b\text{-}reg} \\ &= 314.0 - 6.1 = 307.9 \\ &= 307.9. \end{aligned}$$

df/The apportionment of degrees of freedom follows the same pattern. The value of $[adj]SS_{wg(Y)}$ that appears in the original ANCOVA starts out with degrees of freedom equal to

$$\begin{aligned} [adj]df_{wg(Y)} &= N_T - k - 1 \\ &= 36 - 3 - 1 \\ &= 32. \end{aligned}$$

When you remove $SS_{b\text{-}reg}$ from [adj]$SS_{wg(Y)}$, you are also removing the number of degrees of freedom that go along with it, which is

$$\begin{aligned} df_{b\text{-}reg} &= k-1 \\ &= 3-1 \\ &= 2 \end{aligned}$$

So the degrees of freedom for $SS_{Y(remainder)}$ is

$$\begin{aligned} df_{Y(remainder)} &= [adj]df_{wg(Y)} - df_{b\text{-}reg} \\ &= (N_T - k - 1) - (k - 1) \\ &= 32 - 2 = 30. \end{aligned}$$

As a computational expedient, you can also

calculate [adj]$df_{wg(Y)}$ as $N_T - 2k$.

With these components at hand we can then calculate the F-ratio for homogeneity of regression as

$$F = \frac{MS_{b\text{-reg}}}{MS_{Y(\text{remainder})}} = \frac{SS_{b\text{-reg}} / df_{b\text{-reg}}}{SS_{Y(\text{remainder})} / df_{Y(\text{remainder})}}$$

$$= \frac{6.1/2}{307.9/30}$$

$$= 0.30 \text{ [with df } = 2,30]$$

df denominator	*df numerator* 1	2	3
30	4.17	3.32	2.92
	7.56	5.39	4.51

Here on the right is the relevant portion, though with such a tiny F-ratio one need hardly bother to consult it. Any value of F equal to or smaller than 1.0 will be non-significant. At any rate, the slopes of the three separate group regression lines do not significantly differ from b_{wg}=+1.61. Hence the assumption of homogeneity of regression is satisfied.

Encyclopaedia

of

STATISTICAL THEORY

Encyclopaedia of STATISTICAL THEORY

Volume 2

Dr. Rakesh Kumar Pandey

ANMOL PUBLICATIONS PVT. LTD.
NEW DELHI - 110 002 (INDIA)

ANMOL PUBLICATIONS PVT. LTD.

Regd. Office: 4360/4, Ansari Road, Daryaganj,
New Delhi-110 002 (India)
Ph.: 23278000, 23261597

Branch Office: No. 1015, Ist Main Road, BSK IIIrd Stage
IIIrd Phase, IIIrd Block,
Bangalore-560 085 (India)
Tel.: 080-41723429
Visit us at: www.anmolpublications.com

Encyclopaedia of Statistical Theory

First Edition, 2009

ISBN 978-81-261-4109-8 (Set)

PRINTED IN INDIA

Printed at Mehra Offset Press, Delhi.

Contents

Preface

Encyclopaedia of Statistical theory defines a statistic as a function of a sample where the function itself is independent of the sample's distribution: the term is used both for the function and for the value of the function on a given sample. A statistic is distinct from an unknown statistical parameter, which is not computable from a sample. A key use of statistics is as estimators in statistical inference, to estimate parameters of a distribution given a sample. For instance, the *sample mean* is a statistic, while the *population mean* is a parameter.

Statisticians often contemplate a parameterized family of probability distributions, any member of which could be the distribution of some measurable aspect of each member of a population, from which a sample is drawn randomly. Statistics are used to describe the basic features of the data gathered from an experimental study in various ways. A descriptive Statistics is distinguished from inductive statistics. They provide simple summaries about the sample and the measures. Together with simple graphics analysis, they form the basis of virtually every quantitative analysis of data. In addition to showing how to apply statistical procedures, the authors explain how to meaningfully describe real sets of data, what the statistical tests mean in terms of their practical applications, how to evaluate the validity of the assumptions behind statistical tests, and what to do when statistical assumptions have been violated.

Author

List of Symbols

$\pm$ $\not\subset$ $\subset$ $\supset$ $\subseteq$ $\in$ $\notin$ ∂

$\neq$ $\equiv$ $\approx$ $\angle$ ∇ $\sum$ Π Ω

ϕ α β χ ε γ η λ

μ ν $\perp$ π θ ρ σ τ

$\Re$ ω ξ ψ ζ Υ $\exists$ $\int$

$\iint$ $\iiint$ $\oint$ $\times$ $\sqrt{\ }$ $\therefore$ $\because$ Δ

$\geq$ $\leq$ $\Rightarrow$ $\Downarrow$ $\Leftrightarrow$ ∞ $\propto$ $\mathbb{R}$

$\mathbb{Z}$ $\mathbb{N}$ $\mathbb{C}$ $\mathbb{Q}$

Chapter 1

Principles of Measurement

ELEMENTARY CONCEPTS IN STATISTICS

Statistics is a mathematical science pertaining to the collection, analysis, interpretation or explanation, and presentation of data. It is applicable to a wide variety of academic disciplines, from the physical and social sciences to the humanities. Statistics are also used for making informed decisions.

Statistical methods can be used to summarize or describe a collection of data; this is called descriptive statistics. In addition, patterns in the data may be modeled in a way that accounts for randomness and uncertainty in the observations, and then used to draw inferences about the process or population being studied; this is called inferential statistics. Both descriptive and inferential statistics comprise applied statistics. There is also a discipline called mathematical statistics, which is concerned with the theoretical basis of the subject.

The idea that statistics branched off from mathematics is a widely held misconception. Some place an undue emphasis on the relationship, but the two disciplines are very different.

The purpose of descriptive statistics is to communicate information, while inferential statistics is used to reach conclusions and deductions that possibly explain the data. Both of these together make up applied statistics. There is also a discipline called mathematical statistics, which is concerned with the theoretical basis of the subject.

Mathematical Statistics

Mathematical statistics uses probability theory and other branches of mathematics to study statistics from a purely mathematical standpoint.

Mathematical statistics deals with gaining information from data. In practice, data often contain some randomness or uncertainty. Statistics handles such data using methods of probability theory.

Statistics is divided into:

1. *Descriptive statistics:* The part of mathematical statistics that describes data, i.e. summarises the data and their typical properties

2. Inferential statistics - the part of mathematical statistics that draws conclusions from data, i.e. checks whether the data fulfill some condition and gives guarantees on the involved uncertainty.

Mathematical statistics is the theoretical basis for many practices in applied statistics.

Descriptive Statistics

Descriptive Statistics are used to describe the basic features of the data in a study. They provide simple summaries about the sample and the measures. Together with simple graphics analysis, they form the basis of virtually every quantitative analysis of data. Various techniques that are commonly used are classified as:

1. Graphical description in which we use graphs to summarize data.
2. Tabular description in which we use tables to summarize data.
3. Summary statistics in which we calculate certain values to summarize data.

In general, statistical data can be described as a list of *subjects* or *units* and the data associated with each of them. Although most research uses many data types for each *unit*, we will limit ourselves to just one data item each for this simple introduction.

We have two objectives for our summary:

1. We want to choose a statistic that shows how different *units* seem similar. Statistical textbooks call the solution to this objective, a *measure of central tendency.*
2. We want to choose another statistic that shows how they differ. This kind of statistic is often called a *measure of statistical variability.*

When we are summarizing a quantity like length or weight or age, it is common to answer the first question with the arithmetic mean, the median, or the mode. Sometimes, we choose specific values from the cumulative distribution function called quantiles.

The most common measures of variability for quantitative data are the variance; its square root, the standard deviation; the range; interquartile range; and the average absolute deviation (average deviation).

One important use of descriptive statistics is to summarize a collection of data in a clear and understandable way. For example, assume a psychologist gave a personality test measuring shyness to all 2500 students attending a small college. How might these measurements be summarized? There are two basic methods: numerical and graphical. Using the numerical approach one might compute statistics such as the mean and standard deviation. These statistics convey information about the average degree of shyness and the degree to which people differ in shyness. Using the graphical approach one might create a stem and leaf display and a box plot. These plots contain detailed information about the distribution of shyness scores.

Graphical methods are better suited than numerical methods for identifying patterns in the data. Numerical approaches are more precise and objective.

Since the numerical and graphical approaches compliment each other, it is wise to use both.

Statistical Inference

Inferential statistics or statistical induction comprises the use of statistics to make inferences concerning some unknown aspect of a population. It is distinguished from descriptive statistics.

Statistical inference concerns the problem of inferring properties of an unknown distribution from data generated by that distribution. The most common type of inference involves approximating the unknown distribution by choosing a distribution from a restricted family of distributions. Generally the restricted family of distributions is specified parametrically.

USES OF STATISTICS

One of the potential shortfalls of anecdotal data is that they are idiosyncratic. Just as the congressional staffer told me her father received a better education from the high school they both attended than she did, We could have easily received a higher quality education than my father did. Statistics allow researchers to collect information, or data, from a large number of people and then summarize their typical experience. Do *most* people receive a better or worse education than their parents?

Statistics allow researchers to take a large batch of data and *summarize* it into a couple of numbers, such as an average. Of course, when a large amount of data is summarized into a single number, a lot of information is lost, including the fact that different people have very different experiences. So it is important to remember that, for the most part, statistics do not provide useful information about each individual's experience. Rather, researchers generally use statistics to make *general* statements about a population. Although personal stories are often moving or interesting, it is often important to understand what the *typical* or *average* experience is. For this, we need statistics.

Statistics are also used to reach conclusions about general differences between groups. For example, suppose that in our family, there are four children, two men and two women. Suppose that the women in our family are taller than the men. This personal experience may lead me to the conclusion that women are generally taller than men. Of course, we know that, on average, men are taller than women.

The reason we know this is because researchers have taken large, random samples of men and women and compared their average heights. Researchers are often interested in making such comparisons: Do cancer patients survive longer using one drug than another? Is one method of teaching children to read more effective than another? Do men and women differ in their enjoyment of a certain movie? To answer these questions, we need to collect data from randomly selected samples and compare these data using statistics. The results we get from such comparisons are often more trustworthy than the simple observations people make from non-random samples, such as the different heights of men and women in our family.

Statistics can also be used to see if scores on two variables are related and to make predictions. For example, statistics can be used to see whether smoking cigarettes is related to the likelihood of developing lung cancer. For years, tobacco companies argued that there was no relationship between smoking and cancer. Sure, some people who smoked developed cancer.

But the tobacco companies argued that:

1. Many people who smoked never developed cancer,
2. Many people who smoke tend to do other things that may lead to cancer development, such as eating unhealthy foods and not exercising. With the help of statistics in a number of studies, researchers were finally able to produce a preponderance of evidence indicating that, in fact, there is a relationship between cigarette smoking and cancer. Because statistics tend to focus on overall patterns rather than individual cases, this research did not suggest that *everyone* who smokes would develop cancer.

Rather, the research demonstrated that, on average, people have a greater chance of developing cancer if they smoke cigarettes than if they do not with a moment's thought, you can imagine a large number of interesting and important questions that statistics about relationships can help you answer. Is there a relationship between self-esteem and academic achievement? Is there a relationship between the appearance of criminal defendants and their likelihood of being convicted? Is it possible to predict the violent crime rate of a state from the amount of money the state spends on drug treatment programmes?

If we know the fathers' height, how accurately can we predict sons' height? These and thousands of other questions have been examined by researchers using statistics designed to determine the relationship between variables in a population.

Here is where we stand at this very first step in the enterprise. Off in the distance is the vision of an elaborate and potentially very useful structure that we want to build. But before we can begin building it, we must first sort out the raw materials, which are the logical and conceptual counterparts of bricks and boards, mortar and nails, trowels and hammers. In the present chapter we begin by sorting out the most essential raw materials of all, which are those that pertain to the elemental process of measurement. At first glance, much of what is presented in this chapter is likely to seem more or less familiar, and some of it might even appear rather obvious and common-sensical. If you are just beginning your study of the subject, however, please bear with it and read the chapter carefully through without skipping or skimming, for there are subtleties and complexities here that have very far-reaching implications.

The watchword for this chapter will be the somewhat ugly but very expressive term GIGO, borrowed from the language of computer engineers and programmers. GIGO, which is an acronym for "Garbage In, Garbage Out," is the computer mavin's update of the old adage "You can't make a silk purse out of a sow's ear." Computers are marvelous devices. They can crunch numbers and process information with astonishing speed and accuracy. But no matter how fast or full of multi-megabyte RAM a computer might be, what it gives you by way of output will only be as sensible as what you give it by way of input. If nonsensical garbage is what you put in, then nonsensical garbage is also all that will come out. It will, of course, be very thoroughly processed nonsense. It might even seem elegant and profound. But it is nonsense all the same.

The point is equally applicable to the study of statistics. Statistical methods are basically instruments for processing information. The information that they process is numerical in nature and derives from one or another of several forms of measurement. The various statistical procedures that we will eventually be examining make differing assumptions about the particular kinds of measures you are feeding into them. Feed them the kind of information that they assume they are getting, and the results they give you in return will provide a firm basis for drawing rational conclusions. Feed them information that violates their assumptions, and they will still patiently process it and crank out a result. But that result will be nonsense, and so too will be any conclusion you might draw from it. It might, of course, be very impressive and elegant-looking nonsense, with all the trappings of scientific profundity But it will be nonsense all the same. Garbage In, Garbage Out. This is why it is so very important that you begin your study of the subject with a clear understanding that there are several different forms of measurement, each with its own set of properties, strengths, and limitations.

FUNDAMENTAL FORMS OF MEASUREMENT

At the level of fundamentals, all of the forms of measurement that you have ever encountered or are ever likely to encounter come down to a process of either counting, ordering, or sorting.

Counting.

(*i*) Take a metal tape measure to your desk, count off the number of inches or centimeters it takes to get from the left edge to the right edge, and you are in effect measuring the desk with respect to the general characteristic of width.

(*ii*) Step dry and naked onto a bathroom scale, count off the number of pounds or kilograms that the instrument registers in result of that action, and you are in effect measuring your body with respect to the general characteristic of weight. ~~ Procedures of this general type, based on counting-off the units on some standard measurement scale, constitute what is by far the most commonly understood referent of the term "measurement." The other two types of procedures, based on ordering and sorting, require a somewhat broader understanding of the concept.

Ordering.

The meaning of the term "ordering" in this context is not "placing an order" but rather "Arranging in order."

1. Make a list of the things you recurrently worry about—money, grades, family, friends, personal attractiveness, the degradation of the environment, the meaning of life, or whatever they might be—arranging them in the rank order of the strength of the hold they have upon your mind (strongest, second strongest, third strongest, and so on), and you are in effect measuring each, in relation to all the others, with respect to the general dimension of personal worrisomeness.
2. A variation on this theme would be to rate each of the worrisome items on, for example, a 5-point scale, with "1" and "5" respectively representing what you judge to be

the lowest and highest degrees of worrisomeness, and "2" through "4" representing the gradations of worrisomeness that range between these two extremes.

Sorting

The very simplest form of measurement—so simple, indeed, that your first impulse will probably be to resist seeing it as being any kind of "measurement" at all—involves sorting items into categories. Do, however, try to resist that impulse, for this really is a kind of measurement, and a very useful one at that.

1. Sort the individuals who have applied for admission to a certain college into the two categories *female* and *male,* and you are in effect measuring each applicant with respect to the general characteristic of gender.
2. Sort them into the three categories *admitted, not admitted,* and *wait-listed,* and you are in effect measuring each with respect to the general characteristic of admission status.
3. Perform both of these sortings concurrently, and you are in effect measuring each student with respect to $2 \times 3 = 6$ possible categories of cross-classification: female *and* admitted, male *and* admitted, female *and* not admitted, male *and* not admitted, and so on.

These three forms are not just different versions of measurement. They are different *types.* Like the bricks, boards, nails, and mortar that might be used in the construction of a house, they are made of different stuff, with quite different properties and quite different uses to which they can properly be put. It is exceedingly important for the student of statistics to understand the properties of these three forms of measurement, because these properties are among the main determinants not only of the kinds of mathematical and statistical procedures that can be legitimately applied to a set of measures, but also of the kinds of conclusions that can be meaningfully drawn from the application. Before we get into the details of this point, however, we must first make a brief detour to lay out a distinction and some terminology that applies to all three forms of measurement.

VARIABLES AND VARIATES

In statistical parlance, the general property that is being measured through one or another of the three processes that we have outlined is spoken of as a variable, and any particular measured instance of that property is spoken of as a variate:

Thus, for the first example of measurement by counting, in which we take a tape measure to the width of a desk, the variable is desk width and the variate would the measured width of this, that, or the other particular desk. In which we step on a bathroom scale and record the number of pounds or kilograms, the variable is body weight, and the variate would be the measured weight of this, that, or the other particular body.-

For the example in which you rank-order your recurrent worries, the variable is the relative degree of worrisomeness, as assessed by you subjectively, and each item in the list of worries, according to whether it is ranked as first, second, third, and so on, would be a variate instance of that variable. If you were instead to rate your worries on a 5-point scale, the variable would still be relative degree of worrisomeness, as assessed by you subjectively, though the variates would now be the rating-scale values ("1", "2", "3", etc.) that you assign to each of the particular items on the list.-

For the first example of measurement by sorting, the variable is gender, and each applicant sorted as female or male would represent a variate instance of that variable. The variable is admission status and each applicant sorted into one or another of the three categories would represent a variate instance of that variable. In the third sorting example there are two variables concurrently, gender and admission status, and each applicant cross-classified as female and admitted, male and admitted, and so on, would represent a bivariate instance of both variables together.-

The term "variable" implies that the results of the measurement process are capable of *varying* from one time to another or from one item to another. Thus, the categorical measurement of gender among a mixed group of human subjects will vary from one subject to another between the two possible outcomes, *female* and *male;* rank-order measurements of worrisomeness will vary not only according to the specific items that are being ranked, but also according to who is doing the ranking; and measurements of desk width will of course vary from one desk to another. The opposite of a variable is a constant. The classical example of a constant is the value of pi, which is the ratio of the circumference of a circle to its diameter. Precisely measure this ratio with respect to any circle at all, and it will come out to an absolutely unvarying value of 3.14159... .

If this detour on variables and variates has seemed chiefly a matter of belaboring the meanings of words, please bear in mind that the terminology of a technical subject is no mere decoration. Its function is to provide a way of saying things compactly and efficiently, and the sooner you achieve a mastery of it, the sooner your mind will be free to ascend to higher levels. At any rate, we now conclude the detour and return to the task at hand, which is to describe the properties of the three forms of measurement and examine the implications of these properties. We begin with the process that involves counting, since that is the form of measurement with which you are probably already most familiar. It is also the form of measurement that permits the most powerful forms of mathematical and statistical analysis.

ELEMENTARY CONCEPTS IN STATISTICS

In this introduction, we will briefly discuss those elementary statistical concepts that provide the necessary foundations for more specialized expertise in any area of statistical data analysis. The selected topics illustrate the basic assumptions of most statistical methods and/or have been demonstrated in research to be necessary components of one's general understanding of the "quantitative nature" of reality. Because of space limitations, we will focus mostly on the functional aspects of the concepts discussed and the presentation will be very short. Further information on each of those concepts can be found in statistical textbooks.

WHAT ARE VARIABLES

In statistics, variables refer to measurable attributes, as these typically vary over time or between individuals. Variables can be discrete (taking values from a finite or countable set), continuous (having a continuous distribution function), or neither. Temperature is a continuous variable, while the number of legs of an animal is a discrete variable. This concept of a variable is widely used in the natural, medical and social sciences.

In causal models, a distinction is made between *"independent variables"* and *"dependent variables"*, the latter being expected to vary in value in response to changes in the former. In other words, an independent variable is presumed to potentially affect a dependent one. In experiments, independent variables include factors that can be altered or chosen by the researcher independent of other factors.

For example, in an experiment to test whether or not the boiling point of water changes with altitude, the altitude is under direct control and is the *independent* variable, and the boiling point is presumed to depend upon it and is therefore the *dependent* variable. The collection of results from an experiment, or information to be used to draw conclusions, is known as data. It is often important to consider which variables to allow for, or to directly control or eliminate, in the design of experiments.

While independent variables can refer to quantities and qualities that are under experimental control, they can also include extraneous factors that influence results in a confusing or undesired manner.

In general, if strongly confounding variables exist that can substantially affect the result, then this makes it more difficult to interpret the results. For example, a study into the incidence of cancer with age will also have to take into account *variables* such as income (poorer people may have less healthy lives), location (some cancers vary depending on diet and sunlight), stress and lifestyle issues (cancer may be related to these more than age), and so on. Failure to at least consider these factors can lead to grossly inaccurate deductions. For this reason, controlling unwanted variables is important in research.

Variables are things that we measure, control, or manipulate in research. They differ in many respects, most notably in the role they are given in our research and in the type of measures that can be applied to them.

A symbol or name that stands for a value. For example, in the expression

$$x + y$$

x and *y* are variables.

Measurement Scales

Variables differ in "how well" they can be measured, i.e., in how much measurable information their measurement scale can provide. There is obviously some measurement error involved in every measurement, which determines the "amount of information" that we can obtain. Another factor that determines the amount of information that can be provided by a variable is its "type of measurement scale." Specifically variables are classified as (a) nominal, (b) ordinal, (c) interval or (d) ratio.

1. Nominal variables allow for only qualitative classification. That is, they can be measured only in terms of whether the individual items belong to some distinctively different categories, but we cannot quantify or even rank order those categories. For example, all we can say is that 2 individuals are different in terms of variable A (e.g., they are of different race), but we cannot say which one "has more" of the quality represented by the variable. Typical examples of nominal variables are gender, race, colour, city, etc.
2. Ordinal variables allow us to rank order the items we measure in terms of which has less and which has more of the quality represented by the variable, but still they do not allow us to say "how much more." A typical example of an ordinal variable is the socioeconomic status of families. For example, we know that upper-middle is higher than middle but we cannot say that it is, for example, 18% higher. Also this very distinction between nominal, ordinal, and interval scales itself represents a good example of an ordinal variable. For example, we can say that nominal measurement provides less information than ordinal measurement, but we cannot say "how much less" or how this difference compares to the difference between ordinal and interval scales.
3. Interval variables allow us not only to rank order the items that are measured, but also to quantify and compare the sizes of differences between them. For example, temperature, as measured in degrees Fahrenheit or Celsius, constitutes an interval scale. We can say that a temperature of 40 degrees is higher than a temperature of 30 degrees, and that an increase from 20 to 40 degrees is twice as much as an increase from 30 to 40 degrees.
4. Ratio variables are very similar to interval variables; in addition to all the properties of interval variables, they feature an identifiable absolute zero point, thus they allow for statements such as x is two times more than y. Typical examples of ratio scales are measures of time or space. For example, as the Kelvin temperature scale is a ratio scale, not only can we say that a temperature of 200 degrees is higher than one of 100 degrees, we can correctly state that it is twice as high. Interval scales do not have the ratio property. Most statistical data analysis procedures do not distinguish between the interval and ratio properties of the measurement scales.

Correlational vs. Experimental Research

Most empirical research belongs clearly to one of those two general categories. In correlational research we do not (or at least try not to) influence any variables but only measure them and look for relations (correlations) between some set of variables, such as blood pressure and cholesterol level. In experimental research, we manipulate some variables and then measure the effects of this manipulation on other variables; for example, a researcher might artificially increase blood pressure and then record cholesterol level. Data analysis in experimental research also comes down to calculating "correlations" between variables, specifically, those manipulated and those affected by the manipulation. However, experimental data may potentially provide qualitatively better information: Only experimental data can conclusively demonstrate causal relations between variables. For example, if we found that whenever we change variable A then variable B changes, then we can conclude that "A influences B." Data from correlational research can only be "interpreted" in causal terms based on some theories that we have, but correlational data cannot conclusively prove causality.

Dependent vs. Independent Variables

Independent variables are those that are manipulated whereas dependent variables are only measured or registered. This distinction appears terminologically confusing to many because, as some students say, "all variables depend on something." However, once you get used to this distinction, it becomes indispensable.

The terms dependent and independent variable apply mostly to experimental research where some variables are manipulated, and in this sense they are "independent" from the initial reaction patterns, features, intentions, etc. of the subjects. Some other variables are expected to be "dependent" on the manipulation or experimental conditions. That is to say, they depend on "what the subject will do" in response. Somewhat contrary to the nature of this distinction, these terms are also used in studies where we do not literally manipulate independent variables, but only assign subjects to "experimental groups" based on some pre-existing properties of the subjects. For example, if in an experiment, males are compared with females regarding their white cell count (WCC), Gender could be called the independent variable and WCC the dependent variable.

RELATIONS BETWEEN VARIABLES

Regardless of their type, two or more variables are related if in a sample of observations, the values of those variables are distributed in a consistent manner. In other words, variables are related if their values systematically correspond to each other for these observations. For example, Gender and WCC would be considered to be related if most males had high WCC and most females low WCC, or vice versa; Height is related to Weight because typically tall individuals are heavier than short ones; IQ is related to the Number of Errors in a test, if people with higher IQ's make fewer errors.

WHY RELATIONS BETWEEN VARIABLES ARE IMPORTANT

Generally speaking, the ultimate goal of every research or scientific analysis is finding relations between variables. The philosophy of science teaches us that there is no other way of representing "meaning" except in terms of relations between some quantities or qualities; either way involves relations between variables. Thus, the advancement of science must always involve finding new relations between variables. Correlational research involves measuring such relations in the most straightforward manner. However, experimental research is not any different in this respect. For example, the above mentioned experiment comparing WCC in males and females can be described as looking for a correlation between two variables: Gender and WCC. Statistics does nothing else but help us evaluate relations between variables. Actually, all of the hundreds of procedures that are described in this manual can be interpreted in terms of evaluating various kinds of inter-variable relations.

TWO BASIC FEATURES OF EVERY RELATION BETWEEN VARIABLES

The two most elementary formal properties of every relation between variables are the relation's (a) magnitude (or "size") and (b) its reliability (or "truthfulness").

1. Magnitude (or "size"). The magnitude is much easier to understand and measure than reliability. For example, if every male in our sample was found to have a higher WCC than any female in the sample, we could say that the magnitude of the relation between the two variables (Gender and WCC) is very high in our sample. In other words, we could predict one based on the other (at least among the members of our sample).
2. Reliability (or "truthfulness"). The reliability of a relation is a much less intuitive concept, but still extremely important. It pertains to the "representativeness" of the result found in our specific sample for the entire population. In other words, it says how probable it is that a similar relation would be found if the experiment was replicated with other samples drawn from the same population. Remember that we are almost never "ultimately" interested only in what is going on in our sample; we are interested in the sample only to the extent it can provide information about the population. If our study meets some specific criteria (to be mentioned later), then the reliability of a relation between variables observed in our sample can be quantitatively estimated and represented using a standard measure (technically called p-value or statistical significance level.

WHAT IS "STATISTICAL SIGNIFICANCE" (P-VALUE)

The statistical significance of a result is the probability that the observed relationship (e.g., between variables) or a difference (e.g., between means) in a sample occurred by pure chance ("luck of the draw"), and that in the population from which the sample was drawn, no such relationship or differences exist. Using less technical terms, one could say that the statistical significance of a result tells us something about the degree to which the result is "true" (in the sense of being "representative of the population"). More technically, the value of the p-value represents a decreasing index of the reliability of a result. The higher the p-value, the less we can believe that the observed relation between variables in the sample is a reliable indicator of the relation between the respective variables in the population. Specifically, the p-value represents the probability of error that is involved in accepting our observed result as valid, that is, as "representative of the population." For example, a p-value of .05 (i.e.,1/20) indicates that there is a 5% probability that the relation between the variables found in our sample is a "fluke." In other words, assuming that in the population there was no relation between those variables whatsoever, and we were repeating experiments like ours one after another, we could expect that approximately in every 20 replications of the experiment there would be one in which the relation between the variables in question would be equal or stronger than in ours. (Note that this is not the same as saying that, given that there IS a relationship between the variables, we can expect to replicate the results 5% of the time or 95% of the time; when there is a relationship between the variables in the population, the probability of replicating the study and finding that relationship is related to the statistical power of the design. In many areas of research, the p-value of .05 is customarily treated as a "border-line acceptable" error level.

HOW TO DETERMINE THAT A RESULT IS "REALLY" SIGNIFICANT

There is no way to avoid arbitrariness in the final decision as to what level of significance will be treated as really "significant." That is, the selection of some level of significance, up to which the results will be rejected as invalid, is arbitrary. In practice, the final decision usually depends on whether the outcome was predicted a priori or only found post hoc in the course of many analyses and comparisons performed on the data set, on the total amount of consistent supportive evidence in the entire data set, and on "traditions" existing in the particular area of research. Typically, in many sciences, results that yield $p \leq .05$ are considered borderline statistically significant but remember that this level of significance still involves a pretty high probability of error (5%). Results that are significant at the $p \leq .01$ level are commonly considered statistically significant, and $p \leq .005$ or $p \leq .001$ levels are often called "highly" significant. But remember that those classifications represent nothing else but arbitrary conventions that are only informally based on general research experience.

STATISTICAL SIGNIFICANCE AND THE NUMBER OF ANALYSES PERFORMED

Needless to say, the more analyses you perform on a data set, the more results will meet "by chance" the conventional significance level. For example, if you calculate correlations between ten variables (i.e., 45 different correlation coefficients), then you should expect to find by chance that about two (i.e., one in every 20) correlation coefficients are significant at the $p \leq .05$ level, even if the values of the variables were totally random and those variables do not correlate in the population. Some statistical methods that involve many comparisons, and thus a good chance for such errors, include some "correction" or adjustment for the total number of comparisons. However, many statistical methods (especially simple exploratory data analyses) do not offer any straightforward remedies to this problem. Therefore, it is up to the researcher to carefully evaluate the reliability of unexpected findings. Many examples in this manual offer specific advice on how to do this; relevant information can also be found in most research methods textbooks.

STRENGTH VS. RELIABILITY OF A RELATION BETWEEN VARIABLES

We said before that strength and reliability are two different features of relationships between variables. However, they are not totally independent. In general, in a sample of a particular size, the larger the magnitude of the relation between variables, the more reliable the relation. Assuming that there is no relation between the respective variables in the population, the most likely outcome would be also finding no relation between those variables in the research sample. Thus, the stronger the relation found in the sample, the less likely it is that there is no corresponding relation in the population. As we see, the magnitude and significance of a relation appear to be closely related, and we could calculate the significance from the magnitude and vice-versa; however, this is true only if the sample size is kept constant, because the relation of a given strength could be either highly significant or not significant at all, depending on the sample size.

If there are very few observations, then there are also respectively few possible combinations of the values of the variables, and thus the probability of obtaining by chance a combination of those values indicative of a strong relation is relatively high. Consider the following illustration. If we are interested in two variables (Gender: male/female and WCC: high/low) and there are only four subjects in our sample (two males and two females), then the probability that we will find, purely by chance, a 100% relation between the two variables can be as high as one-eighth. Specifically, there is a one-in-eight chance that both males will have a high WCC and both females a low WCC, or vice versa. Now consider the probability of obtaining such a perfect match by chance if our sample consisted of 100 subjects; the probability of obtaining such an outcome by chance would be practically zero. Let's look at a more general example. Imagine a theoretical population in which the average value of WCC in males and females is exactly the same. Needless to say, if we start replicating a simple experiment by drawing pairs of samples (of males and females) of a particular size from this population and calculating the difference between the average WCC in each pair of samples, most of the experiments will yield results close to 0. However, from time to time, a pair of samples will be drawn where the difference between males and females will be quite different from 0. How often will it happen? The smaller the sample size in each experiment, the more likely it is that we will obtain such erroneous results, which in this case would be results indicative of the existence of a relation between gender and WCC obtained from a population in which such a relation does not exist.

Example. "Baby Boys to Baby Girls Ratio"

Consider the following example from research on statistical reasoning. There are two hospitals: in the first one, 120 babies are born every day, in the other, only 12. On average, the ratio of baby boys to baby girls born every day in each hospital is 50/50. However, one day, in one of those hospitals twice as many baby girls were born as baby boys. In which hospital was it more likely to happen? The answer is obvious for a statistician, but as research shows, not so obvious for a lay person: It is much more likely to happen in the small hospital. The reason for this is that technically speaking, the probability of a random deviation of a particular size (from the population mean), decreases with the increase in the sample size.

Why Small Relations can be Proven Significant only in Large Samples

The examples in the previous paragraphs indicate that if a relationship between variables in question is "objectively" (i.e., in the population) small, then there is no way to identify such a relation in a study unless the research sample is correspondingly large. Even if our sample is in fact "perfectly representative" the effect will not be statistically significant if the sample is small. Analogously, if a relation in question is "objectively" very large (i.e., in the population), then it can be found to be highly significant even in a study based on a very small sample. Consider the following additional illustration. If a coin is slightly asymmetrical, and when tossed is somewhat more likely to produce heads than tails (e.g., 60% vs. 40%), then ten tosses would not be sufficient to convince anyone that the coin is asymmetrical, even if the outcome obtained (six heads and four tails) was perfectly representative of the bias of the coin. However, is it so that 10 tosses is not enough to prove anything? No, if the effect in question were large enough, then ten tosses could be quite enough. For instance, imagine now that the coin is so asymmetrical that no matter how you toss it, the outcome will be heads. If you tossed such a coin ten times and each toss produced heads, most people would consider it sufficient evidence that something is "wrong" with the coin. In other words, it would be considered convincing evidence that in the theoretical population of an infinite number of tosses of this coin there would be more heads than tails. Thus, if a relation is large, then it can be found to be significant even in a small sample.

Can "no relation" be a Significant Result?

The smaller the relation between variables, the larger the sample size that is necessary to prove it significant. For example, imagine how many tosses would be necessary to prove that a coin is asymmetrical if its bias were only .000001%! Thus, the necessary minimum sample size increases as the magnitude of the effect to be demonstrated decreases. When the magnitude of the effect approaches 0, the necessary sample size to conclusively prove it approaches infinity. That is to say, if there is almost no relation between two variables, then the sample size must be almost equal to the population size, which is assumed to be infinitely large. Statistical significance represents the probability that a similar outcome would be obtained if we tested the entire population. Thus, everything that would be found after testing the entire population would be, by definition, significant at the highest possible level, and this also includes all "no relation" results.

How to Measure the Magnitude (Strength) of Relations between Variables

There are very many measures of the magnitude of relationships between variables which have been developed by statisticians; the choice of a specific measure in given circumstances depends on the number of variables involved, measurement scales used, nature of the relations, etc. Almost all of them, however, follow one general principle: they attempt to somehow evaluate the observed relation by comparing it to the "maximum imaginable relation" between those specific variables. Technically speaking, a common way to perform such evaluations is to look at how differentiated are the values of the variables, and then calculate what part of this "overall available differentiation" is accounted for by instances when that differentiation is "common" in the two (or more) variables in question. Speaking less technically, we compare "what is common in those variables" to "what potentially could have been common if the variables were perfectly related."

Let us consider a simple illustration. Let us say that in our sample, the average index of WCC is 100 in males and 102 in females. Thus, we could say that on average, the deviation of each individual score from the grand mean (101) contains a component due to the gender of the subject; the size of this component is 1. That value, in a sense, represents some measure of relation between Gender and WCC. However, this value is a very poor measure, because it does not tell us how relatively large this component is, given the "overall differentiation" of WCC scores. Consider two extreme possibilities:

1. If all WCC scores of males were equal exactly to 100, and those of females equal to 102, then all deviations from the grand mean in our sample would be entirely accounted for by gender. We would say that in our sample, gender is perfectly correlated with WCC, that is, 100% of the observed differences between subjects regarding their WCC is accounted for by their gender.
2. If WCC scores were in the range of 0-1000, the same difference (of 2) between the average WCC of males and females found in the study would account for such a small part of the overall differentiation of scores that most likely it would be considered negligible. For example, one more subject taken into account could change, or even reverse the direction of the difference. Therefore, every good measure of relations between variables must take into account the overall differentiation of individual scores in the sample and evaluate the relation in terms of (relatively) how much of this differentiation is accounted for by the relation in question.

Common "General Format" of Most Statistical Tests

Because the ultimate goal of most statistical tests is to evaluate relations between variables, most statistical tests follow the general format that was explained in the previous paragraph. Technically speaking, they represent a ratio of some measure of the differentiation common in the variables in question to the overall differentiation of those variables. For example, they represent a ratio of the part of the overall differentiation of the WCC scores that can be accounted for by gender to the overall differentiation of the WCC scores. This ratio is usually called a ratio of explained variation to total variation. In statistics, the term explained variation does not necessarily imply that we "conceptually understand" it. It is used only to denote the common variation in the variables in question, that is, the part of variation in one variable that is "explained" by the specific values of the other variable, and vice versa.

How The "Level of Statistical Significance" is Calculated

Let us assume that we have already calculated a measure of a relation between two variables (as explained above). The next question is "how significant is this relation?" For example, is 40% of the explained variance between the two variables enough to consider the relation significant? The answer is "it depends." Specifically, the significance depends mostly on the sample size. As explained before, in very large samples, even very small relations between variables will be significant, whereas in very small samples even very large relations cannot be considered reliable (significant). Thus, in order to determine the level of statistical significance, we need a function that represents the relationship between "magnitude" and "significance" of relations between two variables, depending on the sample size. The function we need would tell us exactly "how likely it is to obtain a relation of a given magnitude (or larger) from a sample of a given size, assuming that there is no such relation between those variables in the population." In other words, that function would give us the significance (p) level, and it would tell us the probability of error involved in rejecting the idea that the relation in question does not exist in the population. This "alternative" hypothesis (that there is no relation in the population) is usually called the null hypothesis. It would be ideal if the probability function was linear, and for example, only had different slopes for different sample sizes. Unfortunately, the function is more complex, and is not always exactly the same; however, in most cases we know its shape and can use it to determine the significance levels for our findings in samples of a particular size. Most of those functions are related to a general type of function which is called normal.

Why The "Normal Distribution" is Important

The "Normal distribution" is important because in most cases, it well approximates the function that was introduced in the previous paragraph. The distribution of many test statistics is normal or follows some form that can be derived from the normal distribution. In this sense, philosophically speaking, the Normal distribution represents one of the empirically verified elementary "truths about the general nature of reality," and its status can be compared to the one of fundamental laws of natural sciences. The exact shape of the normal distribution (the characteristic "bell curve") is defined by a function which has only two parameters: mean and standard deviation.

A characteristic property of the Normal distribution is that 68% of all of its observations fall within a range of ±1 standard deviation from the mean, and a range of ±2 standard deviations includes 95% of the scores. In other words, in a Normal distribution, observations that have a standardized value of less than -2 or more than +2 have a relative frequency of 5% or less. (Standardized value means that a value is expressed in terms of its difference from the mean, divided by the standard deviation.) If you have access to *STATISTICA*, you can explore the exact values of probability associated with different values in the normal distribution using the interactive Probability Calculator tool; for example, if you enter the Z value (i.e., standardized value) of 4, the associated probability computed by *STATISTICA* will be less than .0001, because in the normal distribution almost all observations (i.e., more than 99.99%) fall within the range of ±4 standard deviations. The animation below shows the tail area associated with other Z values.

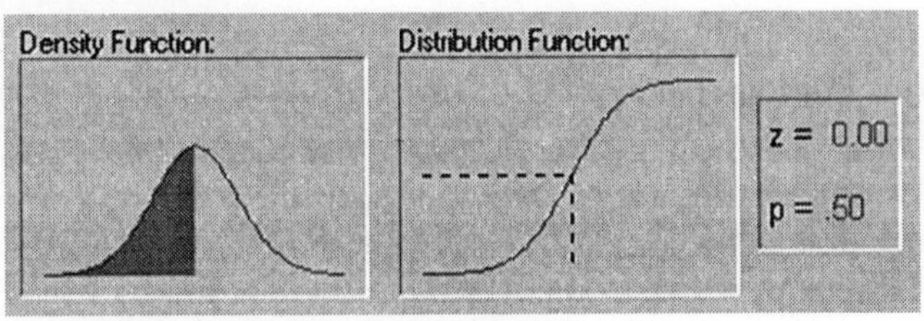

Illustration of How The Normal Distribution is Used in Statistical Reasoning (induction).

Where pairs of samples of males and females were drawn from a population in which the average value of WCC in males and females was exactly the same. Although the most likely outcome of such experiments (one pair of samples per experiment) was that the difference between the average WCC in males and females in each pair is close to zero, from time to time, a pair of samples will be drawn where the difference between males and females is quite different from 0. How often does it happen? If the sample size is large enough, the results of such replications are "normally distributed" (this important principle is explained and illustrated in the next paragraph), and thus knowing the shape of the normal curve, we can precisely calculate the probability of obtaining "by chance" outcomes representing various levels of deviation from the hypothetical population mean of 0. If such a calculated probability is so low that it meets the previously accepted criterion of statistical significance, then we have only one choice: conclude that our result gives a better approximation of what is going on in the population than the "null hypothesis" (remember that the null hypothesis was considered only for "technical reasons" as a benchmark against which our empirical result was evaluated). Note that this entire reasoning is based on the assumption that the shape of the distribution of those "replications" (technically, the "sampling distribution") is normal. This assumption is discussed in the next paragraph.

Are All Test Statistics Normally Distributed?

Not all, but most of them are either based on the normal distribution directly or on distributions that are related to, and can be derived from normal, such as t, F, or Chi-square. Typically, those tests require that the variables analysed are themselves normally distributed in the population, that is, they meet the so-called "normality assumption." Many observed variables actually are normally distributed, which is another reason why the normal distribution represents a "general feature" of empirical reality. The problem may occur when one tries to use a normal distribution-based test to analyse data from variables that are themselves not normally distributed. In such cases we have two general choices. First, we can use some alternative "nonparametric" test; but this is often inconvenient because such tests are typically less powerful and less flexible in terms of types of conclusions that they can provide. Alternatively, in many cases we can still use the normal distribution-based test if we only make sure that the size of our samples is large enough. The latter option is based on an extremely important principle which is largely responsible for the popularity of tests that are based on the normal function. Namely, as the sample size increases, the shape of the sampling distribution (i.e., distribution of a statistic from the sample; this term was first used by Fisher, 1928a) approaches normal shape, even if the distribution of the variable in question is not normal. This principle is illustrated in the following animation showing a series of sampling distributions (created with gradually increasing sample sizes of: 2, 5, 10, 15, and 30) using a variable that is clearly non-normal in the population, that is, the distribution of its values is clearly skewed.

However, as the sample size (of samples used to create the sampling distribution of the mean) increases, the shape of the sampling distribution becomes normal. Note that for n=30, the shape of that distribution is "almost" perfectly normal. This principle is called the central limit theorem (this term was first used by Pσlya, 1920; German, "Zentraler Grenzwertsatz").

How Do We Know The Consequences of Violating The Normality Assumption?

Although many of the statements made in the preceding paragraphs can be proven mathematically, some of them do not have theoretical proofs and can be demonstrated only empirically, via so-called Monte-Carlo experiments. In these experiments, large numbers of samples are generated by a computer following predesigned specifications and the results from such samples are analysed using a variety of tests. This way we can empirically evaluate the type and magnitude of errors or biases to which we are exposed when certain theoretical assumptions of the tests we are using are not met by our data. Specifically, Monte-Carlo studies were used extensively with normal distribution-based tests to determine how sensitive they are to violations of the assumption of normal distribution of the analysed variables in the population. The general conclusion from these studies is that the consequences of such violations are less severe than previously thought. Although these conclusions should not entirely discourage anyone from being concerned about the normality assumption, they have increased the overall popularity of the distribution-dependent statistical tests in all areas of research.

MEASUREMENT BY COUNTING

If you wish to measure the width of your desk, you take a tape measure to it and count off the number of inches or centimeters. If you wish to measure the outdoor temperature at the present moment, you take a thermometer outdoors and count off the number of degrees Fahrenheit or degrees Celsius. If you are sitting in a classroom and wish to measure the number of students in the room, you count them. We will speak of this type of measurement as standard scalar measurement, since each individual instance of it results in a numerical value that refers to a point on some particular standard measurement *scale*—inches, centimeters, degrees Fahrenheit, degrees Celsius, pints, liters, bushels, grams, ounces, light years, volts, ohms, and so on. If you count up the number of students in a classroom as one, two, three, four, etc., and run out of students to count after reaching 29, this numerical value of 29 represents a point on the most general and familiar standard measurement scale of all, the scale of cardinal numbers.

When measurement involves simply counting out the number of a set of items or events according to the series of cardinal numbers—one, two, three, four, etc.—the scale of measurement is spoken of as an absolute scale. All other commonly recognized measurement scales are *relative* in the sense that they are designed to measure not the absolute number of items or events but rather the magnitude of some particular attribute—length, width, weight, temperature, velocity, electrical potential, etc.—relative to the units of some particular scale that has been designed, or has evolved, for taking the measure of that attribute.-

Standard scalar forms of measurement can also be sorted out according to whether they are inter convertible. The desk at which I am presently sitting is 60 inches wide when measured with a yard stick and 152.4 centimeters wide when measured with a meter stick. The actual width of the desk is of course the same in both cases. Inches and centimeters are simply two different and inter convertible ways of measuring it. Each inch is equal to 2.54 centimeters, and each centimeter is equal to 0.3937 inches. Similarly, the outdoor temperature on a pleasant June morning is 61.5 degrees Fahrenheit and 16.4 degrees Celsius. The actual degree of warmness is the same in both cases; it is simply being measured by two different scales. In general, any two scales of measurement that measure the same general property (length, warmness, heaviness, volume, velocity, etc.) and can be systematically translated back and forth into each other's terms are said to be commensurate; otherwise they are incommensurate.

Thus, inches and centimeters are commensurate scales of measurement, as are degrees Fahrenheit and degrees Celsius; whereas inches (or centimeters) and degrees Fahrenheit (or Celsius) are incommensurate scales of measurement. -

EQUAL INTERVAL SCALES

Suppose, now, that we were actually to measure the width of a desk in inches, the current outdoor temperature in degrees Fahrenheit, and the number of students in a classroom by counting them off as one, two, three, four, and so on. Examine these various measures closely and you will see there is one general property, apart from the aspect of counting, that all three have in common. It is that the scales of measurement to which they refer—inches of width, degrees Fahrenheit of temperature, and number of students—are all delineated by equal intervals between their successive units of measurement. Thus, the one-inch interval that separates 23 inches from 24 inches is precisely the same size as the one-inch interval that separates 24 from 25 inches, 30 from 31 inches, or 43 from 44 inches. By extension, the three-inch interval that separates 20 from 23 inches is precisely the same size as any other three-inch interval along the scale; further, any three-inch interval is precisely three times as large as any one-inch interval, but only half as large as any six-inch interval; and so on. The same is true for degrees Fahrenheit and student headcount, as well as for virtually any other commonly recognized scale of measurement that comes down at the level of fundamentals to a process of counting—centimeters, degrees Celsius, volts, ohms, miles per hour, miles per gallon, quarts, bushels, pecks, acres, dollars, square feet, and cubic centimeters, to mention but a few. Measurement scales of this general type are spoken of as *equal interval scales.*

If a measurement scale possesses this property of equal intervals, it is then possible and meaningful to take two or more measures from that scale and perform the simple arithmetic operations of addition and subtraction. For example:

$$\begin{aligned} \text{5 inches} + \text{3 inches} &= \text{8 inches} \\ 45.3°\text{F} - 5.0°\text{F} &= 40.3°\text{F} \\ \text{5 students} - \text{3 students} + \text{10 students} &= \text{12 students} \end{aligned}$$

Because they allow these primary arithmetic operations of addition and subtraction, equal interval scales also permit the calculation of various secondary measures that describe the aggregate properties of what we will be speaking of later as *samples* and *distributions.* An example of such an aggregate measure, with which you are certainly already familiar, is the simple arithmetic average. Suppose that the high temperature on three successive winter days is 31.7°F, 36.4°F, and 29.0°F. Given that the temperature scale of degrees Fahrenheit has equal intervals, we could then meaningfully conclude that the average high temperature for those three days is

$$\frac{31.7°\text{F} + 36.4°\text{F} + 29.0°\text{F}}{\text{3 days}} = 32.37°\text{F}$$

These, however, would not be legitimate or meaningful operations for measures based on scales that have unequal intervals, such as the decibel scale of sound intensity or the Richter scale of earthquake intensity. On the Richter scale, for example, each succeeding unit represents a 10-fold increase in intensity. Thus, a quake that registers 5 on the scale is 10 times stronger than one that registers 4; a quake that registers 6 is 10 times stronger than one that registers 5, thus 100 times stronger than one that registers 4, 1,000 times stronger than one that registers 3; and so on. It would therefore make no sense at all to say that three earthquakes measuring 3, 4, and 8 on the scale have an average magnitude of 5. It is of course certainly possible to calculate the raw average of the numbers 3, 4, and 8, and that average will certainly end up as $(3+4+8)/3 = 5$. But it would not mean anything. In order to calculate a rigorously meaningful average for any set of numerical values, it is essential that the numerical values be based on a measurement scale that has equal intervals.

But for now, back to inches, degrees Fahrenheit, and cardinal numbers. Examine these three standard measurement scales more closely and you will find there is one further property possessed by inches and degrees Fahrenheit that is not shared by the headcount of students, and another property possessed by inches and student headcount that is not shared by degrees Fahrenheit. Although these properties will perhaps seem rather obvious, they have implications that are non-obvious and quite far-reaching.

CONTINUOUS SCALES VERSUS DISCRETE SCALES

For inches and degrees Fahrenheit versus student headcount, the distinction is between continuous scales of measurement versus discrete scales of measurement or, alternatively, between continuous variables versus discrete variables. If you are counting up the number of students in a classroom it is possible to have, for example, either 16 or 17 students, but it would be utter nonsense to end up with a count of 16.75 students, for there is simply no such thing as three-quarters of a student. The same is true when you count up the number of any other set of items that come only in discrete, indivisible units. In cases of this sort the variable (which is the number of items in the set) is discrete, and so too is the integer scale of cardinal numbers—one, two, three, four, etc.—by which the variable is measured. If you are measuring the width of an object in inches, on the other hand, the possible outcomes are not restricted to integer values such as 12 inches, 13 inches, 14 inches, and so on. It is entirely possible to come out with a measure of width that falls somewhere between two successive integer values: for example, 12.47 inches. The same is true for degrees Fahrenheit, pints, grams, seconds, cycles per second, microfarads, and any number of other delineations that would normally be spoken of as a scale of measurement. The common property of such continuous scales is that their units of measurement are in principle infinitely divisible, so that any particular measure taken on them could potentially be drawn out to as many other decimal places as one might care to take it. In reality, of course, there are often practical limits on the precision of the measurement process. With an ordinary household thermometer, for example, it would not be practically possible to measure the current outdoor temperature as 61.0034709102°F. Nonetheless, the temperature of any particular object or location *could* potentially have this or any other multi-decimal value, if only we had a sufficiently precise instrument for measuring it.

RATIO SCALES VERSUS NON-RATIO SCALES

For inches and student headcount versus degrees Fahrenheit, the distinction is between measurement scales for which the point designated as zero represents an absolute zero of the quantity that is being measured, versus measurement scales whose designated zero is only an arbitrary point that happens to be called "zero." Thus, zero inches represents an absolute absence of width. Zero students represents an absolute absence of students. Zero degrees Fahrenheit, however, does not represent an absolute absence of warmth, for it is obviously possible to have measures of temperature that fall *below* zero on the Fahrenheit scale. The same is true of the Celsius temperature scale, whose arbitrary zero merely marks the point at which water freezes under normal barometric pressure. Zero degrees on the Kelvin scale of temperature measurement, on the other hand, *does* mark an absolute absence of temperature, which is to say, an absolute absence of warmth and an absolute absence of the molecular motion from which the quality of warmth derives. In brief, zero degrees Kelvin, which corresponds to about –273°C and –460°F, is as cold as it gets, and there is nothing colder.

So if you are ever out in a temperature of –18° Fahrenheit, you can comfort yourself with the thought that it is still about 442° Fahrenheit warmer than it could be. On the other hand, you will never under any circumstances find a temperature of –18° Kelvin, or even –0.000,018° Kelvin, even if you go to the farthest reaches of interstellar space. Scales of measurement that have an absolute zero point can yield negative values, such as –18° Kelvin, only in two special kinds of cases, both involving the concept of *polarity* or *directionality.*

Scales of measurement that have both equal intervals *and* absolute zero points are spoken of as ratio scales, for the simple reason that they permit the meaningful calculation of ratios. If you find, for example, that object A is 5 inches wide and object B is 15 inches wide, it is legitimate and meaningful to conclude that object B is three times as wide as object A, or alternatively, that object A is only one-third as wide as object B. Similarly, it makes sense to say that 15 students are three times as many as 5 students, and that 5 students are only one-third as many as 15 students. If the high temperatures on two successive winter days are 5°F and 15°F, on the other hand, it makes no sense at all to conclude that the second day is three times as warm as the first–because the zero point from which 5°F and 15°F are starting out is only an arbitrary marker on a scale that potentially extends all the way down to about —460°F. In order to make such ratio judgments concerning temperatures we would have to use a scale, such as the Kelvin scale, whose zero point does mark an absolute zero level of temperature.

COMPOUND MEASURES ON EQUAL INTERVAL SCALES

There are two desks in my office. The width of the larger is 60 inches, and the width of the smaller is 48.5 inches. The sum of these two widths is 108.5 inches, their average is 54.25 inches, and the difference between them is either 11.5 inches or –11.5 inches, depending on whether we subtract the smaller number from the larger or the larger from the smaller. Each of these three quantitative facts is also a *measure,* specifically, a compound measure based on two component measures. In science and in human affairs in general, you will find many compound measures of these general types, all owing their existence to the fact that equal interval scales readily permit addition, subtraction, and the calculation of averages. A familiar family of examples includes that whole range of measures that involve the relational word "per"—miles per gallon, miles per hour, dollars per hour, cycles per second, calories per gram, bushels per acre, and so on. Here are some general precepts for determining the scale properties that pertain to such compound measures.

Sums

The general principle is that the sum of a set of equal interval measures will have the same scale properties as the component measures on which the sum is based. Thus, the sum of a set of measures expressed in inches is itself a measure of inches, and it will have the same scale properties as the original scale of inches (continuous, equal interval, and ratio). The sum of a set of measures of temperature expressed in degrees Fahrenheit is itself a measure of degrees Fahrenheit, and it will have the same scale properties as the original scale of degrees Fahrenheit (continuous, equal interval, and non-ratio). If you count up the number of students in each of three sections of a statistics course, each count is a measure on the scale of cardinal numbers. The sum of the counts will also be a measure on the scale of cardinal numbers, and it will have the same scale properties (discrete, equal interval, and ratio).

Averages

The same principle holds for the average of a set of equal interval measures, with one exception. Thus, the average of a set of equal interval ratio scale measures (e.g., inches) will have the properties of a ratio scale, and the average of a set of equal interval non-ratio measures (e.g., degrees Fahrenheit) will have the properties of a non-ratio scale. The exception is that any average of two or more equal interval measures will have the properties of a *continuous* scale, even though the original component measures themselves might be discrete. Thus, if three sections of a statistics course have 27, 31, and 30 students, respectively, the average number of students in these sections is 29.33..., that is, 29 and one-third "students per class." This of course does not mean that some hapless student is divided into three parts; it is simply the way the arithmetic of the situation happens to work out.

Differences

The principle also holds for the differences between equal interval measures, but again with one exception. Thus, the difference between two ratio measures will also be a ratio measure; the difference between two discrete measures will also be a discrete measure; and the difference between two continuous measures will also be a continuous measure. The exception here is that the difference between two equal interval measures will have the properties of a *ratio* scale, even if the original component measures belong to a non-ratio scale. The reason for this is that it is always possible to end up with a difference between two equal interval measures of absolutely zero, even though the two measures themselves belong to a scale, such as degrees Fahrenheit of temperature, that does not have an absolute zero point. Thus, if you were to measure the temperature at three locations as A = 40°F, B = 50°F, and C = 60°F, it would of course make no sense to say that B is 50/40 = 1.25 times as great as A, nor that C is 60/50 = 1.2 times as great as B. However, it would make perfectly good sense to say that the *difference* between C and A (60–40 = 20) is twice as large as the difference between B and A (50–40 = 10).

A SCORECARD FOR EQUAL INTERVAL SCALES OF MEASUREMENT

This roster of players is not nearly so complex as it might at first seem. For equal interval scales of measurement there are basically two dimensions of classification—ratio versus non-ratio and discrete versus continuous–which yield a total of four possible cross classifications:

non-ratio and discrete;
ratio and discrete;
non-ratio and continuous; and
ratio and continuous

And it is really only the last three that you need to keep track of, as it is difficult to imagine any practically useful scales of measurement that could fall into the first cross-classification category of 'non-ratio and discrete.' This is because any scale delineated by discrete equal intervals is basically a scale for counting the number of discrete, indivisible units of some particular type, be they students in a classroom or hydrogen atoms in a cubic meter of interstellar space, and in any such enumeration it is always a logical possibility to end up with a count of absolutely zero. In any event, bear in mind that the more far-reaching distinction is the one between ratio scales and non-ratio scales, which depends upon whether the scale in question does or does not have an absolute zero point. All equal interval scales, whether ratio or non-ratio, permit the mathematical operations of addition and subtraction, as well as the calculation of averages. But statements of *relative magnitude,* such as "A is three times as large as B," or "B is one-third as large as A," are permitted only by equal interval scales that have an absolute zero point.

RANK-ORDER SCALES AND RATING SCALES

The temperature inside my office at a certain moment on a winter afternoon is 67.2°F, and the outdoor temperature at the same moment is 22.8°F. The equal interval property of the Fahrenheit scale permits you to conclude not only that it is warmer inside than out, but also that the magnitude of the difference is exactly 44.4°F. The same is true of equal interval scales in general: first, they allow us to sort out the relationships of 'greater than' and 'less than'; and second, they allow us to specify just how much greater-than or less-than one measure is in comparison with another.

In general, the relationships of 'greater than' and 'less than' are spoken of as ordinal relationships, so named because they specify the *order* of size, quantity, or magnitude among the particular items that are being measured. Although in principle any measurement scale at all that delineates such ordinal relationships could be described as an ordinal scale, the designation is superfluous for equal interval scales, since these are all ordinal scales necessarily. In practice, the term "ordinal scale" is confined to measurement scales that measure *only* ordinal relationships. Suppose, for example, that I did not have a thermometer. I could judge accurately enough that it is warmer inside than outside. I might even be able to judge that the indoor temperature is "slightly warmer," "moderately warmer," or "quite a lot warmer" than the outdoor temperature. But unless I had an extraordinarily well developed talent for judging temperature differences in fine-grained detail, I would not be in a position to specify *precisely* how much warmer the one is than the other.

If you ever have a choice between measuring a variable with an equal interval scale (which is *also* ordinal) or an ordinal scale (which is *merely* ordinal), choose the equal interval scale, for that is obviously the stronger form of measurement, since it contains a greater amount of information about the item or items that are being measured. As it happens, however, there are many kinds of variables for which measurement by an equal interval scale is either not possible or not practicable, and for which one therefore simply does not have the choice. For example, suppose we were interested in measuring the importance of the following four life goals for each of the students in your class:

- Enjoying a good family life,
- Enjoying good friends,
- Achieving career success,
- Having plenty of money. For better or worse, there is no standard meter stick, no clever electronic instrument, no mental calipers nor anything of the sort that will give us equal interval measures of the variable described by the phrase "importance of life goals." The same is true of any variable that in the final analysis comes down to a matter of preference, attitude, opinion, or judgment. Perhaps someday some really ingenious researcher will come along to invent the requisite equal interval mental calipers. In the meantime, no matter how deft or subtle we might be in attempting to measure such variables, all we will ever end up with is measurement on an ordinal scale.

RANK-ORDER SCALES

The most obvious form of ordinal measurement is rank-ordering. In order to get the feel of this form of measurement, please take a moment to arrange the items that were just mentioned.

- Enjoying a good family life,
- Enjoying good friends,
- Achieving career success, and
- Having plenty of money

In accordance with how important a role they play in your aims and hopes for what your life will be ten or fifteen years from now. Give the most important item the rank of "1", the second most important item the rank of "2", and so on. And then, having done it, reflect upon whether this particular measure of your life priorities, incomplete though it certainly is, would be capable of telling an outside observer anything about you that is worth knowing. Assuming that you have performed the rank-ordering seriously and honestly, I think you will agree that the answer is yes. If your rank-ordering ended up as

1. Family; 2. Friends; 3. Career; 4. Moncy

it says one thing about you; and if it ended up as

1. Money; 2. Career; 3. Friends; 4. Family

it surely says something quite different. Clearly this form of measurement is quite a lot weaker than the measurement of inches of width, pounds of weight, and so forth, but is a potentially useful form of measurement all the same. The key to using it well is to understand its limitations.

The central limitation of rank-order scales is simply that the successive points on such scales, which in all cases translate down to the ordinal succession "first," "second," "third," and so forth, are not intrinsically separated by equal intervals. Thus, for the hypothetical student who ranks the life-goal items as

1. Family; 2. Friends; 3. Career; 4. Money

there is no reason whatever to suppose that the difference in importance between family and friends is of the same size as the difference in importance between friends and career, or between career and money. Similarly, for any two students whose rankings came out as

	1	2	3	4
Student A:	Family	Friends	Career	Money
Student B:	Family;	Friends;	Career;	Money

we could certainly say for both of them that enjoying good friends appears to be more important than achieving career success, but there would be no reason at all to suppose that the magnitude of the preference was the same for the first student as for the second. It would be entirely possible that the first student regarded friends as only marginally more important than career, while the second considered friends to be quite a lot more important than career.

The far-reaching consequence of this limitation is that mathematical operations involving addition, subtraction, and the calculation of averages can be applied to data that derive from a rank-order scale only in certain special circumstances, which in turn severely restricts the kinds of statistical analyses that can be legitimately performed upon such data. We will describe these special circumstances in later chapters as they become relevant. Until then, just bear in mind that while rank-order scales are capable of effectively delineating the ordinal relationships of 'greater than' and 'less than,' that is all they are capable of doing.

RATING SCALES

A useful but potentially problematical variation on the theme of ordinal scales involves measurement by way of assigning numerical ratings. The usefulness of it stems from the fact that a rating scale is capable of providing somewhat more information than a rank-ordering. To illustrate, consider the two hypothetical students mentioned in the preceding section who had both ranked the four life-importance items, from most important to least important, as:

family > friends > career > money

Suppose that they were now to rate these same items on a scale from zero to 4, with the units of the scale defined as

zero = completely unimportant
1 = slightly important
2 = moderately important
3 = quite important
4 = exceedingly important

To allow them to make fine-grained distinctions, we will stipulate that they can break the scale into fractional units if they so wish; thus, in addition to using the integer values zero, 1, 2, 3, and 4, they can also rate items as, for example, 0.5, 2.6, 3.4, etc. When we mentioned the rank-orderings of these two hypothetical students in the preceding section, we indicated it would be entirely possible that the first student regarded friends as only marginally more important than career, while the second considered friends to be quite a lot more important than career. Here is a set of ratings for them that would correspond to this possibility:

	Family	Friends	Career	Money
Student A:	4.0 >	3.8 >	3.6 >	3.4
Student B:	4.0 >	3.5 >	2.0 >	1.0

As you can see, the rating scale would still reflect the same bare ordinal relationships of greater than and less than, but it would also potentially reveal a texture in these relationships that goes beyond the merely ordinal. Both students rate family and friends quite high on the scale, but whereas student A also rates career and money within the range between quite important and exceedingly important, student B places them down toward the bottom of the scale.

The critical difference between rank-ordering and rating is that, whereas the points on a rank-ordering scale (first, second, third, etc.) delineate ordinal relationships and nothing other than ordinal relationships, the successive points on a rating scale represent intervals, in something of the same way that the scale points on a yard stick (6 inches, 7 inches, 8 inches, etc.) or a thermometer (66°F, 67°F, 68°F, etc.) represent intervals. Hence the ability of a rating scale to take us somewhat beyond the merely ordinal.

Please note well, however, that the operative word here is "somewhat." Rating scales can take us somewhat beyond the merely ordinal—but that does not mean that they take us anywhere near where an equal interval scale would take us. When rating scales are represented pictorially, they usually end up looking something like this

Completely unimportant	*Slightly important*	*Moderately important*	*Quite important*	*Exceedingly important*
0	1	2	3	4

which is, of course, highly suggestive of an equal interval scale. The resemblance, however, is superficial and entirely illusory; for while it is true that the rating scale is divided up into intervals, these intervals are not intrinsically *equal.* For student A, there is no reason at all to suppose that the distance between 4 and 3 on the rating scale is the same as the difference between 3 and 2, or between 2 and 1, or between 1 and zero. The same is true for student B and for anyone else who might be rating items on this or any other rating scale. Moreover, there is no reason to suppose that the difference between 4 and 3 for student A is the same as the difference between 4 and 3 for student B; and so on.

The implication is the same as described earlier in connection with rank-ordering. Basic mathematical operations involving addition, subtraction, and the calculation of averages can be applied to data that derive from a rating scale only in certain special circumstances, which in turn severely restricts the kinds of statistical analyses that can be legitimately performed upon such data. Here again, we will describe these special circumstances in later chapters as they become relevant.

Until then, please keep clearly in mind that while rating scales do go somewhat beyond the merely ordinal delineations provided by rank-ordering, they still do not possess the properties of equal interval scales, and thus they do not easily permit the kinds of mathematical operations that are so readily permitted by equal interval scales. This caution is all the more important for rating scales, since their superficial resemblance to an equal interval scale can easily beguile the mind into thinking that the measurement is stronger than it really is.

A familiar example of this tendency to beguile is the standard letter grade scale of A, A–, B+, B, B–, etc., commonly used for the measurement of academic achievement. Clearly it is only a rating scale. The various letters are really only shorthand ways of saying "excellent, superior, top-drawer" for an A, "not quite a full A, but still very good" for an A–, and so on down the line for B+, B, B–, C+, C, etc.

And yet, once the merely ordinal verbal phrases are translated into letters, and the practice becomes enshrined in academic tradition, how very easy it is to drift into the fallacy of supposing that letter grades have the properties of an equal interval scale—to suppose, for example, that the difference between an A and a B is of the same magnitude as the difference between a B and a C, or between a C and a D. In most academic institutions this fallacy is even given official sanction by way of further translating the letter grades into numerical values known as *quality points.* At my own institution the conventional translation is as follows:

A	A–	B+	B	B–	C+	C	C–	D+	D	D–	F
4.0	3.7	3.3	3.0	2.7	2.3	2.0	1.7	1.3	1.0	0.7	0

Now this certainly *looks* like an equal interval scale, and one, moreover, that is capable of making some rather fine distinctions. But here again the resemblance is only superficial and illusory. The letter grade system is fundamentally an ordinal rating scale, and merely exchanging the letters for numbers will not make it anything more than an ordinal rating scale.

MEASUREMENT BY SORTING: CATEGORICAL MEASUREMENT

Measurement of this type involves examining each item in a set of items to determine whether its properties match the defining criteria of this, that, or the other category. Thus, the measure of gender has two possible categorical outcomes, *female* and *male.* Sort the students in your classroom according to the defining criteria of these two categories, and the result for each student is a categorical measure of gender. Sort the same students into the four categories *freshman, sophomore, junior,* and *senior,* and the result for each student is a categorical measure of academic class.

In some cases of categorical measurement, such as the division of gender into *female* and *male* or the division of religious affiliation into *Catholic, Protestant, Jewish, Moslem,* etc., there are no intrinsic relationships of 'greater than' or 'less than' among the categories. It would be nonsense, for example, to say that females have more gender than males, or vice versa. Equally nonsensical would be the suggestion that the category *Catholic,* in and of itself, denotes a greater or lesser degree of religious affiliation than the category *Protestant.* Categories of this type are spoken of as nominal categories (from the Latin *nomen,* meaning "name").

In other cases, however, such as the delineation of academic class into *freshman, sophomore, junior,* and *senior,* the categories do have intrinsic quantitative relationships of 'greater than' or 'less than' . The category of *sophomore* represents a higher academic class than the category of *freshman; junior* represents a higher academic class than *sophomore;* and so on. Another fairly obvious example would be the sorting of college faculty into the categories *instructor, assistant professor, associate professor,* and *full professor.* Categories of this type are spoken of as ordinal categories, so named because there is a quantitative order that exists among them.

At first glance, categorical measurement is likely to seem rather inconsequential. True, you can measure each person in a group with respect to the categorical variable of gender–but then what? Clearly it would make no sense to say "one unit of gender plus two units of gender equals three units of gender," for there is simply no such thing as a unit of gender. It would be equally bizarre to speak of the average gender of a group of persons, or of the average of maleness or femaleness, because there is nothing here that you can take an average *of.* Categorical measurement is sometimes described of as measurement on a "nominal scale," but that is actually a misnomer, since there is really no *scale* involved here at all. It is simply a matter of sorting items into categories.

Note, however, that even the most purely nominal and non-quantitative form of categorical measurement is capable of producing data that do have a distinctly quantitative dimension. Sort the students in your classroom into the two strictly nominal categories *female* and *male,* and you are not only taking the categorical measure of gender for each student; you are also forming the basis for measuring *how many* students there are that fall within each of the two categories. Thus, in a class of 29 students you might find by enumeration that there are 15 females and 14 males. As we noted earlier, anytime you count up the number of discrete items in a set (females, males, Catholics, Protestants, etc.) you are using the scale of cardinal numbers, which, since it is both an equal interval scale and a ratio scale, readily permits the basic mathematical operations of addition, subtraction, multiplication, and division. Hence, even though categorical measurement starts out as a rather weak form of measurement, it is capable of yielding secondary measures that are quite strong. There are certain cautions that do need to be kept in mind with respect to categorical measurement.

An especially useful variation on the theme of categorical measurement is the situation in which you are cross-categorizing items according to two or more dimensions of classification concurrently. This, in fact, is what we did earlier for Table when we cross-classified equal interval scales of measurement according to the two dimensions, non-ratio versus ratio and discrete versus continuous. Similarly, you could sort the students in a classroom concurrently according to any two or more applicable dimensions of categorical measurement, such as gender, academic class, ethnic identity, liberal versus conservative, geographical origins, and so on. The scientific utility of such *bivariate* (two variables) or *multivariate* (three or more variables) cross-classification is that it provides a basis for determining whether the two or more categorical variables tend to be systematically associated with each other. Here as well is a point that will be examined further in later chapters.

TAKING THE MEASURE OF THINGS: BEYOND THE OBVIOUS

The next time you take an examination in a course, you can console yourself with the thought that you are participating in the process of measurement. Your instructor wishes to assess your knowledge and understanding of the subject matter; to that end, he or she will give you a set of tasks to perform; and the degree to which you perform well on those tasks will then be taken as the measure of your knowledge and understanding.

And now for a practice mini-exam. When your instructor measures your knowledge and understanding of the subject matter in this fashion, which of the following forms of measurement does it most closely resemble:

1. The measurement of the width of a desk with a tape measure,
2. The measurement of the number of students in a room by counting them, or
3. The measurement of temperature with a conventional bulb and glass-tube thermometer?

The answer is that it most closely resembles and here is the reason for it. When you measure the width of an object by holding a tape measure right up against it, you are measuring width *directly*.

Similarly, when you are measuring the number of students in a room, you are *directly* counting them off as one, two, three, four, etc. When you measure temperature with a thermometer, on the other hand, what you are measuring directly is not temperature itself, but rather the *effect* that temperature has upon the fluid that is inside the glass tube. As the temperature rises or falls, the fluid in the tube expands or contracts. Hence, the column of fluid rises and falls in accordance with increases and decreases of temperature. Similarly for the exam: what is being measured directly is not your knowledge and understanding of the subject, but rather the effect that knowledge and understanding are presumed to have upon your performance on the exam. Your instructor's assumption is that performance on the exam will be relatively high for those students whose mastery of the subject is strong, and relatively low for those whose mastery of it is weak. But even if the exam is the best and fairest test that any instructor has ever devised, the fact remains that your score on the exam is a direct measure only of how well you perform on the exam. As a measure of the knowledge and understanding that are presumed to underlie the performance, it is strictly indirect.

Scientific researchers are very often in the same position as your instructor. They can get nowhere at all without taking the measure of the variables that are of interest to them; and as many of these variables cannot be measured directly, ways must be found of measuring them indirectly. Indeed, a very considerable part of scientific progress over the past several centuries has resulted from precisely this task of devising indirect ways of measuring variables that cannot be measured directly.

Although indirect measurement has played a prominent role in the development of the hard-core physical sciences, such as physics and chemistry, it is especially important in those realms of scientific research that are concerned with what goes on inside living organisms. For not only is it often the case that the internal states and processes of an organism cannot be directly measured–in many instances they cannot even be directly observed. Take, for example, the internal state known as fear. Now there is certainly a sense in which I can directly observe my own state of fear. If called upon to do so, I could even quantify it somewhat by rating it is 'slight fear', 'moderate fear', 'intense fear', and so on. But I have no such direct access to the fear that you or any other human being might experience, nor to the fear that any non-human organism might experience. In those instances, the only way I could measure fear would be indirectly by way of measuring the effect that fear has upon this, that, or the other aspect of the behavior or physiology of the organism. In fact, if I were to ask you or some other verbal organism to rate the fear on an ordinal scale, that too would be only an indirect measure, for in this case the only thing I would be measuring directly would be the effect that the fear (in conjunction with the request) has upon this particular momentary aspect of your verbal behavior. The same general point applies to any presumed internal state or process of an organism that cannot be directly observed by the investigator. These include a very large part of what is of interest to the behavioral sciences, such as psychology, as well as a considerable portion of what is of interest nowadays to the biological sciences–in general, mental states, mental processes, perception, intelligence, emotion, motivation, the organismic states produced by learning, conditioning, habituation, and the like, to mention but a few.

Although the indirectness of a measure certainly does not disqualify it from being useful, it does raise two flags of caution. Actually, these are flags that should be raised, at least tacitly, every time you come across any kind of measurement at all, even if it seems to be entirely direct and straight-forward. The first has to do with what is conventionally spoken of as the *reliability* of a measure, and the second pertains to what is described as the *validity* of a measure.

RELIABILITY

The dimensions of my solid wooden desk do not vary substantially from one time to another nor from one place to another; and, providing that I avoid extremes of temperature, neither do the dimensions of my steel tape measure. Hence, if the width of the desk measures out at precisely 60.0 inches on one occasion, it is virtually certain to measure out at precisely 60.0 inches on all other occasions as well. Thus, if I were to take three successive measures of the width of my desk at one-hour intervals (at normal room temperature), the three results would almost certainly be: 60.0 inches, 60.0 inches, and 60.0 inches. This is Scenario 1. Now consider two alternative scenarios. Scenario 2. My tape measure is made of a cloth fabric that stretches slightly in accordance with the amount of tension that is placed upon it. Each time I use it I inadvertently apply varying amounts of tension, with the result that the three measure of desk width come out as: 60.5 inches, 60.0 inches, and 60.3 inches. Scenario 3. My tape measure is made of a highly elastic substance that stretches quite a lot when tension is applied to it. In addition, it undergoes a quite large degree of expansion and contraction in response to small moment-to-moment fluctuations in normal room temperature. The three measures in this scenario come out as: 3.0 inches, 1,098.7 inches, and 63.2 inches.

Here in a nutshell is the meaning of the concept of the reliability of measurement: if you repeatedly measure the same item by the same process, and the result is precisely the same on each occasion, the measurement process is highly reliable; if the results are somewhat variable but consistent, it is relatively reliable; and if they are variable and wildly inconsistent, then it is highly unreliable.

VALIDITY

Obviously, a procedure of measurement is worth bothering with only in the degree that it is reliable, in the sense that we have just defined. But equally obvious is that reliability, in and of itself, is not sufficient. Here is a classic example. In the early decades of the nineteenth century there developed a would-be method of psychological assessment known as phrenology, which claimed that important variables of personality and mentality could be measured by mapping out the contours of a person's skull. Thus, the protuberance or depression of the skull in one location was taken as a measure of self-esteem (the greater the protuberance, the higher the self-esteem, and vice versa); elsewhere, it was a measure of cautiousness, benevolence, combativeness, or of some other item in a rather long list of such attributes. There was even a bump on the head that was alleged to measure a person's capacity for reverence.

But preposterous though these measures of the phrenologists might seem, the fact is that they were very highly reliable. Three or five or ten separate measures of the contours of a person's skull made by phrenologist A would all end up virtually identical. Phrenologist B, mapping the same skull, would also end up with virtually identical results. The shortcoming of the phrenologists' measures was not that they were unreliable, but that they were not measuring what they were supposed to be measuring. As direct measures of the depressions and protuberances of the skull they were perfectly fine. But as indirect measures of the dimensions of personality and mentality, they were simply humbug.

In brief, the measures of the phrenologists had plenty of reliability—but as they were not really measuring what they purported to be measuring, they had no validity. The basic meaning of this concept is probably already familiar to you through your experience with school or college examinations of the type described earlier. You have in all likelihood taken some exams that seemed to you to be poor assessments of a student's mastery of the subject, even though you personally might have done well on them; and you have no doubt taken other exams that seemed to you to be good assessments, even though you personally might have done less well on them than you would have liked. In the language of students, it is the distinction between a "fair" exam and an "unfair" exam. A fair exam is one that actually measures what it purports to measure, namely, the student's knowledge and understanding of the subject matter; and an unfair exam is one for which the student's score substantially reflects something other than knowledge and understanding, for example, the student's ability to spot and deal with trick questions, to remember picayune details mentioned by the instructor or the textbook *en passant,* or to adhere to some particular theoretical or ideological party line favored by the instructor.

Chapter 2

Distributions

In its simplest form, a distribution is just a list of the individual measures that are taken on some particular variable. Suppose, for example, that 12 students in a statistics course have achieved the following scores on their first exam, arranged in order from lowest to highest:

61, 69, 72, 76, 78, 83, 85, 85, 86, 88, 93, 97

Take a close look at the texture of interrelationships among these scores–in particular, how they spread out and how they cluster together—and you are in effect examining their distribution. It is a very simple concept, made even simpler by the fact that distributions are very easy to visualize.

Indeed, in most cases a single glimpse of a graphic representation of a distribution will tell you more about it than several minutes of staring at a bare list of numbers. One very simple form of graphic representation is shown below in Figure.

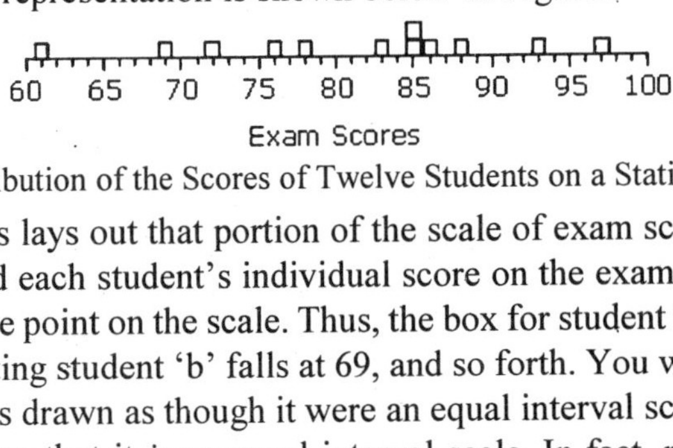

Fig. Distribution of the Scores of Twelve Students on a Statistics Exam

The horizontal axis lays out that portion of the scale of exam scores that includes all 12 of the listed values, and each student's individual score on the exam is represented by a box placed at the appropriate point on the scale. Thus, the box for student 'a' is placed at 61 on the scale, the box representing student 'b' falls at 69, and so forth. You will note that the scale of exam scores in Figure is drawn as though it were an equal interval scale. For purposes of this example, we will assume that it *is* an equal interval scale. In fact, most of what we will be saying in this chapter pertains to distributions that derive from equal interval scales of measurement. It is also possible to speak of distributions in connection with ordinal scale and categorical measurement, but that is something we will save for later.

The type of graph is useful when you are interested in conveying detailed information about each and every measure in the distribution, though with larger numbers of measures it can become quite cumbersome. Besides, for most practical statistical purposes your interest is not so much in the individual identity of your measures as in the overall shape and texture of the distribution that they compose.

The frames of Figure show three other forms of graphic representation that typically serve these purposes much more effectively.

2.2 Frequency Histogram

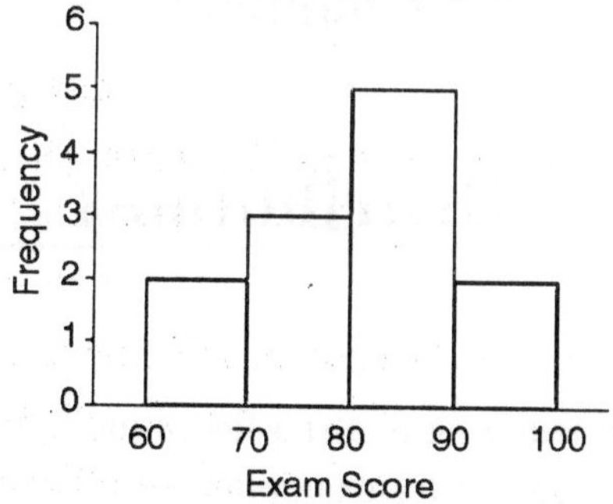

The histogram shown in Figure represents the frequencies by the heights of the columns that extend upward from the horizontal baseline. Thus, the column for the first interval extends two units upward on the left-hand vertical scale to represent the fact that this interval has two cases falling within it; the column for the second interval extends three units upward on the vertical scale to indicate that it has three cases falling within it; and so on.

2.2b. Frequency Polygon

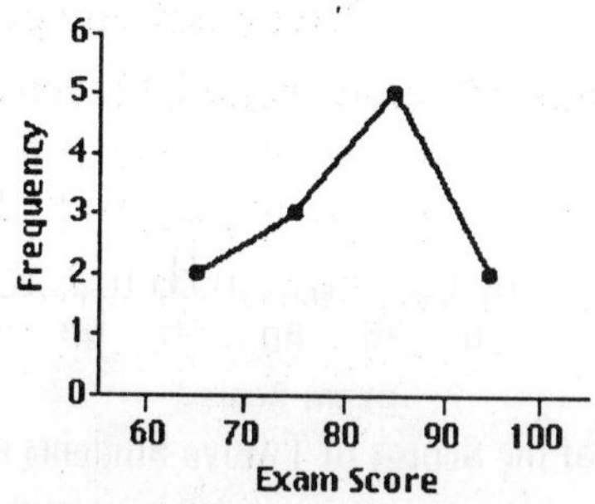

If you were to place a dot at the middle top of each column in the histogram, connect the dots with straight lines, and then erase the columns, you would have the frequency polygon shown in Figure.

2.2c. Smoothed Frequency Polygon

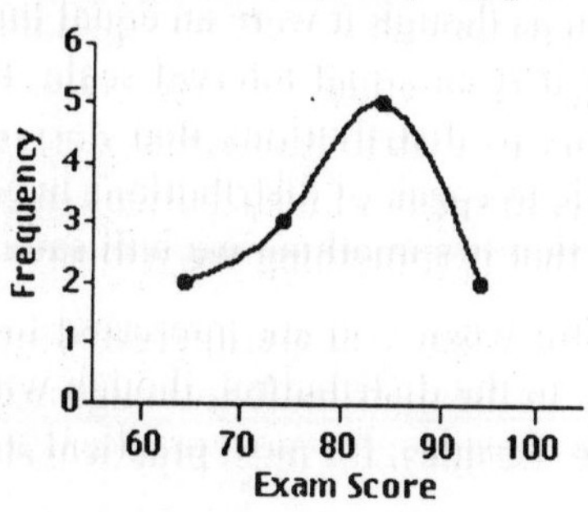

Draw the frequency polygon as a smooth curve and you have the smoothed frequency polygon shown.

All three are representing the same basic facts, though each in a somewhat different format.

The process of constructing these graphs begins by breaking the scale of exam scores into intervals and then counting up the number of individual measures that fall within each interval. The resulting aggregate information can then be displayed in terms of either absolute frequencies (the actual number of cases within each interval) or relative frequencies (percentages or proportions). For the present illustration we display it only in terms of absolute frequencies. The horizontal axis of the three graphs represents the intervals that the scale of exam scores is broken into. For present purposes we simply break the scale into the four conventional academic grade-score intervals:

1. >60 and <70 (equal to or greater than 60 and less than 70)
2. >70 and <80 (equal to or greater than 70 and less than 80)
3. >80 and <90 (equal to or greater than 80 and less than 90)
4. >90 (equal to or greater than 90)

The following table shows the number of exam scores that fall into each of these intervals, and the adjacent image frame will show each of these three ways of graphing the distribution as you click the links beneath the frame. If you are new to this type of subject matter, you might want to take a moment with each graph to figure out exactly what information it is representing.

Exam Score in Interval	*Interval*	*Frequency*
61 69	>60 <70	2
72 76 78	>70 <80	3
83 85	>80 <90	5
85 86 88		
93 97	>90	2

A number of smoothed polygons of the type shown in Figure as you encounter them, please keep in mind that this is a kind of special purpose graphic format usually reserved either for describing abstract theoretical distributions, of which you will see much more in subsequent chapters, or else for providing a global overview of the shape of a distribution without getting too bogged down in details. For most other purposes (for researchers, reporting the results of research; for students, doing exams and homework assignments) the convention is to represent the frequency aspects of a distribution with either a histogram or a regular (i.e., non-smoothed) polygon.

THE PARAMETERS OF DISTRIBUTIONS

The term *parameter* comes from two Greek words, *para* ('beside,' 'along side of') and *metron* (a unit of measurement), that combine to convey the meaning of a fixed limit or boundary. When you define something (from the Latin *definire,* 'to limit'), you are in effect laying out its limits or boundaries. Hence, the parameters of a distribution are its defining characteristics.

The department of psychology at a fairly large institution is concurrently offering six sections of the introductory statistics course. Although these sections are expected to cover the same general range of subject matter, they are otherwise independent of each other, taught by different instructors who are free to choose their own text materials and construct their own exams. Figure portrays the distributions of exam scores for these six sections, each in the form of a smoothed frequency polygon. In each graph, the horizontal axis delineates the exam-score intervals, while the vertical axis is scaled in terms of the percentages of students whose scores fell within the various intervals. Note that these horizontal axes now show an additional interval on the left, defined as >50 and <60, to accommodate the regrettable fact that several students among the six sections ended up with grades between 50 and 59.99... . Please take a moment to examine and compare the shapes of these six distributions before going on to the commentary that follows.

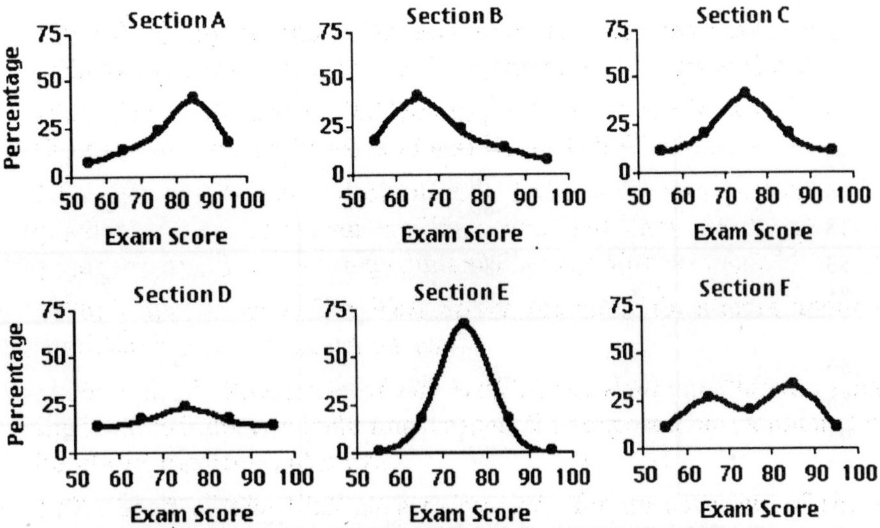

Fig. Distributions of Exam Scores in Six Sections of a Statistics Course

The main parameters of a distribution are generally spoken of as

- *Central tendency,*
- *Variability,*
- *Skew,*

- *Kurtosis,* and
- *Modality.*

Central tendency and variability will need to be examined separately in considerable detail, since a large part of the work we will doing in this webtext requires that these two parameters be both clearly understood and precisely measured. The remaining three parameters in the list also require a clear understanding, though for most practical statistical work it will not be necessary to measure them in detail. We will begin with a general conceptual introduction to the parameters of skew, kurtosis, and modality, and then shift gears to examine central tendency and variability in greater depth and detail.

SKEW

The term *skew* comes from an Old French word meaning to shun or avoid. It is the same linguistic ancestor from which we get the English words 'eschew,' which also means to shun or avoid, and 'askew,' which conveys the meaning of lopsided or tilted off to one side. A skewed distribution is therefore one that is askew, lopsided, tending to shun or avoid one or the other of the extremes of the range within which it falls.

Examine the first two graphs in Figure and you will see that the distributions of exam scores for Sections A and B are both conspicuously skewed, though in opposite directions. In Section A the exam scores tend to cluster toward the higher end of the range and taper off toward the lower end, whereas in Section B they tend to cluster toward the lower end of the range and taper off toward the higher end. In general, a distribution that is lopsidedly heavy at the higher end of the range and light at the lower end (e.g., Section A) is described as a *negatively skewed* distribution, while one that is heavy at the lower end and light at the higher end (e.g., Section B) is spoken of as a *positively skewed* distribution. Or to put it in pictorial terms, a negatively skewed distribution is one whose elongated tail extends to the left (the low or "negative") end of the range, while a positively skewed distribution is one whose elongated tail extends to the right (the high or "positive") end of the range.

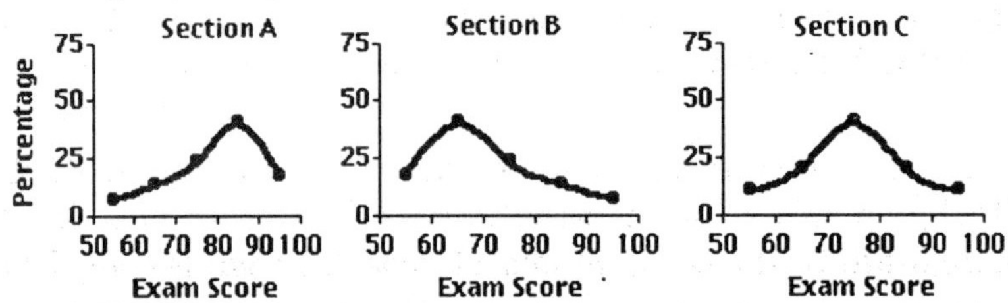

An unskewed distribution, on the other hand, is one that is not lopsided in either the one direction or the other; typically it has two tails that trail off in both directions symmetrically. An example of an unskewed distribution is the third graph in Figure showing the symmetrical distribution of exam scores in Section C.

KURTOSIS

Kurtosis (from a Greek word meaning 'curvature' or 'convex') refers to whether the shape of a distribution is relatively short and flat, or tall and slender, or somewhere in-between those two extremes. If it is short and flat, like the distribution shown for Section D, it is described as *platykurtic* (flat-curved); if it is tall and slender, like the distribution shown for Section E, it is spoken of as *leptokurtic* (slender-curved); and if it is neither the one extreme nor the other, like the distribution shown for Section C, it is described as *mesokurtic* (medium-curved). In a platykurtic distribution the individual measures are spread out fairly uniformly across their range, whereas in a leptokurtic distribution they tend to cluster compactly at some particular point in the range.

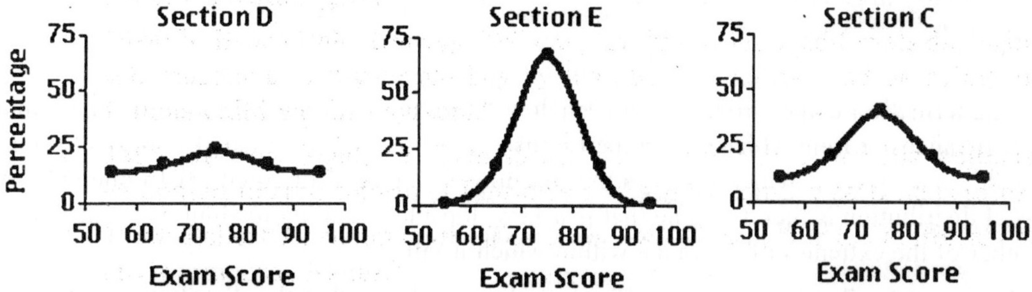

Thus, the exam scores for Section D are distributed fairly uniformly among the five intervals, with only a slight tendency to cluster and form a peak in the vicinity of the third interval (70 to 79.99...). The scores for Section E, on the other hand, have a quite pronounced tendency to cluster in the third interval, with only a few scores falling outside this interval. In the mesokurtic distribution illustrated by Section C the clustering is more moderate than in the leptokurtic distribution, and the curve as it falls away from the peak is more tapering than in the platykurtic distribution. A class containing approximately equal numbers of students whose mastery of the subject matter is very strong, quite strong, moderately strong, not so strong, and so on, would tend to produce the platykurtic distribution shown for Section D, while a class containing a large majority of students who all have approximately the same level of mastery would tend to produce a leptokurtic distribution of the kind shown for Section E.

MODALITY

Modality refers simply to the number of distinct peaks, or areas of cluster, that appear within a distribution, with each such peak being spoken of as a *mode.* A distribution with only one distinct peak is described as *unimodal:* if it has two distinct peaks it is spoken of as *bimodal;* three peaks, *trimodal;* and so on. Most distributions that you are likely to encounter will be either unimodal or bimodal. Only rarely will you find distributions that have three or more distinct peaks.

The distributions for Sections A, B, C, D, and E are all unimodal distributions with varying degrees of skew and kurtosis. The plot for Section F, on the other hand, portrays a bimodal distribution with one mode in the vicinity of the second interval (60 to 69.99...) and another in the vicinity of the fourth interval (80 to 89.99...). When the two modes of a bimodal distribution are of unequal prominence, as in the present case, they are distinguished from one another by the designations *major* and *minor*. Thus, the more prominent peak on the right is the major mode of the distribution, while the less prominent one on the left is the minor mode.

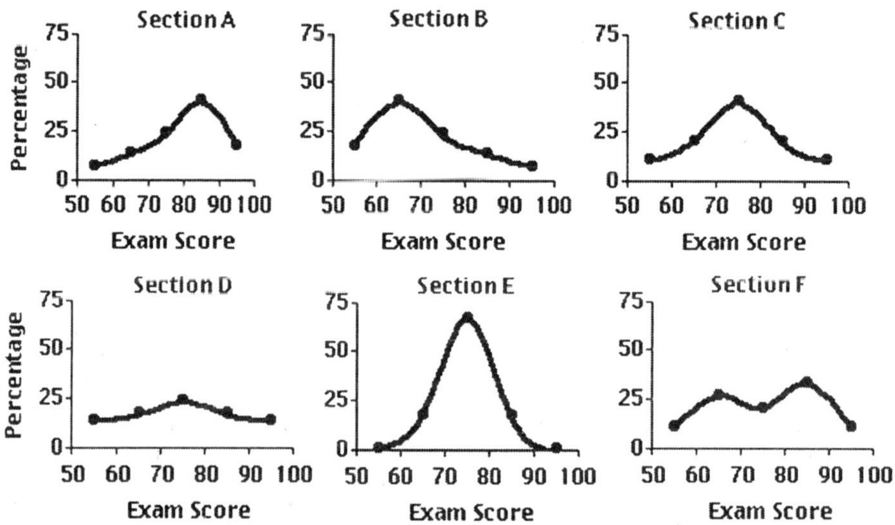

Whenever you find bimodality in a distribution, the first thing to consider is the possibility that what you are really looking at is two distributions mixed together. Suppose, for example, that you were to plot the frequency distribution of the heights of 100 male elementary school pupils. If it were to happen that 60 of your subjects were fifth graders and the other 40 were third graders, your distribution would almost certainly end up as bimodal. Indeed, its general outlines would probably look quite a lot like the distribution of exam scores for Section F, with the minor mode produced by the height distribution of the smaller number of younger and shorter third graders and the major mode produced by the height distribution of the larger number of older and taller fifth graders. Similarly, suppose that Section F is composed of two fairly distinct groups of students, the larger group quite well prepared for the exam and the smaller group not so well prepared. The larger well prepared group would produce the major mode of the distribution, while the smaller not so well prepared group would produce the minor mode.

CENTRAL TENDENCY AND VARIABILITY

Although skew, kurtosis, and modality are each important and indispensable concepts for understanding the general nature of distributions, the parameters with which you will need to become most intimately acquainted are those of central tendency and variability. By way of brief preview, central tendency refers to the tendency of the individual measures in a distribution to cluster together toward some point of aggregation, while variability describes the contrary tendency for the individual measures to disperse or spread out away from each other. Basically, they are refined variations on the complementary themes of closeness versus separation, similarity versus difference. Variability refers to the overall separations and differences that exist among the individual measures in a distribution, while central tendency refers to their closeness and similarity.

Among the six exam score distributions portrayed in Figure, the clearest illustrations of central tendency and variability are found in the symmetrical and unimodal plots for Sections C, D, and E. The platykurtic distribution for Section D shows a fairly weak tendency to cluster in the region of the third interval and a quite strong complementary tendency to disperse away from that point of central tendency. The leptokurtic distribution for Section E, on the other hand, shows a quite strong tendency to cluster in the vicinity of the third interval and only a weak tendency to disperse away from it.

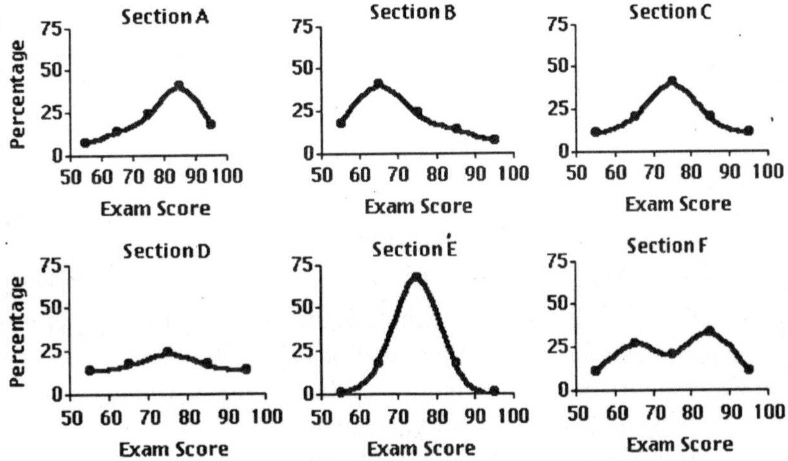

The mesokurtic distribution for Section C falls in-between these two extremes, showing both a moderate tendency to cluster and a moderate tendency to disperse. The remaining three plots in Figure also illustrate central tendency and variability, although they do so somewhat more complexly. If most of the dispersion within a distribution falls asymmetrically to either the left or the right side of the region of cluster, the result is a skewed distribution such as the ones shown for Sections A and B. Even more complex is the bimodal distribution for Section F, for here you have two distinct regions of clustering and a considerable degree of overall dispersion resulting from the separation of those regions.

MEASURES OF CENTRAL TENDENCY

Potentially, you could have two quite different types of measures of central tendency. The first would measure the *strength* of the tendency for measures to cluster together, and the second would measure the *location* where the clustering tends to occur. Typically when we speak of measures of central tendency, it is the second type of measure that we are talking about—the measure of location. The measure of the strength of the tendency, on the other hand, is actually the flip-side of the measures of variability that we will be examining in the section that follows this one.

The three most commonly used measures of the location of central tendency are the mode, the median, and the mean. In brief, the mode is the point or region within a distribution where the largest number of individual measures congregate, the median is the midpoint of all the individual measures, and the mean is the arithmetic average of all the individual measures.

As a rule, the only time you will find the mode and median to be precisely coincident with the mean is when the distribution is unimodal and perfectly symmetrical. In skewed distributions the mean, median, and mode will tend to be separated from one another, with the mean falling toward the tail of the skew, the mode falling away from the tail, at the peak, and the median falling somewhere in-between. Thus, in a positively skewed distribution the mean will be to the right, the mode to the left, and the median in-between, while in a negatively skewed distribution the mean will be to the left, the mode to the right, and the median in-between. These relationships among the three measures of central tendency are shown below in Figure.

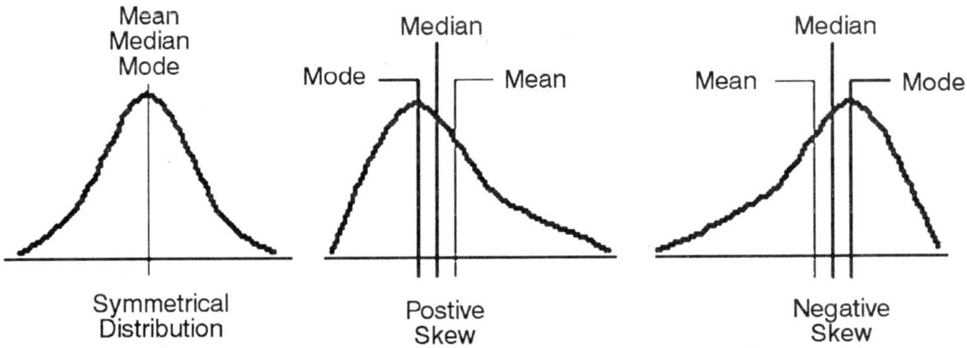

Fig. Relationships of the Mean, Median, and Mode in Symmetrical and Skewed Distributions

If your purposes in assessing the central tendency of a distribution are purely descriptive, your best procedure would be to examine not just the mode or the median or the mean separately, but all three together. But once you step beyond those limited purposes into the much broader realm of analytical and inferential statistics, the mode and the median become virtually useless The reason for this is that measures of mode and median do not have the properties of an equal interval scale of measurement—this is true even if the data on which they are based do have the properties of an equal interval scale—hence they cannot meaningfully be subjected to any further mathematical operations. In brief, they are both mathematical dead ends. The mean, on the other hand, providing it is based on data that derive from an equal interval scale of measurement, will itself have the properties of an equal interval scale and thus *can* be subjected to further mathematical operations.

The arithmetic mean of a distribution is simply the sum of all the values in the distribution divided by the number of values. Thus, the mean of a distribution consisting of the values 1, 2, 3, 4, and 5 is

$$\text{mean} = \frac{\text{sum of all values}}{\text{number of values}}$$

$$= \frac{1+2+3+4+5}{5} = \frac{15}{5} = 3.0.$$

You can think of the mean as the balance point within a distribution, a kind of center of gravity to which each measure in the distribution contributes in proportion to its size. Thus a distribution consisting of the values 1, 2, 2, and 3 would balance, as shown below, at (1+2+2+3)/4 = 2.0.

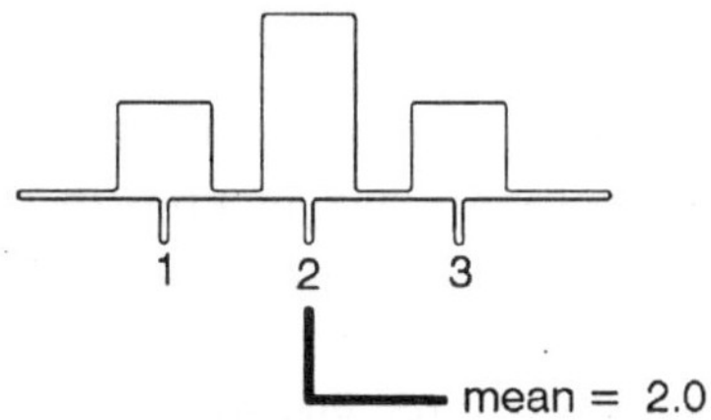

Replace any of the four values in this distribution with some other value, and the balance point shifts accordingly. In the following illustration we replace the 3, first with a 5 and then with a 9. For the first replacement the balance point shifts from 2.0 to 2.5, and for the second it shifts even further to 3.5.

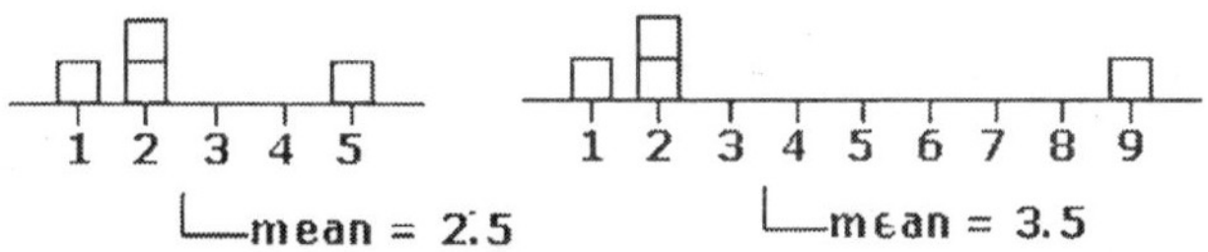

And please take special note of the tidy proportionality of these shifts. In this particular distribution there are four values, each contributing 1/4 = 25% to the determination of the mean. Increase any particular value by one point (one unit of measurement), and you will increase the mean by 0.25 units; increase the value by two points, and you will increase the mean by 0.50 units; and so on. Conversely, decrease any value in the distribution by a certain amount and you will decrease the mean by 25% of that amount. In a distribution composed of three values, the contribution of each individual value would be 1/3 = 33.3%; in one composed of 10 values it would be 1/10 = 10%; and so on.

And now for a formula and an introduction to some conventional statistical notation. When you calculate the mean of a distribution you are performing two separate computational steps: first, you are taking the sum of all the individual values in the distribution; and then you are dividing that sum by the total number of values in the distribution. In the formulaic description of these two steps that we gave earlier, notice that the second step is expressed not in words but by an abstract symbol:

$$\text{mean} = \frac{\text{sum of all values}}{\text{number of values}}.$$

The horizontal line between "sum of all values" and "number of values" is a conventional, concise symbolic way of saying something that would otherwise be very cumbersome to say, especially if it needed to be said over and over again in a variety of different situations: namely, "take whatever is above the line and divide it by whatever is below it." As it happens, you have a very long-standing familiarity with this particular symbolic notation, as you do with its substitutes "χ" and "/ ", as well as with its cousins "+", "–", and "*x*", and so your eye and mind see immediately precisely what operation it is telling you to perform. The formula that follows replaces each of the other terms and operations involved in the calculation of the mean with symbols that are equally concise. Please do not be intimidated by these symbols, for there is nothing at all elusive or arcane about them. It is simply a matter of familiarization. After a while their meanings will leap out at you as though you had been using them all your life.

FORMULA FOR THE CALCULATION OF THE ARITHMETIC MEAN

Definition of Terms

N = the number of individual values in the distribution.

X_i This is a symbol referring to the N individual values of your distribution in the abstract. Each value in the distribution is spoken of as a variate instance of the variable X. Thus, the first value in the distribution would be X_1; the second would be X_2; and so on to the last value, X_N.

Σ This symbol (upper-case Greek letter 'sigma') does not represent a numerical value, but rather, like "+" and "–", a mathematical operation that is to be performed upon certain designated numerical values. It is a computational signpost saying "calculate the sum of these values"; hence its conventional name, the *summation sign.* Thus for a distribution of size $N = 4$, consisting of the values 6, 9, 12, and 15, the expression X_i would be equivalent to $6+9+12+15 = 42$.

M We will use the bold-face letter M to represent the arithmetic mean of a distribution. Thus M_X would be the mean of a set of X values, M_Y would be the mean of a set of Y values, and so on. [Note that hardcopy statistics textbooks will often represent the mean of X (Y, Z, etc.) by writing the letter X (Y, Z, etc.) with a horizontal bar over it.]

Formula$_T$.

$$\text{mean} = \frac{\text{sum of all values}}{\text{number of values}}.$$

Substitute the symbolic expressions for the verbal phrases in this structure, and you have$_T$

$$M_x = \frac{X_i}{{}_T N_T}.$$

The best way to get a feeling for what a formula *does* is to apply it to a specific numerical example. Here again is the distribution of 12 exam scores mentioned earlier.

X_1	X_2	X_3	X_4	X_5	X_6	X_7	X_8	X_9	X_{10}	X_{11}	X_{12}
61	69	72	76	78	83	85	85	86	88	93	97

Add these $N = 12$ scores together and you will find that their sum comes out to 973. Substitute this value into the formula for the mean, and you have$_T$

$$M_X = \frac{X_i}{_T N_T} = \frac{973}{12} = 81.08.$$

It is as simple as that.

Actually, the value of the mean that you would calculate in this example would not come out to precisely 81.08, but rather to 81.08333.... Almost always when you calculate a mean you will end up having to do some rounding. However, do be careful not to round it down too much, because the mean often enters into subsequent calculations that can be thrown considerably askew if it is rounded excessively. In general, it is good practice to calculate the mean out to at least one decimal place beyond the number of decimal places contained in the original data. Thus, if the original data have no decimal places, as in the present example, carry the mean out to at least one decimal place; if they have one decimal place, carry it out to two; if they have two decimal places, carry it out to three; and so on.

MEASURES OF VARIABILITY

One very easy measure of the variability that exists within a distribution is the range, which is simply the distance between the lowest score in the distribution and the highest. Equally easy is the interquartile range, which is the distance between the lowest and highest of the middle 50 percent of the scores in the distribution. Thus for our 12 exam scores, the range extends from a low of 61 to a high of 97, and the interquartile range extends from a low of 76 to a high of 86.

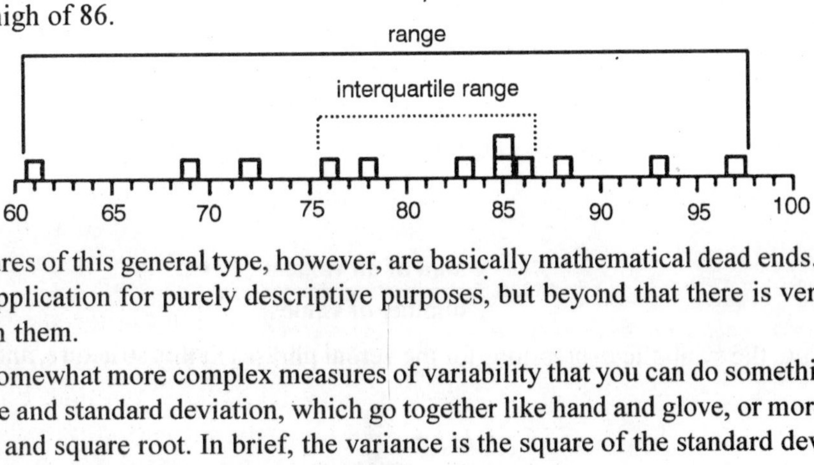

Measures of this general type, however, are basically mathematical dead ends. They have a limited application for purely descriptive purposes, but beyond that there is very little you can do with them.

Two somewhat more complex measures of variability that you can do something with are the variance and standard deviation, which go together like hand and glove, or more precisely, like square and square root. In brief, the variance is the square of the standard deviation, and the standard deviation is the square root of the variance.

variance = (standard deviation)^2 standard deviation = sqrt[variance]	*Note:* Owing to the limitations of HTML coding for a web document, we will not be able to use the conventional radical sign to indicate "square root of." We will instead be using the notation "sqrt." Thus sqrt[variance] means "the square root of the variance," sqrt[16] means 'the square root of 16," and so forth.

They are both ways of measuring variability, based on the realization that the dispersion within a distribution—the overall amount by which the individual values of X_i differ among themselves—is in direct proportion to the aggregate amount by which they differ or deviate from their collective mean. For each particular value of X_i within a distribution, the distance between it and the mean can be arrived at through a simple process of subtraction. By convention, X_i is placed to the left of the subtraction sign, the mean is placed to the right, and the result of the operation is spoken of as a deviate (noun). Thus

$$\text{deviate} = X_i - M_X.$$

In and of themselves, these deviate measures are useless as an aggregate measure of variability, because they will always sum up to zero (within the limits of rounding errors), no matter how much or little variability the distribution might actually contain. The way to get around this sum-to-zero problem is to take each deviate score and square it. Then add up all the squared deviates to arrive at a quantity known as the sum of squared deviates, conveniently abbreviated as *SS*.

$$\text{squared deviate} = (X_i - M_X)^2$$
$$\text{sum of squared deviates: } SS = (X_i - M_X)^2$$

The quantity calculated as *SS* will come out to zero only when the actual amount of variability within the distribution is zero. In all other cases it will come out as something larger than zero, in an amount precisely determined by the degree of variability that actually exists within the distribution.

The variance is then simply the average of the squared deviates, and the standard deviation is the square root of that average. [But note the special caution below.] By convention, the variance and standard deviation of a distribution are symbolized as s^2 and s, respectively. Thus

Variance:

$$s^2 = \frac{\Sigma(X_i - M_X)^2}{N} = \frac{SS}{N}.$$

Standard Deviation:

$$s = \text{sqrt}\left[\frac{(X_i - M_X)^2}{N}\right] = \text{sqrt}\left[\frac{SS}{N}\right].$$

An Important Caution Concerning the Calculation of Variance and Standard Deviation

Textbooks in statistics will often use N–1 in the denominator of the formulas for the variance and standard deviation, rather than simply N, as I have done. Therein hangs a tale. The formulas that I have given, which use N in the denominator, provide the actual, unadorned, plain-vanilla variance and standard deviation for any particular set or *sample* of X_i values. When you use N-1 in the denominator, what you are actually doing is using the particular sample of X_i values to *estimate* the variance and standard deviation of the entire *population* of X_i values from which the particular sample comes. You will see in later chapters that the technical reasons for this substitution derive from the theory of probability. The technical distinction between samples and populations will be explained later in the present chapter.

Thus, for a simple distribution consisting of the X_i values 1, 2, 3, 4, and 5:

	deviate = $X_i (X_i - M_X)$	Squared deviate $(X_i - M_X)^2$	sum of squared deviates: $SS = 10$ variance: $s^2 = SS/N = 10/5 = 2$
1	1–3 = –2	4	standard deviation:
2	2–3 = –1	1	s = square_root(variance)
3	3–3 = 0	0	= square_root(2)
4	4–3 = +1	1	= +1.41
5	5–3 = +2-4	4	
		sum = 10	

You will sometimes find the variance (s^2) referred to as the mean square, and the standard deviation (s) referred to as the root mean square. The former is a shorthand way of saying "the mean of the squared deviates," which is what the variance basically is, and the latter is a shorthand way of saying "the square root of the mean of the squared deviates," which is what the standard deviation basically is.

At any rate, once you know the values of N and SS for a distribution, the remaining calculations for variance and standard deviation are utterly simple and straightforward. The only potential complexity is that it will often prove rather laborious to calculate the value of SS using the formula I have just given you:

$$SS = (X_i - M_X)^2$$

This construction is typically spoken of as a conceptual formula, since it allows you to see the structure of exactly what you are doing when you calculate SS. In effect it is saying: take the difference between each value of X_i and the mean of the distribution, square each of those differences, and then add them all up.

Obviously this structural visibility is an advantage for anyone who is just starting out in statistics, for it helps to keep the underlying logic of the calculation clearly in view. But it also has the disadvantage that it can be quite tedious and cumbersome to use in actual practice. In the calculation of SS, and thus of the variance and standard deviation, it is often preferable to use the following computational formula. In general, a computational formula is one that is algebraically equivalent to, and hence yields the same result as the corresponding conceptual formula, more easily, though without the advantage of clearly showing the underlying logic of the calculation. Incidentally, this computational formula for SS will also do the job more precisely, since it minimizes rounding errors. First some additional items of symbolic notation:

X_i^2 The square of any particular value of X_i within the distribution. If $X_i = 3$, then $X_i^2 = 3 \text{x} 3 = 9$; if $X_i = 4$, then $X_i^2 = 4 \text{x} 4 = 16$; and so on.

ΣX_i^2 The sum of all the squared X_i values within the distribution. If the values of X_i within a distribution are 3, 4, and 5, the sum of the squared X_i values is $\Sigma X_i^2 = 3^2 + 4^2 + 5^2 = 9 + 16 + 25 = 50$.

ΣX_i The sum of the original (unsquared) values of X_i within a distribution. If the values of X_i within a distribution are 3, 4, and 5, the sum of the original (unsquared) squared X_i values is simply $\Sigma X_i = 3 + 4 + 5 = 12$.

$(\Sigma X_i)^2$ The square of the value of ΣX_i. That is, calculate the sum of the original (unsquared) values of X_i within the distribution, and then square that sum. For a distribution consisting of the X_i values 3, 4, and 5, $(\Sigma X_i)^2$ is equivalent to $(3+4+5)^2 = 12^2 = 144$. Please note very carefully that $(\Sigma X_i)^2$ is not the same thing as X_i^2. For $(\Sigma X_i)^2$ you are summing up all the original values of X_i and then squaring that sum, while for ΣX_i^2 you are squaring each individual value of X_i and then taking the sum of all those squared values.

And then the formula:

$$SS = \Sigma X_i^2 \frac{(\Sigma X_i)^2}{{}^2N^2}.$$

Thus, for our distribution of 12 exam scores:

X_i	X_i^2	
61	3721	*sum of squared deviates:*
69	4761	$= 80{,}063 - [(973^2)/12]$
72	5184	$= 80{,}063 - 78894.08$
76	5776	$= 1168.92$
78	6084	*variance:*
83	6889	$s^2 = 1168.92/12 = 97.41$
85	7225	*standard deviation:*
85	7225	s = sqrt[variance]
86	7396	= sqrt[97.41]
88	7744	= +9.87
93	8649	
97	9409	
Sums	$X_i =$ 973	$\Sigma X_i^2 =$ 80,063

Although the variance and standard deviation are equally valid measures of variability, the standard deviation is by far the more easily visualized and intuitively comprehended, because it is the one that is expressed in the same units of measurement as the original values of X_i of which the distribution is composed. When you calculate the standard deviation of our distribution of exam scores, the resulting value of s also refers to the scale of exam scores.

The variance (s^2), on the other hand, would refer to *squared* exam scores, which do not readily lend themselves to graphic representation nor intuitive understanding. This is not to suggest that the variance is inferior. It is, in fact, a very useful measure of variability in its own right, and in some respects it is even more useful than the standard deviation, notwithstanding that one cannot easily draw a picture of it. But that is a point best saved until later.

At any rate, as we have just calculated, our distribution of exam scores has a standard deviation of $s = +9.87$. Figure will give you an idea of what this rather abstract quantity–"plus or minus 9.87"–is saying about the distribution. Recalling that the mean of the distribution is 81.08, move one standard deviation to the right of the mean (M_X+1s), and you end up at 81.08+9.87 = 90.95 on the exam-point scale. Move one standard deviation to the left of the mean (M_X–1s), and you end up at 81.08–9.87 = 71.21. For many types of distributions, the range between +1 and –1 standard deviation tends to encompass about two-thirds of all the individual values of X_i. Within the present distribution it is exactly two-thirds.

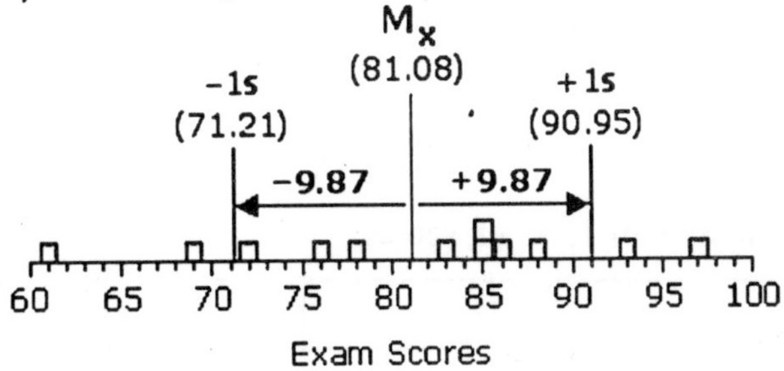

Fig. Graphic Representation of +1 and —1 Standard Deviation.

In effect, "plus or minus" one standard deviation represents a range within the distribution, centered upon the mean. Unlike the analogous interquartile range, however, the range between –1s and +1s is a measure of variability defined by *all* of the values of X_i within the distribution, each in proportion to its distance from the mean of the distribution. Thus, a compact distribution will have a relatively narrow range between –1s and +1s, while a more dispersed distribution will have a wider range, in proportion to the degree of its dispersion. Either way, you can in general expect approximately two-thirds of the values of X_i within a distribution to fall within the range between –1s and +1s, providing the distribution is not extremely skewed. For an extremely skewed distribution, it will tend to be more than two-thirds.

SAMPLES, POPULATIONS, AND THEORETICAL DISTRIBUTIONS

Up to this point we have been speaking of distributions as though they were all cut from the same fabric. In fact they are cut from several different kinds of fabric, and the closer we get to the subject matter of inferential statistics, the more critically important it becomes to understand exactly what general type of distribution you are dealing with in any particular context. The most general distinction among distributions pertains to whether they are empirical or theoretical.

An empirical distribution is one composed of some set of variates–that is, values of X_i– that have either been observed or are capable in principle of being observed. If you were to measure the level of serum cholesterol of 100 adult Canadian males, the resulting 100 values of X_i would constitute an empirical distribution. If you were to speak of the distribution of serum cholesterol levels among adult Canadian males in general, that too would be an empirical distribution, even though you might not have observed all or even most of the multitudinous X_i values of which this distribution is composed.

A theoretical distribution, on the other hand, is one that is derived from certain basic facts, principles, or assumptions, by logical and mathematical reasoning involving a more or less complex sequence of conditional statements of the general form "If such-and-such is true, then so-and-so must also be true." In general, the procedures of inferential statistics begin with one or more empirical distributions and conclude by making reference to a theoretical probability distribution.

THEORETICAL DISTRIBUTION

Lest the term theoretical conjure up images in your mind of ivory-tower impracticality, please be assured that our interest here is no mere intellectual exercise. In fact, theoretical distributions of the type that allow us to make rational judgments concerning probability are what lie at the very heart of the whole enterprise of inferential statistics. They are what make inferential statistics inferential, rather than merely descriptive; they are also what make the study of statistical methodology worth the bother. Although we will not be able to say much about theoretical probability distributions until we have covered some basic concepts of probability, there is one particular species of theoretical distribution that we will be referring to before that time, so you will need to have at least a general sense of it. Actually, it is a form of distribution that you have probably already encountered in some previous context, perhaps under its popular and unofficial title, the bell-shaped curve. Its official name is the normal distribution.

The general outlines of the normal distribution. Although the curve that you see in this graph looks as though it might be a smoothed polygon outlining an empirical frequency distribution, it is in fact a mathematical abstraction, generated by a formula known as the normal probability density function. You will not need to be able to generate it yourself, nor will you need to do any of the higher mathematics required to work out its detailed properties and implications. That has already been done for us, long ago, by the pioneers of statistical theory. For present purposes, you can think of this graph as representing an abstract frequency distribution.

The vertical axis in the graph delineates relative frequencies, which could be scaled as either percentages or proportions, and the horizontal axis represents units of z—that is, distances from the mean of the distribution, with each unit of z equal to one standard deviation. The mean of the distribution (z=0) stands precisely at its center, and plus-and-minus one standard deviation (+1z and –1z) fall precisely at the points where the curved outline of the distribution on either side of the mean changes from convex to concave. The range between +1 and –1 standard deviation encompasses approximately two-thirds (68.26%) of the total distribution, with equal portions (34.13%) falling between the mean and –1 standard deviation and between the mean and +1 standard deviation. The remaining 31.74% falls in equal portions of 15.87% to the left of –1 standard deviation and to the right of +1 standard deviation. These are only some of the more immediately visible properties of the normal distribution.

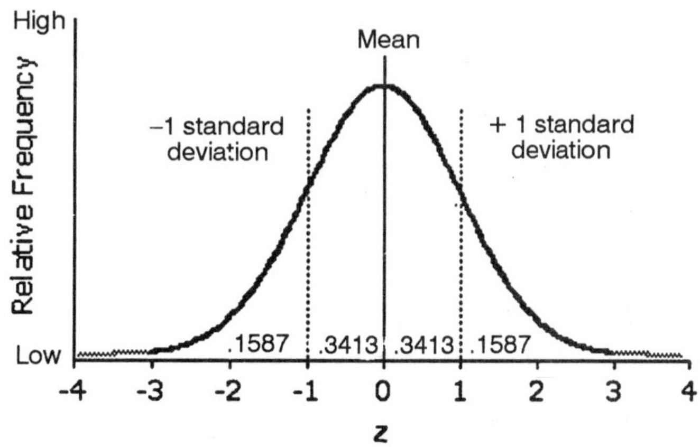

Fig. The Theoretical Normal Distribution

This particular theoretical distribution first arose in the eighteenth and nineteenth centuries in connection with the attempt to specify the probabilities, or odds, that are involved in certain games of chance. At first it was purely theoretical and of no particular interest to anyone apart from gamblers and mathematicians. But with the passage of time it became increasingly clear that the general shape of the theoretical distribution tends to be closely approximated by the distributions of a very large number of real-world empirical variables. So here you have it in the proverbial nutshell–a mathematical abstraction inspired by frivolous games of chance, which nonetheless turns out to be applicable to a very wide range of scientifically interesting real-world phenomena. You will see numerous examples of the applicability of this and other theoretical probability distributions as we proceed.

EMPIRICAL DISTRIBUTIONS: POPULATIONS AND SAMPLES

The particular point on which the applicability of these theoretical distributions is focused is in the relationship between populations and samples. The statistical distinction between populations and samples is basically the distinction between all and some. If you were to measure all instances of some particular variable, the result would be a population distribution. If you were to measure only a limited subset of the instances of that variable, and take those measures to be representative of the entire set of instances, the result would be a sample distribution. Thus, if our distribution of N=12 exam scores comes from a class that has exactly 12 students in it, and if we are interested in nothing other than how this particular class of 12 students performed on this particular exam, then what we are dealing with is the distribution of the entire population of the instances in which we happen to be interested.

Suppose, however, that the class actually has 60 students in it, and that the instructor, after grading the first 12 exams, takes a moment to examine the distribution of this limited subset of scores in order to get a sense of how the 60 students in general did on the exam. In this case, the exam scores of all 60 students (even though most of the exams have not yet been graded) would constitute the population distribution, and the 12 exam scores of the subset would constitute a sample distribution that the instructor takes to be representative of this population.

Alternatively, suppose that the instructor, who plans to give slightly modified versions of this particular exam to other classes over the next several years, examines the distribution of all 60 exam scores in the current class to get a sense of how students in general will do on the exam in subsequent years. In this case the 60 exam scores are not a population, but rather a sample taken to be representative of a considerably larger population, namely, the as-yet nonexistent scores of an as-yet undetermined number of students who will be taking the exam in subsequent editions of the course.

The distinction is important enough to bear repeating: a population includes all instances of the particular variable in which you happen to be interested, irrespective of whether all these instances have actually been observed; while a sample includes only a limited subset of the population, selected in such a way as to insure that it is representative of the totality of the population. And while we are at it, please note carefully that this technical statistical sense of population is not limited to populations of people. You can also have populations of cats, rats, mice, lice, jonquils, junipers, paramecia, erythrocytes, galaxies, water molecules, hydrogen atoms, and electrons. Indeed, any type of empirical fact at all can be thought of as constituting a population, providing that the reference is to all instances of it. Thus, you could speak not only of the population of tree squirrels in a locality, but also of the population of tree squirrel nests, or of the populations of heights or sizes of tree squirrel nests.

When the great 18th century chemist Lavoisier showed precisely how hydrogen and oxygen combine to form water, he was not interested just in this, that, or the other particular sample of hydrogen, or oxygen, or water, but in the entire population of water molecules and the entire population of instances–past, present, and future–in which hydrogen and oxygen might combine in certain proportions and under certain conditions of temperature and barometric pressure. When Newton, about a century earlier, examined the effect of earth-moon gravitational interactions upon oceanic tides, he was not interested just in this, that, or the other particular sample of gravitational effects, but in the entire population of gravitational interactions.

Although the scope of scientific research has broadened very considerably since the days of Newton and Lavoisier, the central aim and strategy remain the same. The overarching aim of the various sciences is to figure out the general facts and principles that pertain to their respective domains of phenomena. But the populations of phenomena to which these general facts and principles refer are typically too large or otherwise inaccessible to be examined in their totality, so they are instead examined by way of systematic sampling. The essential task of inferential statistics is to determine what can reasonably be concluded about an entire domain of phenomena–a population–on the basis of having examined only a limited sample of instances drawn from that domain. As you will see later in considerable detail, the key to accomplishing this task is provided by the theory of probability.

Up to this point we have been like a runner doing warm-up exercises, preparing to run, but not yet actually running. Now the warm-up period is over and the race is about to begin. At first the effort might seem more strenuous, but once you find your stride, I think you will discover that the running flows much more smoothly than the warm-up. By performing some full-blown statistical analyses in connection with a concept that you have certainly already encountered elsewhere under the label of correlation. Historically, this was one of the first forms of statistical analysis to be developed (ca. 1846–1896), and it remains to this day one of the most often and productively employed statistical instruments of scientific research.

THE NORMAL DISTRIBUTION

The normal distribution is a concept with which most people have some familiarity, although they often have never heard of the term. A more familiar name for the normal distribution is the bell curve, because a normal distribution forms the shape of a bell. The normal distribution is extremely important to statistics, and has some specific characteristics that make it so useful. In this chapter it is derscribed what a normal distribution is and why it is so important to researchers. Then some of the features of the normal distribution, and of sampling is discussed, in more depth.

Normal distributions are a family of distributions that have the same general shape. They are symmetric with scores more concentrated in the middle than in the tails. Normal distributions are sometimes described as bell shaped.

Examples of normal distributions are shown below. Notice that they differ in how spread out they are. The area under each curve is the same. The height of a normal distribution can be specified mathematically in terms of two parameters: the mean (μ) and the standard deviation (σ).

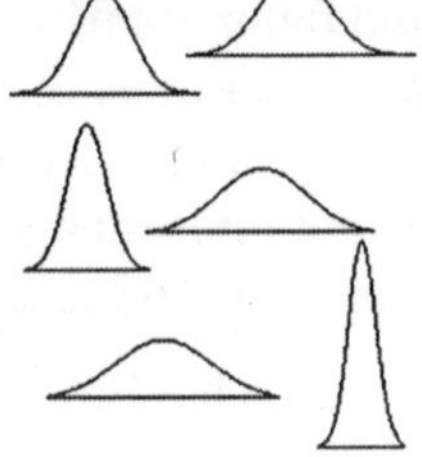

Characteristics of the Normal Distribution: If you take a look at the normal distribution shape presented in Figure, you may notice that the normal distribution has three fundamental characteristics. First, it is symmetrical, meaning that the upper half and the lower half of the distribution are mirror images of each other. Second, the mean, median, and mode are all in the same place, in the centre of the distribution (i.e., the top of the bell curve). Because of this second feature, the normal distribution is highest in the middle, it is unimodal, and it curves downward toward the top and bottom of the distribution. Finally, the normal distribution is asymptotic, meaning that the upper and lower tails of the distribution never actually touch the baseline, also known as the *X* axis.

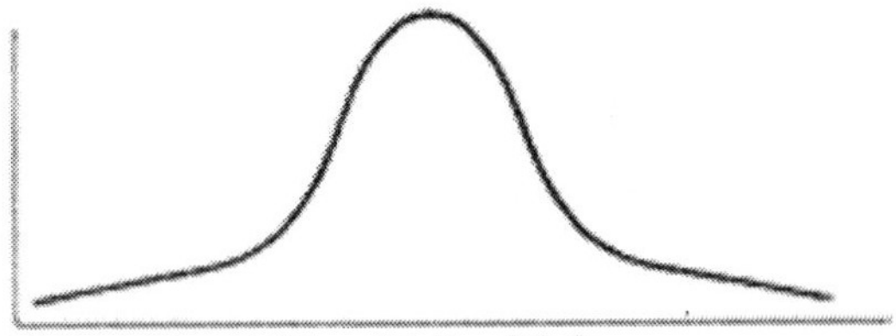

The normal distribution.

MATHEMATICAL FORMULA FOR HEIGHT OF A NORMAL DISTRIBUTION

The height (ordinate) of a normal curve is defined as:

$$\frac{1}{\sqrt{2\pi\sigma^2}} e^{\frac{-(X-\mu)^2}{2\sigma^2}}$$

where μ is the mean and σ is the standard deviation, π is the constant 3.14159, and e is the base of natural logarithms and is equal to 2.718282.

x can take on any value from -infinity to +infinity.

f(*x*) is very close to 0 if x is more than three standard deviations from the mean (less than -3 or greater than +3).

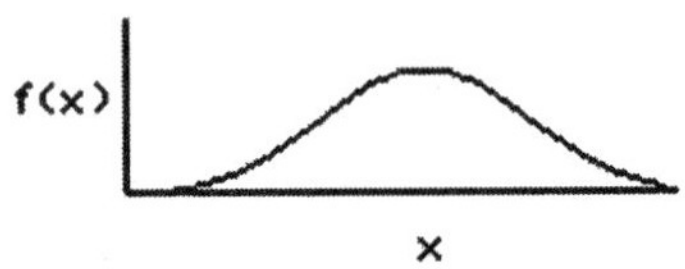

STANDARD NORMAL DISTRIBUTION

The standard normal distribution is a normal distribution with a mean of 0 and a standard deviation of 1. Normal distributions can be transformed to standard normal distributions by the formula:

$$z = \frac{X-\mu}{\sigma}$$

where X is a score from the original normal distribution, μ is the mean of the original normal distribution, and σ is the standard deviation of original normal distribution. The standard normal distribution is sometimes called the z distribution. A z score always reflects the number of standard deviations above or below the mean a particular score is. For instance, if a person scored a 70 on a test with a mean of 50 and a standard deviation of 10, then they scored 2 standard deviations above the mean. Converting the test scores to z scores, an X of 70 would be:

$$z = \frac{70 - 50}{10} = 2$$

So, a z score of 2 means the original score was 2 standard deviations above the mean. Note that the z distribution will only be a normal distribution if the original distribution (X) is normal.

Applying the formula

$$z = \frac{X - \mu}{\sigma}$$

will always produce a transformed distribution with a mean of zero and a standard deviation of one. However, the shape of the distribution will not be affected by the transformation. If X is not normal then the transformed distribution will not be normal either. One important use of the standard normal distribution is for converting between scores from a normal distribution and percentile ranks.

Areas under portions of the standard normal distribution are shown below. About 0.68 (0.34 + 0.34) of the distribution is between -1 and 1 while about 0.96 of the distribution is between -2 and 2.

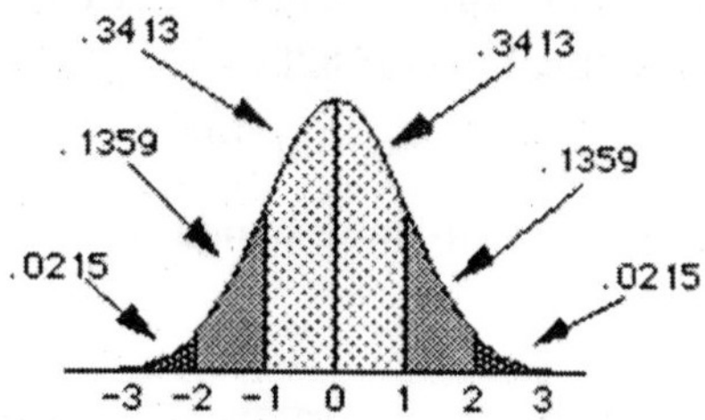

WHY IS THE NORMAL DISTRIBUTION SO IMPORTANT?

These three characteristics of the normal distribution are each critical in statistics because they allow us to make good use of probability statistics. When researchers collect data from a sample, sometimes all they want to know about are characteristics of the sample. For example, if we wanted to examine the eating habits of 100 first-year college students, we would just select 100 students, ask them what they eat, and summarize our data.

These data might give me statistics such as the average number of calories consumed each day by the 100 students in our sample, the most commonly eaten foods, the variety of foods eaten, and so on. All of these statistics simply *describe* characteristics of our sample, and are therefore called descriptive statistics. Descriptive statistics generally are used only to describe a specific sample. When all we care about is describing a specific sample, it does not matter whether the scores from the sample are normally distributed or not.

Many times, however, researchers want to do more than simply describe a sample. Sometimes, they want to know what the exact probability is of something occurring in their sample just due to chance. For example, if the average student in our sample consumes 2,000 calories a day, what are the chances, or probability, of having a student in the sample who consumes 5,000 calories a day? In addition, researchers often want to be able to make inferences about the population based on the data they collect from their sample. To determine whether some phenomenon observed in a sample represents an actual phenomenon in the population from which the sample was drawn, inferential statistics are used.

For example, if we find that the men in our sample eat an average of 200 more calories a day than the women, we may want to know the likelihood, or probability, that this gender difference that we have in our sample represents a real difference between the larger populations of men and women in their first year of college. To calculate these probabilities, we need to rely on the normal distribution, because the characteristics of the normal distribution allow statisticians to generate exact probability statistics. In the next section, we will briefly explain how this works.

It is important to note that the normal distribution is what is known in statistics as a theoretical distribution. That is, one rarely, if ever, gets a distribution of scores from a sample that forms an exact, normal distribution. Rather, what you get when you collect data is a distribution of scores that may or may not approach a normal, bell-shaped curve.

Because the theoretical normal distribution is what statisticians used to develop the probabilities discussed earlier, a distribution of scores that is not normal may be at odds with these probabilities. Therefore, there are a number of statistics that begin with the assumption that scores are normally distributed. When this assumption is violated (i.e., when the scores in a distribution are not normally distributed), there can be dire consequences.

The most obvious consequence of violating the assumption of a normal distribution is that the probabilities associated with a normal distribution are not valid. For example, if you have a normal distribution of scores on some variable (e.g., IQ test scores of adults in the United States), you can use the probabilities based on the normal distribution to determine exactly what percentage of the scores in the distribution will be 120 or higher on the IQ test. But suppose the scores in our distribution do not form a normal distribution. Suppose, for some reason, we have an unusually large number of high scores (e.g., over 120) and an unusually small number of low scores (e.g., below 90) in our distribution.

If this were the case, when we use probability estimates based on the normal distribution, we would underestimate the actual number of high scores in our distribution and overestimate the actual number of low scores in our distribution.

SAMPLING ISSUES

The problem of violating the assumption of normality becomes most problematic when our sample is not an adequate representation of our population. In most social science research, we do not have access to an entire population. Therefore, we usually need to rely on a sample for our data, and we must hope that our sample represents our population. ofcourse, to select a sample that represents our population, we must begin by very carefully defining our population.

For example, if we conduct a study in California to examine students' knowledge of world geography, and we select a sample of 500 fifth-grade students from two elementary schools in Los Angeles, what is our population? Do they represent all fifth graders in those two schools? Or all fifth graders in Los Angeles? Or all fifth graders in public schools in Los Angeles? Or in the state? Or in the country?

It turns out that this is a very difficult question to answer. There are all kinds of reasons why a sample may not be a good representation of the larger population that we think it is supposed to represent. Returning to our sample of 500 fifth graders from Los Angeles, suppose that we are trying to make the claim that this sample represents all fifth grade students in Los Angeles public schools. Why might this not be true? For one thing, the two schools from which these students were selected may be different from the typical Los Angeles public elementary school. They may have a larger number of wealthy students than the average school, for example. Or they may spend an unusually small amount of time discussing geography in their fifth grade classrooms. Or they may have an inordinate number of diplomats' children.

On the other hand, these students in our sample may be an excellent representation of the larger population of fifth graders in Los Angeles public schools. Even if they may differ in their wealth or number of students in the classroom or what they eat for breakfast, as long as they do not differ in some *systematic* way *on the variable of interest* (i.e., geography knowledge), there is no reason to believe that this sample does not represent this population.

So the problem of defining our target population is somewhat tricky. Once we have done this, the task becomes one of selecting a sample that has the best chance of representing our population, if possible. One method researchers have adopted is called random sampling. Random sampling occurs if members of the sample to be examined in the study are selected *at random* from the population. We should note that the term *at random* has a somewhat different meaning in statistics than in everyday language. It does not mean haphazardly.

Rather, it means that every member of the population has an equal chance of being selected for inclusion in the sample. Although this may sound easy, in many cases it is quite difficult. For example, if we define our population as all adults in the United States, how could we make sure that every adult has an equal chance of being selected for our sample? We cannot simply select names out of a phonebook, because not everyone has a phone and those who do not own a phone are more likely to be poor than those who do. We cannot simply select names off of voter registration lists because not everyone is registered to vote, and those who are registered are generally better educated than those who are not. As we can see, it can be difficult to get a truly random sample from a population.

A second method of sampling sometimes used by researchers is called representative sampling. In this method, the researcher selects one or more traits to use in selecting a sample from the population. For example, if we are interested in looking at issues of gender and ethnicity, we may try to select a sample that represents the national population (i.e., a sample that is 52% female, 11% African American, etc.). Unfortunately, this can be a costly and time-consuming way of selecting a sample, so many researchers decide against representative sampling. A third, quite common method of selecting a sample is called convenience sampling. In this method, a sample is selected because there was relatively easy access to the members of the sample. For example, if we were to conduct a study of student achievement, we may select a school that is fairly close to where we live or work so that we have less difficulty selecting the data. In addition, human beings have the option of declining participation in a study, so we must select a school that agrees to participate in our study.

Once we find a relatively close school that agrees to participate in our study, we may be working with a group of wealthy Lithuanian immigrants who are quite different from the larger population of students in our city, state, and country. With convenience sampling, it is often difficult to define the target population before identifying the sample. Rather, a sample is selected and its characteristics are noted (e.g., race, socioeconomic status, any unique or special characteristics) so that the population that the sample represents can later be inferred.

The Relationship Between Sampling Method and the Normal Distribution

The relationship between the normal distribution and sampling methods is as follows. The probabilities generated from the normal distribution depend on (a) the shape of the distribution, and (b) the idea that the sample is not somehow systematically different from the population. If we select a sample randomly from a population, we know that this sample may not look the same as another sample of equal size selected randomly from the same population. But any differences between our sample and other random samples of the same size selected from the same population would differ from each other randomly, not systematically. In other words, our sampling method was not biased such that we would continually select a sample from one end of our population (e.g., the more wealthy, the better educated, the higher achieving) if we continued using the same method for selecting our sample. Contrast this with a convenience sampling method. If we only select schools that are near our home or work, we will continually select schools with similar characteristics. For example, if we live in the Bible Belt, our sample will probably be biased in that our sample will more likely hold fundamentalist religious beliefs than the larger population of schoolchildren. Now if this characteristic is not related to the variable we are studying (e.g., achievement), then it may not matter that our sample is biased in this way. But if this bias is related to our variable of interest, then we may have a problem.

SKEW AND KURTOSIS

A distribution is skewed if one of its tails is longer than the other. The first distribution shown has a positive skew. This means that it has a long tail in the positive direction. The distribution below it has a negative skew since it has a long tail in the negative direction. Finally, the third distribution is symmetric and has no skew. Distributions with positive skew are sometimes called "skewed to the right" whereas distributions with negative skew are called "Skewed to the left."

Distributions with positive skew are more common than distributions with negative skews.

One example is the distribution of income. Most people make under $40,000 a year, but some make quite a bit more with a small number making many millions of dollars per year. The positive tail therefore extends out quite a long way whereas the negative tail stops at zero.

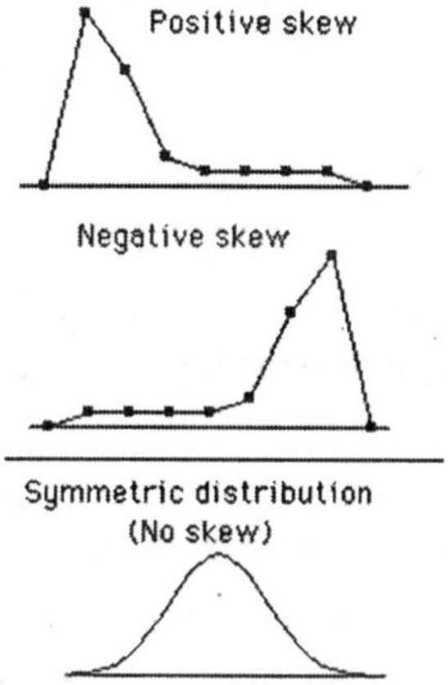

For.a more psychological example, a distribution with a positive skew typically results if the time it takes to make a response is measured. The longest response times are usually much longer than typical response times whereas the shortest response times are seldom much less than the typical response time. A histogram of the author's performance on a perceptual motor task in which the goal is to move the mouse to and click on a small target as quickly as possible is shown below. The X axis shows times in milliseconds.

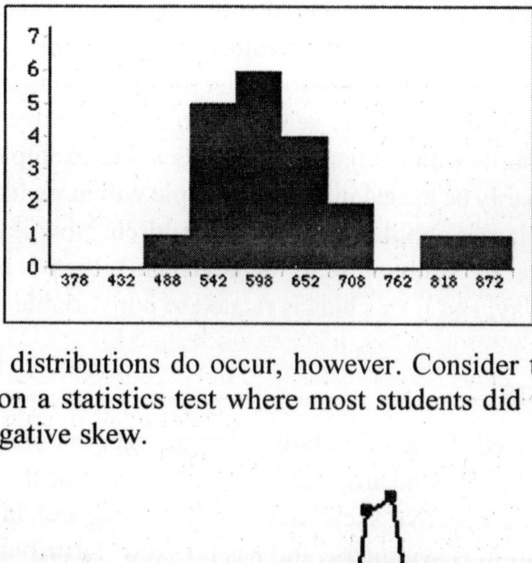

Negatively skewed distributions do occur, however. Consider the following frequency polygon of test grades on a statistics test where most students did very well but a few did poorly. It has a large negative skew.

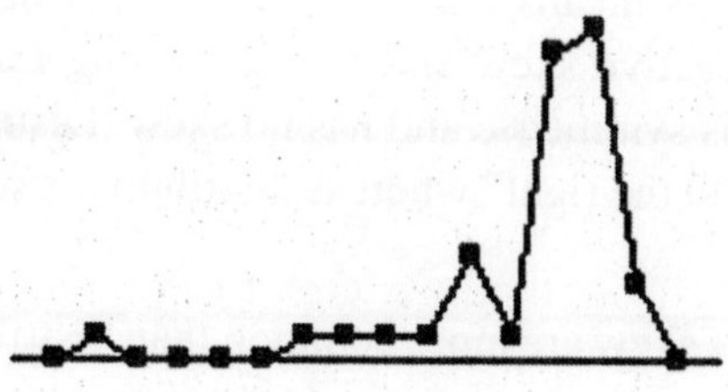

Skew can be calculated as:

$$skew = \frac{\sum (X-\mu)^3}{N\sigma^3}$$

where μ is the mean and σ is the standard deviation.

The normal distribution has a skew of 0 since it is a symmetric distribution.

As a general rule, the mean is larger than the median in positively skewed distributions and less than the median in negatively skewed distributions. There are counter examples. For example it is not uncommon for the median to be higher than the mean in a postively skewed bimodal distribution or with discrete distributions.

Two characteristics used to describe a distribution of scores are skew and kurtosis. When a sample of scores is not normally distributed (i.e., the bell shape), there are a variety of shapes it can assume. One way a distribution can deviate from the bell shape is if there is a bunching of scores at one end and a few scores pulling a tail of the distribution out toward the other end. If there are a few scores creating an elongated tail at the higher end of the distribution, it is said to be positively skewed. If the tail is pulled out toward the lower end of the distribution, the shape is called negatively skewed.

As you might have guessed, skewed distributions can distort the accuracy of the probabilities based on the normal distribution. For example, if most of the scores in a distribution occur at the low end with a few scores at the higher end (positively skewed distribution), the probabilities that are based on the normal distribution will underestimate the actual number of scores at the lower end of this skewed distribution and overestimate the number of scores at the higher end of the distribution. In a negatively skewed distribution, the opposite pattern of errors in prediction will occur.

Kurtosis refers the shape of the distribution in terms of height, or flatness. When a distribution has a peak that is higher than that found in a normal, bell-shaped distribution, it is called leptokurtic. When a distribution is flatter than a normal distribution, it is called platykurtic. Because the normal distribution contains a certain percentage of scores in the middle area (i.e., about 68% of the scores fall between 1 standard deviation above and 1 standard deviation below the mean), a distribution that is either patykurtic or letpkurtic will likely have a different percentage of scores near the mean than will a normal distribution. Specifically, a letptokurtic distribution will probably have a greater percentage of scores closer to the mean and fewer in the upper and lower tails of the distribution whereas a platykurtic distribution will have more scores at the ends and fewer in the middle than will a normal distribution.

Applying Normal Distribution Probabilities to A Non-normal Distribution

To illustrate some of the difficulties that can arise when we try to apply the probabilities that are generated from using the normal distribution to a distribution of scores that is skewed, we present a distribution of sixth-grade students& scores on a measure of self-esteem. In these data, 677 students completed a questionnaire that included four items designed to measure students' overall sense of self-esteem. Examples of these questions include "On the whole, We are satisfied with ourself" and "We feel we have a number of good qualities." Students responded to each of these four questions using a 5-point rating scale with. "1 = not at all true" and "5 = very true." Students' responses on these four items were then averaged, creating a single self-esteem score that ranged from a possible low of 1 to a possible high of 5. The frequency distribution for this self-esteem variable is presented in Figure.

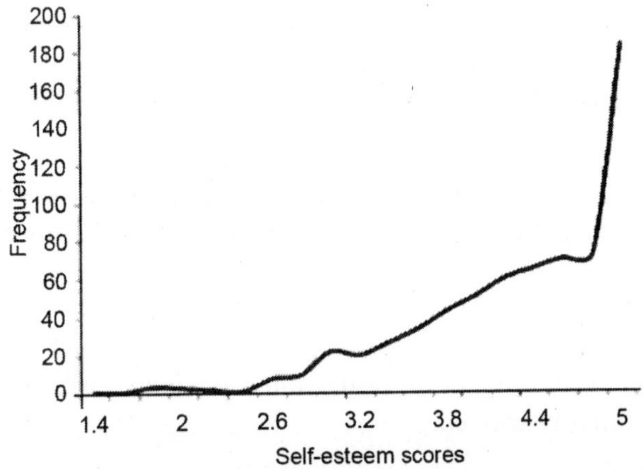

Figure. Frequency Distribution for Self-esteem Scores.

As we can see, the distribution of scores presented in Figure does not form a nice, normal, bell-shaped distribution. Rather, most of the students in this sample scored at the high end of the distribution and a long tail extends out toward the lower end of the scale. This is a classic, negatively skewed distribution of scores. The happy part of this story is that most of the students in this sample appear to feel quite good about themselves. The sad part of the story is that some of the assumptions of the normal distribution are violated by this skewed distribution. Let's take a look at some specifics.

One of the qualities of a normal distribution is that it is symmetrical, with an equal percentage of the scores between the mean and 1 standard deviation below the mean as there are between the mean and 1 standard deviation above the mean. In other words, in a normal distribution, there should be about 34% of the scores within 1 standard deviation above the mean and 34% within 1 standard deviation below the mean.

In our distribution of self-esteem scores presented earlier, the mean is 4.28 and the standard deviation is .72. A full 50% of the distribution falls between the mean and 1 standard deviation above the mean in this group of scores. So, although we might predict that about 16% of our distribution will have scores more than 1 standard deviation above the mean in a normal distribution, in our skewed distribution of self-esteem scores, we can see that there are no students with scores more than 1 standard deviation above the mean. In the coming we present a more thorough discussion of how to use the normal distribution to calculate standard deviation units and percentile scores in a normal distribution.

As this example demonstrates, the probabilities that statisticians have generated using the normal distribution may not apply well to skewed or otherwise non-normal distributions of data. This should not lead you to believe, however, that non-normal distributions of scores are worthless. In fact, even if you have a non-normal distribution of scores in your sample, these scores can create normal sampling distributions for use in inferential statistics. What is perhaps most important to keep in mind is that a non-normal distribution of scores may be an indication that your sample may differ in important and systematic ways from the population that it is supposed to represent.

When making inferences about a population based on a sample, be very careful to define the population precisely and to be aware of any biases you may have introduced by your method of selecting your sample. The normal distribution can be used to generate probabilities about the likelihood of selecting an individual or another sample with certain characteristics (e.g., distance from the mean) from a population. If your sample is not normal and your method of selecting the sample may be systematically biased to include those with certain characteristics (e.g., higher than average achievers, lower than average income, etc.), then the probabilities of the normal distribution may not apply well to your sample.

The theoretical normal distribution is a critical element of statistics primarily because many of the probabilities that are used in inferential statistics are based on the assumption of normal distributions. Statisticians use these probabilities to determine the probability of getting certain statistics and to make inferences about the population based on the sample. Even if the data in a sample are not normally distributed, it is possible that the data in the population from which the sample was selected may be normally distributed. It is described how the normal distribution, through the use of z scores and standardization, is used to determine the probability of obtaining an individual score from a sample that is a certain distance away from the sample mean.

STANDARDIZATION AND Z SCORES

If we know the mean and standard deviation of a distribution of scores, we have enough information to develop a picture of the distribution. Sometimes researchers are interested in describing individual scores within a distribution. Using the mean and the standard deviation, researchers are able to generate a standard score, also called a *z* score, to help them understand where an individual score falls in relation to other scores in the distribution.

Through a process of standardization, researchers are also better able to compare individual scores in the distributions of two separate variables. Standardization is simply a process of converting each score in a distribution to a z score. A z score is a number that indicates how far above or below the mean a given score in the distribution is in standard deviation units. So standardization is simply the process of converting individual raw scores in the distribution into standard deviation units.

Suppose that you are a college student taking final exams. In our biology class, you take your final exam and get a score of 65 out of a possible 100. In your statistics final, you get a score of 42 out of 200. On which exam did you get a "better" score? The answer to this question may be more complicated than it appears. First, we must determine what we mean by "better." If better means percentage of correct answers on the exam, clearly you did better on the biology exam.

But if your statistics exam was much more difficult than your biology exam, is it fair to judge your performance solely on the basis of percentage of correct responses? A more fair alternative may be to see how well you did compared to other students in your classes. To make such a comparison, we need to know the mean and standard deviation of each distribution. With these statistics, we can generate z scores.

Suppose the mean on the biology exam was 60 with a standard deviation of 10. That means you scored 5 points above the mean, which is half of a standard deviation above the mean (so higher than the average for the class). Suppose further that the average on the statistics test was 37 with a standard deviation of 5. Again, you scored 5 points above the mean, but this represents a full standard deviation over the average. Using these statistics, on which test would you say you performed better? To fully understand the answer to this question, let's examine standardization and *z* scores in more depth.

STANDARDIZATION AND Z SCORES IN DEPTH

It is often difficult to compare two scores on two variables when the variables were measured using different scales. The biology test was measured on a scale from 1 to 100, whereas the statistics exam used a scale from 1 to 200. When variables have such different scales of measurement, it is almost meaningless to compare the raw scores (i.e., 65 and 42) on these exams. Instead, we need some way to put these two exams on the same scale, or to *standardize* them.

One of the most common methods of standardization used in statistics is to convert raw scores into standard deviation units, or z scores. The formula for doing this is very simple and is presented in Table.

As you can see from the formulas in Table, to standardize a score (i.e., to create a z score), you simply subtract the mean from an individual raw score and divide this by the standard deviation. So if the raw score is above the mean, the z score will be positive whereas a raw score that is less than the mean will produce a negative z score. When an entire distribution of scores is standardized, the average (i.e., mean) z score for the standardized distribution will always be 0, and the standard deviation of this distribution will always be 1.0.

Formula for calculating a *z* score.

$$z = \frac{raw\,score - mean}{s\tan dard\,deviation}$$

or

$$z = \frac{X - \mu}{\sigma}$$

or

$$z = \frac{X - \overline{X}}{s}$$

When X = raw score

μ = population mean

σ = standard deviation

$\overline{X}$ = sample mean

s = sample standard deviation

Interpreting z Scores

z scores tell researchers instantly how large or small an individual score is relative to other scores in the distribution. For example, if we know that one of our students got a z score of "1.5 on an exam, we would know that student scored 1.5 standard deviations below the mean on that exam. If another student had a z score of .29,1 would know the student scored .29 standard deviation units above the mean on the exam.

Now, with the appropriate cautions in mind, let's consider a couple more uses of z scores and standardization. One of the most handy features of z scores is that, when used with a normally distributed set of scores, they can be used to determine percentile scores. That is, if you have a normal distribution of scores, you can use z scores to discover which score marks the 90 percentile of a distribution (i.e., that raw score at which 10% of the distribution scored above and 90% scored below). This is because statisticians have demonstrated that in a normal distribution, a precise percentage of scores will fall between the mean and 1 standard deviation above the mean. Because normal distributions are perfectly symmetrical, we know that the exact same percentage of scores that falls between the mean and 1 standard deviation *above* the mean will also fall between the mean and 1 standard deviation *below* the mean. In fact, statisticians have determined the precise percentage of scores that will fall between the mean and any z score (i.e., number of standard deviation units above or below the mean).

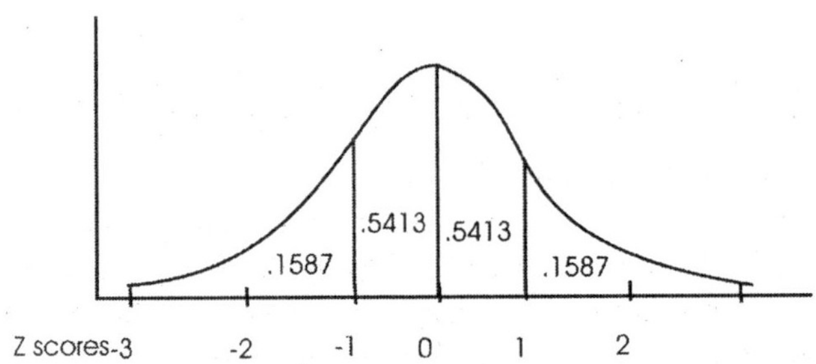

Fig. The standard normal distribution

Let us consider that in a normal distribution the mean always marks the exact centre of the distribution, you know that the mean is the spot in the distribution in which 50% of the cases fall below and 50% fall above. With this in mind, it is easy to find the score in a distribution that marks the 90th percentile, or any percentile, for that matter. In Figure, we can see the percentage of scores in a normal distribution that fall between different *z* score values. This figure contains the *standard normal distribution.*

Let us consider an example. Suppose we know that the average SAT-math score for White males is 517, with a standard deviation of 100, and forms a normal distribution. In this distribution, we already know that the score that marks the 50th percentile is 517. Suppose we wanted to know the score that marks the 90th percentile. To find this number, we have to follow a series of simple steps.

Step 1. Using a *z* score table from a traditional statistios textbook, find the *z* score that marks the 90th percentile. To do this, we need to remember that the 90th percentile is 40 percentile points above the mean, so we are looking for the *z* score at which 40% of the distribution falls between the mean and this *z* score. An alternative method is to find the *z* score at which 10% of the distribution falls above, because the 90th percentile score divides the distribution,into sections with 90% of the score falling below this point and 10% falling above, z score tables in traditional statistics textbooks always provide at least one of these scores (i.e., the percentage of the distribution that falls between the mean and the z score or the percentage that falls above the z score), and often both. In the current example, the z score that corresponds with the 90th percentile is 1.28. So z = 1.28.

Step 2. Convert this z score back into the original unit of measurement. Remember that the SAT-math test is measured on a scale from 0 to 800. We now know that the mean for White males who took the test in 1989 was 517, and that 90th percentile score of this distribution is 1.28 standard deviations above the mean (because z = 1.28). So what is the actual SAT-math score that marks the 90th percentile? To answer this, we have to convert our z score from standard deviation units into raw score units and add this to the mean. The formula for doing this is

$$X = \mu + (z)(\sigma)$$

In this equation, *X* is the raw score we are trying to discover, M, is the average score in the distribution, *z* is the *z* score we found, and a is the standard deviation for the distribution. Plugging our numbers into the formula, we find that

x=517 + (1.28)(100) x=517+128 *X=645*

Step 3. Now we can wrap words around our result and answer our original question. When doing this, it is often helpful to use the original question when stating our finding, as follows:

Question: What is the score that marks the 90th percentile of the distribution of White male students' SAT-math scores in 1989?

Answer: The score of 645 marks the 90th percentile of the distribution of White male students' SAT-math scores in 1989. This *z* score, percentile score, and the corresponding raw score are depicted in Figure.

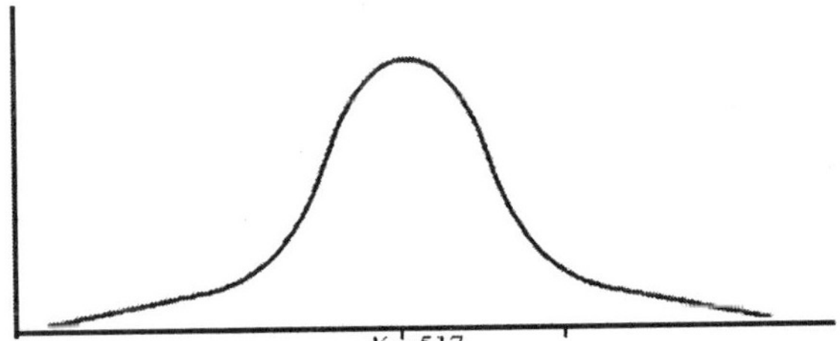

Finding the score that marks the 90th percentile of the distribution.

Just as we can use *z* scores to find the raw score that marks a certain percentile in a distribution, we can also use *z* scores to help us convert a known raw score into a percentile score. For example, if we know that a student in our distribution has a score of 425 on the SAT-math test, we want to know the percentage of the distribution that scored above and below 425. This is the type of conversion that has happened when students' standardized test scores are published in the local newspaper using percentiles under headlines such as "California Students Score in 45th Percentile on National Test!" Similarly, when a proud parent exclaims "My Johnny is in the top 10% in height for his age group." a conversion from a raw score to a percentile score has taken place, with the help of a z score. Here's how it's done:

Step 1 : We must begin by converting the raw score into a *z* score. In our example, the raw score is 425 *(X* = 425). To convert this into a z score, we simply recall our mean (m = 517) and our standard deviation (a= 100) and then plug these numbers into the z score formula: z =425-517/100 z =-88/100 z =.88 Step 2: Now that we have a z score, we need to look in the nearest z score table to find either the percentage of the normal distribution that falls between the mean and a z score of -.88 or the percentage of the distribution that falls below a z score of -.88. Notice that we are dealing with a negative z score in our example. Most z score tables only report positive z scores, but because normal distributions are symmetrical, the percentage of the distribution that fall between the mean and z are identical whether the z score is positive or negative. Similarly, the percentage that falls above a positive z score is identical to the percentage that falls below a negative z score. Our z score table tells me that 31% of the normal distribution of scores will fall between the mean and a z score of-.88 and 19% will fall below a z score of-.88.

Step 3: To wrap words around this result, we must begin with the recollection that in our example, a z score of -.88 corresponds with a raw score of 425 on the SAT-math test among the White males who took the test in 1989. So, to wrap words around our result, we would say that "A score of 425 on the SAT-math test marks the 19th percentile of the distribution of test scores among White males in 1989."

The examples just presented represent two handy uses of z scores for understanding botl an entire distribution of scores and individual scores within that distribution. Standardizeo scores are used in a variety of statistics, and are perhaps most helpful for comparing scores that are measured using different scales of measurement. It is difficult to compare two scores that are measured on different scales (e.g., height and weight) without first converting them into a common unit of measurement. Standardizing scores is simply this process of conversion. In the final section of this chapter, we present and briefly describe two distributions of scores described by both raw scores and z scores.

Examples: Comparing Raw Scores and Z Scores

To illustrate the overlap between raw scores and standardized z scores,we first present data from a sample of elementary and middle school students from whom we collected data a few years ago. gave these students a survey to assess their motivational beliefs and attitudes about a standardized achievement test they were to take the following week. One of the items on the survey read "The ITBS test will measure how smart we are." Students responded to this question using an 8-point scale with "1 = Strongly Disagree" and "8 = Strongly Agree." The frequency distribution is presented in Figure. This distribution has a mean of 5.38 and a standard deviation of 2.35.

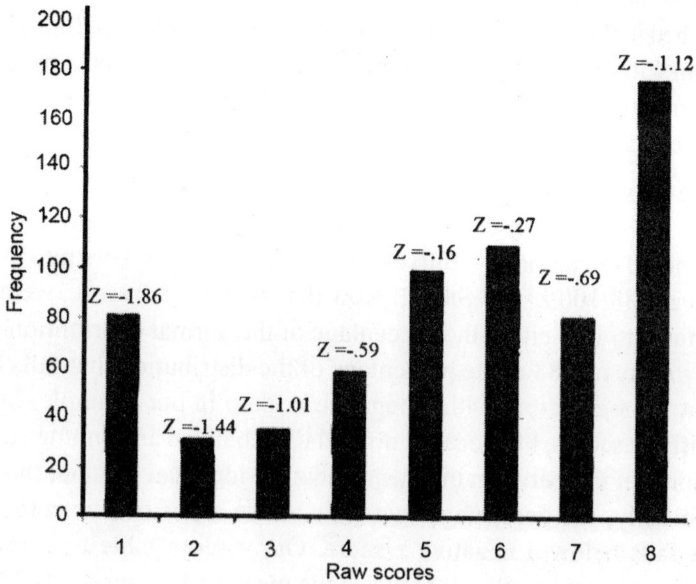

Fig. Frequency Distribution for "The Test will show how Smart we are" Item.

As you can see, this is not a normal, bell-shaped distribution. This distribution has a sort of odd shape where there is the hint of a normal distribution in Scores 2 through 7 but then there are "spikes" at the ends, particularly at the high end. The result is an asymmetrical distribution. If you compare the z scores on top of each column with the raw scores at the bottom of each column, you can see how these scores are related to each other. For example, we can see that all of the raw scores of 5 or lower have negative z scores. This is because the mean of a distribution always has a z score of 0, and any raw scores below the mean will have negative z scores. In this distribution, the mean is 5.38, so all raw scores of 5 and below have negative z scores and all raw scores of 6 or above have positive z scores.

Another feature of this distribution that is clearly illustrated by the z scores is that there is a larger range of scores below the mean than there are above the mean. This is fairly obvious, because the mean is well above the midpoint on this scale. The highest scores in this distribution are just a little more than 1 standard deviation above the mean (z = 1.12), whereas the lowest scores are nearly 2 standard deviations below the mean (z = –1.86). Finally, the inclusion of standard deviation scores with each raw score allows us to immediately determine how many standard deviations away from the mean a particular raw score falls. For example, we can see that a student who had a raw score of 3 on this variable scored just about exactly 1 standard deviation below the mean (z = –1.01).

For our second example, we have chosen a variable with a much smaller standard deviation. Using the same 8-point scale described earlier, students were asked to respond to the item "we think it is important to do well on the ITBS test." Students overwhelmingly agreed with this statement, as the mean (7.28) and relatively small standard deviation (1.33) revealed. The frequency distribution for the scores on this item is presented in Figure.

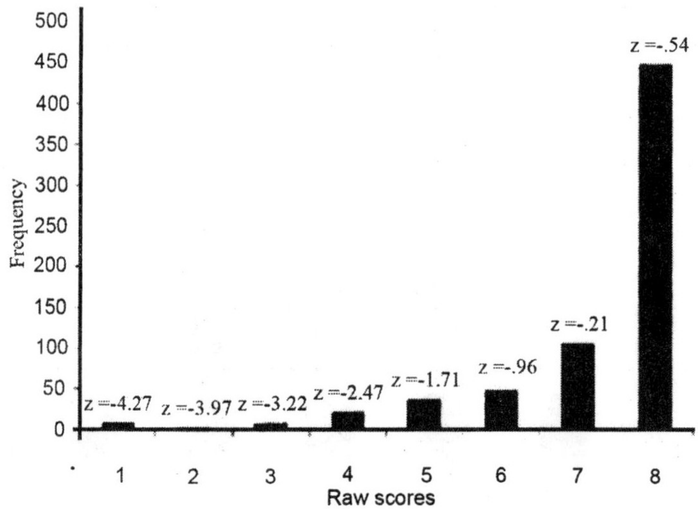

Fig. Frequency Distribution for "Important to do well" Item.

In this graph, we can see that the distribution is highly skewed, with most students circling the number 8 on the scale. Because so many students answered similarly, the standard deviation is quite small, with only a relatively few scores at the lower end of the distribution. The small standard deviation coupled with the high mean create a situation where very low scores on the scale have extremely small z scores. For example, the few students with a raw score of 1 on the scale (n = 7) had z scores of –4.72, indicating that these students were more than 4 standard deviations below the mean. Those students with the highest score on the scale were only about half a standard deviation above the mean because, with such a high mean, it was impossible to get a score very far above the mean.

The two examples provided previously both illustrate the relation between z scores and raw scores for distributions that are skewed. In both of these distributions, the means were above the midpoint on the scale, and subsequently there was a greater range of z scores below the mean than there were above the mean. Such is not the case when the distribution of scores are normally distributed. To illustrate this, we use data from a different data set. We used surveys to measure a sample of high school students' motivational goals in school. One goal that we measured is known as a performance-approach goal. This goal reflects a concern, or a desire, to outperform classmates and peers for the sake of demonstrating superior ability. The items on the survey were measured using a scale from 1 to 5 (1 = "not at all true" and 5 = "very true"). The frequency distribution is presented in Figure.

This distribution of scores had a mean of 3.00 and a standard deviation of .92. As we can see, the data are quite normally distributed. When the data are normally distributed, we would expected most of our cases to have z scores at or near zero, because in a normal distribution most of the cases are near the mean. Also notice that as we move farther away from the mean (i.e., z scores over 2.0 or less than –2.0), there are fewer cases. In a normal distribution, then, the probability of finding a particular z score becomes smaller as the value of the z score moves further away from zero. As Figures illustrate, this is not always the case in skewed distributions.

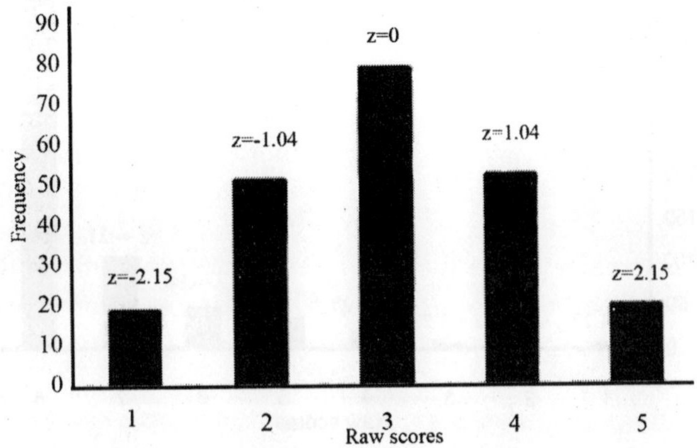

Figure. Frequency Distribution for Performance-approach Goals.

z scores provide a handy way of interpreting where a raw score is in relation to the mean. They also demonstrate that the range of possible z scores depends heavily on where the mean falls in the range of possible scores as well as the size of the standard deviation for the distribution. We can use *z* scores to quickly and easily determine where an individual score in a distribution falls relative to other scores in the distribution, either by interpreting the z score in standard deviation units or by calculating percentile scores. We can also use z score to determine how unusual a given score in a distribution is (i.e., the probability of obtaining an individual score of that size when selecting the individual at random). We will use the information that we have learned about the mean, standard deviation, normal distributions, z scores, and probability to understand one of the most important concepts in statistics: the standard error.

STANDARD ERRORS

The concept of standard error is one that many students of statistics find confusing when they first encounter it. In all honesty, there are many students, and many researchers, who never fully grasp the concept. We are convinced that many people have problems with understanding standard errors because they require a bit of a leap into the abstract and because, with the advent of computer programmes, it is possible to lead a long and productive research life without having to think about or analyse a standard error for years at a time.

Therefore, many researchers choose to gloss over this abstract concept. This is a mistake. We hold this opinion because, as a teacher of statistics, We have learned that when one is able to truly understand the concept of standard error, many of our most beloved inferential statistics (*t* tests, ANOVA, regression coefficients, correlations) become easy to understand. So let me offer this piece of advice: Keep trying to understand the contents of this chapter, and other information you get about standard errors, even if you find it confusing the first or second time you read it. With a little effort and patience, you can understand standard errors and many of the statistics that rely on them.

WHAT IS A STANDARD ERROR?

There are two answers to this question. First, there is the technical answer, which is the definition of a standard error. A standard error is, in effect, the standard deviation of the sampling distribution of some statistic (e.g., the mean, the difference between two means, the correlation coefficient, etc.). We realise that this makes no sense until you know what a sampling distribution is, and it is explained this in the next section of this chapter. For now, it is recommend that we repeat the definition to yourself 10 times: "The standard error is, in effect, the standard deviation of the sampling distribution of some statistic." The second answer is that the standard error is the denominator in the formulas used to calculate many inferential statistics.

In the following chapters, you will see the standard error as the denominator in many formulas. This is because the standard error is *the measure of how much* random *variation we would expect from samples of equal size drawn from the same population.* Again, look at the preceding sentence, think about it, and rest assured that it is explained in more detail in the next few pages.

The Conceptual Description of the Standard Error of the Mean

To begin this more detailed discussion of standard errors, it is introduced that the esoteric component of the concept. This is the section that you may need to read several times to let it sink in. Although there are standard errors for all statistics, we will focus on the standard error of the mean. When we think of a distribution of scores, we think of a certain number of scores that are plotted in some sort of frequency graph to form a distribution. In these distributions, each case has a score that is part of the distribution. Just as these simple frequency distributions are plotted, or graphed, we can also plot distributions of sample means. Imagine that we wanted to find the average shoe size of adult American women. In this study, the population we are interested in is all adult American women.

But it would be expensive and tedious to measure the shoe size of all adult American women. So we select a sample of 100 women, at random, from our population. At this point, it is very important to realise that our sample of 100 women may or may not look like the typical American woman (in terms of shoe size). When we select a sample at random, it is possible to get a sample that represents an extreme end of the population (e.g., a sample with an unusually large average shoe size). If we were to throw our first sample of women back into the general population and chose another random sample *of the same size* (i.e., 100), it is possible that this second sample may have an average shoe size that is quite different from our first sample.

Once you realise that different random samples of equal size can produce different mean scores on some variable (e.g., different average shoe sizes), the next step in this conceptual puzzle is easy: If we were to take 1,000 different random samples of women, each of 100, and compute the average shoe size of each sample, these 1,000 sample means would form their own distribution. This distribution would be called the sampling distribution of the mean.

To illustrate this concept, let's consider an example with a small population (TV = 5). Suppose our population consists of five college students enrolled in a seminar on statistics. Because it is a small seminar, these five students represent the entire population of this seminar. These students each took the final exam that was scored on a scale from 1 to 10, with lower scores indicating poorer performance on the exam. The scores for each student are presented in Table, arranged in descending order according to how well they did on the exam.

Population of students' scores on final exam.

Students	*Score of Final Exam*
Students 1	3
Students 2	6
Students 3	6
Students 4	7
Students 5	9

If we were to select a random sample of two students from this population ($n = 2$), we might get student 2 and student 5. This sample would have a mean of 7.5 because [(6 + 9) + 2 = 7.5]. If we were to put those two students back into the population and randomly select another sample of 2,1 might get Student 4 and Student 5. This sample would have a mean of 8 because [(7 + 9) 2 = 8]. We put those students back into the population and randomly select another sample of 2, such as Students 1 an 3. This sample would have a mean of 4.5. As we can see, just by virtue of whom is included in each random sample we select from our population, we get different sample means. Now if we were to repeat this process of randomly selecting samples of two students from our population, calculating their mean, and returning the members of the sample to the population (called sampling with replacement), eventually we would get a distribution of sample means that would look something like the distribution presented in Figure. As we can see, these means form a distribution. This example illustrates how random samples of a given size selected from a population will produce a distribution of sample means, eventually forming a sampling distribution of the mean.

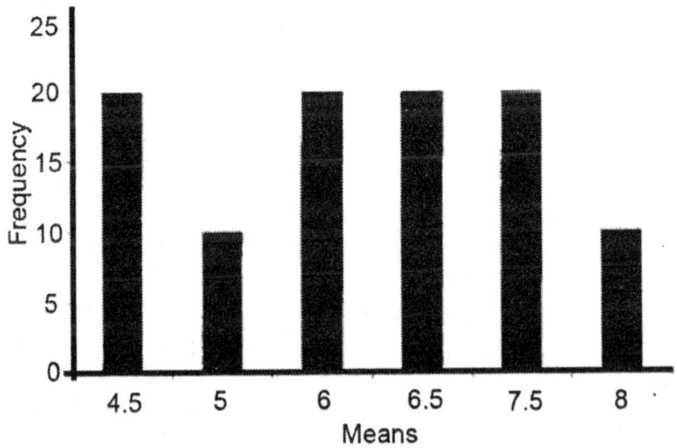

Fig. Sampling Distribution of the Mean.

Just as the other distributions we have discussed have a mean and a standard deviation, this sampling distribution of the mean also has these characteristics. To distinguish a sampling distribution from a simple frequency distribution, the mean and standard deviation of the sampling distribution of the mean have special names. The mean of the sampling distribution of the mean is called the expected value.

It is called the expected value because the mean of the sampling distribution of the means is the same as the population mean. When we select a sample from the population, our best guess is that the mean for the sample will be the same as the mean for the population, so our *expected* mean will be the population mean. The standard deviation of the sampling distribution of the mean is called the standard error. So the standard error is simply the standard deviation of the sampling distribution.

The final step in understanding the concept of standard error of the mean is to understand what this statistic tells us. You will remember that the standard deviation tells us the average difference, or deviation, between an individual score in the distribution and the mean for the distribution.

The standard error of the mean provides essentially the same information, except it refers to the average difference between the expected value (i.e., the population mean) and an individual sample mean. So one way to think about the standard error of the mean is that it tells us how confident we should be that a sample mean represents the actual population mean.

Phrased another way, the standard error of the mean provides a measure of how much *error* we can expect when we say that a sample mean represents the mean of the larger population. That is why it is called a standard *error*.

HOW TO CALCULATE THE STANDARD ERROR

Most of the time, researchers do not draw 1,000 samples of equal size from the population and then figure out the mean and standard deviation of this distribution of sample means. In fact, most of the time, researchers collect data from only a single sample, and then use this sample to make inferences about the population from which the sample was drawn. How can we make inferences about a larger population on the basis of a single sample?

To make such inferences about the population from a single sample, researchers must use what they know about their sample to make educated guesses, or estimates, about the population. demonstrate this concept using the shoe-size example mentioned earlier. Suppose that we have a random sample of 100 women. Now if this sample were truly selected at random (i.e.; every adult woman in the United States had an equal chance of being selected), our most logical assumption would be that this sample represents the larger population accurately.

Therefore, we would have to assume that the mean shoe size of our sample (suppose it is 6) is also the mean shoe size of the larger population. Of course, we cannot know if this is true. In fact, as discussed earlier, there is good reason to believe that our sample may not represent our population well. But if the only information we have about U.S. adult women's shoe size comes from our sample of 100 women, our *best guess* about what the larger population of women looks like must be that they are similar to this sample of 100 women. Now we are faced with a critical question: When we guess that the population of women in the United States has an average shoe size of 6 (based on our sample average), how much *error* can we *expect* to have in this estimation? In other words, what is the *standard error?*

To answer this question we must examine two characteristics of our sample. First, how large is our sample? The larger our sample, the less error we should have in our estimate about the population. This makes sense because the larger our sample, the more our sample should look like our population, and the more accurate our estimates of our population will be. If there are 100 million women in the United States and we use a sample of 50 million to predict their average shoe size, we would expect this prediction to be more accurate than a prediction based on a sample of 100 women. Therefore, the larger our sample, the smaller our standard error.

The second characteristic of our sample that we need to examine is the standard deviation. Remember that the standard deviation is a measure of how much variation there is in the scores in our sample. If the scores in our sample are very diverse (i.e., a lot of variation, a large standard deviation), we can assume that the scores in our population are also quite diverse. So, if some women in our sample have size 2 shoes and others have size 14 shoes, we can assume there is also a pretty large variety of shoe sizes in our population. On the other hand, if all of the women in our sample have shoe sizes of either 5, 6, or 7, we can assume that most of the women in the larger population have an equally small variety of shoe sizes.

Of course, these assumptions about the population may not be true, we must rely on them because this is all the information we have. So, the larger the sample standard deviation, the greater the assumed variation of scores in the population, and consequently the larger the standard error of the mean.

Formulas for calculating the standard error of the mean.

$$\sigma_{\bar{x}} = \frac{\sigma}{\sqrt{n}}$$

or

$$s_{\bar{x}} = \frac{s}{\sqrt{n}}$$

Where σ is the standard deviation for the population

s is the sample estimate of the standard deviation

n is the size of the sample

An examination of the formula for calculating the standard error of the mean reveals the central role of the sample standard deviation (or population standard deviation, if known) and the sample size in determining the standard error. The formula is simply the standard deviation of the sample or population divided by the square root of *n,* the sample size. As with all fractions, as the numerator gets larger, so does the resulting standard error. Similarly, as the size of the denominator decreases, the resulting standard error increases.

Small samples with large standard deviations produce large standard errors, because these characteristics make it more difficult to have confidence that our sample accurately represents our population. In contrast, a large sample with a small standard deviation will produce a small standard error, because such characteristics make it more likely that our sample accurately represents our population.

THE CENTRAL LIMIT THEOREM

Simply put, the central limit theorem states that as long as you have a reasonably large sample size (e.g., $n > 30$), the sampling distribution of the mean will be normally distributed, even if the distribution of scores in your sample is not. In earlier chapters, it is discussed distributions that were not in the shape of a nice, normal, bell curve. What the central limit theorem proves is that even when you have such a non-normal distribution in your sample, the sampling distribution of the mean will form a nice, normal, bell-shaped distribution as long as you have at least 30 cases in your sample. Even if you have fewer than 30 cases in your sample, the sampling distribution of the mean will probably be near normal if you have at least 10 cases in your sample. Even in our earlier example where we had only two cases per sample, the sampling distribution of the mean had the beginning of a normal shape.

Although we do not concern ourselves here with why the central limit theorem works, you need to understand why the veracity of this theorem is so important. A number of statistics rely on probabilities that are generated from normal distributions. For example, we may want to know whether the average IQ test scores of a sample of 50 adults in California is different from the larger population of adults. If our sample has an average IQ test score of 110, and the national average is 100, we can see that our sample average differs from the population average by 10 points. Is 10 points a meaningful difference or a trivial one? To answer that question, we must be able to discover the probability of getting a difference of 10 points by random chance. In other words, if we were to select another random sample of 50 adults from California and compute their average IQ test score, what are the odds that they will have an average that is 10 points higher than the national average of 100? To determine this probability, we must have a normal distribution of sample means, or a normal sampling distribution of the mean. The central limit theorem indicates that as long as we have a sample size of at least 30, our sampling distribution of the mean will be normal.

The Normal Distribution and *t* Distributions: Comparing *z* Scores and *t* Values

To determine the probability of randomly selecting an individual case with a particular score on some variable from a population with a given mean on that variable. We did this by converting the raw score into a *z* score. Now that we know how to compute a standard error, we can use z scores again to determine the probability of randomly selecting a sample with a particular mean on a variable from a population with a given mean on the same variable. We can also use the family of *t* distributions to generate *t* values to figure out the same types of probabilities. To explain this, we will begin by comparing the normal distribution with the family of *t* distributions.

The normal distribution is a theoretical distribution with a bell shape and is based on the idea of population data. We also know that the probabilities associated with z scores are associated with the normal distribution. In addition, we know that a standard deviation derived from sample data is only an *estimate* of the population standard deviation. Because the formula for calculating the sample standard deviation has n - 1 in the denominator, we also know that the smaller the sample, the less precisely the sample standard deviation estimates the population standard deviation. Finally, we know that the standard error formula is based partly on the standard deviation.

When we put all of this information together, we end up with a little bit of the dilemma. If we can use the standard error to generate z scores and probabilities, and these z scores and probabilities are based on the normal distribution, what do we do in those cases where we are using sample data and we have a small sample? Won't our small sample influence our standard error? And won't this standard error influence our z scores? Will our z scores and probabilities be accurate if we have a small sample? Fortunately, these concerns have already be addressed by brains larger than mine. It turns out that the normal distribution has a close family of relatives: the family of *t* distributions.

These distributions are very much like the normal distribution, except the shape of *t* distributions is influenced by sample size. With large samples (e.g., > 120), the shape of the *t* distribution is virtually identical to the normal distribution. As sample size decreases, however, the shape of the t distribution becomes flatter in the middle and higher on the ends. In other words, as sample size decreases, there will be fewer cases near the mean and more cases away from the mean, out in the tails of the distribution. Like the normal distribution, t distributions are still symmetrical.

Just as we use the z table to find probabilities associated with the normal distribution, we use the table of t values to find probabilities associated with the t distributions. Along the left column there are numbers in ascending order. These are degrees of freedom and they are directly related to sample size. To use this table, you simply calculate a t value (using basically the same formula that you use to find a z score) and then, using the appropriate degrees of freedom, figure out where your t value falls to determine the probability of finding a t value of that size. Whenever you don't know the population standard deviation and must use an estimate from a sample, it is wise to use the family of t distributions. Here is an example to illustrate these ideas. We used this formula to calculate a z score from a raw score:

$$z = \frac{raw\ score - mean}{s\tan dard\ deviation}$$

or

$$z = \frac{X - \mu}{\sigma}$$

where X = raw score

μ = population mean

σ = standard deviation

The formula for converting a sample mean into a z score is almost identical, except the individual raw score is replaced by the sample mean and the standard deviation is replaced by the standard error. In addition, if we do not know the population standard deviation, the standard deviation estimate from the sample must be used and we are computing a t value rather than a z score. These formulas are found in Table.

Now, suppose that we know that the average American man exercises for 60 minutes a week. Suppose, further, that we have a random sample of 144 men and that this sample exercises for an average of 65 minutes per week with a standard deviation of 10 minutes. What is the probability of getting a random sample of this size with a mean of 65 if the actual population mean is 60 *by chance?* To answer this question, we compute a t value:

$$t = \frac{65 - 60}{10/\sqrt{144}}$$

$$t = \frac{5}{.83}$$

$$t = 6.02$$

The row with oo degrees of freedom, we can see that the probability of getting a *t* value of this size *by chance* with a sample of this size is less than .001. Notice that if we had calculated a z score rather than a *t* score (i.e., if the *population* standard deviation had been 10), our z value would have been the same (i.e., z = 6.02) and our probability, would have been less than .00003. The normal distribution (associated with z scores) and the *t* distributions are virtually identical when the sample size is larger than 120.

Finally, to illustrate the difference between the *t* distributions and the normal distribution, suppose that our sample size had been 25 rather than 144. We would have calculated the *t* value just as we did before, but our standard error would be different (because our sample size is smaller), thereby producing a smaller *t* value:

$$t = \frac{65-60}{10/\sqrt{25}}$$

$$t = \frac{5}{2}$$

$$t = 2.50$$

Now, looking at our table of *t* values with 24 degrees of freedom, we find that the probability of getting a *t* value of this size is just about .02. Notice that if we had our larger sample size of 144, the probability of getting a *t* value of 2.50 would have been closer to .01.

z score and *t* value formulas.

when σ is known

$$z = \frac{\textit{sample mean} - \textit{population mean}}{\textit{s tan dard error}}$$

$$z = \frac{\overline{X} - \mu}{\sigma \overline{x}}$$

when σ is not known

$$t = \frac{\textit{sample mean} - \textit{population mean}}{\textit{s tan dard error}}$$

$$z = \frac{\overline{X} - \mu}{s_{\bar{x}}}$$

Where μ = population mean

σ^2 = standard error

$\overline{X}$ = sample mean

$s_{\bar{x}}$ = sample of the standard error

So when the sample size is large, the normal distribution and the *t* distribution are virtually identical. But as our sample size decreases, the *t* distribution changes and so do the probabilities associated with it. When the population standard deviation is known, the normal distribution can be used. But when the population standard deviation is not know, or the sample size is small, the family of *t* distributions should be used. Now we can turn our attention the how standard errors are used in other statistics.

THE USE OF STANDARD ERRORS IN INFERENTIAL STATISTICS

Some type of standard error is used in every inferential statistic, including those discussed in this book *(t* tests, ANOVA, regression, etc.). In inferential statistics, we want to know whether something that we observe in our sample represents a similar phenomenon in the larger population from which the sample was drawn. For example, if we compare the average weight of a sample 100 men to that of a sample of 100 women and find that, on average, men in our sample weigh 60 pounds more than women in our sample, we may want to know whether we should conclude that, on average, men in the larger population weigh more than women in the population. Similarly, if we find a correlation of $r = .45$ between height and weight in our sample of 100 men, we might want to know whether this relationship between height and weight in our sample means there is probably a relationship between these two variables in the larger population of men. To answer these questions, we need to use standard errors.

In many inferential statistics formulas, we need to see whether the phenomenon we observed in our samples is large or small relative to our standard error. A standard error is a measure of the average amount of variance, or difference, we can expect from different samples of the same size selected from a population. So, the question we are asking with many inferential statistics is whether some statistic we see in our sample is big or small compared to the amount of variance (or error) we would expect if we had randomly selected a *different* sample of the same size. This question can be summarized with the following fraction:

$$\frac{size\ of\ sample\ statistic}{s\tan dard\ error}$$

As an illustration, let us return to the example comparing the average weight of men and women. We already know that, in our samples, the difference between the average weight of men and women was 60 pounds. The statistic that we am interested in here is the *difference* between the two means (i.e., the average weight of men and the average weight of women). If we were to select two different samples of the same size from the populations of men and women and find the difference in those two sample means, we would probably find a difference that was either larger or smaller than the difference we found in the comparison of the first two samples. If we kept selecting different samples and compared their means, we would eventually get a sampling distribution of the differences between the means, and this sampling distribution would have a standard error. Suppose that the standard error of this sampling distribution was 10. Let's plug that standard error into our fraction formula presented earlier:

$$\frac{sample\ statistic}{s\tan dard\ error} = \frac{60}{60}$$

The difference between our two sample means is six times larger than the difference we would expect to find just due to random sampling error. This suggests that the difference between our two sample means is probably not due to chance. Using a table of probabilities based on the *t* distribution, we can calculate the exact probability of getting a ratio this large (i.e., 60:10, or 6:1). So, to summarize, the standard error is often used in inferential statistics to see whether our sample statistic is larger or smaller than the average differences in the statistic we would expect to occur by chance due to differences between samples. We now discuss some examples to demonstrate the effect of sample size and the standard deviation on the size of the standard error of the mean.

Example: Sample Size and Standard Deviation Effects on The Standard Error

To illustrate the effect that sample size and standard deviation have on the size of the standard error of the mean, let's take a look at a variable from a set of data we collected a few years ago. The purpose of the study was to examine students' motivational beliefs about standardized achievement tests. We examined whether students thought it was important to do well on the standardized test they were about to take in school, whether they had anxiety about the test, whether they expected to do well on the test, whether they thought of themselves as good test takers, and so on.

One of the goals of the study was to compare the motivational beliefs of elementary school students with those of middle school students. The sample for the study included 137 fifth graders in elementary school with 536 seventh and eighth graders in middle school. Suppose we wanted to know the standard error of the mean on the variable "We expect to do well on the test" for each of the two groups in the study, the elementary school students and the middle school students. To calculate these standard errors, we would need to know the standard deviation for each group on our variable and the sample size for each group. These statistics are presented in Table.

Standard Deviation and Sample Action

	Standard Dev.	*Sample Size*	*Standard Dev.*	*Sample Size*
Expect to do with	1.34	1.37	1.46	5.36

A quick glance at the standard deviations for each group reveals that they are very similar (s = 1.38 for the elementary school sample, s = 1.46 for the middle school sample). However, because there is quite a large difference in the size of the two samples, we should expect somewhat different standard errors of the mean for each group. Which group do you think will have the larger standard error of the mean?

To find the standard error of the mean, we simply need to divide the standard deviation by the square root of the sample size. For the elementary school sample, we need to divide 1.38 by the square root of 137. The square root of 137 = 11.70. When we divide 1.38 by 11.70 we get .12. So the standard error of the mean for the elementary sample is .12. Following the same procedure for the middle school sample, we find that the standard error of the mean for this group will equal 1.46 divided by the square root of 546. The square root of 546 = 23.37. When we divide 1.46 by 23.37 we get .06.The standard error of the mean for the middle school sample (s.e. = .06) is half the size of the standard error of the mean for the elementary school sample (s.e. = .12). Because the standard deviations were roughly equal for these two groups, virtually all of the difference in their standard errors is attributable to differences in sample size.

To illustrate the effect of the standard deviation on the size of the standard error, let's take a look at a second variable from this study, students' scores on the verbal portion of the standardized achievement tests. Scores on this portion of the test range from a possible low of 0 to a possible high of 100. In the elementary school sample, the standard deviation on this variable was 23.81. The sample size is still 137. To find the standard error of the mean, we must divide 23.81 by the square root of 137 (which we know from our previous example is 11.70. And, 23.81 divided by 11.70 equals 2.04. So the standard error of the mean in this example is 2.04. When we compare this number with the standard error of the mean for the elementary school sample on the "Expect to do well on the test" variable (s_x = .12), we see that the larger standard deviation for the test score variable created a much larger standard error, even though the sample size remained the same, 137.

As these examples demonstrate, the size of the standard error of the mean depends on the size of the standard deviation and the size of the sample. As sample size increases, the standard error of the mean decreases. As the size of the standard deviation increases, the size of the standard error of the mean increases as well. Remember that the standard error is generally used in the denominator of the formulas statisticians use to calculate inferential statistics. Therefore, smaller standard errors will produce larger statistics (because smaller denominators produce larger overall numbers than larger denominators do when the numerators are equal). Larger statistics are more likely to be statistically significant. Therefore, all else being equal, larger sample sizes are more likely to produce statistically significant results because larger sample sizes produce smaller standard errors.

Standard error is often a difficult concept to grasp the first time it is encountered (or the second or the third). Because it is such a fundamental concept in inferential statistics, however, we encourage you to keep trying to make sense of both the meaning and the usefulness of standard errors. Standard errors can be used to determine probabilities of sample statistics (such as the mean) in much the same way that we used standard scores to determine probabilities associated with individual scores. Because of the usefulness of standard errors in determining probabilities, standard errors play a critical role in determining whether a statistic is statistically significant. Because standard errors are influenced by sample size, statistical significance will also be influenced by these sample characteristics. The issue of statistical significance, and the effects of sample size on statistical significance, are discussed in more depth.

Chapter 3

Linear Correlation and Regression

Correlation and regression refer to the relationship that exists between two variables, X and Y, in the case where each particular value of X_i is paired with one particular value of Y_i. For example: the measures of height for individual human subjects, paired with their corresponding measures of weight; the number of hours that individual students in a statistics course spend studying prior to an exam, paired with their corresponding measures of performance on the exam; the amount of class time that individual students in a statistics course spend snoozing and daydreaming prior to an exam, paired with their corresponding measures of performance on the exam; and so on.

Fundamentally, it is a variation on the theme of *quantitative functional relationship*. The *more* you have of this variable, the *more* you have of that one. Or conversely, the *more* you have of this variable, the *less* you have of that one. Thus: the more you have of height, the more you will tend to have of weight; the more that students study prior to a statistics exam, the more they will tend to do well on the exam.

Or conversely, the greater the amount of class time prior to the exam that students spend snoozing and daydreaming, the less they will tend to do well on the exam. In the first kind of case (the *more* of this, the *more* of that), you are speaking of a *positive* correlation between the two variables; and in the second kind (the *more* of this, the *less* of that), you are speaking of a *negative* correlation between the two variables.

Correlation and regression are two sides of the same coin. In the underlying logic, you can begin with either one and end up with the other. We will begin with correlation, since that is the part of the correlation-regression story with which you are probably already somewhat familiar.

CORRELATION

Here is an introductory example of correlation, taken from the realm of education and public affairs. If you are a college student in the United States, the chances are that you have a recent and perhaps painful acquaintance with an instrument known as the Scholastic

Achievement Test (SAT, formerly known as Scholastic Aptitude Test), annually administered by the College Entrance Examination Board, which purports to measure both academic achievement at the high school level and aptitude for pursuing further academic work at the college level. As those of you who have taken the SAT will remember very well, the letter informing you of the results of the test can occasion either great joy or great despair. What you probably did not realize at the time, however, is that the letter you received also contributed to the joy or despair of the commissioner of education of the state in which you happened that year to be residing.

This is because every year the College Entrance Examination Board publicly announces the state-by-state average scores on the SAT, and every year state education officials rejoice or squirm over the details of this announcement, according to whether their own state averages appear near the top of the list or near the bottom.

The presumption, of course, is that state-by-state differences in average SAT scores reflect underlying differences in the quality and effectiveness of state educational systems.

And sure enough, there are substantial state-by-state differences in average SAT scores, year after year after year. The differences could be illustrated with the SAT results for any particular year over the last two or three decades, since the general pattern is much the same from one year to another. We will illustrate the point with the results from the year 1993, because that was the year's sample of SAT data examined in an important research article on the subject.

Among the states near the top of the list in 1993 (verbal and math SAT averages combined) were Iowa, weighing in at 1103; North Dakota, at 1101; South Dakota, at 1060, and Kansas, at 1042. And down near the bottom were the oft-maligned "rust belt" states of the northeast: Connecticut, at 904; Massachusetts, at 903; New Jersey, at 892; and New York, more that 200 points below Iowa, at 887.

You can easily imagine the joy in DesMoines and Topeka that day, and the despair in Trenton and Albany. For surely the implication is clear: The state educational systems in Iowa, North Dakota, South Dakota, and Kansas must have been doing something right, while those in Connecticut, Massachusetts, New Jersey, and New York must have been doing something not so right.

Before you jump too readily to this conclusion, however, back up and look at the data from a different angle. When the College Entrance Examination Board announces the annual state-by-state averages on the SAT, it also lists the percentage of high school seniors within each state who took the SAT.

This latter listing is apparently offered only as background information–at any rate, it is passed over quickly in the announcement and receives scant coverage in the news media. Take a close look at it, however, and you will see that the background it provides is very interesting indeed. Here is the relevant information for 1993 for the eight states we have just mentioned. See if you detect a pattern.

State	*Percentage taking SAT*	*Average SAT score*
Iowa	5	1103
North Dakota	6	1101
South Dakota	6	1060
Kansas	9	1042
Connecticut	88	904
Massachusetts	81	903
New Jersey	76	892
New York	74	887

Mirabile dictu! The four states near the top of the list had quite small percentages of high school seniors taking the SAT, while the four states near the bottom had quite large numbers of high school seniors taking it. I think you will agree that this observation raises some interesting questions. For example: Could it be that the 5% of Iowa high school seniors who took the SAT in 1993 were the *top* 5%? What might have been the average SAT score for Connecticut if the test in that state had been taken only by the top 5% of high school seniors, rather than by (presumably) the "top" 88%? You can no doubt imagine any number of variations on this theme.

Figure shows the relationship between these two variables—percentage of high school seniors taking the SAT versus average state score on the SAT—for all 50 of the states. Within the context of correlation and regression, a two-variable coordinate plot of this general type is typically spoken of as a scatterplot or scattergram. Either way, it is simply a variation on the theme of Cartesian coordinate plotting that you have almost certainly already encountered in your prior educational experience. It is a standard method for graphically representing the relationship that exists between two variables, *X* and *Y*, in the case where each particular value of X_i is paired with one particular value of Y_i.

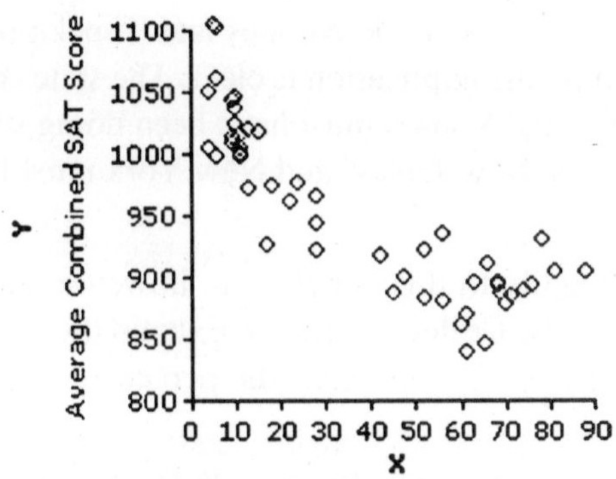

Fig. Percentage of High School Seniors Taking the SAT versus Average Combined State SAT Scores: 1993

For the present example, designating the percentage of high school seniors within a state taking the SAT as *X*, and the state's combined average SAT score as *Y*, we would have a total of N=50 paired values of X_i and Y_i. Thus for Iowa, X_i=5% would be paired with Y_i=1103; for

Massachusetts, X_i=81% would be paired with Y_i=903; and so on for all the other 50 states. The entire *bivariate* list would look like the following, except that the abstract designations for X_i and Y_i would of course be replaced by particular numerical values.

State	X_i *Percentage taking SAT*	Y_i *Average SAT score*
1_i	X_1	Y_1
2_i	X_2	Y_2
$::::_i$	$::::_i$	$::::_i$
49_i	X_{49}	Y_{49}
50_i	X_{50}	Y_{50}

The next step in bivariate coordinate plotting is to lay out two axes at right angles to each other. By convention, the horizontal axis is assigned to the *X* variable and the vertical axis to the *Y* variable, with values of *X* increasing from left to right and values of *Y* increasing from bottom to top.

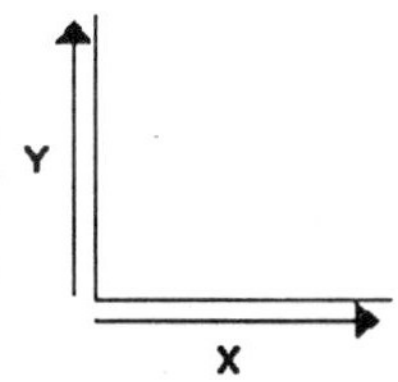

A further convention in bivariate coordinate plotting applies only to those cases where a causal relationship is known or hypothesized to exist between the two variables. In examining the relationship between two causally related variables, the independent variable is the one that is capable of *influencing* the other, and the dependent variable is the one that is capable of being *influenced* by the other. For example, growing taller will tend to make you grow heavier, whereas growing heavier will have no systematic effect on whether you grow taller. In the relationship between human height and weight, therefore, height is the independent variable and weight the dependent variable. The amount of time you spend studying before an exam can affect your subsequent performance on the exam, whereas your performance on the exam cannot retroactively affect the prior amount of time you spent studying for it. Hence, amount of study is the independent variable and performance on the exam is the dependent variable.

In the present SAT example, the percentage of high school seniors within a state who take the SAT can conceivably affect the state's average score on the SAT, whereas the state's average score in any given year cannot retroactively influence the percentage of high school seniors who took the test. Thus, the percentage of high school seniors taking the test is the independent variable, *X*, while the average state score is the dependent variable, *Y*. In cases of this type, the convention is to reserve the *X* axis for the independent variable and the *Y* axis for the dependent variable. For cases where the distinction between "independent" and "dependent" does not apply, it makes no difference which variable is called *X* and which is called *Y*.

In designing a coordinate plot of this type, it is not generally necessary to begin either the *X* or the *Y* axis at zero. The *X* axis can begin at or slightly below the lowest observed value of X_i, and the *Y* axis can begin at or slightly below the lowest observed value of Y_i.

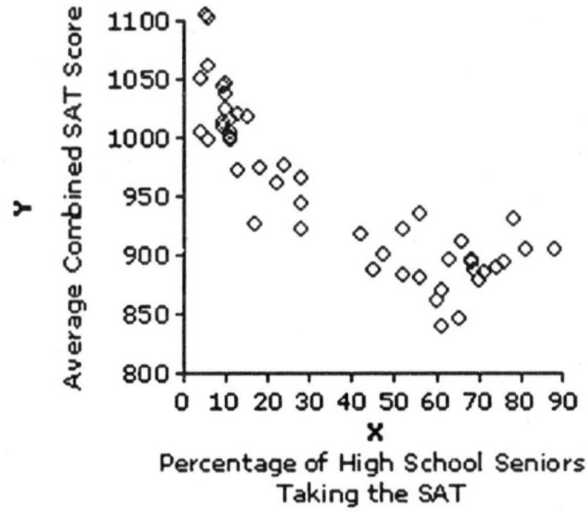

In Figure the X axis does begin at zero, because any value much larger than that would lop off the lower end of the distribution of X_i values; whereas the Y axis begins at 800, because the lowest observed value of Y_i is 838.

At any rate, the clear message of Figure is that states with relatively low percentages of high school seniors taking the SAT in 1993 tended to have relatively high average SAT scores, while those that had relatively high percentages of high school seniors taking the SAT tended to have relatively low average SAT scores. The relationship is not a perfect one, though it is nonetheless clearly visible to the naked eye. The following version of Figure will make it even more visible. It is the same as shown before, except that now we include the straight line that forms the best "fit" of this relationship.We will return to the meaning and derivation of this line a bit later.

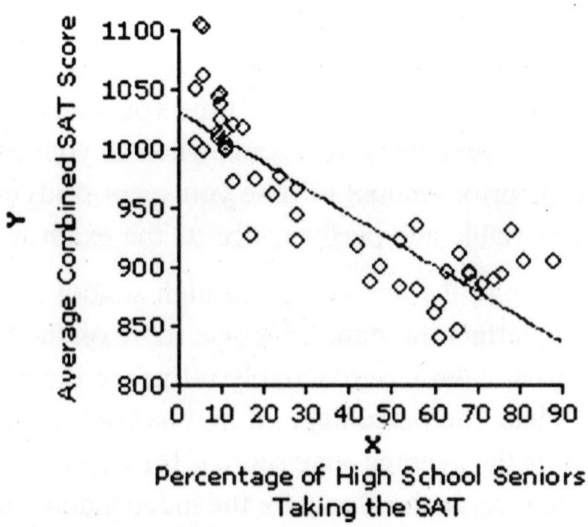

Actually, in this particular example there are two somewhat different patterns that the 50 state data points could be construed as fitting. The first is the pattern delineated by the solid downward slanting straight line, and the second is the one marked out by the dotted and mostly downward sloping curved line that you will see if you click the line labeled "Toggle!"

A relationship that can be described by a straight line is spoken of as *linear* (short for 'rectilinear'), while one that can be described by a curved line is spoken of as *curvilinear.* We will touch upon the subject of curvilinear correlation in a later chapter. Our present coverage will be confined to linear correlation.

Figure illustrates the various forms that linear correlation is capable of taking. The basic possibilities are:

1. Positive correlation;
2. Negative correlation;
3. Zero correlation. In the case of zero correlation, the coordinate plot will look something like the rather patternless jumble shown in Figure, reflecting the fact that there is no systematic tendency for X and Y to be associated, either the one way or the other. The plot for a positive correlation, on the other hand, will reflect the tendency for high values of X_i to be associated with high values of Y_i, and vice versa; hence, the data points will tend to line up along an upward slanting diagonal. The plot for negative correlation will reflect the opposite tendency for high values of X_i to be associated with low values of Y_i, and vice versa; hence, the data points will tend to line up along a downward slanting diagonal.

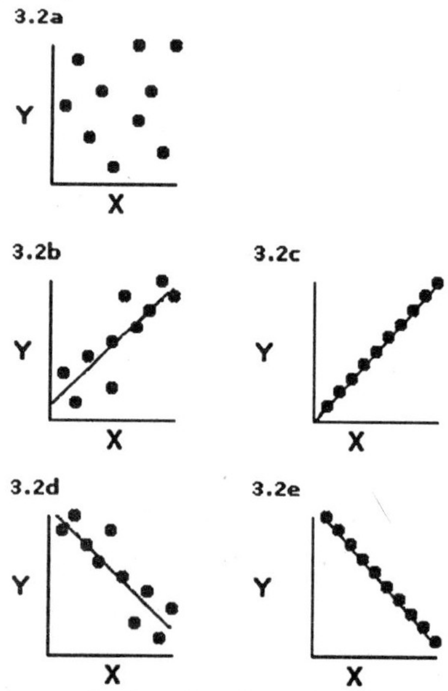

The limiting case of linear correlation, is when the data points line up along the diagonal like beads on a taut string. This arrangement, typically spoken of as *perfect* correlation, would represent the maximum degree of linear correlation, positive or negative, that could possibly exist between two variables. In the real world you will normally find perfect linear correlations only in the realm of basic physical principles; for example, the relationship between voltage and current in an electrical circuit with constant resistance. Among the less tidy phenomena of the behavioral and biological sciences, positive and negative linear correlations are much more likely to be of the "imperfect" types.

THE MEASUREMENT OF LINEAR CORRELATION

The primary measure of linear correlation is the Pearson product-moment correlation coefficient, symbolized by the lower-case Roman letter r, which ranges in value from $r=+1.0$ for a perfect positive correlation to $r=-1.0$ for a perfect negative correlation. The midpoint of its range, $r=0.0$, corresponds to a complete absence of correlation. Values falling between $r=0.0$ and $r=+1.0$ represent varying degrees of positive correlation, while those falling between $r=0.0$ and $r=-1.0$ represent varying degrees of negative correlation.

A closely related companion measure of linear correlation is the coefficient of determination, symbolized as r^2, which is simply the square of the correlation coefficient. The coefficient of determination can have only positive values ranging from $r^2=+1.0$ for a perfect correlation (positive or negative) down to $r^2 = 0.0$ for a complete absence of correlation. The advantage of the correlation coefficient, r, is that it can have either a positive or a negative sign and thus provide an indication of the positive or negative *direction* of the correlation. The advantage of the coefficient of determination, r^2, is that it provides an equal interval and ratio scale measure of the *strength* of the correlation. In effect, the correlation coefficient, r, gives you the true direction of the correlation (+ or –) but only the square root of the strength of the correlation; while the coefficient of determination, r^2, gives you the true strength of the correlation but without an indication its direction. Both of them together give you the whole works.

We will examine the details of calculation for these two measures in a moment, but first a bit more by way of introducing the general concepts. Figure shows four specific examples of r and r^2, each produced by taking two very simple sets of X and Y values, namely

$$X_i = \{1, 2, 3, 4, 5, 6\} \text{ and } Y_i = \{2, 4, 6, 8, 10, 12\}$$

and pairing them up in one or another of four different ways. In Example I they are paired in such a way as to produce a perfect positive correlation, resulting in a correlation coefficient of $r=+1.0$ and a coefficient of determination of $r^2=1.0$. In Example II the pairing produces a somewhat looser positive correlation that yields a correlation coefficient of $r=+0.66$ and a coefficient of determination of $r^2= 0.44$.

For purposes of interpretation, you can translate the coefficient of determination into terms of percentages (i.e., percentage=$r^2\times100$), which will then allow you to say such things as, for example, that the correlation in Example I ($r^2=1.0$) is 100% as strong as it possibly could be, given these particular values of X_i and Y_i, whereas the one in Example II ($r^2=0.44$) is only 44% as strong as it possibly could be.

Alternatively, you could say that the looser positive correlation of Example II is only 44% as strong as the perfect one shown in Example I. The essential meaning of "strength of correlation" in this context is that such-and-such percentage of the variability of Y is associated with (tied to, linked with, coupled with) variability in X, and vice versa. Thus, for Example I, 100% of the variability in Y is coupled with variability in X; whereas, in Example II, only 44% of the variability in Y is linked with variability in X.

FIGURE. FOUR DIFFERENT PAIRINGS OF THE SAME VALUES OF *X* AND *Y*

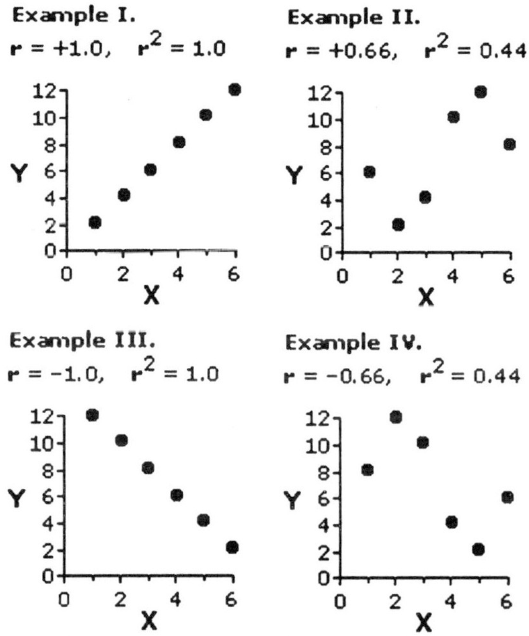

The correlations shown in Examples III and IV are obviously mirror images of the ones just described. For Example III the six values of X_i and Y_i are paired in such a way as to produce a perfect negative correlation, which yields a correlation coefficient of $r = -1.0$ and a coefficient of determination of r^2=1.0. In Example IV the pairing produces a looser negative correlation, resulting in a correlation coefficient of $r = -0.66$ and a coefficient of determination of r^2= 0.44. Here again you can say for Example III that 100% of the variability in *Y* is coupled with variability in *X*; whereas for Example IV only 44% of the variability in *Y* is linked with variability in *X*. You can also go further and say that the perfect positive and negative correlations in Examples I and III are of equal strength (for both, r^2=1.0) but in opposite directions; and similarly, that the looser positive and negative correlations in Examples II and IV are of equal strength (for both, r^2=0.44) but in opposite directions.

To illustrate the next point in closer detail, we will focus for a moment on the particular pairing of X_i and Y_i values that produced the positive correlation shown in Example II of Figure.

Pair	X_i	Y_i
a	1	6
b	2	2
c	3	4
d	4	10
e	5	12
f	6	8

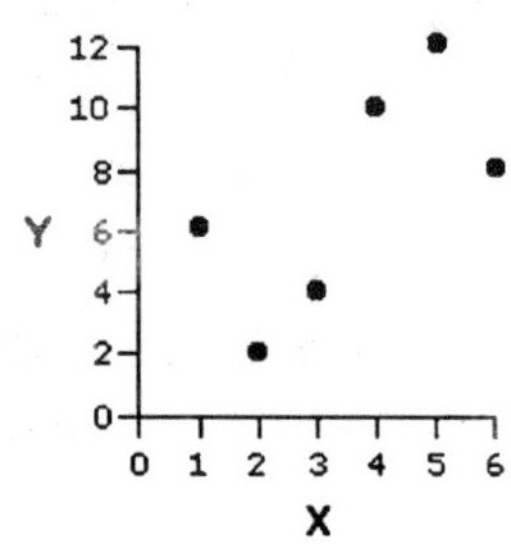

When you perform the computational procedures for linear correlation and regression, what you are essentially doing is defining the straight line that best fits the bivariate distribution of data points, as shown in the following version of the same graph. This line is spoken of as the regression line, or line of regression, and the criterion for "best fit" is that the sum of the squared vertical distances between the data points and the regression line must be as small as possible.

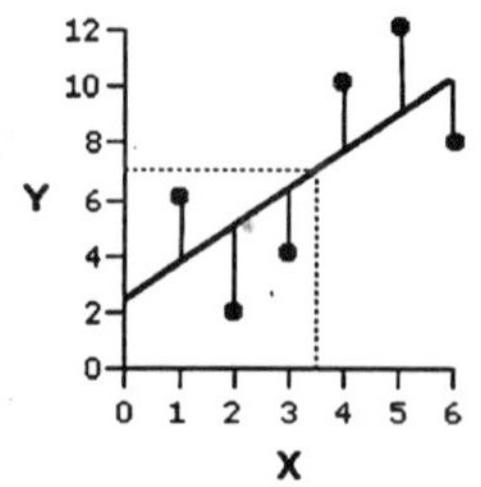

As it happens, this line of best fit will in every instance pass through the point at which the mean of X and the mean of Y intersect on the graph. In the present example, the mean of X is 3.5 and the mean of Y is 7.0. Their point of intersection occurs at the convergence of the two dotted gray lines.

The details of this line–in particular, where it begins on the Y axis and the rate at which it slants either upward or downward–will not be explicitly drawn out until we consider the regression side of correlation and regression. Nonetheless, they are implicitly present when you perform the computational procedures for the correlation side of the coin. As indicated above, the slant of the line upward or downward is what determines the sign of the correlation coefficient (r), positive or negative; and the degree to which the data points are lined up along the line, or scattered away from it, determines the strength of the correlation (r^2).

You have already encountered the general concept of *variance* for the case where you are describing the variation that exists among the variate instances of a single variable. The measurement of linear correlation requires an extension of this concept to the case where you are describing the co-variation that exists among the *paired* bivariate instances of two variables, X and Y, together. We have already touched upon the general concept. In positive correlation, high values of X tend to be associated with high values of Y, and low values of X tend to be associated with low values of Y. In negative correlation, it is the opposite: high values of X tend to be associated with low values of Y, and low values of X tend to be associated with high values of Y. In both cases, the phrase "tend to be associated" is another way of saying that the variability in X tends to be coupled with variability in Y, and vice versa–or, in brief, that X and Y tend to vary *together*. The raw measure of the tendency of two variables, X and Y, to co-vary is a quantity known as the covariance. As it happens, you will not need to be able to calculate the quantity of covariance in and of itself, because the destination we are aiming for, the calculation of r and r^2, can be reached by way of a simpler shortcut. However, you will need to have at least the general concept of it; so keep in mind as we proceed through the next few paragraphs that covariance is a measure of the degree to which two variables, X and Y, co-vary.

In its underlying logic, the Pearson product-moment correlation coefficient comes down to a simple ratio between

- The amount of covariation between X and Y that is actually observed,
- The amount of covariation that *would* exist if X and Y had a perfect (100%) positive correlation.

Thus

$$r = \frac{\text{observed covariance}}{\text{maximum possible positive covariance}}.$$

As it turns out, the quantity listed above as "maximum possible positive covariance" is precisely determined by the two separate variances of X and Y. This is for the simple reason that X and Y can co-vary, together, only in the degree that they vary separately. If either of the variables had zero variability (for example, if the values of X_i were all the same), then clearly they could not co-vary. Specifically, the maximum possible positive covariance that can exist between two variables is equal to the *geometric mean* of the two separate variances.

For any N numerical values, *a*, *b*, *c*, etc., the geometric mean is the nth root of the product of those values. Thus, the geometric mean of a and b would be the square root of axb; the geometric mean of *a*, *b* and *c* would be the cube root of axbxc; and so on.

So the structure of the relationship now comes down to

$$r = \frac{\text{observed covariance}}{\text{sqrt}[(\text{variance}X) \times (\text{variance}Y)]}.$$

Recall that "sqrt" means "the square root of."

Although in principle this relationship involves two variances and a covariance, in practice, through the magic of algebraic manipulation, it boils down to something that is computationally much simpler. In the following formulation you will immediately recognize the meaning of SS_X, which is the sum of squared deviates for X; by extension, you will also be able to recognize SS_Y, which is the sum of squared deviates for Y.

In order to get from the formula above to the one below, you will need to recall that the variance (s^2) of a set of values is simply the average of the squared deviates: SS/N.

The third item, SC_{XY}, denotes a quantity that we will speak of as the sum of co-deviates; and as you can no doubt surmise from the name, it is something very closely akin to a sum of squared deviates. SS_X is the raw measure of the variability among the values of X_i; SS_Y is the raw measure of the variability among the values of Y_i; and SC_{XY} is the raw measure of the co-variability of X and Y together.

$$r = \frac{SC_{XY}}{\text{sqrt}[SS_X \times SS_Y]}.$$

To understand this kinship,

For any particular item in a set of measures of the variable X,

$\text{deviate}_X = X_i - M_X$.

Similarly, for any particular item in a set of measures of the variable Y,

$\text{deviate}_Y = Y_i - M_Y$.

As you have probably already guessed, a co-deviate pertaining to a particular *pair* of XY values involves the deviate$_X$ of the X_i member of the pair and the deviate$_Y$ of the Y_i member of the pair. The specific way in which these two are joined to form the co-deviate is

$$\text{co-deviate}_{XY} = (\text{deviate}_X) \text{ X } (\text{deviate}_Y)$$

And finally, the analogy between a co-deviate and a squared deviate:

For a value of X_i, the squared deviate is

$$(\text{deviate}_X) \times (\text{deviate}_X)$$

For a value of Y_i it is $(\text{deviate}_Y) \times (\text{deviate}_Y)$

And for a pair of X_i and Y_i values, the co-deviate is

$$(\text{deviate}_X) \times (\text{deviate}_Y)$$

This should give you a sense of the underlying concepts. Just keep in mind, no matter what particular computational sequence you follow when you calculate the correlation coefficient, that what you are fundamentally calculating is the ratio

$$r = \frac{\text{observed covariance}}{\text{maximum possible positive covariance}}$$

which, for computational purposes, comes down to

$$r = \frac{SC_{XY}}{\text{sqrt}[SS_X \times SS_Y]}.$$

Now for the nuts-and-bolts of it. Here, once again, is the particular pairing of X_i and Y_i values that produced the positive correlation. But now we subject them to a bit of number-crunching, calculating the square of each value of X_i and Y_i, along with the cross-product of each X_iY_i pair. These are the items that will be required for the calculation of the three summary quantities in the above formula: SS_X, SS_Y, and SS_{XY}.

Pair	X_i	Y_i	X_i^2	Y_i^2	X_iY_i
a	1	6	1	36	6
b	2	2	4	4	4
c	3	4	9	16	12
d	4	10	16	100	40
e	5	12	25	144	60
f	6	8	36	64	48
sums	21	42	91	364	170

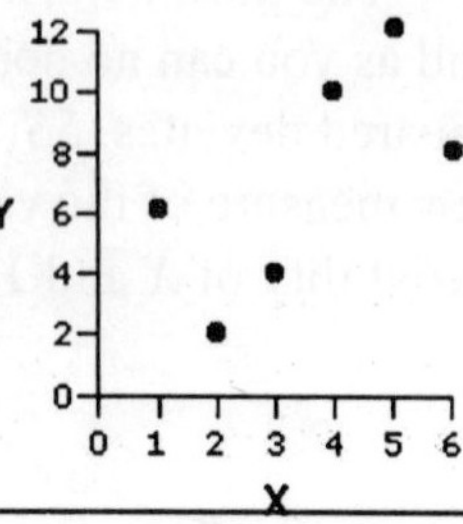

SS_X : sum of squared deviates for X_i values$_T$

The sum of squared deviates for a set of X_i values can be calculated according to the computational formula$_T$

$$SS_x = \Sigma X_i^2 - \frac{(\Sigma X_i)^2}{N}.$$

In the present example,

$$N = 6 \text{ [because there are 6 values of } X_i]$$
$$X_i^2 = 91$$
$$X_i - 21$$
$$(X_i)^2 = (21)^2 = 441$$

Thus:

$$SS_X = 91–(441/6) = 17.5$$

SS_Y: sum of squared deviates for Y_i values

Similarly, the sum of squared deviates for a set of Y_i values can be calculated according to the formula$_T$

$$SS_y = \Sigma Y_i^2 - \frac{(\Sigma Y_i)^2}{N}.$$

In the present example,$_T$

$$N = 6 \text{ [because there are 6 values of } Y_i]$$
$$Y_i^2 = 364$$
$$Y_i = 42$$
$$(Y_i)^2 = (42)^2 = 1764$$

Thus:

$$SS_Y = 364–(1764/6) = 70.0$$

SC_{XY}: Sum of co-deviates for paired values of X_i and Y_i

A moment ago we observed that the sum of co-deviates for paired values of X_i and Y_i is analogous to the sum of squared deviates for either of those variables separately. You will probably be able to see that this analogy also extends to the computational formula for the sum of co-deviates:

$$SC_{xy} = \Sigma(X_iY_i) - \frac{(\Sigma Y_i)(\Sigma Y_i)}{N}.$$

Again, for the present example,$_T$

$$N = 6 \text{ [because there are 6 } X_iY_i \text{ pairs]}$$
$$\Sigma X_i = 21$$
$$\Sigma Y_i = 42$$
$$(\Sigma X_i)(\Sigma Y_i) = 21 \times 42 = 882$$
$$\Sigma(X_iY_i) = 170$$

Thus:

$$SC_{XY} = 170–(882/6) = 23.0$$

Once you have these preliminaries,$_T$

$$SS_X = 17.5,\ SS_Y = 70.0, \text{ and } SC_{XY} = 23.0$$

you can then easily calculate the correlation coefficient as

$$r = \frac{SC_{XY}}{\text{sqrt}[SS_X \times SS_Y]}$$

$$= \frac{23.0}{\text{sqrt}[17.5 \times 70.0]} = +0.66$$

and the coefficient of determination as

$$r^2 = (+0.66)^2 = 0.44$$

To make sure you have a solid grasp of these matters, please take a moment to work your way through the details of Table, which will show the data and calculations for each of the examples of Figure. Recall that each example starts out with the same values of X_i and Y_i; they differ only with respect to how these values are paired up with one another.

The calculations we have just worked were performed on some very small and simple sets of data. Larger, more complex data sets will of course require something more laborious– but the general principles and specific computational procedures are precisely the same, either way. Consider, for example, the data set we referred to at the beginning of the chapter, pertaining to the correlation between the percentage of high school seniors taking the SAT versus average state score on the SAT.

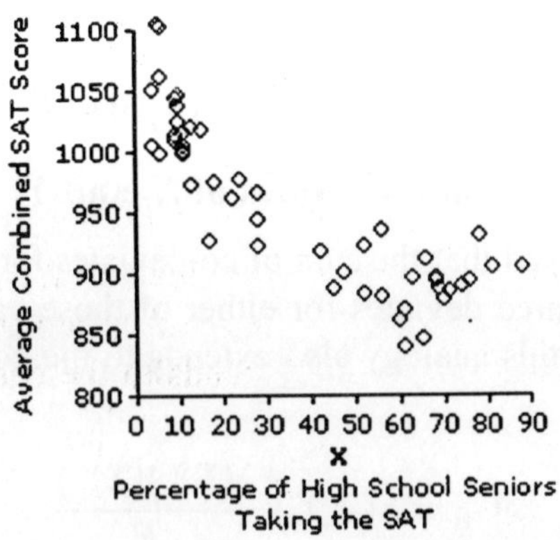

Table 3.2 shows the details of calculation for this data set. As you will see, it requires quite a large number of separate operations, many of which result in multi-digit numerical values. There was a time in the not too distant past when students of statistics had to perform calculations of this sort armed with nothing but paper, pencil, and patience, and it was a very laborious enterprise indeed. But that was then, and this is now. With a fairly inexpensive pocket calculator and a little practice in using it to its full advantage, you can perform complex calculations of this sort with a speed that would have made an earlier generation of statistics students weep with envy. With a computer spreadsheet, and again a little practice, you can do it even faster. With pre-packaged computer software, you can do it with as little time and effort as it takes you to enter the paired values of X_i and Y_i.

At any rate, once you perform the operations required to arrive at the following sums, all the rest is simple and straightforward:

Sums of				
X_i	Y_i	X_i^2	Y_i^2	X_iY_i
1,816	47,627	102,722	45,598,101	1,650,185

Given these results from the preliminary number-crunching, you could then easily calculate

$$SS_X = 36,764.88$$
$$SS_Y = 231,478.42$$
$$SC_{XY} = -79,672.64$$
$$r = -0.86$$
$$r^2 = 0.74$$

THE INTERPRETATION OF CORRELATION

The interpretation of an observed instance of correlation can take place at two quite distinct levels. The first of these involves a fairly conservative approach that emphasizes the observed fact of covariation and does not go very far beyond this fact. The second level of interpretation builds on the first, but then goes beyond it to consider whether the relationship between the two correlated variables is one of *cause and effect.* The latter is a potentially more fruitful approach to interpretation, but also a potentially more problematical one.

CORRELATION AS COVARIATION

When you find two variables to be correlated, the fundamental meaning of that fact is that the particular paired instances of X_i and Y_i that you have observed tend to *co-vary.* The positive or negative sign of r, the coefficient of correlation, indicates the direction of covariation, and the magnitude of r^2, the coefficient of determination, provides an equal interval and ratio measure of the degree of covariation. Thus, when we find a correlation coefficient of $r = -.86$ for our 1993 SAT data, the fundamental meaning of this numerical fact is that the particular paired instances of X_i and Y_i listed in Table 3.2 show some degree of covariation, and that the direction of that covariation is negative, or inverse. When we square r to get the coefficient of determination, $r^2 = 0.74$, the fundamental meaning of this numerical fact is that the degree of covariation is 74%. That is, 74% of the variance of the Y variable is coupled with variability in the X variable; similarly, 74% of the variance of the X variable is associated with variability in the Y variable. Conversely, you could say that 26% of the variance of the Y variable is *not* coupled with variability in the X variable, and similarly that 26% of the variance of the X variable is *not* associated with variability in the Y variable.

The basic concepts involved in this bare-bones covariation interpretation are illustrated in the following diagram. Each of the full circles represents 100% of the variance of either X or Y. In the case of zero correlation there is no tendency for X and Y to co-vary; and thus, as illustrated by the two separated circles at the top, there is zero overlap between the variability of X and the variability of Y. Any non-zero correlation (positive or negative) will reflect the fact that X and Y do tend to co-vary; and the greater the degree of covariation, as measured by r^2, the greater the overlap.

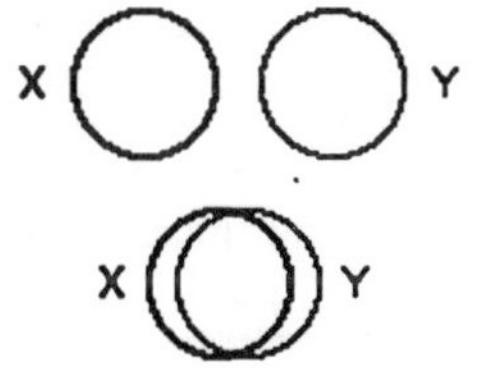

The bottom two circles illustrate the overlap for our observed SAT correlation of $r = -0.86$, and more generally for any instance of correlation where r is either -0.86 or $+0.86$. The area of overlap represents the fact that 74% of the variance of Y is coupled with variability in X, and vice versa; and the areas of non-overlap represent the fact that 26% of the variance of Y is unrelated to variability in X, just as 26% of the variance of X is unrelated to variability in Y. This non-overlapping portion of the variance of either X or Y is spoken of as the residual variance. In general, the proportion of the variance of either variable that is coupled with variability in the other is given by r^2, and the proportion of residual variance for either variable is given by $1-r^2$.

If we were examining correlation simply as a mathematical abstraction, this interpretation is all we would really need. Correlation is covariation, covariation is correlation, and all the rest is just a matter of filling in the details. The next level of interpretation ventures beyond the safe and tidy realm of mathematical abstraction and asks the question: What (if anything) does an observed correlation between two variables have to do with empirical reality? *Ex nihilo nihil fit.* Freely translated, it means that nothing comes from nowhere, so everything must come from somewhere. Granted that correlation is covariation. The question is, where does the covariation come from?

THE QUESTION OF CAUSE AND EFFECT

Correlation is a tool, and any tool, if misused, is capable of doing harm. Use a hammer the wrong way, and you will smash your thumb. Use correlation the wrong way, by jumping too quickly, glibly, and simple-mindedly to inferences about cause and effect, and you will arrive at conclusions that are false and misleading. The risk is so great that many statistics instructors and textbooks actively discourage students from even thinking about causal relationships in connection with correlation. Usually it takes the form of a caution: "You cannot infer a causal relationship solely on the basis of an observed correlation." Occasionally, it almost sounds like an eleventh commandment: "Thou shalt not infer causation on the basis of correlation!" The caution is correct. The commandment is much overstated.

If two variables are systematically related to each other as cause and effect, then variation in the cause will produce corresponding variation in the effect, and the two will accordingly show some degree of correlation. There is, for example, a fairly high positive correlation between human height and weight, and for obvious reasons. The taller a person is, the greater the basic mass of the body; and for those inclined to corpulence, the more room there is on the frame of a taller body for adding additional mass. In brief, height and weight are related to each other as cause and effect, and the correlation between the two variables *reflects* this causal relationship. Alternatively, you could say that the causal relationship between height and weight *produces* the observed correlation.

But the fact that causal relationships between variables can produce correlations does not entail that a causal relationship lies behind each and every instance of correlation. An observed correlation tells you nothing more than that the two variables co-vary. In some cases the covariation does reflect a causal relationship between the variables, and in other cases it does not. The trick is in determining which is which. An observed correlation between two variables does give you grounds for considering the *possibility* of a causal relationship—but that possibility must then be carefully and cautiously weighed in the balance of whatever other information you might have about the nature of the two variables.

Whenever you find two variables, X and Y, to be correlated, the basic possibilities concerning the question of cause and effect are the following:

Possibility 1. By the time you complete the chapters of this webtext, you will have raised the general question of Possibility 1 so often that it will seem as though you were born with it. When you sample the events of nature and observe a pattern, it may be that the pattern of the sample reflects a corresponding pattern in the entire population from which the sample is drawn. But then again, it may be that the pattern observed in the sample is only a fluke, the result of nothing more than mere chance coincidence. This, of course, is the general question of *statistical significance,* and once you get past the current chapter, there will be scarcely a page in this text that does not refer to it in one way or another. We will broach the question in somewhat greater detail toward the end of this chapter. Meanwhile, suffice it to say that before you even begin thinking about the issue of cause and effect, you first need to determine whether it is reasonable to assume that the observed correlation comes from anything other than mere chance coincidence. The remaining possibilities presuppose that this determination has been validly made in the affirmative.

Possibility 2. First, of course, is the possibility that there *is* a causal relationship between X and Y, either direct or indirect, such that variation in X produces variation in Y

$$X \cdots\cdots > Y$$

or alternatively, such that variation in Y produces variation in X

$$X < \cdots\cdots Y$$

In the latter case, you would do well to switch the X and Y labels of your variables; for as we noted earlier, the convention is to reserve "Y" for the dependent variable (the effect) and "X" for the independent variable (the cause).

At any rate, in seeking to determine whether an observed XY correlation betokens the existence of a causal relationship, the logical first step would be to rule out the possibility that it reflects something other than a causal relationship between X and Y. Remember, we are assuming here that the question of statistical significance has already been answered in the affirmative. The observed correlation is assumed to come from something more than mere chance coincidence; and if that something is not a causal relationship between *X* and *Y*, then what else could it possibly be?

Possibility 3. If you examine the records of the city of Copenhagen for the ten or twelve years following World War II, you will find a strong positive correlation between.

- The annual number of storks nesting in the city;
- The annual number of human babies born in the city. Jump too quickly to the assumption of a causal relationship, and you will find yourself saddled with the conclusion either that storks bring babies or that babies bring storks.

Or consider this one. If you examine the vital statistics of any country over a period of years, you will find a virtually perfect positive correlation between.

- The annual number of live male births;
- The annual number of live female births. Do baby boys bring baby girls, or is it the other way around?

In both of these examples what you have is a situation where two variables end up as correlated, not because one is influencing the other, but rather because both are influenced by a third variable, *Z*, that is not being taken into account. That is, the causal relationship here is not *X*····>*Y* or *X*<····*Y*, but rather

$$Z \cdots\begin{matrix} \cdots > X \\ \cdots > Y \end{matrix}.$$

For the male-female births example, the third variable is quite simply the overall annual birth rate. More babies are born in some years than in others. But no matter what the birth rate in any given year, the proportions of male and female births tend to remain fairly constant, with male births slightly outnumbering female births. (In the United States for recent decades it has been in the vicinity of 51.25% males and 48.75% females.) Thus, a relatively high birth rate will bring with it relatively high numbers of both male and female births, and a relatively low birth rate will bring relatively low numbers of both male and female births.

$$\text{total number of births} \cdots\begin{matrix} \cdots > \text{number of male births} \\ \cdots > \text{number of female births} \end{matrix}.$$

In short, the annual numbers of male and female births are correlated with each other only because they are both correlated with fluctuations in the annual birth rate.

The third variable for the correlation between storks and babies does not leap off the page quite so conspicuously, but it is there all the same. During the ten or twelve years following World War II, the populations of most western European cities steadily grew as a result of migrations from surrounding rural areas. There was also that spurt of fecundity known as the post-war baby boom. Here is how it worked out for the city of Copenhagen, which is also home to annually fluctuating numbers of storks. As population increased, there were more people to have babies, and therefore more babies were born. Also as population increased, there was more building construction to accommodate it, which in turn provided more nesting places for storks; hence increasing numbers of storks.

····>more buildings
····> ····>increasing numbers of storks
increasing population ···
····>more baby makers
····>····>increasing numbers of babies

Notice in this kind of situation that it makes no sense to speak of *X* as the independent variable and *Y* as the dependent variable. For in fact, *X* and *Y* are both independent of each other and dependent on variable *Z*.

INTERPRETING THE SAT CORRELATION

So what shall we make, in this context, of our observed correlation between

X = percentage of high school seniors within a state taking the SAT, and

Y = the state's average combined score on the SAT

which you will recall was measured as r=—0.86 and r^2=0.74? Question one: Is the observed correlation statistically significant–i.e., is it unlikely to have occurred by mere chance coincidence? For the moment I will have to ask you to take my word that it is. Through procedures that we will examine later, you will see that the mere-chance likelihood of finding a correlation this strong for a bivariate sample of size N=50 is very tiny indeed. Question two: Is there anything other than a straightforward *X*·····>*Y* causal relationship that could plausibly account for the observed correlation? Is it possible, for example, that *X* and *Y* are correlated with each other only because they are both being influenced by some third variable, *Z*? Alternatively, is it possible that *X* and *Y* are influencing each other reciprocally? I think you will agree that the possibility of reciprocal influence is unlikely, as it is difficult to imagine how a state's average score on the SAT in a given year could retroactively influence the percentage of high school seniors in the state who took the test. The possibility of a third variable, *Z*, cannot be quite so quickly ruled out, though it is not immediately obvious just what *Z* might be. Certainly there are other variables that play a rȯle in the situation, but that does not necessarily mean that they are influencing *X* and *Y* *separately,* according to the paradigm

····> *X* [I.e., *Z*····>*X* and *Z*····>*Y*
Z····
····> *Y* but not *X*····>*Y* or *Y*····>*X*]

For all of the possible candidates for *Z* that I can think of (economic factors, demographic factors, geographic factors, etc.), the scenario is one in which *Z* would influence *first X*, and *then Y* through the mediation of *X*, according to the paradigm

Z····>*X*····>*Y* [I.e., *Z*····>*X* and *X*····>*Y*]

Here is one fairly obvious example. Few if any high school seniors take the SAT for the sheer fun of it. Those who take it do so because they are applying to colleges that require the

SAT. In some states it is a smaller percentage who apply to such colleges, hence a smaller percentage who take the SAT; and in other states it is a larger percentage who apply to such colleges, hence a larger percentage who take the SAT. Z is the state percentage of seniors applying to colleges that require the SAT; X is the state percentage of seniors who take the SAT; and the positive correlation that we would surely find between these two variables, if we were to measure it, would clearly betoken a relationship of cause and effect (Z···>X).

In any event, from everything we know about the two primary variables, X and Y, in this situation, the possibility of a straightforward X···>Y causal relationship is an entirely plausible one. Imagine two states, A and B, whose respective percentages of high school seniors taking the SAT are A=5% and B=65%. Now the 5% of seniors in state A that happen to take the SAT may not represent precisely the top 5% in that state, but surely they are more likely to represent the top 10 or 15% than the top 60 or 70%. On the other hand, there is no way at all that the 65% who take the test in state B could come mostly from the top 10 or 15% of the high school seniors in that state, nor even from the top 25%. In fact, nearly a quarter of them could not possibly come even from the top 50%. So here is our presumed causal connection in a nutshell: A smaller percentage of students taking the SAT will tend to represent the upper portion of the scale of academic prowess, and will thus tend to produce a higher state average SAT score; whereas a larger percentage of students taking the test will tend to include not only students near the top of the scale, but also others not so near the top, and will thus tend to produce a lower state average SAT score.

Assuming that the X···>Y causal relationship is as I have described it, we could then link up again with the covariation interpretation of correlation and observe the following (recall that $r = -0.86$ and $r^2 = 0.74$). Of the total variability that exists among the 50 states with respect to average SAT scores, 74% is associated with variability in the percentages of high school seniors taking the SAT. In effect, 74% of the state-by-state differences in average SAT scores are *explained* by the fact that different states have different percentages of high school seniors taking the test. And that is 74% that does not need to be explained by presumed quality differences among state educational systems–unless you fancy the rather far-fetched notion that the most effective educational systems are those that produce the smallest percentages of students applying to colleges that require the SAT. Assuming that this latter notion is as preposterous as it seems, the maximum proportion of state-by-state variability in average SAT scores that could conceivably be due to differences among state educational systems is the 26% that is *not* explained by state-by-state differences in the percentage of seniors taking the SAT. And please note that this is only the maximum *possible* proportion. It could well turn out that measurable differences among state educational systems account for only a fraction of the 26% residual variance of Y, if indeed they account for any part of it at all.

REGRESSION

The appearance of the term *regression* at this point (literally, backward movement) is something of an historical accident. It could just as easily have been called *progression.* The basic concept is the same we found for correlation, though now it has added into it the visual imagery of movement– essentially, of two things, two variables, moving together. As indicated earlier, correlation and regression are two sides of the same statistical coin. When you measure the linear correlation of two variables, what you are in effect doing is laying out a straight line that best fits the average "together-movement" of these two variables. That line is spoken of as the line of regression, and its utility is not only as a device for helping us to visualize the relationship between the two variables. It can also serve very usefully as a basis for making rational *predictions.*

To illustrate, consider again our 1993 SAT correlation. Assuming that the negative correlation for that year is likely to occur in subsequent years, you are now in a position to predict a state's average SAT score for some subsequent year, before the results are reported, simply on the basis of knowing the percentage of students within the state who take the SAT that year.

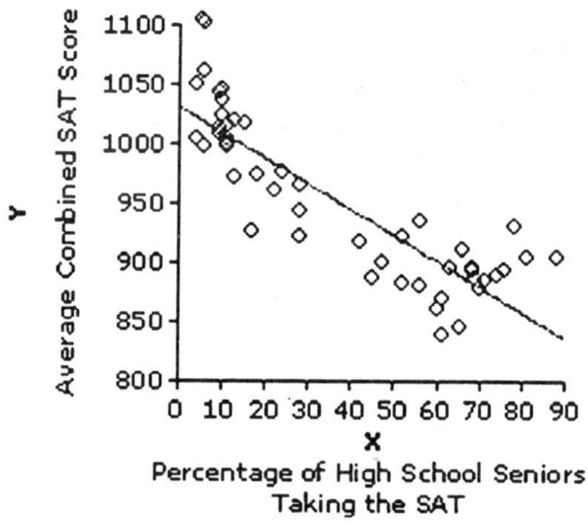

If 10% of the high school seniors within a state take the SAT, it is a fairly safe bet that the average combined SAT score for that state will be somewhere in the vicinity of 1,010–perhaps a bit higher or lower, but in any event somewhere in the vicinity. If 70% of the high school seniors in some other state take the SAT, it is a fairly safe bet that the average for that state will be nowhere near 1,010, but rather somewhere in the vicinity of 880. Regression analysis provides a rational foundation for making such predictions; it also provides a basis for specifying precisely what we mean by "somewhere in the vicinity."

As we noted earlier, when you perform the computational procedures for linear correlation and regression, what you are essentially doing is defining the straight line that best fits the bivariate distribution of data points. The criterion for "best fit" is that the sum of the squared vertical distances between the data points and the regression line must be as small as possible. The slant of the resulting line will correspond to the direction of correlation (upward, +; downward, –); and the tightness of the data points around the line will correspond to the strength of the correlation. You can think of the regression line as representing the average relationship that exists between *X* and *Y*, as observed within this particular sample.

The location and orientation of the regression line are defined by two quantities, spoken of as regression constants, that can be easily derived from the results of calculations already performed in Table. These are

a = the point at which the line crosses the Y axis (the 'intercept'); and

b = the rate at which the line angles upward or downward along the X axis (the 'slope').

The computational formulas for these two quantities are quite simple and can be introduced without elaborate comment:

For the slope:

$$b = \frac{SC_{XY}}{SS_X}$$

and for the intercept:

$$a = M_Y - bM_X$$

Before we perform these calculations for the SAT data, I think it might be useful to illustrate the process with a simpler data set. For this purpose, consider yet again the pairing of X_i and Y_i values that produced the positive correlation shown in Example II of Figure.

Pair	X_i	Y_i
a	1	6
b	2	2
c	3	4
d	4	10
e	5	12
f	6	6
means	3.5	7.0

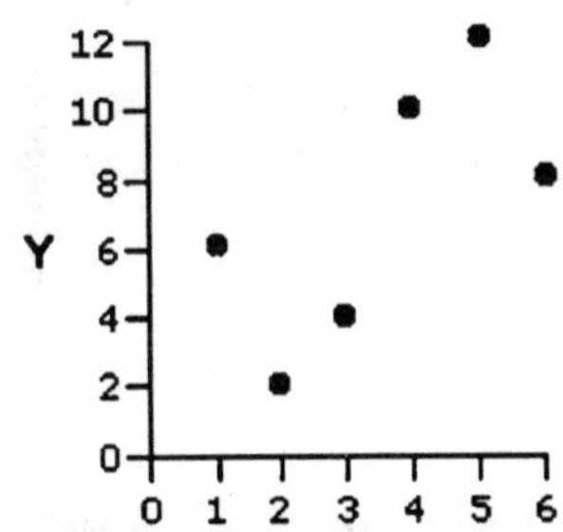

SS_X = 17.5
SS_Y = 70.0
SC_{XY} = 23.0

Given these previously calculated values:

Slope:

$$b = \frac{SC_{XY}}{SS_X} = \frac{23.0}{17.5} = +1.31$$

intercept:

$$a = M_Y - bM_X = 7.0 - [1.31(3.5)] = 2.4.$$

In the following graph I show the same figure that appears above, but now constructed in such a way as to emphasize the intercept and slope of the regression line. The intercept, shown on the left-hand side of the graph, is the point at which the regression line crosses the vertical Y axis–providing that the Y axis is lined up with the point on the horizontal axis where X is equal to zero. (Be careful with this, because scatter plots do not always begin the X axis at X=0.) The slope of the regression line is indicated by the green pattern in the graph that looks like a flight of stairs. What this pattern shows is that for each increase of one unit in the value of X, the value of Y increases by 1.31 units. Thus, when X is equal to zero, Y is equal to the intercept, which is 2.4; when X=1.0, Y is equal to the intercept plus 1.31 (i.e., 2.4+1.31=3.71); when X=2.0, Y is equal to the intercept plus 2.62 (i.e., 2.4+2.62=5.02); and so on.

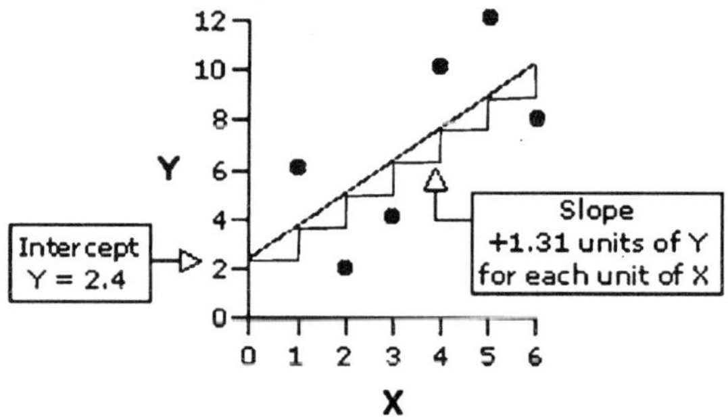

Now we perform the same calculations for the data set of our 1993 SAT correlation. In Table 3.2 we have already arrived at the summary values

mean of X = 36.32 X = percentage of high school seniors taking the SAT

mean of Y = 952.54 Y = state average combined SAT score

SS_X = 36,764.88

SS_Y = 231,478.42

SC_{XY} = – 79627.64

Given these values, the slope of the regression line can be calculated as

$$b = \frac{SC_{XY}}{SS_X} = \frac{-79627.64}{36{,}764.88} = -2.17$$

and the intercept as

$$a = M_Y - bM_X$$
$$= 952.54 - [-2.17(36.32)] = 1031.35.$$

For this data set, the regression line intercepts the vertical axis at the point where Y is equal to 1031.35, and then slants downward (–) 2.17 units of Y for each unit of X. Thus, when X is equal to zero, Y is equal to 1031.35; when X=10, Y is equal to the intercept minus 2.17×10 (i.e., 1031.35–21.7=1009.65); when X=20, Y is equal to the intercept minus 2.17x20 (i.e., 1031.35–43.4=987.95); and so on.

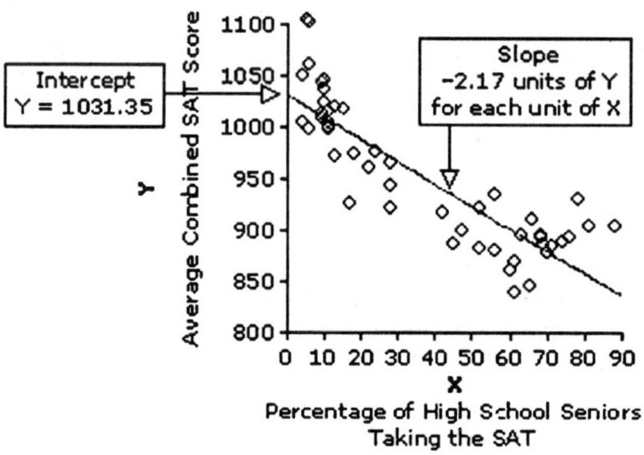

These are the mechanics of regression in a nutshell; now to the logic and strategy of prediction. If the observed correlation between two variables, *X* and *Y*, proves to be statistically significant—unlikely to have occurred through mere chance coincidence—the rational presumption is that it pertains not just to this particular sample of X_iY_i pairs, but to the relationship between *X* and *Y* in general. And once you know the relationship between *X* and *Y* in general, you are then in a position to figure out the value of Y_i that is likely to be associated with any particular newly observed value of X_i. The procedure for making such a prediction is illustrated pictorially below.

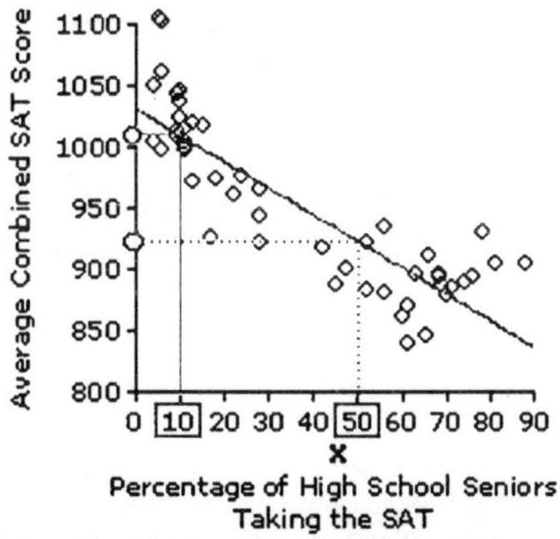

From the observed correlation in this 1993 sample, we infer that the general relationship between *X* and *Y* can be described by a regression line that has an intercept of a=1,031.35 and a slope of $b = -2.17$. Suppose, now, that for some subsequent year a certain state has $X_i = 10\%$ of its high school seniors taking the SAT. If you wanted to predict Y_i, the average SAT score for that state, the obvious way to proceed would be to start with the observed value of $X_i = 10\%$, go straight up to the line of regression, and then turn left to see where you end up on the *Y* axis. That will be your predicted value of Y_i, which as you can see from the graph is something quite close to *Y*=1,010. For $X_i = 50\%$, on the other hand, the predicted value is in the vicinity of $Y = 925$.

In practice, of course, the predicted values of Y_i are not arrived at graphically, but through calculation. For any particular observed linear correlation between two variables, *X* and *Y*, the value of Y_i to be predicted on the basis of a newly observed value of X_i is given by the following formula. Please note, however, that this version of the formula is only preliminary. There is something we will need to add to it a bit later

$$\text{predicted } Y_i = a + b\,X_i.$$

Try this formula out with a few different values of X_i and you will see that it is arriving mathematically, hence more precisely, at the same result that would be reached through the graphical method shown above. The formula does it by starting at a, the point at which the regression line intercepts the *Y* axis, and then moving up or down the *Y* axis (depending on the direction of the correlation) one unit of slope (*b*) for each unit of *X*

for $$X_i = 10\%$$

predicted $Y_i = 1{,}031.35 + (-2.17 \times 10)$
$= 1{,}009.65$

and for $X_i = 50\%$

predicted $Y_i = 1{,}031.35 + (-2.17 \times 50)$
$= 922.85.$

Now we are of course not claiming for either of these cases that the actual values of Y_i will fall *precisely* at the points we have calculated. All we can rationally assert is that actual values of Y_i for the case where $X_i = 10\%$ will tend to approximate the predicted regression-line value of 1,009.65; that actual values of Y_i for the case where $X_i = 50\%$ will tend to approximate the predicted regression-line value of 922.85; and so on for any other values of X_i that fall within the range of X_i values observed within the sample. It will probably be intuitively obvious to you that the strength of this "tendency to approximate" will be determined by the strength of the correlation observed within the original sample: The stronger the observed correlation, the more closely the actual values of Y_i will tend to approximate their predicted values; and conversely, the weaker the correlation, the greater will be the tendency of the actual values of Y_i to deviate from their predicted values. A moment ago I indicated that the formula for a predicted value of Y_i

$$\text{predicted } Y_i = a + bX_i$$

needs to have something added to it. What needs to be added is a measure of probable error, something that reflects the strength of the observed correlation, hence the strength of the tendency for actual values of Y_i to approximate their predicted values. Although the full conceptual background for this step will not be available until we have covered some basic concepts of probability, it is possible at this point to convey at least a practical working knowledge of it. Within the context of linear regression, the measure of probable error is a quantity spoken of as the standard error of estimate. Essentially, it is a kind of standard deviation. Here again is the scatter plot for the 1993 SAT correlation.

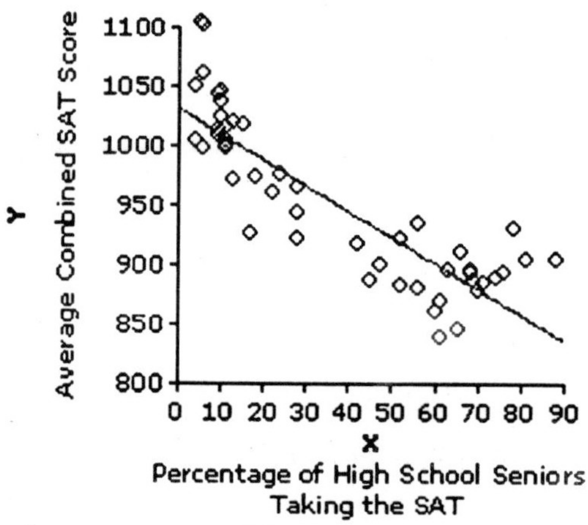

In your mind's eye, please try to envision a green line extending straight up or straight down from each of the blue data points to the red regression line. Each of these imaginary green lines is a measure of the degree to which the associated data point *deviates* (along the Y axis) from the regression line. Square each of these distances, then take the sum of those squares, and you will have a sum of squared deviates. In statistical parlance, each deviate (the imaginary green line) is spoken of as a residual, so the sum of their squares can be denoted as the sum of squared residuals, which we will abbreviate as $SS_{residual}$. At any rate, divide this sum of squared deviates (residuals) by N, and you will have a variance. Take the square root of that variance, and you will have a standard deviation.

As it happens, the sum of squared residuals can be arrived at mathematically through the simple formula

$SS_{\text{residual}} = SS_Y \times (1 - r^2)$ Recall that r^2 is the proportion of variability in Y that is associated with variability in X, and that $1-r^2$ is the proportion (residual) that is not associated with variability in X. Multiplying SS_Y by $1-r^2$ therefore gives you the amount of SS_Y that is residual, "left over," not accounted for by the correlation between X and Y.

For the 1993 SAT example, this yields

$$\begin{aligned} SS_{\text{residual}} &= 231{,}478.42 \times (1-0.86^2) \\ &= 60{,}184.38. \end{aligned}$$

Divide this quantity by N, and you will have the residual variance of Y:

$$60{,}184.38/50=1{,}203.69.$$

Take the square root of it, and you will have the standard deviation of the residuals:

$$\text{sqrt}[1{,}203.69] = \pm 34.69.$$

This standard deviation of the residuals is almost, but not quite, equivalent to the standard error of estimate. The difference is that the quantity we have just calculated is purely descriptive–it pertains only to this particular sample of paired X_iY_i values–whereas the standard error of estimate aims to reach beyond the sample into the realm of events as yet unobserved. This extension–from the particular sample of X_iY_i pairs to the relationship between X and Y in general—is achieved through the simple expedient of dividing $SS_{residual}$ by N–2 rather than by N. The rationale for this N–2 denominator will have to wait until a later chapter. For the moment, suffice it to say that the standard error of estimate, which we will abbreviate as SE, is given by the formula

$$SE = \text{sqrt}[(SS_{\text{residual}})/(N-2)].$$

For the present example, our standard error of estimate is therefore

$$SE = \text{sqrt}[60{,}184.38 / (50-2)] = \pm 35.41.$$

In brief: On the basis of what we have observed within our sample of X_iY_i pairs, we estimate that if the regression line of the sample were to be applied to the entire population of X_iY_i pairs, the Y residuals of the population would have a standard deviation somewhere very close to±35.41.

The next version of the SAT scatter plot shows how all of this applies to the task of prediction. A parallel line drawn 35.41 units of Y above the regression line will give you +1 standard error of estimate; one drawn 35.41 units of Y below the regression line will give — 1 standard error of estimate; and the inference is that the range between +1SE and –1SE will include approximately two-thirds of all the X_iY_i pairs within the population. Thus, when you predict an unknown value of Y_i according to the formula

$$\text{predicted } Y_i = a + bX_i$$

the true value of Y_i has about a two-thirds chance of falling within plus-or-minus 35.41 points of your predicted value, that is, within plus-or-minus 1 standard error of estimate. In making predictions of this type, the convention is to state the predicted value not simply as

$$\text{predicted } Y_i = a + b\,X_i$$

but rather as 'predicted Y' plus-or-minus 1 standard error of estimate. That is

$$\text{predicted } Y_i = a + b\,X_i \pm \text{SE}.$$

Thus, our predicted state average SAT scores for the cases where 10% and 50% of a state's high school seniors take the test are, in their full form

$$\text{for } X_i = 10\%$$

$$\text{predicted } Y_i = 1{,}031.35+(-2.17 \times 10)\pm 35.41$$

$$= 1{,}009.65\pm 35.41$$

$$\text{and for } X_i = 50\%$$

$$\text{predicted } Y_i = 1{,}031.35+(-2.17 \times 50)\pm 35.41$$

$$= 922.85\pm 35.41.$$

That is, for $X_i = 10\%$ we predict that the corresponding value of Y_i has a two-thirds chance of falling between Y=974.24 and Y=1,045.06; for $X_i = 50\%$, we predict that the corresponding value of Y_i has a two-thirds chance of falling between Y=887.44 and Y=958.26; and so on. Providing that the sample is adequately representative of the relationship between X and Y in general, we can expect approximately two-thirds of the entire 'population' of X_iY_i pairs to fall within the range defined by plus-or-minus 1 standard error of estimate, and only about one-third to fall outside that range. Hence, any particular prediction of the general form

$$\text{predicted } Y_i = a + b\,X_i \pm \text{SE}$$

will have about a two-thirds chance of catching the true value of Y_i in its net and only a one-third chance of missing it. Another way of expressing this concept is in terms of confidence. For a linear-regression prediction of this general form, you can be about two-thirds confident that the true value of Y_i will fall within ±1SE of the predicted value.

But the proof, as they say, is in the pudding. If you examine the SAT data for any testing year subsequent to 1993, you will find that about two-thirds of the actual values of Y_i do in fact fall within the range defined by the regression line of the 1993 sample, plus-or-minus 1SE. Hence any particular prediction of the form

$$\text{predicted } Y_i = a + b\,X_i \pm \text{SE}$$

would have had about a two-thirds chance of falling within the net.

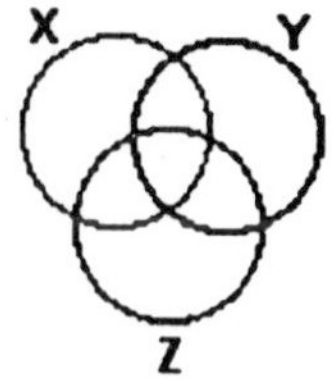

The first question to be asked of an observed correlation is whether it comes from anything other than mere chance coincidence. It is now time to take that question up in greater depth; however, as it is a question whose implications extend far beyond the confines of correlation and regression.$_T$

PARTIAL CORRELATION

Suppose you were to measure each of *N* subjects on each of three variables, *X*, *Y*, and *Z*, and find the following correlations:

X versus *Y*:	$r_{XY} = +.50$	$r^2_{XY} = .25$
X versus *Z*:	$r_{XZ} = +.50$	$r^2_{XZ} = .25$
Y versus *Z*:	$r_{YZ} = +.50$	$r^2_{YZ} = .25$

For the moment, focus on the value of r^2, which in each case (for this streamlined hypothetical example) is equal to .25. What this means is that for each pair of variables–*XY*, *XZ*, and *YZ*–the covariance, or variance overlap, is 25%. As illustrated in the following diagram, 25% of the variability of *X* overlaps with variability in *Y*; 25% of the variability of *X* overlaps with variability in *Z*; and 25% of the variability of *Y* also overlaps with variability in *Z*.

Note as well that there is one region where all three of the variability circles overlap. The meaning of this three-way overlap is that a certain amount of the correlation found between any two of the variables is tied in with the correlation that each of those two has with the third. Thus, of the 25% variance overlap found between *X* and *Y*, approximately half (judging by the naked eye) is tied in with the overlaps that exist between *XZ* and *YZ*. Similarly for the 25% overlap between *X* and *Z*, where about half is bound up with the overlaps for *XY* and *YZ*. And similarly as well for the 25% overlap of *YZ*, where about half is tied up with the overlaps for *XY* and *XZ*.

Partial correlation is a procedure that allows us to measure the region of three-way overlap precisely, and then to remove it from the picture in order to determine what the correlation between any two of the variables would be (hypothetically) if they were not each correlated with the third variable. Alternatively, you can say that partial correlation allows us to determine what the correlation between any two of the variables would be (hypothetically) if the third variable were held constant. The partial correlation of *X* and *Y*, with the effects of *Z* removed (or held constant), would be given by the formula

$$r_{XY\cdot Z} = \frac{r_{XY} - (r_{XZ})(r_{YZ})}{\text{sqrt}[1 - r^2_{XZ}] \times \text{sqrt}[1 - r^2_{YZ}]}$$

which for the present example would work out as

$$r_{XY \cdot Z} = \frac{50-(.50)(.50)}{\text{sqrt}[1-25] \times \text{sqrt}[1-.25]}$$

$$r_{XY \cdot Z} = +.33$$

Hence $$r^2_{XY \cdot Z} = .11.$$

The same general structure would apply for calculating the partial correlation of X and Z, with the effects of Y removed:

$$r_{YZ \cdot Y} = \frac{r_{XZ}-(r_{XY})(r_{YZ})}{\text{sqrt}[1-r^2_{XY}] \times \text{sqrt}[1-r^2_{YZ}]}$$

and for for calculating the partial correlation of Y and Z, with the effects of X removed:

$$r_{YZ \cdot X} = \frac{r_{YZ}-(r_{XY})(r_{XZ})}{\text{sqrt}[1-r^2_{XY}] \times \text{sqrt}[1-r^2_{XZ}]}.$$

Here is the apparatus of partial correlation applied to a real-life example. The Wechsler Adult Intelligence Scale (WAIS) is a device often used to measure "intelligence" beyond the years of childhood. Among its several sub-scales are three labeled as C, A, and V. The "C" stands for "comprehension," which chiefly reflects the test-taker's ability to comprehend the meanings and implications of written passages. The "A" refers to the test-taker's ability to perform tasks that require arithmetic ability. And the "V" stands for "vocabulary," which as you might imagine is a measure that increases or decreases in accordance with the breadth of the test-taker's vocabulary within the domain of the language in which the test is constructed. The following table shows the correlations typically found among these three sub-scales.

C versus A:	$r_{CA} = +.49$	$r^2_{CA} = .24$
C versus V:	$r_{CV} = +.73$	$r^2_{CV} = .53$
A versus V:	$r_{AV} = +.59$	$r^2_{AV} = .35$

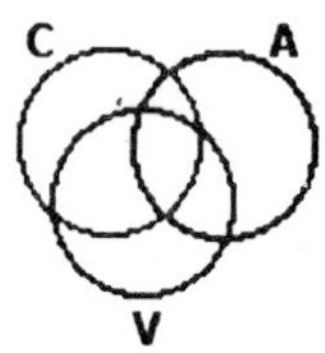

Here the overlaps are less evenly proportioned, although the logic is quite the same. Of the 24% variance overlap that occurs in the relationship between comprehension and arithmetic ability, a substantial portion reflects the fact that both of these variables are correlated with vocabulary. If we were to remove the effects of vocabulary from the relationship between C and A, the resulting partial correlation would be

$$r_{CA \cdot V} = \frac{r_{CA}-(r_{CV})(r_{AV})}{\text{sqrt}[1-r^2_{CV}] \times \text{sqrt}[1-r^2_{AV}]}$$

$$r_{CA \cdot V} = \frac{.49-(.73)(.59)}{\text{sqrt}[1-.53] \times \text{sqrt}[1-.35]}$$

$$r_{CA \cdot V} = \text{-+}.11$$

Hence $$r^2_{CA \cdot V} = .01.$$

In brief: with the effects of vocabulary removed, the correlation between comprehension and arithmetic ability collapses down to hardly anything at all. The practical inference is that if we were to administer the WAIS to a sample of subjects who were homogeneous with respect to breadth of vocabulary, the correlation between their scores on the comprehension and arithmetic sub-scales would prove fairly scant, on the order of r=+.11 and r^2 =.01.

In most cases a partial correlation of the general form $r_{XY \cdot Z}$ will turn out smaller than the original correlation r_{XY}. In those cases where it turns out larger, the third variable, Z, is typically spoken of as a supressor variable on the assumption that it is suppressing the larger correlation that would appear between X and Y if Z were held constant.

Suppose, for example, that a rather cranky professor has just administered an exam in his statistics course, and that for each student in the course we have measures on each of the following three variables:

X = the amount of effort spent on studying for the exam beforehand

Y = the student's score on the exam

Z = a measure of the degree to which the professor inspires fear and trembling in the student

And here are the correlations among the three variables:

X versus Y:	$r_{XY} = +.20$	$r^2_{XY} = .04$
X versus Z:	$r_{XZ} = +.80$	$r^2_{XZ} = .64$
Y versus Z:	$r_{YZ} = -40$	$r^2_{YZ} = .16$

Now isn't it odd that the correlation between X and Y should end up as a scant r_{XY}= +.20 and r^2_{XY}=.04, indicating a mere 4% covariance between the degrees of effort that students put into to the exam and the scores that they receive on it? Examine the other two correlations, however, and you will see that it is not so odd after all. The greater the fear and trembling, the greater the effort that students tend to put into preparing for the exam; hence r_{XZ}= +.80 and r^2_{XZ}=.64. On the other hand, the greater the fear and trembling, the less well students tend to do on the exam, as witness r_{YZ}=–.40 and r^2_{YZ}=.16. Remove the supressing effects of fear and trembling from the equation,

$$r_{XY \cdot Z} = \frac{.20-(.80)(-.40)}{\text{sqrt}[1-.64] \times \text{sqrt}[1-.16]} \qquad \begin{matrix} r_{XY}=.20 & r^2_{XY}=.04 \\ r_{XY \cdot Z}=.95 & r^2_{XY \cdot Z}=.90 \end{matrix}$$

$$r_{XY \cdot Z} = +.95$$

and the correlation between effort and exam score goes from a scant r_{XY}= +.20 to an impressive $r_{XY \cdot Z}$ = +.95. Or alternatively: remove the fear and trembling, and the covariance between effort and exam score goes from a mere 4% to a very substantial 90% ($r^2_{XY \cdot Z}$=.90).

RANK-ORDER CORRELATION

The version of correlation examined in the main body applies to those cases where the values of X and of Y are both measured on an equal- interval scale. It is also possible to apply the apparatus of linear correlation to cases where X and Y are measured on a merely ordinal scale. When applied to ordinal data, the measure of correlation is spoken of as the Spearman rank–order correlation coefficient, typically symbolized as r_s.

Suppose, for example, that two experts, X and Y, were asked to rank N=8 items with respect to some dimension germane to their field of expertise (rank#1=highest, rank#8=lowest). To make it specific, you can imagine two physicians ranking 8 patients with respect to the severity of their disease; two psychotherapists ranking 8 patients with respect to the likelihood of improvement; two wine experts ranking 8 wines from best to worst; two statisticians ranking 8 statistical concepts with respect to their fundamental importance; or whatever else it might be that strikes your fancy.

As a token of my liberal-mindedness – for I am one of those benighted souls who find all wines to taste suspiciously like vinegar – I will use the image of the wine experts. The following table shows the rankings from 1 to 8, best to worst, of two experts, X and Y.

wine	X	Y
a	1	2
b	2	1
c	3	5
d	4	3
e	5	4
f	6	7
g	7	8
h	8	6

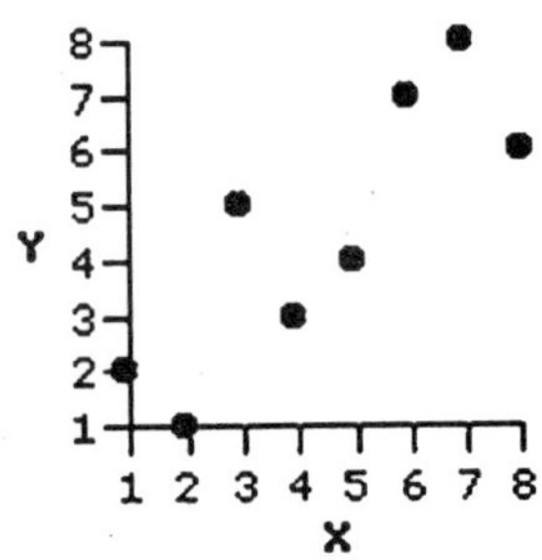

As you can see from the accompanying graph, there is a substantial degree of agreement between the rankings of the two experts. Plug the bivariate values of X and Y into the formulaic,

$$r = \frac{SC_{XY}}{\text{sqrt}\,[SS_X \times SS_Y]}$$

and you will find

$$r = +.83$$

$$r^2 = .69.$$

As it happens, these are exactly the same values you will get when you calculate the Spearman coefficient, r_s. The simple reason for this is that r and r_s are algebraically equivalent in the case where the values of X and Y consist of two sets of N rankings. The only advantage of r_s is that the calculations are easier if you are doing them by hand. [Note, however, that r_s is precisely equal to r only when the rankings within X and Y are the consecutive integer values: 1, 2, 3, and so on, with no ties. With tied ranks there will tend to be discrepancies between r_s and r. If the proportion of tied ranks is fairly large, you would be better advised to plug your rankings for X and Y into the standard formula for r.]

THE SIMPLE FORMULA FOR R_s, FOR RANKINGS WITHOUT TIES

wine	*X*	*Y*	*D*	D^2
a	1	2	–1	1
b	2	1	1	1
c	3	5	–2	4
d	4	3	1	1
e	5	4	1	1
f	6	7	–1	1
g	7	8	–1	1
h	8	6	2	4
N = 8			–ΣD² = 14	

Here is the same table you saw above, except now we also take the difference between each pair of ranks (D=*X*–*Y*), and then the square of each difference. All that is required for the calculation of the Spearman coefficient are the values of *N* and-D^2, according to the formula

$$r_s = 1 - \frac{6\Sigma D^2}{N(N^2 - 1)}.$$

If this formula seems a bit odd to you, you are in good company. Generations of statistics students have been presented with it, and generations have puzzled over such mind- bending questions as: why do you start out with "1" and subtract something from it?; where does that $N(N^2–1)$ in the denominator come from?; and, above all, how does that peculiar "6" get into the numerator?

Here are the answers to these age-old questions in a nutshell.

1. For any set of *N* paired bivariate ranks, the minimum possible value of ΣD^2 occurs in the case of perfect positive correlation. In this case, rank 1 for *X* is paired with rank 1 for *Y*, rank 2 for *X* with rank 2 for *Y*, and so on. Each value of *D* will accordingly be equal to zero, and so too will be the sum of the squared values of *D*.
2. Conversely, the maximum possible value of ΣD^2 occurs in the case of perfect negative correlation. This maximum possible value is in every instance equal to

$$\text{maximum-}\Sigma D2 = \frac{N(N^2 - 1)}{3}.$$

Thus, for *N*=8 with perfect negative correlation:$_T$

Item	*X*	*Y*	*D*	D^2	
a	1	8	–7	49	
b	2	7	–5	25	
c	3	6	–3	9	ΣD² = 168
d	4	5	–1	1	
e	5	4	1	1	8(8²—1)/3 = 168
f	6	3	3	9	
g	7	2	5	25	
h	8 .	1	7	49	

3. The ratio of the observed-D^2 to its maximum possible value will therefore be equal to zero in the case of perfect positive correlation, to +1.0 in the case of perfect negative correlation, and to +.50 in the case of zero correlation.

$$\frac{\Sigma D^2}{N(N^2-1)/3} = \frac{3\Sigma D^2}{N(N^2-1)}$$

Double this ratio, subtract it from 1, and voila! you have a quantity that will be equal to +1.0 in the case of perfect positive correlation, to –1.0 in the case of perfect negative correlation, and to zero in the case of zero correlation.

$$r_s = 1 - \frac{6\Sigma D^2}{N(N^2-1)}.$$

And here, finally, is the calculation of r_s for the example with which we began:

wine	*X*	*Y*	*D*	D^2	
a	1	2	–1	1	$r_s = 1 - \frac{6\Sigma D^2}{N(N^2-1)}$
b	2	1	1	1	
c	3	5	–2	4	
d	4	3	1	1	
e	5	4	1	1	$= 1 - \frac{6 \times 4}{8(8^2-1)}$
f	6	7	–1	1	
g	7	8	–1	1	
h	8	6	2	4	
$N = 8$	$\Sigma D^2 = 14$				$= +.83$
					$r^2_s = .69$

The meanings of r_s and r^2_s in a rank- order correlation are essentially the same as those of r and r^2 in a correlation based on equal- interval data. For the present example, r^2_s=.69 means that the covariance between the *X* and *Y* rankings is 69% as strong as it possibly could be, and the positive sign of r_s=+.83 signals that this covariation occurs along the upward slant, with higher values of *X* tending to be associated with higher values of *Y*, and vice versa. However, I would not recommend taking the parallels much farther than this. In particular, I think it would not make much sense to subject bivariate rankings to the predictive apparatus of linear regression.

CORRELATION AND REGRESSION

We have examined statistics and parameters that describe a single variable at a time, such as the mean, standard deviation, z scores, and standard errors. Although such single-variable statistics are important, researchers are often interested in examining the relations among two or more variables. One of the most basic measures of the association among variables, and a foundational statistic for several more complex statistics, is the correlation coefficient. Although there are a number of different types of correlation coefficients, the most commonly used in social science research is the Pearson product-moment correlation coefficient. Abrief description of three other types of correlations: the point-biserial coefficient, the Spearman rho coefficient, and the phi coefficient.

CORRELATION

Researchers compute correlation coefficients when they want to know how two variables are related to each other. For a Pearson product-moment correlation, both of the variables must be measured on an interval or ratio scale, otherwise known as continuous variables. For example, suppose we wanted to know whether there was a relationship between the amount of time students spend studying for an exam and their scores on the exam. We suspect that the more hours students spend studying, the higher their scores will be on the exam.

But we also suspect that there is not a perfect correspondence between time spent studying and test scores. Some students will probably get low scores on the exam even if they study for a long time, simply because they may have a hard time understanding the material. Indeed, there will probably be a number of students who spend an inordinately long period of time studying for the test precisely *because* they are having trouble understanding the material. On the other hand, there will probably be some students who do very well on the test without spending very much time studying. Despite these "exceptions" to our rule, We will still hypothesize that, *on average,* as the amount of time spent studying increases, so do students' scores on the exam.

There are two fundamental characteristics of correlation coefficients researchers care about. The first of these is the direction of the correlation coefficient. Correlation coefficients can be either positive or negative. A positive correlation indicates that the values on the two variables being analysed move in the same direction. That is, as scores on one variable go up, scores on the other variable go up as well (on average).

Similarly, on average, as scores on one variable go down, scores on the other variable go down. Returning to our earlier example, if there is a positive correlation between the amount of time students spend studying and their test scores, We can tell that, on average, the more time students spent studying, the higher their scores on the test. This is *equivalent* to saying that, on average, the *less* time they spent studying, the *lower* their scores on the test. Both of these represent a *positive* correlation between time spent studying and test scores.

A negative correlation indicates that the values on the two variables being analysed move in *opposite* directions. That is, as scores on one variable go up, scores on the other variable go down, and vice-versa (on average). If there were a negative correlation between the amount of time spent studying and test scores, we would know that, on average, the more time students spend studying for the exam, the *lower* they actually score on the exam. Similarly, with a negative correlation we would also conclude that, on average, the *less* time students spent studying, the *higher* their scores on the exam.

These positive and negative correlations are represented by scattergrams in Figure. Scattergrams are simply graphs that indicate the scores of each case in a sample simultaneously on two variables. For example, in the "Positive Correlation" scattergram in Figure, the first case in the sample studied for 1 hour and got a score of 30 on the exam. The second case studied for 2 hours and scored a 40 on the exam.

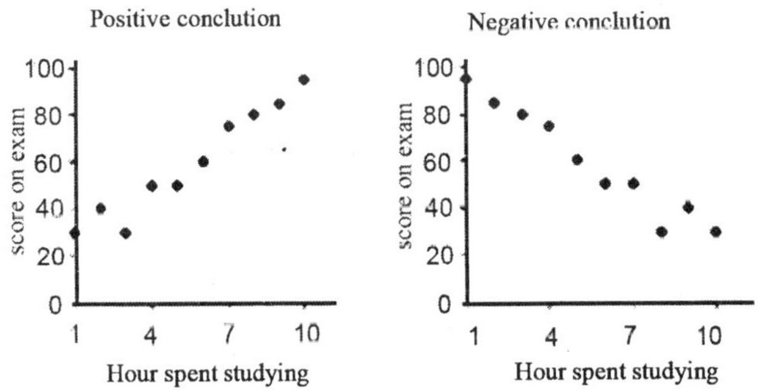

Fig. Examples of Positive and Negative Correlations

The second fundamental characteristic of correlation coefficients is the strength or magnitude of the relationship. Correlation coefficients range in strength from —1.00 to +1.00. A correlation coefficient of .00 indicates that there is no relationship between the two variables being examined. That is, scores on one of the variables are not related in any meaningful way to scores on the second variable. The closer the correlation coefficient is to either –1.00 or +1.00, the stronger the relationship is between the two variables. A perfect negative correlation of –1.00 indicates that for every member of the sample or population, a higher score on one variable is related to a lower score on the other variable. A perfect positive correlation of + 1.00 reveals that for every member of the sample or population, a higher score on one variable is related to higher score on the other variable.

Perfect correlations are never found in actual social science research. Generally, correlation coefficients stay between –.70 and +.70. Some textbook authors suggest that correlation coefficients between –.20 and +.20 indicate a weak relation between two variables, those between .20 and .50 (either positive or negative) represent a moderate relationship, and those larger than .50 (either positive or negative) represent a strong relationship. These general rules of thumb for judging the relevance of correlation coefficients must be taken with a grain of salt. For example, even a "small" correlation between alcohol consumption and liver disease (e.g., +.15) is important, whereas a strong correlation between how much children like vanilla and chocolate ice cream (e.g., +.70) may not be so important.

The scattergrams presented in Figure represent very strong positive and negative correlations (r=.97 and r = ".97 for the positive and negative correlations, respectively; r is the symbol for the sample Pearson correlation coefficient). In Figure, a scattergram representing virtually no correlation between the number of hours spent studying and the scores on the exam is presented. Notice that there is no discernible pattern between the scores on the two variables. In other words, the data presented in Figure reveal that it would be virtually impossible to predict an individual's test score simply by knowing how many hours the person studied for the exam.

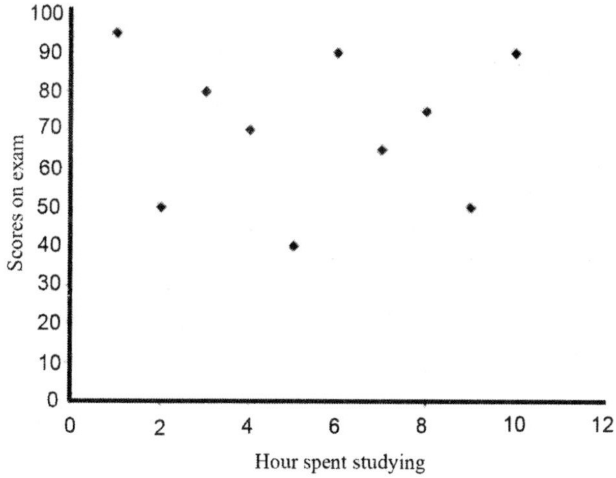

Fig. No Correlation Between Hours Spent Studying and Exam Scores

PEARSON'S CORRELATION

The correlation between two variables reflects the degree to which the variables are related. The most common measure of correlation is the Pearson Product Moment Correlation (called Pearson's correlation for short). When measured in a population the Pearson Product Moment correlation is designated by the Greek letter rho (ρ). When computed in a sample, it is designated by the letter "r" and is sometimes called "Pearson's r." Pearson's correlation reflects the degree of linear relationship between two variables. It ranges from +1 to –1. A correlation of +1 means that there is a perfect positive linear relationship between variables. The scatterplot shown on this page depicts such a relationship. It is a positive relationship because high scores on the X–axis are associated with high scores on the Y–axis.

A correlation of -1 means that there is a perfect negative linear relationship between variables. The scatterplot shown below depicts a negative relationship. It is a negative relationship because high scores on the X-axis are associated with low scores on the Y-axis.

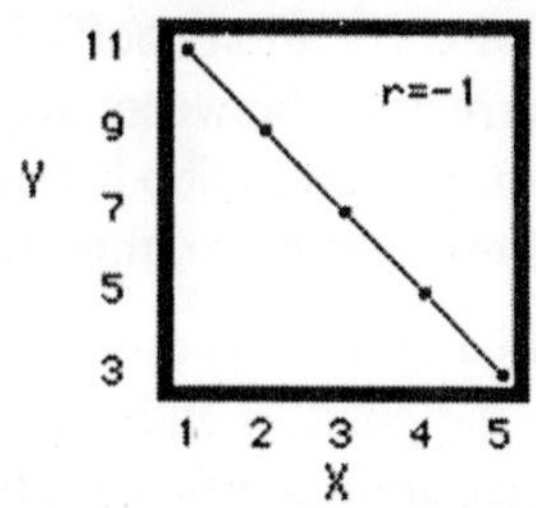

A correlation of 0 means there is no linear relationship between the two variables. The second graph shows a Pearson correlation of 0.

Correlations are rarely if ever 0, 1, or -1. Some real data showing a moderately high correlation are shown on the next page.

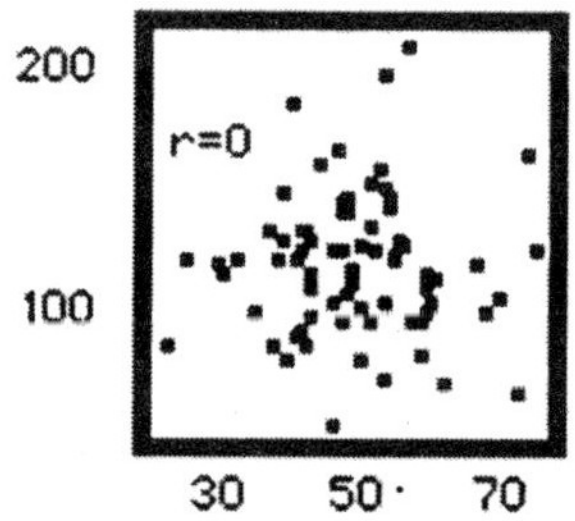

The scatterplot below shows arm strength as a function of grip strength for 147 people working in physically-demanding jobs. The plot reveals a strong positive relationship. The value of Pearson's correlation is 0.63.

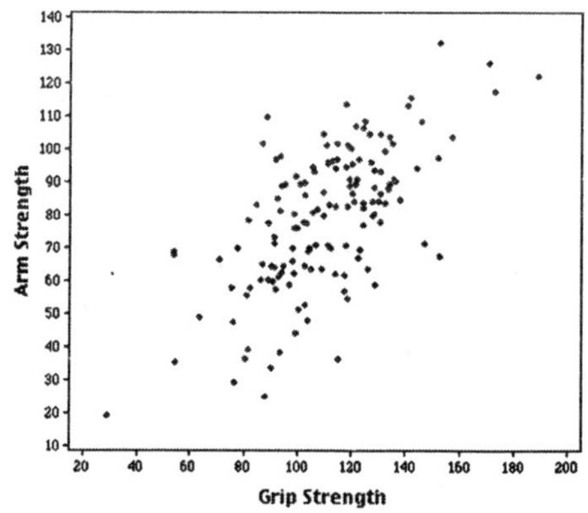

PEARSON CORRELATION COEFFICIENTS

The first step in understanding how Pearson correlation coefficients are calculated is to notice that we are concerned with a sample's scores on *two* variables at the same time. Returning to our previous example of study time and test scores, suppose that we randomly select a sample of five students and measure the time they spent studying for the exam and their exam scores. The data are presented in Table.

Data for Correlation Coefficient

	Hours spent Studying (X Variable)	*Exam Score (Y Variable)*
Student 1	5	80
Student 2	6	85
Student 3	7	70
Student 4	8	90
Student 5	9	85

For these data to be used in a correlation analysis, it is critical that the scores on the two variables *arepaired.* That is, for each student in our sample, the score on the x variable (hours spent studying) is paired with his or her own score on the *Y* variable (exam score). If we want to determine the relation between hours spent studying and exam scores, we cannot pair Student 1's hours spent studying with Student 4's test score. We must match each student's score on the x variable with his or her own score on the *Y* variable. Once we have done this, we can determine whether, on average, hours spent studying is related to exam scores.

Calculating the Correlation Coefficient

There are several different formulas that can be used to calculate Pearson product-moment correlation coefficients. Each of these formulas produces the same result and differ only in their ease of use. In fact, none of them are particularly easy to use. We present one of them here to illustrate the principles of correlation coefficients. Should you find yourself in need of knowing the relation between two variables, it is recommend to use a calculator or statistics programme that generates correlation coefficients for you.

The formula presented in Table requires that you *standardize* your variables. Remember that when you standardize a variable, you are simply subtracting the mean from each score in your sample and dividing by the standard deviation. What this does is provide a z score for each case in the sample. Those members of the sample with scores below the mean will have negative *z* scores, whereas those members of the sample with scores above the mean will have positive z scores.

Definitional formula for Pearson correlation

$$r = \frac{\Sigma\left(z_x z_y\right)}{N}$$

Wehre r = Pearson product moment correlation coefficient

z_x = a z score for variable X

z_y = a paired z score for variable Y

N = the number of pairs of X and Y scores

Notice that this formula looks similar to some of the other formulas that we have already encountered. For example, the denominator is *N*, which is the number of pairs of scores (i.e., the number of cases in the sample). Whenever we divide by *N*, we are finding an *average.* So we know that the correlation coefficient will be an *average* of some kind. But what is it an average of? Now take a look at the numerator. Here, we see that we must find the sum (E) of something. There, we had to find the sum of the squared deviations between each individual score and the mean. But in the formula for computing the correlation coefficient, we have to find the sum of the cross products between the z scores on each of the two variables being examined for each case in the sample. When we multiply each individual's score on one variable with that individual's score on the second variable (i.e., find a cross product), sum those across all of the individuals in the sample, and then divide by *N*, we have an average cross product, and this is known as covariance. If we standardize this covariance, we end up with a correlation coefficient. In the formula provided in Table, we simply standardized the variables before we computed the cross products, thereby producing a standardized covariance statistic, which is a correlation coefficient.

In this formula, notice what is happening. First, we are multiplying the paired z scores together. When we do this, notice that if an individual case in the sample has scores above the mean on each of the two variables being examined, the two z scores being multiplied will both be positive, and the resulting cross product will also be positive. Similarly, if an individual case has scores below the mean on each of the two variables, the two *z* scores being multiplied will both be negative, and the cross product will again be positive. Therefore, if we have a sample where low scores on one variable tend to be associated with low scores on the other variable, and high scores on one variable tend to be associated with high scores on the second variable, then when we add up to products from our multiplications we will end up with a *positive* number. This is how we get a positive correlation coefficient.

Now consider what happens when high scores on one variable are associated with low scores on the second variable. If an individual case in a sample has a score that is higher than the average on the first variable (i.e., a positive z score) and a score that is below the mean on the second variable (i.e., a negative z score), when these two z scores are multiplied together, they will produce a *negative* product. If, for most of the cases in the sample, high scores on one variable are associated with low scores on the second variable, the sum of the products of the z scores [E(zxZy)] will be *negative.* This is how we get a negative correlation coefficient.

WHAT THE CORRELATION COEFFICIENT DOES, AND DOES NOT, TELL US

Correlation coefficients such as the Pearson are very powerful statistics. They allow us to determine whether, on average, the values on one variable are *associated -with* the values on a second variable. This can be very useful information, but people, including social scientists, are often tempted to ascribe more meaning to correlation coefficients than they deserve. Namely, people often confuse the concepts of *correlation* and causation. Correlation (co-relation) simply means that variation in the scores on one variable *correspond* with variation in the scores on a second variable. Causation means that variation in the scores on one variable *cause* or *create* variation in the scores on a second variable.

When we make the leap from correlation to causation, we may be wrong. As an example, a short story is offered, which is heard in one of introductory psychology class. As the story goes, one winter shortly after World War II, there was an explosion in the number of storks nesting in some northern European country. Approximately 9 months later, there was a large jump in the number of babies that were born. Now, the link between storks and babies being what it is, many concluded that this correlation between the number of storks and the number of babies represented a causal relationship. Fortunately, science tells us that babies do not come from storks after all, at least not human babies. However, there is something that storks and babies have in common: Both can be "summoned" by cold temperatures and warm fireplaces. It seems that storks like to nest in warm chimneys during cold winters. As it happens, cold winter nights also foster baby-making behaviour. The apparent cause-and-effect relationship between storks and babies was in fact caused by a third variable: a cold winter.

For a more serious example, we can look at the relationship between SAT scores and first-year college grade point average. The correlation between these two variables is about .40. Although these two variables are moderately correlated, it would be difficult to argue that higher SAT scores *cause* higher achievement in the first year of college. Rather, there is probably some other variable, or set of variables, responsible for this relationship. For example, we know that taking a greater number of advanced math courses in high school is associated with higher SAT scores *and* with higher grades in first-year math courses in college.

The point of these examples is simple: Evidence of a relationship between two variables (i.e., a correlation) does not necessarily mean that there is a causal relationship between the two variables. However, it should also be noted that a correlation between two variables is a *necessary ingredient* of any argument that the two variables are causally related. In other words, we cannot claim that one variable causes another (e.g., that smoking causes cancer) if there is no correlation between smoking and cancer. If we do find a correlation between smoking and cancer, we must rule out other factors before we can conclude that it is smoking that causes cancer.

In addition to the correlation-causation issue, there are a few other important features of correlations worth noting. First, simple Pearson correlations are designed to examine *linear* relations among variables. In other words, they describe average *straight* relations among variables. For example, if you find a positive correlation between two variables, you can predict how much the scores in one variable will increase with each corresponding increase in the second variable. But not all relations between variables are linear. For example, there is a curvilinear relationship between anxiety and performance on a number of academic and nonacademic behaviours. When taking a math test, for example, a little bit of anxiety actually may help performance. However, once a student becomes too nervous, this anxiety can interfere with performance. We call this a curvilinear relationship because what began as a positive relationship between performance and anxiety at lower levels of anxiety becomes a negative relationship at higher levels of anxiety. This curvilinear relationship is presented graphically in Figure. Because correlation coefficients show the *average* relation between two variables, when the relationship between two variables is curvilinear, the correlation coefficient can be quite small, suggesting a weaker relationship than may actually exist.

	Hours spent Studying (X Variable)	*Exam Score (Y Variable)*
Student 1	0	95
Student 2	2	895
Student 3	4	100
Student 4	7	95
Student 5	10	100

Another common problem that arises when examining correlation coefficients is the problem of truncated range. This problem is encountered when the scores on one or both of the variables in the analysis do not have much variety in the distribution of scores, possibly due to a ceiling or floor effect. For example, suppose that we gave a sample of students a very easy test with a possible high score of 100. Then suppose we wanted to see if there was a correlation between scores on our test and how much time students spent studying for the test. Suppose we got the following data, presented in Table.

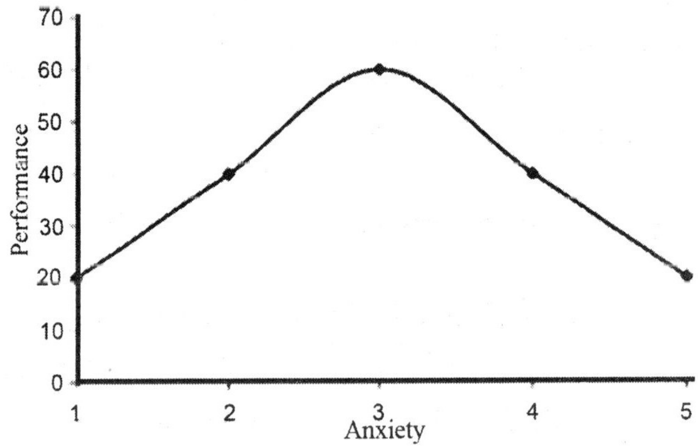

Fig. A curvilinear relationship.

In this example, all of our students did well on the test, whether they spent many hours studying for it or not. Because the test was too easy, a ceiling effect may have occurred, thereby truncating the range of scores on the exam. Although there may be a relationship between how much time students spend studying and their knowledge of the material, our test was not sensitive enough to reveal this relationship. The weak correlation that will be produced by the data in Table may not reflect the true relationship between how much students study and how much they learn.

STATISTICALLY SIGNIFICANT CORRELATIONS

When researchers calculate correlation coefficients, they often want to know whether a correlation found in sample data represents the existence of a relationship between two variables in the population from which the sample was selected. In other words, they want to test whether the correlation coefficient is statistically significant. To test whether a correlation coefficient is statistically significant, the researcher begins with the null hypothesis that there is absolutely no relationship between the two variables in the population, or that the correlation coefficient in the population equals zero. The alternative hypothesis is that there, in fact, is a statistical relationship between the two variables in the population, and that the population correlation coefficient is *not* equal to zero. So what we are testing here is whether our correlation coefficient is statistically significantly different from 0. These two competing hypothesis can be expressed with symbols:

$$H_{\circ} : \rho = 0$$

$$H_1 : \rho \neq 0$$

Where ρ is rho, the population correlation coefficient.

The *t* distribution is used to test whether a correlation coefficient is statistically significant. Therefore, we must conduct a *t* test. As with all *t* tests, the *t* test that we use for correlation coefficients involves a ratio, or fraction. The numerator of the fraction is the difference between two values. The denominator is the standard error. When we want to see whether a sample correlation coefficient is significantly different from zero, the numerator of the *t* test formula will be the sample correlation coefficient, *r*, minus the hypothesized value of the population correlation coefficient (p), which in our null hypothesis is zero. The denominator will be the standard error of the sample correlation coefficient:

$$t = \frac{r - \rho}{s_{r.}}$$

where ***r*** is the sample correlation coefficient,

ρ is the population correlation coefficient,

Fortunately, with the help of a little algebra, we do not need to calculate s_r to calculate the *t* value for correlation coefficients. First, for the sake of knowledge, let me present the formula for s_r:

$$s_r = \sqrt{(1 - r^2) \div (N - 2)}$$

where r^2 is the correlation coefficient squared and *N* is the.number of cases in the sample.

The formula for calculating the *t* value is

$$t = (r)\sqrt{\frac{N-2}{1-r^2}}$$

To illustrate this formula in action, let's consider an example. Some research suggests that there is a relationship between the number of hours of sunlight people are exposed to during the day and their mood. People living at extreme northern latitudes, for example, are exposed to very little sunlight in the depths of winter days and weeks without more than a few hours of sunlight per day. There is some evidence that such sunlight deprivation is related to feelings of depression and sadness. In fact, there is even a name for the condition: seasonal affective disorder, or SAD. To examine this relationship for ourself, 100 people is random selected from various regions of the world, measure the time from sunrise to sunset on a given day where each person lives, and get a measure of each person's mood on a scale from 1 to 10 (1="very sad, " 10="very happy"). Because the members of our sample live at various latitudes, the number of daylight hours will vary. If we conduct our study in January, those participants living in the north will have relatively short days whereas those living in the south will have long days.

Suppose that we were to compute a Pearson correlation coefficient with these data and found that the correlation between number of sunlight hours in the day and scores on the mood scale was r = .25. Is this a statistically significant correlation? To answer that question, we must find a t value associated with this correlation coefficient and determine that probability of obtaining a t value of this size by chance. In this example,

$$t = (.25)\sqrt{\frac{100-2}{1-.25^2}}$$

$$t = (.25)\sqrt{\frac{98}{1-.25^2}}$$

$$t = (.25)\sqrt{\frac{98}{1-.0625}}$$

$$t = (.25)\sqrt{\frac{98}{1-.9375}}$$

$$t = (.25)\sqrt{104.53}$$

$$t = (.25)\ 10.22$$

$$t = 1.56,\ df = 98$$

To see whether this t value is statistically significant. There we can see that, because our degrees of freedom equals 98, we must look at t values in both the $df = 60$ row and the $df = 120$ row. Looking at the $df = 60$ row, we can see that a t value of 2.56 has a probability of between .01 and .02 (for a two-tailed test). We get the same results when looking in the $df = 120$ row. Therefore, we conclude that our p value is between .01 and .02. If our alpha level is the traditional .05, we would conclude that our correlation coefficient is statistically significant. In other words, we would conclude that, on the basis of our sample statistic, in the larger population of adults the longer the daylight hours, the better their mood, on average. We could convey all of that information to the informed reader of statistics by writing "We found a significant relationship between number of daylight hours and mood (r = .25, $t_{(98)} = 2.56, p < .05$)."

This example also provides a good opportunity to once again remind you of the dangers of assuming that a correlation represents a causal relationship between two variables. Although it may well be that longer days cause the average adult to feel better, these data do not prove it. An alternative causal explanation for our results is that shorter days are also associated with *colder* days whereas longer days are generally associated with *warmer* days. It may be the case that *warmth* causes better moods whereas the lack of warmth causes depression and sadness. If people had warm, short days they might be just as happy as if they had warm, long days. So remember: Just because two variables are correlated, it does not mean that one causes the other.

THE COEFFICIENT OF DETERMINATION

Although correlation coefficients give an idea of the strength of the relationship between two variables, they often feel a bit nebulous. If you get a correlation coefficient of .40, is that a strong relationship? Fortunately, correlation coefficients can be used to obtain a more concrete-feeling statistic: the coefficient of determination. Even better, it is easy to calculate.

When we want to know if two variables are related to each other, we are really asking a somewhat more complex question: Are the variations in the scores on one variable somehow associated with the variations in the scores on a second variable? Put another way, a correlation coefficient tells us whether we can know anything about the scores on one variable if we already know the scores on a second variable. In common statistical vernacular, what we want to be able to do with a measure of association, like a correlation coefficient, is be able to *explain* some of the variance in the scores on one variable with the scores on a second variable. The coefficient of determination tells us how much of the variance in the scores of one variable can be understood, or explained, by the scores on a second variable.

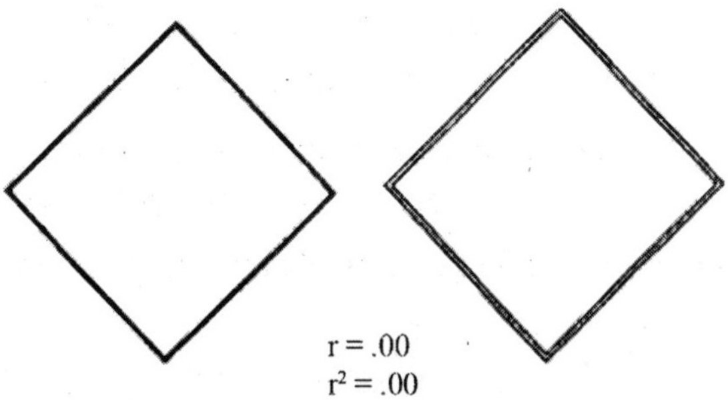

Fig. Uncorrelated Variables

One way to conceptualize explained variance is to understand that when two variables are correlated with each other, they *share* a certain percentage of their variance. Consider an example. If we have a sample of 10 people, and we measure their height and their weight, we've got 10 scores on each of two variables. Assuming that our 10 people differ in how tall they are, there will be some total amount of variance in their scores on the height variable. There will also be some total amount of variance in their scores on the weight variable, assuming that they do not all weigh the same amount. These total variances are depicted in Figure as two full squares, each representing 100% of the variance in their respective variables. Notice how they do not overlap.

When two variables are related, or correlated, with each other, there is a certain amount of shared variance between them. In Figure, the two squares are not touching each other, suggesting that all of the variance in each variable is independent of the other variable. There is no overlap. But when two variables are correlated, there is some *shared* variance. The stronger the correlation, the greater the amount of shared variance, and the more variance you can explain in one variable by knowing the scores on the second variable. The precise percentage of shared, or explained, variance can be determined by squaring the correlation coefficient. This squared correlation coefficient is known as the coefficient of determination. Some examples of different coefficients of determination are presented in Figure. The stronger the correlation, the greater the amount of shared variance, and the higher the coefficient of determination. It is still important to remember that even though the coefficient of determination is used to tell us how much of the variance in one variable can be explained by the variance in a second variable, coefficients of determination do not necessarily indicate a *causal* relationship between the two variables.

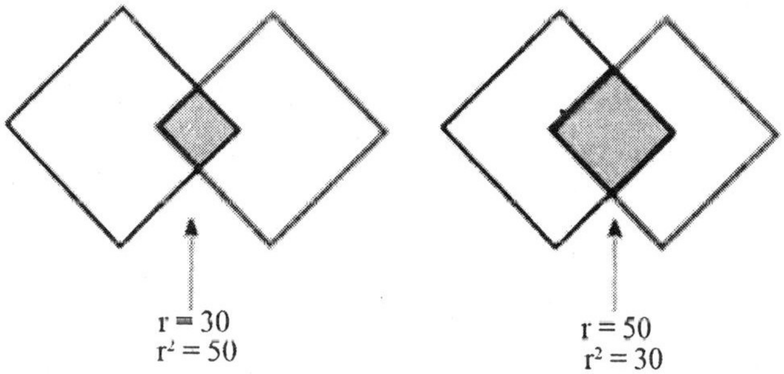

Fig. Examples of Different Coefficients of Determination

A BRIEF WORD ON OTHER TYPES OF CORRELATION COEFFICIENTS

Although Pearson correlation coefficients are probably the most commonly used and reported in the social sciences, they are limited by the requirement that both variables are measured on interval or ratio scales. Fortunately, there are methods available for calculating the strength of the relationship between two variables even if one or both variables are not measured using interval or ratio scales. In this section we briefly describe three of these "other" correlation coefficients. It is important to note that all of these statistics are very similar to the Pearson correlation coefficient and each produces a correlation coefficient that is similar to the Pearson *r*. They are simply specialized versions of the Pearson correlation coefficient that can be used when one or both of the variables are not measured using interval or ratio scales.

Point Biserial

When one of our variables is a continuous variable (i.e., measured on an interval or ratio scale) and the other is a two-level categorical (a.k.a. nominal) variable (also known as a dichotomous variable), we need to calculate a point-biserial correlation coefficient. This coefficient is a specialized version of the Pearson correlation coefficient discussed. For example, supposed we wanted to know whether there was a relationship between whether a person owns a car (yes or no) and their score on a written test of traffic rule knowledge, such as the tests one must pass to get a driver's license. In this example, we are examining the relation between one categorical variable with two categories (whether one owns a car) and one continuous variable (one's score on the driver's test). Therefore, the point-biserial correlation is the appropriate statistic in this instance.

Phi

Sometimes researchers want to know whether two dichotomous variables are correlated. In this case, we would calculate a phi coefficient (Φ), which is another specialized version of the Pearson *r*. For example, suppose we wanted to know whether gender (male, female) was associated with whether one smokes cigarettes or not (smoker, non smoker). In this case, with two dichotomous variables, we would calculate a phi coefficient. (Note: Those readers familiar with chi-square analysis will notice that two dichotomous variables can also be analysed using chi-square.)

Spearman Rho

Sometimes data are recorded as ranks. Because ranks are a form of ordinal data, and the other correlation coefficients discussed so far involve either continuous (interval, ratio) or dichotomous variables, we need a different type of statistic to calculate the correlation between two variables that use ranked data. The Spearman rho, a specialized form of the Pearson *r*, is appropriate. For example, many schools use students' grade point averages (a continuous scale) to rank students (an ordinal scale). In addition, students' scores on standardized achievement tests can be ranked. To see whether a students' rank in their school is related to their rank on the standardized test, a Spearman rho coefficient can be calculated.

EXAMPLE: THE CORRELATION BETWEEN GRADES AND TEST SCORES

Student achievement can be measured in a variety of ways. One common method of evaluating students is to assign them letter grades. These grades can be converted into numbers (e.g., A=4, B = 3, etc.). In high school, students grades across all of their classes (e.g., mathematics, science, social studies, etc.) can be combined into an average, thereby creating a grade point average (GPA), which is measured on a continuous, interval scale ranging from a possible low of 0 to a possible high of 4.33 (if the school gives grades of A+). Because grades are assigned by teachers, they are sometimes considered to be overly subjective. That is, different teachers may assign different grades to the same work. Similarly, there are some individual teachers that may give different grades to two students who produce the same quality of work. To ostensibly overcome such subjectivity, another form of assessment, the standardized test, was created. With this type of assessment, all students of a given grade level answer the same questions and their responses are scored by computer, thereby removing the human element and its subjectivity.

Some argue that standardized tests of ability and teacher-assigned grades generally measure the same thing. That is, for the most part, bright students will score well both on the test and in their grades. Others argue that standardized tests of ability and teacher-assigned grades really measure somewhat different things. Whereas standardized tests may measure how well students answer multiple-choice questions, teachers have the benefit of knowing students, and can take things like students' effort, creativity, and motivation into account when assigning grades. The first step in discovering which of these two viewpoints is more accurate is to see how strongly grades and test scores are related to each other. If there is a very strong correlation between the two, then both grades and test scores may in fact be measuring the same general trait. But if the two scores are only moderately correlated, perhaps they really do measure separate constructs. By constructs, we mean the actual "thing" that we are trying to measure. In the preceding example, if grades and test scores are strongly correlated, we could argue that both of these measures represent some underlying construct, such as "intelligence" or "academic ability." On the other hand, if these two variables are not strongly correlated, they each may represent different things, or constructs.

Recently data is collected from a sample of 314 eleventh-grade students at a high school in California. Among the data we collected were their cumulative GPA (i.e., their GPA accumulated from the time they began high school until the time the data were collected). In addition, we gave students the Naglieri Nonverbal Test of Abilities (NNTA; Naglieri, 1996), a nonverbal test of general mental reasoning and critical thinking skills. To see if there was a statistically significant correlation between these two measures of ability, we used the SPSS statistical software programme to calculate a correlation coefficient and *a p* value. The SPSS printout from this analysis is presented in Table.

The results presented in Table provide several pieces of information. First, there are three correlation coefficients presented. The correlations on the diagonal show the correlation between a single variable and itself. Therefore, the first correlation coefficient presented reveals that GPA is correlated with itself perfectly (r=1.0000). Because we always get a correlation of 1.00 when we correlate a variable with itself, these correlations presented on the diagonal are meaningless. That is why there is not a *p* value reported for them. The numbers in the parentheses, just below the correlation coefficients, report the sample size. There were 314 eleventh-grade students in this sample. The correlation coefficient that is *off the* diagonal is the one we're interested in. Here, we can see that students' GPA was moderately correlated with their scores on the Naglieri test $(r = .4291)$. This correlation is statistically significant, with *a p* value of less than .0001 $(p < .0001)$.

SPSS printout of correlation analysis.

	GPA	*Naglieri*
GPA	1.0000	
	(314)	
	p = .	
Naglieri	.4291	1.0000
	(314)	(314)
	p = .000	p = .

To gain a clearer understanding of the relationship between GPA and Naglieri test scores, we can calculate a coefficient of determination. We do this by squaring the correlation coefficient. When we square this correlation coefficient (.4291 × .4291=.1841), we see that GPA explains a little bit more than 18% of the variance in the Naglieri test scores. Although this is a substantial percentage, it still leaves more than 80% of the ability test scores unexplained. Because of this large percentage of unexplained variance, we must conclude that teacher-assigned grades reflect something substantially different from general mental reasoning abilities and critical thinking skills, as measured by the Naglieri test.

Correlation coefficients, in particular Pearson correlation coefficients, provide a way to determine both the direction and the strength of the relationship between two variables measured on a continuous scale. This index can provide evidence that two variables are related to each other, or that they are not, but does not, in and of itself, demonstrate a *causal* association between two variables. We also were introduced to the concepts of *explained* or *shared variance* and the coefficient of determination. Determining how much variance in one variable is shared with, or explained by, another variable is at the core of all of the statistics that are discussed. In particular, correlation coefficients are the precursors to the more sophisticated statistics involved in multiple regression. We examine *t* tests, which allow us to look at the association between a two-category independent variable and a continuous dependent variable.

REGRESSION

The concept of correlation was introduced. Correlation involves a measure of the degree to which two variables are related to each other. A closely related concept, coefficient of determination. This statistic provides a measure of the strength of the association between two variables in terms of percentage of variance explained. Both of these concepts are present in regression.

Regression is a very common statistic in the social sciences. One of the reasons it is such a popular technique is because it is so versatile. Regression, particularly multiple regression, allows researchers to examine how variables are related to each other, the strength of the relations, and relative predictive power of several independent variables on a dependent variable, and the unique contribution of one or more independent variables when controlling for one or more covariates. It is also possible to test for interactions in multiple regression. With all of the possible applications of multiple regression, it is clear that it is impossible to describe all of the functions of regression.

SIMPLE VERSUS MULTIPLE REGRESSION

The difference between simple and multiple regression is similar to the difference between one-way and factorial ANOVA. Like one-way ANOVA, simple regression analysis involves a single independent, or predictor variable and a single dependent, or outcome variable. This is the same number of variables used in a simple correlation analysis. The difference between a Pearson correlation coefficient and simple regression analysis is that whereas the correlation does not distinguish between independent and dependent variables, in a regression analysis there is always a designated predictor variable and a designated dependent variable. That is because the purpose of regression analysis is to make *predictions* about the value of the dependent variable given certain values of the predictor variable.

This is a simple extension of correlation analyses. If we are interested in the relationship between height and weight, for example, we could use simple regression analysis to answer this question: If we know a man's height, what would we *predict* his weight to be? Of course, the accuracy of our prediction will only be as good as our correlation will allow, with stronger correlations leading to more accurate predictions. Therefore, simple linear regression is not really a more powerful tool than simple correlation analysis. But it does give me another way of conceptualizing the relation between two variables, a point we elaborate on shortly. The real power of regression analysis can be found in multiple regression.

Like factorial ANOVA, multiple regression involves models that have two or more predictor variables and a single dependent variable. For example, suppose that, again, we are interested in predicting how much a person weighs (i.e., weight is the dependent variable). Now, suppose that in addition to height, we know how many minutes of exercise the person gets per day, and how many calories a day he consumes. Now I've got three predictor variables (height, exercise, and calories consumed) to help me make an educated guess about the person's weight. Multiple regression analysis allows me to see, among other things, (a) how much these three predictor variables, as a group, are related to weight, (b) the strength of the relationship between each predictor variable and the dependent variable *while controlling for the other predictor variables in the model,* (c) the *relative* strength of each predictor variable, and (d) whether there are interaction effects between the predictor variables. As you can see, multiple regression is a particularly versatile and powerful statistical technique.

VARIABLES USED IN REGRESSION

As with correlation analysis, in regression the dependent and independent variables need to be measured on an interval or ratio scale. Dichotomous (i.e., categorical variables with two categories) predictor variables can also be used. There is a special form of regression analysis, logit regression, that allows us to examine dichotomous dependent variables, but this type of regression. We limit our consideration of regression to those types that involve a continuous dependent variable and either continuous or dichotomous predictor variables.

REGRESSION IN DEPTH

Regression, particularly simple linear regression, is a statistical technique that is very closely related to correlations. In fact, when examining the relationship between two continuous (i.e., measured on an interval or ratio scale) variables, either a correlation coefficient or a regression equation can be used. Indeed, the Pearson correlation coefficient is nothing more than a simple linear regression coefficient that has been standardized.

The benefits of conducting a regression analysis rather than a correlation analysis are (a) regression analysis yields more information, particularly when conducted with one of the common statistical software packages, and (b) the regression equation allows us to think about the relation between the two variables of interest in a more intuitive way. Whereas the correlation coefficient provides us with a single number (e.g., $r = .40$), which we can then try to interpret, the regression analysis yields a formula for calculating the predicted value of one variable when we know the actual value of the second variable. Here's how it works.

The key to understanding regression is to understand the formula for the regression equation. So we begin by presenting the most simple form of the regression equation, describe how it works, and then move on to more complicated forms of the equation. In Table, the regression equation used to find the predicted value of *Y,* along with definitions of the components, are presented.

Table: The regression equation

$$\hat{Y} - bX + a$$

Where $\hat{y}$ is the predicted value of the $\hat{y}$ variable

b is the unstandardized regression coefficient, or the slope

a is intercepe (i.e., the point where regression line intercepes the $\hat{y}$ axis)

In simple linear regression, we begin with the assumption that the two variables are linearly related. In other words, if the two variables are actually related to each other, we assume that every time there is an increase of a given size in value on the *X* variable (called the predictor or independent variable), there is a corresponding increase (if there is a positive correlation) or decrease (if there is a negative correlation) *of a given size in* the *Y* variable (called the dependent, or outcome, or criterion variable). In other words, if the value of *X* increases from a value of 1 to a value of 2, and *Y* increases by 2 points, then when *X* increases from 2 to 3, we would predict that the value of *Y* would increase another 2 points.

To illustrate this point, let's consider the following set of data. Suppose we wanted to know whether there was a relationship between the amount of education people have and their monthly income. Education level is measured in years, beginning with kindergarten and extending through graduate school. Income is measured in thousands of dollars. Suppose that we randomly select a sample of 10 adults and measure then: level of education and their monthly income, getting the data provided in Table.

When we look at these data, we can see that, in general, monthly income increases as the level of education increases. This is a general, rather than an absolute, trend because in some cases a person with more years of education makes less money per month than someone with less education (e.g., Case 10 and Case 9, Case 6 and Case 5). So although not every person with more education makes more money, *on average* more years of education are associated with higher monthly incomes. The correlation coefficient that describes the relation of these two variables is r=.83, which is a very strong, positive correlation.

Table: Income and Education Level Data

	Education Level (X) *(In thousands)*	*Monthly income(Y)*
Case 1	6 years	1
Case 2	8 years	1.5
Case 3	11 years	1
Case 3	12 years	2
Case 4	12 years	4
Case 6	13 years	5
Case 7	14 years	5
Case 8	16 years	6
Case 9	16 years	10
Case 10	21 years	8
Mean	12.9	4.1
Standard Deviation	4.25	3.32

If we were to plot these data on a simple graph, we would produce a scatterplot, such as the one provided in Figure. In this scatterplot, there are 10 data points, one for each case in the study. Note that each data point marks the spot where education level (the *X* variable) and monthly income (the *Y* variable) meet for each case. For example, the point that has a value of 10 on the *Y* axis (income) and 16 on the *X axis* (education level) is the data point for the 10th case in our sample.

These 10 data points in our scatterplot reveal a fairly distinct trend. Notice that the points rise somewhat uniformly from the lower left corner of the graph to the upper right corner. This shape is a clear indicator of the positive relationship (i.e., correlation) between education level and income. If there had been a perfect correlation between these two variables (i.e., *r* " 1.0), the data points would be aligned in a perfectly straight line, rising from lower left to upper right on the graph. If the relationship between these two variables were weaker (e.g., *r*=.30), the data points would be more widely scattered, making the lower-left to upper-right trend much less clear.

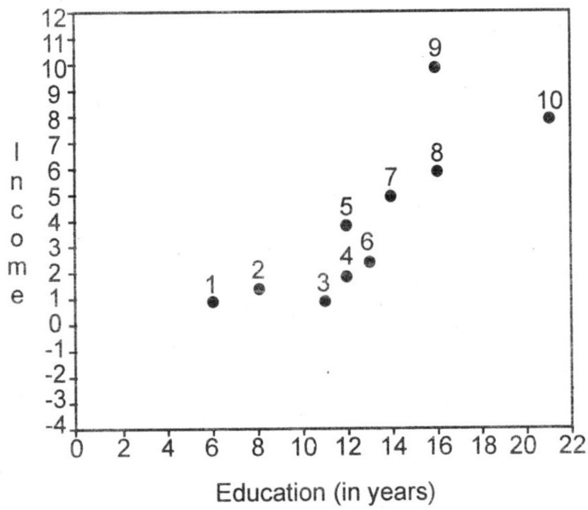

Fig. Scatterplot for Education and Income

With the data provided in Table, we can calculate all of the pieces of the regression equation. The regression equation allows us to do two things. First, it lets us find predicted values for the *Y* variable for any given value of the *X* variable. In other words, we can predict a person's monthly income if we know how many years of education he or she has. Second, the regression equation allows us to produce the regression line. The regression line is the basis for linear regression and can help us understand how regression works. There are a number of different types of regression formulas, but the most commonly used is called ordinary least squares regression, or OLS. OLS is based on an idea that we have seen before: the *sum of squares*. If you wanted to, you could draw a number of straight lines that bisect the data points presented in the scatterplot in Figure. For example, you could draw a horizontal line that extends out from the number 5 on the *Y* axis. Similarly, you could draw a straight line that extends down from the number 10 on the *Y* axis to the number 25 on the *X* axis.

No matter how you decided to draw your straight line, notice that at least some of the data points in the scatterplot will not fall exactly *on* the line. Some or all will fall above the line, some may fall directly on the line, and some or all will fall below the line. Any data point that does not fall directly on the line will have a certain amount of distance between the point and the line.

Now if you were to calculate the distance between the data point and the line you have drawn, and then square that distance, you would have a *squared deviation* for that point. If you calculated the squared deviation for each data point that did not fall on the line, and added all of these squared deviations together, you would end up with the *sum of squared deviations,* or *sum of squares*.

Now here is the key: The sum of the squared deviations, or sum of squares, will differ depending on where you draw your line. In any scatterplot, there is only *one* line the produces the *smallest* sum of squares. This line is known as the line of *least squares,* and this is the regression line. So, the reason this type of regression is called ordinary *least squares* regression is because in this type of regression, the regression line represents the straight line that produces the *smallest* sum of squared deviations from the line. This regression line represents the *predicted* values of *Y* at any given value of *X*. Of course, when we predict a value of *Y* for a given value of *X*, our prediction may be off.

This error in prediction is represented by the distance between the regression line and the actual data point(s) in the scatterplot. To illustrate how this works, we first need to calculate the properties of the regression line (i.e., its slope and intercept). Then, we draw this regression line into the scatterplot, and you can see how well it "fits" the data (i.e., how close the data points fall to the regression line).

If you take a look at the formula for the regression equation in Table, you will that there are four components: (a) *Y* is the predicted value of the 7 variable, (b) *b* is the unstandardized regression coefficient, and is also the slope of the regression line, (c) *X* is the value of the *X* variable, and (d) *a* is the value of the intercept (i.e., where the regression line crosses the *Y* axis). Because *Y* is the value produced by regression equation, let's save that one for last. And because *X* is just a given value on the *X* variable, there is not really anything to work out with that one. So let's take a closer look at *a* and *b*.

We cannot calculate the intercept before we know the slope of the regression line, so let's begin there. The formula for calculating the regression coefficient is

$$b = r \times \frac{S_y}{S_x}$$

where *b is the regression coefficient,*
r is the correlation between the X and Y variables,
S_y *is the standard deviation of the Y variable,*
S_x *is the standard deviation of the X variable,*

Looking at the data in Table, we can see that $r = .83$, $s_y = 3.12$, $s_x = 4.25$. When we plug these numbers into the formula, we get the following:

$$b = .83 \times \frac{3.12}{4.25}$$

$$b = (.83) \times (.73)$$

$$b = .61$$

Notice that the regression coefficient is simply the correlation coefficient times the ratio of the standard deviations for the two variables involved. When we multiply the correlation coefficient by this ratio of standard deviations, we are roughly transforming the correlation coefficient into the scales of measurement used for the two variables. Notice that there is a smaller range, or less variety, of scores on our *Y* variable that there is on our *X* variable in this example. This is reflected in the ratio of standard deviations used to calculate *b*.

Now that we've got our *b,* we can calculate our intercept, *a.* The formula for *a* is as follows:

$$a = \overline{Y} - b\overline{X}$$

where $\overline{Y}$ is the average value of *Y*, $\overline{X}$ is the average value of *X*, and *b* is the regression coefficient. When we plug in the values from Table, we find that

$$a = 4.1 - (.61)(12.9)$$

$$a = 4.1 - 7.87$$

$$a = -3.77$$

This value of *a* indicates that the intercept for the regression line is –3.77. In other words, the regression line crosses the *Y* axis at a value of –3.77. In still other words, this intercept tells us that when *X*=0, we wo uld predict the value of *Y* to be –3.77. Of course, in the real world, it is not possible to have a monthly income of negative 3.77 thousand dollars. Such unrealistic values remind us that we are dealing with *predicted* values of *Y*. Given our data, if a person has absolutely no formal education, we would *predict* that person to make a negative amount of money.

Now we can start to fill out our regression equation. The original formula

$$\hat{Y} = a + bX$$

now reads

$$\hat{Y} = -3.77 + .61X$$

It is important to remember that when we use the regression equation to find predicted values of *Y* for different values of *X*, we are not calculating the *actual* value of *Y*. We are only making *predictions* about the value of *Y*. Whenever we make predictions, we will sometimes be incorrect. Therefore, there is bound to be some error *(e)* in our predictions about the values of *Y* at given values of X. The stronger the relationship (i.e., correlation) between our *X* and 7 variables, the less error there will be in our predictions. The error is the difference between the actual, or observed, value of *Y* and the predicted value of *Y*. Because the predicted value of *Y* is simply $a + bX$, we can express the formula for the error in two ways:

$$e = Y - \hat{Y}$$

$$e = Y - a + bX$$

So rather than a single regression equation, there are actually two. One of them, the one presented in Table, is for the *predicted* value of *Y (Y)*. The other one is for the actual, or *observed* value of *Y*. This equation takes into account the errors in our predictions, and is written as $Y{=}bX{+}a + e$.

Now that we've got our regression equation, we can put it to use. First, let's wrap words around it, so that we can make sure we understand what it tells us. Our regression coefficient tells us that "For every unit of increase in *X*, there is a corresponding predicted increase of .61 units in *Y*" Applying this to our variables, we can say that "For every additional year of education, we would *predict* an increase of .61 ($1,000), or $610, in monthly income." We know that the predicted value of *Y* will *increase* when *X* increases, and vice-versa, because the regression coefficient is *positive*. Had it been negative, we would predict a decrease in *Y* when *X* increases.

Next, let's use our regression equation to find predicted values of *Y* at given values of x. For example, what would we predict the monthly income to be for a person with 9 years of formal education? To answer this question, we plug in the value of 9 for the *X* variable and solve the equation:

$$\hat{Y} = -3.77 + .61(9)$$

$$\hat{Y} = -3.77 + 5.59$$

$$\hat{Y} = 1.82$$

So we would predict that a person with 9 years of education would make $1,820 per month, plus or minus our error in prediction *(e)*.

Finally, we can use our regression equation to compute our regression line. We already know, from the value of the intercept, that our regression line will cross the *Y* axis at a value of "3.77. To draw a straight line, all we need to do is calculate one additional point. To make sure we include all of the points in our scatterplot, let's just calculate a predicted value of y for a person with 25 years of education.

$$\hat{Y} = -3.77 + .61(25)$$

$$\hat{Y} = -3.77 + 15.25$$

$$\hat{Y} = 11.48$$

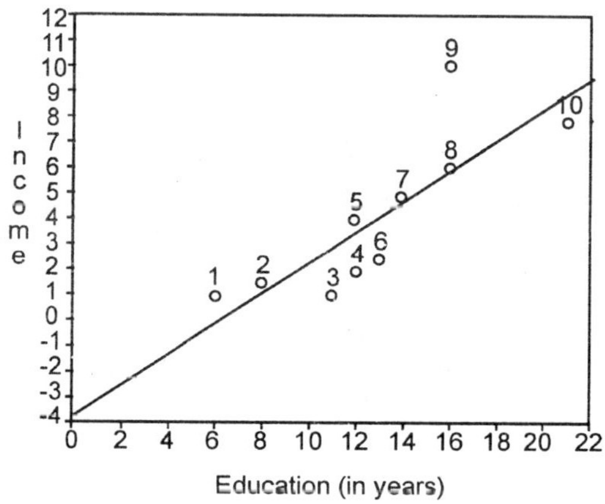

Fig. Scatterplot with Regression Line

If we were to draw a regression line through our scatterplot, using the two points we found from our intercept and the predicted *Y* value, we would get something like the line presented in Figure.

First, notice that our regression line does not accurately predict the actual y values for any of our cases except for Case 8. That data point is precisely on the regression line. For each of the other nine cases, there is some amount of error present in the prediction. In some cases, the amount of error is very little (e.g., Case 7), whereas in others the amount of error is quite large (e.g., Case 9). These errors in prediction are known as residuals. In some cases, our predicted value was greater than our observed value (e.g., Cases 1, 2, 5, 7, and 9).

For these cases, we have overpredicted their income based on their level of education. Such overpredictions produce *negative* residuals (because the residual=observed scores - predicted score). For other cases (Cases 3, 4, 6, and 10) we underpredicted the *Y* value, creating *positive* residuals. Second, notice the distance between each case and the line. When we square each of these distances, and then add them all together, we get the sum of squares. Third, notice that the regression line marks the line where the sum of the squared distances is smallest. To test this, try drawing some other lines and noting the way it increases the overall amount of error in prediction. Finally, notice where the regression line crosses the *Y* axis (the intercept) and the how much higher up the *Y* axis the regression line goes for each increase of one unit value in *X* (the slope). The slope and the intercept will correspond with the values that we found for *b* and a, respectively.

MULTIPLE REGRESSION

Now that we've discussed the elements of simple linear regression, let's move on to a consideration of multiple regression. Despite the impressive qualities of simple linear regression, the plain truth is that when we only have two variables, simple linear regression does not provide much more information than would a simple correlation coefficient. Because of this, you rarely see a simple linear regression with two variables reported in a published study. But multiple regression is a whole different story. Multiple regression is a very powerful statistic that can be used to provide a staggering array of useful information. At this point, it may be worth reminding you that in a short book like this, we only scratch the surface of what multiple regression can do and how it works. The interested reader should refer to one or all of the references listed at the end of this chapter to find more information on this powerful technique. To illustrate some of the benefits of multiple regression, let's add a second *predictor* variable to our example. So far, using the data from Table, we have examined the relationship between education level and income. In this example, education level has been used as *our predictor* or *independent* variable and income has been used as our *dependent* or *outcome* variable. We found that, on average, in our sample one's monthly salary is predicted to increase by \$610 for every additional year of schooling the individual has received.

But there was some error in our predictions, indicating that there are other variables that predict how much money one makes. One such predictor may be the length of time one has been out of school. Because people tend to make more money the longer they have been in the workforce, it stands to reason that those adults in our sample who finished school a long time ago may be making more than those who finished school more recently. Although Case 4 and Case 5 each had 12 years of schooling, Case 5 makes more money than Case 4. Perhaps this is due to Case 5 being in the workforce longer than Case 4.

When we add this second predictor variable to the model, we get the following regression equation:

$$\hat{Y} = a + bX_1 + bX_2$$

where $\hat{y}$ is the predicted value of the dependent variable, X_1 is the value of the first predictor variable, and X_2 is the value of the second predictor variable.

This regression equation with two predictor variables will allow me to examine a number of different questions. First, we can see whether our two predictor variables, combined, are significantly related to, or predictive of, our dependent variable, and how much of the variance our predictor variables explain in our dependent variable. Second, we can test whether each of our predictor variables is significantly related to our dependent variable *when controlling for the other predictor variable.* When we say "controlling for the other predictor variable" we mean that we can examine whether a predictor variable is related to the dependent variable after we partial out, or take away, the portion of the variance in our dependent variable that has already been accounted for by our other independent variable.

Third, we can see which of our two predictor variables is the stronger predictor of our dependent variable. Fourth, we can test whether one predictor variable is related to our dependent variable after controlling for the other predictor variable, thus conducting a sort of ANCOVA. There are many other things we can do with multiple regression, but we will limit our discussion to these four. (Note: Perhaps the most important omission from this discussion is the ability to examine statistical interactions between the predictor variables in multiple regression. Suppose that for our 10 cases in our sample we also measured the number of years that they have been in the workforce, and get the data presented in Table.

Table: Income and Education Level Data

	Education Level (X_1)	*Working* (X_1)	*Monthly income* (T)
Case 1	6 years	10	1
Case 2	8 years	14	1.5
Case 3	11 years	8	1
Case 4	12 years	7	2
Case 5	12 years	20	4
Case 6	13 years	15	2.5
Case 7	14 years	17	5
Case 8	16 years	22	6
Case 9	16	30	10
Case 10	21 years	16	8
Mean	12.9	15	4.1
Standard Deviation	4.25	7.20	3.32
Concielation with income	r=.83	r=.70	

The data presented in Table reveal that both years of education and years in the workforce are positively correlated with monthly income. But how much of the variance in income can these two predictor variables explain *together?* Will years of education still predict income *when -wecontrol for the effects of years in the workforce?*

In other words, after we partial out the portion of the variance in income that is accounted for by years in the workforce, will years of education still be able to help us predict income? Which of these two independent variables will be the stronger predictor of income? And will each make a unique contribution in explaining variance in income?

To answer these questions, we use the SPSS statistical software package to analyse our data. (Note: With only 10 cases in our sample, it is not wise to run a multiple regression. We are doing so for illustration purposes only.

When conducting multiple regression analyses, you should have at least 30 cases plus 10 cases for each predictor variable in the model.) We begin by computing Pearson correlation coefficients for all three of the variables in the model. The results are presented in Table.

Table: Correlation among variables in regression model

	Years of Education	*Years in Workforce*	*Monthly Income*
Years of Education	1,000		
Years in Workforce	.310	1,000	1.00
Monthly Income	826	.995	1.00

These data reveal that both level of education and years in the workforce are both correlated with monthly income (r = .826 and r = .695 for education and workforce with income, respectively). In Table we can also see that there is a small-to-moderate correlation between our two predictors, years of education and years in the workforce (r=.310). Because this correlation is fairly weak, we can infer that both of these independent variables may predict education level.

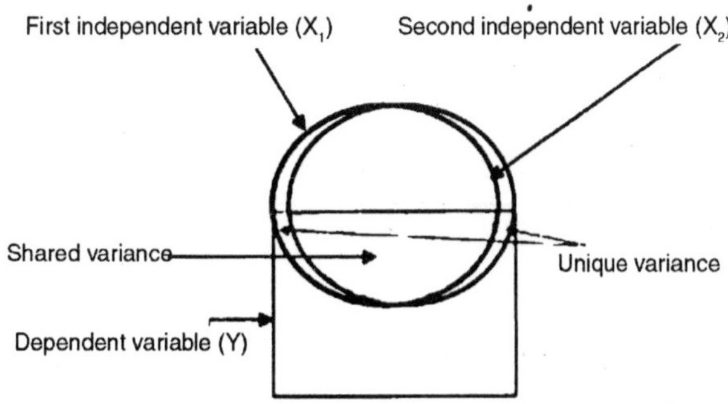

Fig. Shared Variance in Multiple Regression

Remember that in a multiple regression, we've got multiple predictor variables trying to explain variance in the dependent variable. For a predictor variable to explain variance in a dependent variable, it must be *related* to the dependent variable. In our current example, both of our predictor variables are strongly correlated with our dependent variable, so this condition is met.

In addition, for each of our predictor variables to explain a *unique,* or *independent* portion of the variance in the dependent variable, our two predictor variables cannot be too strongly related to *each other.* If our two predictor variables are strongly correlated with each other, then there is not going to be very much unexplained variance *in our predictor variables* left over to explain variance in the dependent variable. For example, suppose that the correlation between scores on a reading test were strongly correlated with scores on a writing test (r = .90).

Now suppose that we wanted to use reading and writing test scores to predict students' grades in the English class. Because reading and writing test scores are so highly correlated with each other, we will probably not explain any more of the variance in English class grades using both predictor variables than if we had just used one or the other. In other words, once we use reading test scores to predict English class grades, adding writing test scores to our regression model will probably not explain any more of the variance in our dependent variable, because reading and writing test scores are so closely related to each other.

This concept is represented graphically in Figure. The shaded area represents shared variance. Notice that the shaded area in the two predictor variables is so large, it is virtually impossible for any of the unshaded areas in each predictor variable to overlap with the dependent variable. These unshaded areas represent the unique variance explaining power of each predictor. You can see that when these unique portions of the predictor variables are small, it is difficult for each predictor to explain a unique portion of the variance in the dependent variable. Strong correlations among predictor variables is called multicollinearity and can cause problems in multiple regression analysis because it can make it difficulty to identify the unique relation between each predictor variable and the dependent variable.

Chapter 4

Statistical Significance

"How often things occur by the merest chance."
—Terence, Phormio

The essential task of inferential statistics is to determine what can reasonably be concluded about an entire domain of phenomena–a population–on the basis of having examined only a limited sample of instances drawn from that domain. A 'sample' is a relatively small window through which the investigator hopes to see the outlines of some larger, more inclusive reality. In some cases, the glimpse provided by the sample might truly represent the larger reality, while in other cases it might misrepresent it, leading the investigator to erroneous conclusions. This latter possibility derives from the fact that the phenomena of nature are often shot through and through with random variability. While the risk of drawing erroneous conclusions from a limited sample is considerable within any domain of phenomena, it is especially great in those domains that pertain to the mentality, behavior, and biology of living organisms.

At any rate, whenever there is random variability inherent in the phenomena under investigation, there is always the possibility that the observed facts result from nothing other than mere chance coincidence; and until that possibility is rigorously assessed, no conclusions can reasonably be drawn from a sample, one way or the other. 'Statistical significance' is the logical and mathematical apparatus by which that assessment is accomplished. This is far and away the most commonly encountered question of inferential statistics within the various fields of scientific research: "Here are some facts that we have extracted from the complex, chance-infused flow of real-world events. Now that we have them, how confident can we be that they signify anything more than mere chance, mere random coincidence–the luck of the scientific draw, so to speak?"

Statistical significance and effect size are two of the most fundamental concepts in quantitative research. Both of these provide indexes of how meaningful the results of statistical analyses are. Despite their frequent appearance in reports of quantitative research (particularly measures of statistical significance), too many researchers lack a clear understanding of what precisely what these concepts mean. The purpose of this chapter is to provide a solid foundation of understanding about statistical significance and effect size for you, the reader, to understand these concepts when you encounter them. Because statistical significance and effect size can be calculated for virtually any statistic, it is not possible in this short chapter to provide instructions on how to determine statistical significance or calculate an effect size across all research situations. Therefore, the focus of this chapter is to describe what these concepts mean and how to interpret them, as well as to provide general information about how statistical significance and effect sizes are determined.

Statistics are often divided into two types: descriptive statistics and inferential statistics, descriptive statistics are those statistics that *describe* the characteristics of a given set of data. For example, if we collect weight data for a group of 30 adults, we can use a variety of statistics to describe the weight characteristics of these 30 adults (e.g., their average, or mean, weight, the range from the lowest to the highest weight, the standard deviation for this group, etc.). Notice that all of these descriptive statistics do nothing more than provide information about this specific group of 30 individuals from whom we collected data.

Although descriptive statistics are useful and important, researchers are often interested in extending their results beyond the specific group of people from whom they have collected data (i.e., their sample, or samples). From their sample data, researchers often want to determine whether there is some phenomenon of interest occurring in the larger population(s) that these samples represent. For example, we may want to know whether, in the general population, boys and girls differ in their levels of physical aggression. To determine this, we could conduct a study in which we measure the physical aggression levels of every boy and girl in America and see whether boys and girls differ. This study would be very costly, however, and very time consuming. Another approach is to select a sample of boys and a sample of girls, measure their levels of physical aggression, see if they differ, and from these sample data *infer* about differences in the larger populations of boys and girls. If we eventually conclude that our results are statistically significant, in essence we am concluding that the differences we observed in the average levels of aggression of the boys and girls in our two samples represents a likelihood that there is also a difference in the average levels of aggression in the populations of boys and girls from which these samples were selected.

As the name implies, *inferential* statistics are always about making inferences about the larger population(s) on the basis of data collected from a sample or samples. To understand how this works, we first need to understand the distinction between a population and a sample and get comfortable with some concepts from probability. Once we have developed an understanding of statistical significance, we can then compare the concepts of statistical significance and practical significance. Briefly, effect size is a measure of how large an observed effect is without regard to the size of the sample. In the earlier example examining levels of aggression, the effect that we are interested in is the difference in boys' and girls' average levels of aggression.

Within the context of correlation, the question of statistical significance concerns the relationship between r, which is the correlation that is observed within a limited sample of X_iY_i pairs, and rho, which is the correlation that exists, in the larger reality beyond the sample, between X and Y in general. The reason why the question arises is that it is possible, by the merest chance coincidence, to obtain rather impressive-looking values of r within a sample, even when the correlation between X and Y in general is rho=0. This is especially true when the size of the sample is small, but decreasingly so as the size of the sample increases. Here is a simple experiment by which you can gain some hands-on experience with the general principle to which this statement is referring. If you repeatedly toss a pair of dice, there is no reason at all to expect that the paired outcomes of the dice would be correlated. Another way of saying it is that the correlation within the entire potential population of paired dice outcomes is rho=0. Now for the experiment. For each of its two parts you will need a pair of dice, preferably of different colors. One of the dice is designated as X and the other as Y.

A few years ago I have had the students in several of my own statistics classes perform these exercises, and in the process collected one batch of 473 correlation coefficients, each based on a sample of N=5 paired dice outcomes, and another batch of 424 correlation coefficients, each based on a sample of N=10 paired dice outcomes. The distributions of these two batches of sample correlation coefficients are shown in Figure. For purposes of illustration, the portions of the two histograms representing sample values of r that deviated from the central value of zero by as much as ±0.50 are marked off and colored in light blue. As you can see, these blue areas of the distributions include nearly 40% of the correlation coefficients based on samples of size N=5, but only about 12% of those based on samples of size N=10.

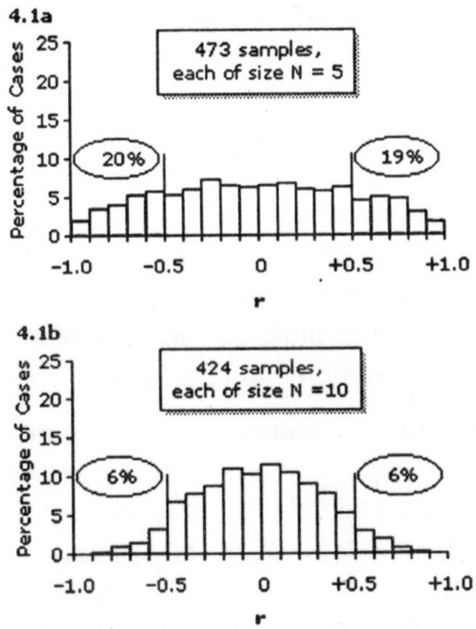

Fig. Observed Distributions of Random Values of r Drawn from Populations for which rho=0

The somewhat jagged outlines of the distributions are explained by the fact that our batches of 473 and 424 samples, although large in terms of time and effort, are still relatively small in comparison with the number of samples that could have been taken, if only my students had been willing to make a life-time commitment to the experiment. Luckily for both them and me, this life-long labor is not really necessary, since the true forms of these distributions of sample values of r are already known. The knowledge of them comes not from anyone actually observing several hundred million paired dice outcomes, but from logical and mathematical reasoning grounded on the basic concepts of probability that we will begin examining in Chapter. Although the process begins with a question that seems highly abstract and hypothetical, the result that comes from it has a very wide range of practical application. Imagine a population of X_iY_i pairs–paired dice outcomes, or anything else–for which the overall correlation is rho=0. If you were to draw a vast number of bivariate samples from that population, each sample of size *N*=5, and calculate the correlation coefficient for each sample, and then lay all of these sample values of *r* out in the form of a relative frequency distribution–what would that distribution look like?

The answer to this question is depicted in Figure, which you will see looks very much like the distribution of 473 sample values of *r* shown in Figure, except that it is smoother and precisely symmetrical. Here again, as a convenient reference point, the portion of the distribution representing sample values of *r* that would deviate from the central value of zero by as much as ±0.50 are marked off and colored in light blue. As indicated, this would include about 40% of the distribution. That is, about 20% of your multitude of sample correlation coefficients would be as large or larger than +0.50 in the positive direction, and about 20% would be as large or larger than –0.50 in the negative direction. Another way of putting it is that any particular one of the samples has about a 20% likelihood of coming out, by mere chance, with a correlation coefficient equal to or greater than +0.50 in the positive direction, and about a 20% likelihood of coming out, by mere chance, with a correlation coefficient equal to or greater than –0.50 in the negative direction. The principle applies not just to the tossing of dice, but to bivariate situations in general where

- The correlation within the entire population is rho=0,
- The samples drawn from the population are of size *N*=5.

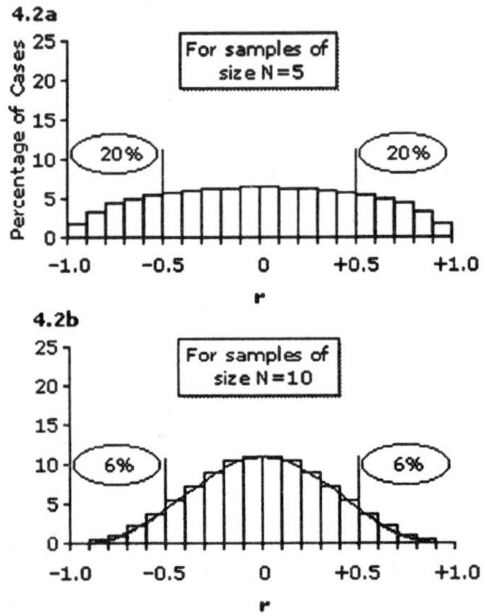

Fig. Theoretical Distributions of Random Values of *r* (*N*=5 and *N*=10) Drawn from Populations for which rho=0

Figure shows the same type of theoretical distribution for the case where the correlation within the entire population is rho=0 and the samples are of size N=10. Here as well, you can see that the theoretical distribution for samples of size N=10 looks very much like the corresponding distribution of 424 actual samples, except that it is smoother and precisely symmetrical. In this situation only about 12% of the sample values of r would deviate from zero by as much as ±0.50. Half of this 12% would fall at or beyond +0.50 in the positive direction, and the other half would fall at or beyond –0.50 in the negative direction. Alternatively, you can say that any particular one of the samples has about a 6% likelihood of coming out, by mere chance, with a correlation coefficient equal to or greater than +0.50 in the positive direction, and about a 6% likelihood of coming out, by mere chance, with a correlation coefficient equal to or greater than –0.50 in the negative direction.

You have probably already noticed that the outline of the distribution shown in Figure looks like a somewhat squat version of a normal distribution. In fact, it is not a normal distribution; though as you can see from the graphs in Figure, below, theoretical distributions of this general type do come closer and closer to the form of a normal distribution as you increase the size of the samples. By the time you reach a sample size of N=30, the shape of the distribution of sample correlation coefficients is virtually identical to that of a normal distribution. Notice also in these two remaining graphs that increasing the size of the samples decreases even further the tendency of sample correlation coefficients to deviate from the zero correlation that exists within the population from which the samples are drawn. Thus, for samples of size N=20, it is only 2.5% of the sample values of r that will deviate from zero by as much as ±0.50, and for samples of size N=30 it is only 0.5%.

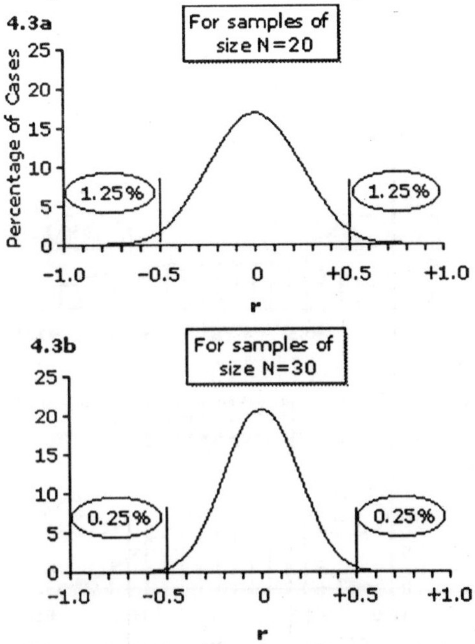

Fig. Theoretical Distributions of Random Values of r(N=20 and N=30) Drawn from Populations for which rho=0

Consider, now, the following hypothetical scenario. A team of medical investigators is exploring the properties of a newly discovered human nutrient, which temporarily goes by the name of nutrient *X*. On the basis of what is known about the chemical properties of nutrient *X*, the investigators hypothesize that it should have the effect of increasing the production of a certain blood component, called factor *Y*, that is known to play an important role in strengthening the body's resistance to infection. To test this hypothesis, the investigators measure average daily intake of nutrient *X* over a period of time for a certain number of normal adult human subjects; and then, at the end of that period they draw a small volume of blood from each subject and measure the number of units of factor *Y*. Upon analyzing their results, they find the correlation between

X: average daily intake of nutrient *X*,

Y: number of units of factor *Y* per milliliter of blood

for this particular sample to be $r = +0.50$. In short, their hypothesis led them to expect a positive correlation between *X* and *Y* within the sample, and that is exactly what they found.

Now comes the question of statistical significance, which boiled down to its barest bones is simply this. Given that impressive-looking correlations can occur within limited samples, even when there is no correlation at all between *X* and *Y* in the larger reality beyond the sample, what confidence can we have that the +0.50 correlation observed in this particular case is not just a fluke of mere chance coincidence?

From what we said just prior to talking about nutrient *X* and factor *Y*, you can see that the answer to this question obviously depends on the size of the sample, which in our description of this hypothetical experiment was left unspecified. If the size of the sample is $N = 5$, then the investigators will have a 20% chance of observing a positive correlation coefficient as large as +0.50 even if the correlation within the entire population is merely rho = 0. With a sample of size N=10 it drops to a 6% chance; for N=20 it falls further to a scant 1.25% chance; and for N=30 it falls even further to a minuscule one-quarter of one percent. Clearly, the larger the size of the investigators' sample, the more confidence you can have that their observed correlation of $r = +0.50$ is not just a fluke of mere chance coincidence.

In most areas of scientific research, the criterion for statistical significance is conventionally set at the 5% level. That is, an observed result is regarded as statistically significant–as something more than a mere fluke–only if it had a 5% or smaller likelihood of occurring by mere chance coincidence. Otherwise, it is regarded as statistically non-significant. We will be examining the concept of statistical significance at much greater length, not only for correlation but for numerous other statistical procedures as well. For whatever immediate need you might have to determine the significance of a correlation coefficient, you will be able to get by with the information presented in Table and in Figure. Table shows the positive or negative values of r that are required for statistical significance at the 5% level for various sample sizes, from N=5 through N=32, and for two different kinds of situations to be described below. Figure 4.4 shows the same information in graphic form, but extended across a wider range of sample sizes, from N=5 to N=100.

Positive or Negative Values of r Required for Statistical Significance at the 5% Level, for Samples of Size N=5 through N=32

For any particular sample size, an observed value of r is regarded as statistically significant at the 5% level if and only if its distance from zero is equal to or greater than the distance of the tabled value of r. Thus, for a sample of size N=20, an observed value of r=+0.40 or r= –0.40 would be significant at the 5% level for a directional hypothesis, but non-significant for a non-directional hypothesis; an observed value of r=+0.44 or r= –0.44 would be significant for both kinds of hypotheses; and an observed value of r=+0.37 or r= –0.37 would be non-significant for both kinds of hypotheses. (Tabled values of r are rounded to two decimal places.)

	Hypothesis			*Hypothesis*	
N	*Directional* $\pm r$	*Non-Directional* $\pm r$	*N*	*Directional* $\pm r$	*Non-Directional* $\pm r$
5	0.81	0.88	19	0.39	0.46
6	0.73	0.81	20	0.38	0.44
7	0.67	0.75	21	0.37	0.43
8	0.62	0.71	22	0.36	0.42
9	0.58	0.67	23	0.35	0.41
10	0.55	0.63	24	0.34	0.40
11	0.52	0.60	25	0.34	0.40
12	0.50	0.58	26	0.33	0.39
13	0.48	0.55	27	0.32	0.38
14	0.46	0.53	28	0.32	0.37
15	0.44	0.51	29	0.31	0.37
16	0.43	0.50	30	0.31	0.36
17	0.41	0.48	31	0.30	0.36
18	0.40	0.47	32	0.30	0.35

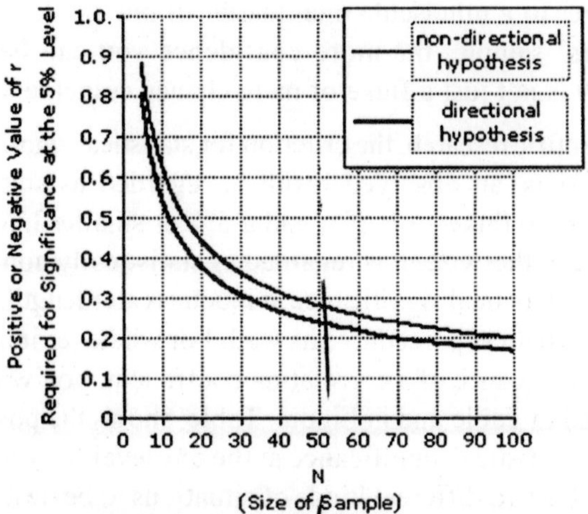

Fig. Values of r Required for Statistical Significance at the 5% Level, for Samples of Size N=5 through N=100

The difference between the two kinds of situations is defined by the investigator's *hypothesis,* which is either directional or non-directional. Within the context of correlation, a directional hypothesis is one that leads the investigator to specify, in advance, one or the other of the following expectations:

Positive Directional Hypothesis: The relationship between X and Y in the general population is positive (the more of X, the more of Y), hence this particular sample of X_iY_i pairs will show a positive correlation;

or

Negative Directional Hypothesis: The relationship between X and Y in the general population is negative (the more of X, the less of Y), hence this particular sample of X_iY_i pairs will show a negative correlation.

A non-directional hypothesis, on the other hand, leads only to the expectation that the correlation between X and Y within the general population might be something other than zero, with no specification of the particular direction in which it might go. Essentially, it is an either-or combination of the two types of directional hypothesis:

Non-directional Hypothesis: The relationship between X and Y in the general population is something other than zero, hence this particular sample of X_iY_i pairs will show a non-zero correlation, either positive or negative, though we have no basis for predicting just which of these it will be.

The important logical difference between these two kinds of situations is that a non-directional hypothesis could potentially be supported by finding either a positive or a negative correlation within the sample, whereas a directional hypothesis could be supported only by finding a correlation within the sample that is in the direction specified; that is, only by finding a positive correlation when the positive direction has been specified, and only by finding a negative correlation when the negative direction has been specified. This logical difference between the two situations entails a different standard of statistical significance. Specifically, for any particular sample size, the value of r required for significance at the 5% level is larger for a non-directional hypothesis than for a directional hypothesis. You need not worry just yet about the detailed rationale of this point. For the moment, it is sufficient to understand the basic distinction between directional and non-directional hypotheses, and to know that the standard of statistical significance for a non-directional hypothesis is more stringent.

In our imaginary study of the relationship between nutrient X and blood factor Y, the investigators specified in advance that they expected the correlation to be positive. Hence, the applicable standard of statistical significance is the one that pertains to directional hypotheses. If the sample in this study was only of size N=5, the required value of r would be +0.81; so the observed correlation of +0.50, falling short of this required value, would be non-significant. It would also fall short of the required values for a sample of size N=6, 7, 8, 9, 10, or 11. For a sample of size N=12, however, the observed value of +0.50 hits the required value dead center; so for a sample of this size, the observed correlation coefficient would be significant precisely at the 5% level. For any sample size larger than N=12, the observed value of +0.50 would be larger than the required value, and so it would be significant even beyond the 5% level. For our earlier SAT example we had a sample of size N=50 and an observed correlation coefficient of r= –0.86. We might conceivably have had some reason to specify a negative directional hypothesis in advance of examining the data–but as we did not, in fact, make that specification, the applicable standard must be the one for a non-directional hypothesis. Here the relevant information is found in Figure.

Start at the horizontal axis where N is equal to 50; go straight up to the red line, which refers to the standard of significance for a non-directional hypothesis; then go straight across to the left to find the required value of r on the vertical axis. That required value is r=±0.28, which means that for a sample of size N=50 and a non-directional hypothesis, an observed correlation coefficient of either r=+0.28 or r= –0.28 would be significant precisely at the 5% level, and that an observed r greater than +0.28 in the positive direction, or greater than –0.28 in the negative direction, would be significant beyond the 5% level. Our actually observed correlation coefficient of r= –0.86 is therefore significant beyond the 5% level. In a later chapter we will describe a much more precise computational method for determining the statistical significance of a correlation coefficient for all possible values of N and for a wider range of levels of significance. At that time you will see that our observed correlation of r= –0.86 for the SAT example is significant not only beyond the 5% level, but also considerably beyond the 1% level.

SAMPLES AND POPULATIONS

The first step in understanding statistical significance is to understand the difference between a sample and a population. This difference has been discussed earlier, both in brief and in more detail. Briefly, a sample is an individual or group from whom or from which data are collected. A population is the individual or group that the sample is supposed to represent. For the purposes of understanding the concept of statistical significance, it is critical that you remember that when researchers collect data from a sample, they are often interested in using these data to make inferences about the population from which the sample was drawn. Statistical significance refers to the likelihood, or probability, that a statistic derived from a *sample* represents some genuine phenomenon in the *population* from which the sample was selected. In other words, statistical significance provides a measure to help us decide whether what we observe in our sample is also going on in the population that the sample is supposed to represent.

One factor that often complicates this process of making inferences from the sample to the population is that in many, if not most, research studies in the social sciences, the population is never explicitly defined. This is somewhat problematic, because when we argue that a statistical result is statistically significant, we are essentially arguing that the result we found in our sample is representative of some effect in the population from which the sample was selected. If we have not adequately defined our population, it is not entirely clear what to make of such a result. For the purposes of this chapter, however, suffice it to say that samples are those individuals or groups from whom or which data are collected, whereas populations are the entire collection of individuals or cases from which the samples are selected.

PROBABILITY

Probability plays a key role in inferential statistics. When it comes to deciding whether a result in a study is statistically significant, we must rely on probability to make the determination.

When we calculate an inferential statistic, that statistic is part of a sampling distribution. We will recall that whenever we select a sample from a population and calculate a statistic from the sample, we have to keep in mind that if we had selected a different sample *of the same size from the same population,* we probably would get a slightly different statistic from the new sample. For example, if we randomly selected a sample of 1,000 men from the population of men in the United States and measured their shoe size, we might find an average shoe size of 10 for this sample. Now, if we were to randomly select a new sample of 1,000 men from the population of men in the United States and calculate their average shoe size, we might get a different mean, such as 9. If we were to select an infinite number of random samples of 1,000 and calculate the average shoe sizes of each of these samples, we would end up with a sampling distribution of the mean, and this sampling distribution would have a standard deviation, called the standard error of the mean. Just as there is a sampling distribution and a standard error of the mean, so there are sampling distributions and standard errors for all statistics, including correlation coefficients, *F* ratios from ANOVA, *t* values from *t* tests, regression coefficients, and so on.

Because these sampling distributions have certain stable mathematical characteristics, we can use the standard errors to calculate the exact probability of obtaining a specific sample statistic, from a sample of a given size, using a specific known or hypothesized population parameter. It's time for an example. Suppose that, from previous research by the shoe industry, we know that the average shoe size for the population of men in the U.S. is a size 9. Because this is the known average for *the population,* this average is *a parameter* and not a statistic. Now suppose we randomly select a sample of 1,000 men and find that their average shoe size is 10, with a standard deviation of 2. Notice that our average for our sample (10) is a *statistic* because it comes from our sample, not our population. With these numbers, we can answer two slightly different but related questions. First, if the average shoe size in the population is really 9, what is the probability of selecting a random sample of 1000 men who have an average shoe size of 10? Second, is the difference between our population mean (9) and our sample mean (10) *statistically significant?* The answer to our first question provides the basis for the answer to our second question.

Notice that if we do not select our sample at random, it would be easy to find a sample of 1,000 men with an average shoe size of 10. We could buy customer lists from shoe stores and select 1,000 men who bought size 10 shoes. Or we could place an advertisement in the paper seeking men who wear a size 10 shoe. But if our population mean is really 9, and our sample is really selected at random, then there is some probability, or *chance,* that we could wind up with a sample of 1,000 men with an average shoe size of 10. In statistics, this *chance* is referred is as random sampling error or random chance.

Back to the example. If our population mean is 9, and our random sample of 1000 men has a mean of 10 and a standard deviation of 2, we can calculate the standard error by dividing the standard deviation by the square root of the sample size.

$$s_{\bar{x}} = 2 \div \sqrt{1000}$$

$$s_{\bar{x}} = 2 \div 31.62$$

$$s_{\bar{x}} = .06$$

Where $s_{\bar{x}}$ = the standard error of the mean

Now that we know the standard error is .06, we can calculate a t value to find the approximate probability of getting a sample mean of 10 *by random chance* if the population mean is really 9. (Note: For sample sizes larger than 120, the t distribution is identical to the normal distribution. Therefore, for large sample sizes, t values and z values, and their associated probabilities, are virtually identical.) t =10"9/06 t =1/06 t =16.67

When using the t distribution to find probabilities, we can simply take the absolute value of t. Once we have our absolute value for t (t = 16.67), we can consult the t table and see that, when the degrees of freedom equals infinity (i.e., greater than 120), the probability of getting a t value of 16.67 is less than .001. In fact, because the critical t value associated with a probability of .001 is only 3.291, and our actual t value is 16.67, we can conclude that the random chance of getting a sample mean of 10 when the population mean is 9 is *much* less than .001. In other words, when we randomly select a sample of 1,000 men and calculate their average shoe size, when we know that the average shoe size of men in the population is 9, we would *expect* to get a sample mean of 10 much less than 1 time in 1,000. With our table of t values, that is as accurate as we can get.

So we have already calculated the probability, or random chance, of finding a sample mean of 10 when the population mean is 9 is very small, less than one in a thousand, when the sample size is 1,000 and is randomly selected. This probability is known as a p value, with the p standing for *probability*. In our current example, we would say that we found *a* $p < .001$, which is the *way p* values are generally reported in research reports and scholarly journals. Now we can turn our attention to the second question: Is the difference between a population mean of 9 and a sample mean of 10 statistically significant? Well, the quick answer is "Yes." The longer answer requires us to delve into the world of hypothesis testing.

HYPOTHESIS TESTING AND TYPE I ERRORS

In the real world of social science research, the model of testing a p value against some set of null and alternative hypotheses is rarely followed. In fact, in most social science research, the researcher collects data from a sample, runs some statistic of interest, and decides that if the probability of finding the statistic by chance (i.e., the p value) is less than .05 (i.e., $p < .05$), the result is statistically significant. But the decision to proclaim any statistic that has a p value less than .05 "statistically significant" is grounded in a tradition of hypothesis testing that is worth reviewing briefly. Note that extended discussions of hypothesis testing can be found in any standard statistics textbook.

The idea hère is simple. Before we decide whether a result is statistically significant, we should establish a standard, or benchmark, *before* we calculate the statistic. To do this, we develop a hypothesis and establish a criterion that we will use when deciding whether to retain or reject our hypothesis. The primary hypothesis of interest in social science research is the null hypothesis (H_0). As the name implies, the null hypothesis always suggests that there will be an *absence* of effect. To illustrate, let us return to the shoe-size example. Recall that we already knew our population average shoe size was 9. Given this, we would expect that if we were to randomly select a sample from that population, and calculate the average shoe size for the sample, that average would also be 9. We might know that there is a *chance* our sample would have a different mean than our population, but our best guess is that our sample would have the same mean as our population. Therefore, our null hypothesis would be that our population mean and our sample mean would not differ from each other (i.e., no effect). We could write this hypothesis symbolically as follows:

$$H_{\circ} : \mu = \overline{X} = 9$$

Where μ represents the population mean

$\overline{X}$ represents the sample mean

Notice that at this point, we have not yet selected our sample of 1,000 men and we have not yet calculated a sample mean. This entire hypothesis building process occurs *a priori.* Of course, where there is one hypothesis (the null), it is always possible to have alternative hypotheses. One alternative to the null hypothesis is the opposite hypothesis. Whereas the null hypothesis is that the sample and population means will equal each other, an alternative hypothesis could be that they will not equal each other. This alternative hypothesis (H_A or E1) would be written symbolically as

$$H_A : \mu \neq \overline{X}$$
$$\overline{X} \neq 9$$

Where μ represents the population mean

$\overline{X}$ represents the sample mean

At this point in the process, we have established our null and alternative hypotheses. You may assume that all we need to do is randomly select our 1000 men, find their average shoe size, and see if it is different from or equal to 9. But, alas, it is not quite that simple. Suppose that we get our sample and find their average shoe size is 9.00001. Technically, that is different from 9, but is it different enough to be considered meaningful? Keep in mind that whenever we select a sample at random from a population, there is always a chance that it will differ slightly from the population. Although our best guess is that our sample mean will be the same as our population mean, we have to remember that it would be almost impossible for our sample to look *exactly* like our population. So our question becomes this: How different does our sample mean have to be from our population mean before we consider the difference meaningful, or *significant.* If our sample mean is just a little different from our population mean, we can shrug it off and say "Well, the difference is probably just due to *random sampling error,* or *chance.*" But how different do our sample and population means need to be before we conclude that the difference is probably *not* due to chance? That's where our alpha level, or Type I error, comes into play.

Sampling distributions and standard errors of these distributions allow us to compute probabilities for obtaining sample statistics of various sizes. When we say "probability" this is, in fact, shorthand for "the probability of obtaining this sample statistic due to *chance* or *random sampling error.*" Given that samples generally do not precisely represent the populations from which they are drawn, we should expect some difference between the sample statistic and the population parameter simply due to the luck of the draw, or random sampling error. Had we reached into our population and pulled out another random sample, we probably would get slightly different statistics again. So some of the difference between a sample statistic, like the mean, and a population parameter will always be due to who we happened to get in our random sample, which is why it is called random sampling error. A statistic like the mean, the sampling distribution of the mean is a normal distribution. So our random sampling method will produce many sample means that are close to the value of the population mean and fewer that are further away from the population mean. The further the sample mean is from the population mean, the less likely it is to occur by chance, or random sampling error.

Before we can conclude that the differences between the sample statistic and the population parameter are probably *not* just due to random sampling error, we have to decide how unlikely the chances are of getting a difference between the statistic and the population parameter just by chance. In other words, before we can *reject the null hypothesis,* we want to be reasonably sure that any difference between the sample statistic and the population parameter is not just due to random sampling error, or chance. In the social sciences, the convention has generally been to set that level at .05. In other words, social scientists have generally agreed that if the probability of getting a difference between the sample statistic and the population parameter is less than 5%, we will reject the null hypothesis and conclude that the differences between the statistic and the parameter are probably not due to chance.

The agreed-upon probability of .05 (symbolized as a = .05) represents the Type I error rate that we, as researchers, are willing to accept *before we conduct our statistical analysis.* Remember that the purpose of our analysis is to determine whether we should retain or reject our null hypothesis. When we decide to reject the null hypothesis, what we are saying in essence, is that we are concluding that the difference between our sample statistic and our population parameter is *not* due to random sampling error. But when we make this decision, we have to remember that it is always possible to get even very large differences just due to random sampling error, or chance. In our shoe-size example, when we randomly select 1,000 men, it is possible that, just due to some fluke, we select 1,000 men with an average shoe size of 17. Now this is extremely unlikely, but it is always possible. We never know what you're going to get when you select a random sample. In our earlier example, where our sample had an average shoe size of 10, we found the probability of getting a sample mean of 10 when our population mean was 9, by chance, was less than one in a thousand. Though unlikely, *it is still possible* that this difference between our sample and population means were just due to chance. So because our *p* value ($p<.001$) is much smaller than our alpha level (a = .05), we will reject the null hypothesis and conclude that our sample mean is actually different from our population mean, that this is probably not just a fluke of random sampling, and that our result is statistically significant. When we reach this conclusion, we may be wrong. In fact, we may be rejecting the null hypothesis, even though the null hypothesis is true.

To summarize, when we do inferential statistics, we want to know whether something that we observe in a sample represents an actual phenomenon in the population. So we set up a null hypothesis that there is no real difference between our sample statistic and our population parameter, and we select an alpha level that serves as our benchmark for helping us decide whether to reject or retain our null hypothesis. If our *p* value (which we get *after* we calculate our statistic) is smaller than our selected alpha level, we will reject the null hypothesis. When we reject the null hypothesis, we are concluding that the difference between the sample statistic and the population parameter is probably not due to chance, or random sampling error. However, when we reach this conclusion, there is always a chance that we will be wrong, having made a Type I error. One goal of statistics is to avoid making such errors, so to be extra safe we may want to select a more conservative alpha level, such as .01, and say that unless our *p* value is even smaller than .01, we will retain our null hypothesis. In our shoe-size example, our p value was much smaller than either .05 or .01, so we reject the null hypothesis and conclude that, for some reason, our sample of 1,000 men had a statistically significantly larger average shoe size than did our general population.

EFFECT SIZE

As an indication of the importance of a result in quantitative research, statistical significance has enjoyed a rather privileged position for decades. Social scientists have long given the "$p<.05$" rule a sort of magical quality, with any result carrying a probability greater than .05 being quickly discarded into the trash heap of "non significant" results. Recently, however, researchers and journal editors have begun to view statistical significance in a slightly less flattering light, recognizing one of its major shortcomings: It is perhaps too heavily influenced by sample size. As a result, more and more researchers are becoming aware of the importance of effect size, and increasingly are including reports of effect size in their work. To determine whether a statistic is statistically significant, we follow the same general sequence regardless of the statistic (z scores, *t* values, *F* values, correlation coefficients, etc.). First, we find the difference between a sample statistic and a population parameter (either the actual parameter or, if this is not known, an hypothesized value for the parameter). Next, we divide that difference by the standard error. Finally, we determine the probability of getting a ratio of that size due to chance, or random sampling error.

The problem with this process is that when we divide the numerator (i.e., the difference between the sample statistic and the population parameter) by the denominator (i.e., the standard error), the sample size plays a large role. In all of the formulas that we use for standard error, the larger the sample size, the smaller the standard error. When we plug the standard error into the formula for determining *t* values, f-values, and *z* scores, we see that the smaller the standard error, the larger these values become, and the more likely that they will be considered statistically significant. Because of this effect of sample size, we sometimes find that even very small differences between the sample statistic and the population parameter can be statistically significant if the sample size is large.

To illustrate this point, let us consider an example with two different sample sizes. Suppose we know that the average IQ score for the population of adults in the United States is 100. Now suppose that I randomly select two samples of adults. One of our samples contains 25 adults, the other 1600. Each of these two samples produces an average IQ score of 105 and a standard deviation of 15. Is the difference between 105 and 100 statistically significant? To answer this question, we need to calculate a *t* value for each sample. The standard error for our sample with 25 adults will be

$$\sigma_{\bar{x}} = 15 \div \sqrt{25} \Rightarrow 15 \div 5 \Rightarrow 3$$

Where $\sigma_{\bar{x}}$ is the standard of the mean

The standard error for our second sample, with 1,600 adults, will be

$$\sigma_{\bar{x}} = 15 \div \sqrt{1600} \Rightarrow 15 \div 40 \Rightarrow .375$$

Plugging these standard errors into our *t* value formulas, we find that the *t* value for the 25 person sample is 105 - 100/3, or 1.67. Looking in our table *of t* distributions we can see that the *p* value for a *t* value of 1.67 is between .10 and .20. The *t* value for the sample with 1,600 adults is 105 - 100/.375, or 13.33, with a corresponding p value *of* $p<.0001$. If we are using an alpha level of .05, then a difference of 5 points on the IQ test would not be considered statistically significant if we only had a sample size of 25, but would be highly statistically significant if our sample size were 1600. Because sample size plays such a big role in determining statistical significance, many statistics textbooks make a distinction between statistical significance and practical significance. With a sample size of 1600, a difference of even 1 point on the IQ test would produce a statistically significant result *(t= I* + .375 => t = 2.67, $p < 01$). However, if we had a very small sample size of 4, even a 15-point difference in average IQ scores would not be statistically significant *(t* = 15 4- 7.50 => $t = 2.00, p > 10$). But is a difference of one point on a test with a range of over 150 points really important in the real world? And is a difference of 15 points not meaningful? In other words, is it a significant difference in the *practical* sense of the word *significant?* One way to answer this question is to examine the effect size.

There are different formulas for calculating the effect sizes of different statistics, but each of these formulas share common features. The formulas for calculating most inferential statistics involve a ratio of a numerator (such as the difference between a sample mean and a population mean in a one-sample *t test)* divided by a standard error. Similarly, most effect size formulas use the same numerator, but divide this numerator by a standard *deviation* rather than a standard error. The trick, then, is knowing how to come up with the appropriate standard deviation to use in a particular effect size formula.

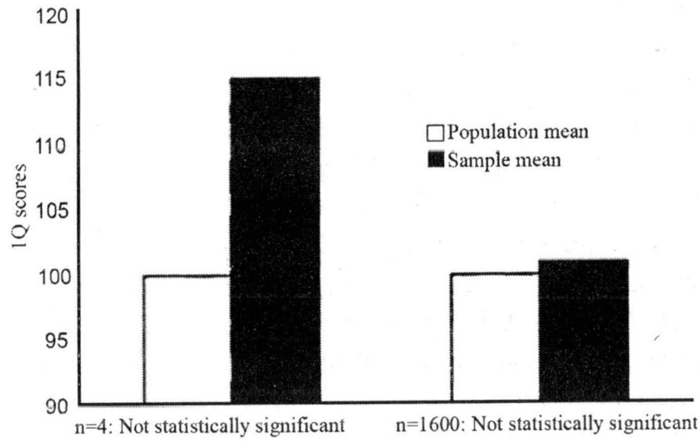

The influence of sample size on statistical significance.

We can examine the simplest form of effect size by returning to our examples using IQ scores. Remember that we have a population with an average IQ score of 100 and a standard deviation of 15. We also had two samples, each with average IQ scores of 105, and one with a sample size of 25 and the other with a sample size of 1,600. Also recall that to find the standard error for calculating our *z* scores, we simply divided the standard deviation by the square root of the sample size. So for the sample with 25 members, our standard error was where ax is the standard error of the mean

$$\sigma_{\bar{x}} = 15 \div \sqrt{25} \Rightarrow 15 \div 5 \Rightarrow 3$$

To calculate an effect size, what we need to do is convert this standard error back into a standard deviation. If we *divide* the standard deviation by the square root of the sample size to find the standard error, we can *multiply* the standard error by the square root of the sample size to find the standard deviation. When we do this, we find that

$$\sigma = 3 \times \sqrt{25} \Rightarrow 3 \times 5 \Rightarrow 15$$

Notice that the standard deviation would be exactly the same if we calculated it for the larger sample size of 1,600, even though the standard error was much smaller for that sample.

$$\sigma = .375 \times \sqrt{1600} \Rightarrow .375 \times 40 \Rightarrow 15$$

Once we have our standard deviation, it is easy to calculate an effect size, which has the symbol *d.* In the IQ example, we could determine the effect size as follows:

$$d = \frac{105 - 100}{15}$$

$$d = \frac{5}{15}$$

$$d = .33$$

Where *d* is the effect of size

The formula for the effect size translates the numerator into standard deviation units. When the numerator represents some sort of difference score (e.g., the difference between two or more group means, the difference between a sample statistic and a population parameter), the effect size will represent that difference in standard deviation units.

This is similar to representing the difference in standard error units, as most inferential statistics do (e.g., t values, F values, correlation coefficients), except that sample size is *eliminated* from the process.

There are no hard and fast rules regarding the interpretation of effect sizes. Some textbook authors suggest that effect sizes smaller than .20 are small, those between .25 and .75 are moderate, and those over .80 are large. When determining whether an effect size is meaningful, it is important to consider what you are testing. If we are comparing the mortality rates of two samples trying two different experimental drugs, even small effect sizes are important, because we are talking about life and death. But if we are comparing two different samples' preference for ice cream flavours, even fairly large effect sizes may have little more than trivial importance in the real world.

The recent push by some researchers to focus more heavily on effect sizes than on statistical significance reminds me that we should conclude this section of the chapter by urging you to take both effect size *and* statistical significance into consideration as you read and conduct research. The exact same effect size was produced with the 2 5-person sample as with the 1,600-person sample. These results suggest that sample size does not matter. In fact, sample size is very important. Stated simply, it is easier to come up with fluke, or *chance* results with smaller sample sizes than with larger sample sizes.

Our tests of statistical significance, which are sensitive to sample size, tell us the probability that our results are just due to random sampling error, or chance. Because larger sample sizes have a better likelihood of representing the populations from which they were selected, the results of studies that use larger sample sizes are more *reliable* than those using smaller sample sizes, if all else is equal (e.g., how the samples were selected, the methods used in the study, etc.). When used together, tests of statistical significance and measures of effect size can provide important information regarding the reliability and importance of statistical results. Of course, our own judgments about the meaning, causes, and consequences of our results are also important factors.

THREE OVERARCHING CONCEPTS

Although sampling distributions often have an intellectual elegance and beauty in and of themselves, their chief function lies in the workaday realm of practical application. They are the devices by which scientific researchers can rationally determine how confident they may be that their observed results reflect anything more than mere chance coincidence.

The general concept of sampling distributions via the particular example of binomial distributions, along with the normal distribution whose form and properties the binomial distributions tend to approximate. Before going on to consider other sampling distributions and their associated tests of statistical significance, we will take a moment to draw out some further details concerning the general concept of statistical significance.

MEAN CHANCE EXPECTATION

All tests of statistical significance involve a comparison between

• An observed value	E.g., an observed correlation coefficient; an observed number of heads in *N* tosses of a coin; an observed number of recoveries in *N* patients; and so on.
• The value that one would expect to find, on average, if nothing other than chance coincidence, mere random variability, were operating in the situation.	E.g., a correlation coefficient of r=0, assuming that the correlation within the general population of paired X_iY_i values is rho=0; 5 heads in 10 tosses of a coin, assuming that the mere chance probability of a head on any particular toss is p=.5; 400 recoveries in 1,000 patients, assuming that the probability of mere chance recovery for any particular patient is p=.4; and so on.

A useful generic label for the second of these values is mean chance expectation, which we can abbreviate as MCE. For binomial probability situations of the sort, MCE is in every case equal to the mean, μ, of the relevant binomial sampling distribution. Thus, for the case of 10 coin tosses, MCE for the number of heads is

$MCE_{(heads)} = \mu = Np$ Where N=10, the number of tosses, and p=.5, the probability of a head on any particular one of the N tosses.

$= 10 \times .5 = 5$

If you actually perform 10 coin tosses, it is of course entirely possible that you might end up with as many as 6 or 7 or 8 heads, or as few as 4 or 3 or 2. Indeed, there is a 1/1,024 chance–small, but nonetheless greater than zero–of ending up with all 10 heads, and an equal 1/1,024 chance of ending up with no heads at all. When we say that MCE for heads in this situation is 5, the reference is not to 5 exactly but to 5 probabilistically. Essentially, it is the concept of a sampling distribution expressed in somewhat different language.

Within the entire range of possible outcomes in this situation–zero heads, 1 head, 2 heads, and so on, up through 10 heads–the closer a particular outcome is to the MCE value of 5, the more likely it is to occur by mere chance coincidence; and conversely, the farther away it falls from the MCE value of 5, the less likely it is to occur by mere chance coincidence.

To help consolidate your grasp of this concept, the following graph shows the outlines of the binomial sampling distribution for this situation, and the table beneath it lists the respective probabilities for each of the 11 possible outcomes.

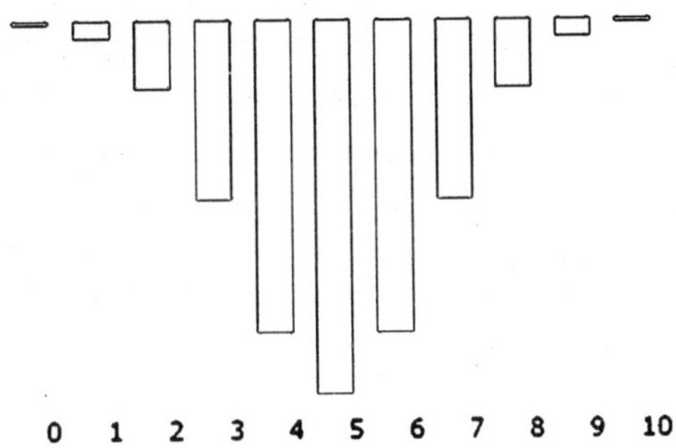

Number of Heads in 10 Tosses of a Coin

Number of Heads	*Probability*
0	.00098
1	.00977
2	.04395
3	.11719
4	.20508
5	.24609
6	.20508
7	.11719
8	.04395
9	.00977
0	.00098

THE NULL HYPOTHESIS AND THE RESEARCH HYPOTHESIS

We will be taking some familiar items of symbolic notation–” = ,” “not=,” “>,” and “<“– and printing them in blue to convey the concepts of random variability and statistical significance. When you encounter these blue symbols, they are to be read as follows:

$a = b$	a and b will not significantly differ; that is, a will be equal to b, within the limits of random variability;
a not= *b*	a and b will significantly differ (there will be a difference between a and b that goes beyond what could be expected on the basis of mere random variability);
$a > b$	a will be significantly greater than b (a will be greater than b in a degree that goes beyond what could be expected on the basis of mere random variability);
$a < b$	a will be significantly smaller than b (a will be smaller than b in a degree that goes beyond what could be expected on the basis of mere random variability).

When performing a test of of statistical significance, it is a useful convention to distinguish between

- The particular hypothesis that the research is seeking to examine; e.g., that the experimental medication has some degree of effectiveness as a treatment of the disease;
- The logical antithesis of the research hypothesis; e.g., that the experimental medication has no effectiveness at all.

The second of these items is commonly spoken of as the null hypothesis, the word "null" in this context having essentially the meaning of "zero." Hence its conventional symbolic notation, H_0, which is H for "hypothesis," with a subscripted zero. The research hypothesis, on the other hand, sometimes also spoken of as the experimental hypothesis or alternative hypothesis, is typically symbolized as H_1. In general, the null hypothesis is to the effect that the observed result–e.g., the number of patient recoveries–will not significantly differ from mean chance expectation; in other words, that it will be equal to MCE, within the limits of random variability:

$$H_0\text{: Observed value } = \text{ MCE}$$

And the research hypothesis is to the effect that there *will* be a significant difference between the observed result and MCE.

$$H_1\text{: Observed value not= MCE}$$

The specific form of the research hypothesis can be either directional or non-directional, depending upon the particular question that is at issue in the research. The version of H_1 that we have just given is non-directional, in that it asserts there will be a difference between the observed result and MCE, but makes no claim about the particular direction of the difference. It is tantamount to saying: There will be a significant difference between the observed result and MCE, though we have no basis for predicting in just which direction the difference will go, and will therefore take a significant difference in either direction as support for our hypothesis. A directional hypothesis, on the other hand, is one that does specify the particular direction of the difference. The investigators did not merely hypothesize that the observed number of patient recoveries would be different from MCE; their specific hypothesis was that it would be greater than MCE:

$$H_1\text{: Observed value} > \text{MCE}$$

Had they instead been examining the effect of sleep deprivation on the outcome of the disease, they might have had reason to expect that the number of recoveries would be smaller than MCE.

$$H_1\text{: Observed value} < \text{MCE}$$

A non-directional hypothesis is essentially an either-or combination of the two forms of directional hypotheses. Thus, when we put forward a non-directional hypothesis of the general type

$$H_1\text{: Observed value not= MCE}$$

what we are really saying is

$$H_1\text{:or} \begin{array}{l} \text{observed value} > \text{MCE} \\ \\ \text{observed value} < \text{MCE} \end{array}$$

The logical equivalent of a non directional research hypothesis.

Although this form of research hypothesis is conventionally spoken of as non-directional, you will see from this breakdown of it that the more accurate label would be bi-directional.

DIRECTIONAL VERSUS NON-DIRECTIONAL RESEARCH HYPOTHESES AND ONE-WAY VERSUS TWO-WAY TESTS OF SIGNIFICANCE

The distinction between a directional and a non-directional hypothesis is not just a matter of using the right words in the right place. It also marks a very important distinction in logic. To illustrate the nature of this distinction, return for a moment to that fanciful mind-over-matter coin-toss example introduced at the beginning; except now we will examine it in two different versions.

Scenario 1. As we described it at the time, our claimant of paranormal powers was aiming to produce "an impressive number of heads, to the point where you will at least be willing to take my claim seriously." Putting him to the test, we toss the coin 100 times and find that 59 of these tosses come up as heads. As the observed number of heads is only 9 greater than the 50 that we would calculate as MCE, you are perhaps not very impressed by this result at first glance. Upon closer examination, however, you find that this particular binomial outcome (with N=100, k=59, p=.5 and q=.5) yields a z-ratio of +1.7 which, is associated with a probability of only P=.0446.

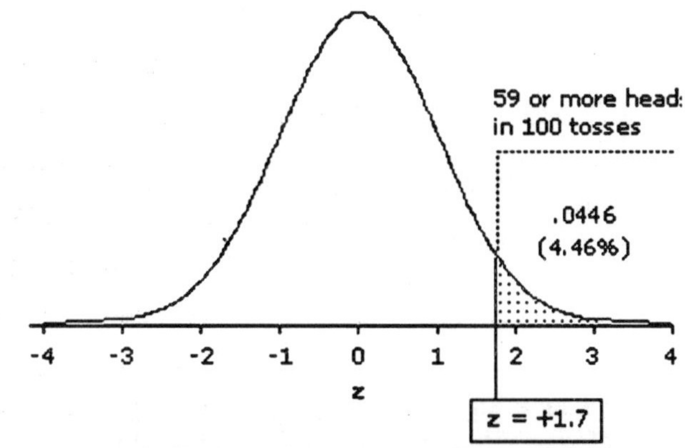

Fig. Location of z=+1.7 in the Unit Normal Distribution

That is: Of all the combinations of heads and tails that might occur by mere chance in 100 tosses of a coin, only 4.46% of them would include as many as 59 heads. So the result in this scenario is significant a bit beyond the conventional .05 level.

Scenario 2. Suppose that our mind-over-matter claimant had instead asserted the following:

"Although the power within me is substantial, I cannot control its direction. Sometimes it prefers heads, sometimes tails, and all that I as its humble instrument can do is go with the flow of it. In the absence of such a power, you would expect the outcomes of any particular set of coin tosses to approximate a 50/50 split of heads and tails. What I will do is produce a split significantly different from the 50/50 that you would expect on the basis of mere chance coincidence–that is, either significantly more than 50% heads or significantly more than 50% tails." -

Putting him to the test, we toss the coin 100 times and find that 59 of these tosses come up as heads.

Although both of these scenarios have a total of 100 tosses and end up with 59 heads instead of the 50 of MCE, the logic of probability in the two cases is quite different. In Scenario 1 the research hypothesis is directional, explicitly stating at the outset that the observed number of heads will be greater than the MCE value of 50:

$$H_1\text{: observed no. of heads} > \text{MCE}$$

while in Scenario 2 it is non-directional, which is to say, bi-directional.

$$H_1 \begin{cases} \text{observed no. of heads} > \text{MCE} \\ \text{or} \\ \text{observed no. of tails} > \text{MCE} \end{cases}$$

So when you assess statistical significance in these two scenarios, you are really asking two quite different questions. In Scenario 1 the question takes the specific directional form

In 100 tosses of a coin, what is the probability of getting an excess of heads this large or larger?

whereas in Scenario 2 it takes the form

In 100 tosses of a coin, what is the probability of getting an excess of either heads or tails this large or larger?

And clearly, the mere-chance probability for "either heads or tails" is greater than for just "heads." Except now we show the probabilities for both sides of the either/or outcome. The probability of ending up with as many as 59 heads is P=.0446, whereas the probability for ending up with as many as 59 heads or as many as 59 tails is twice that amount, P=.0892.

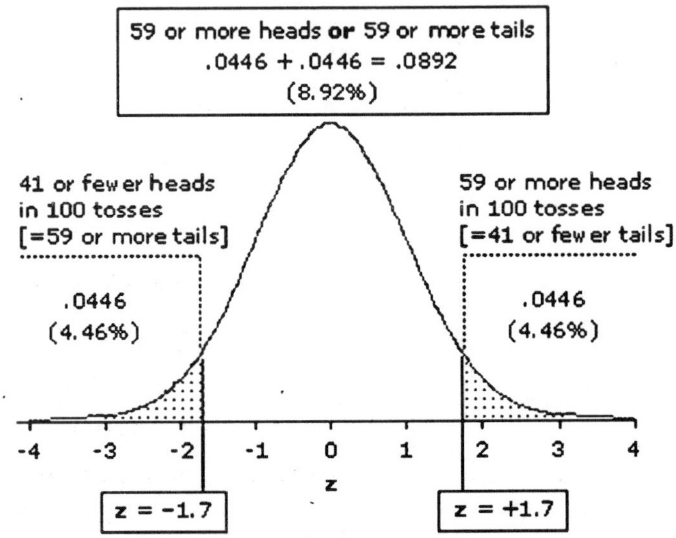

Fig. Locations of z=±1.7 and in the Unit Normal Distribution

Thus, by the conventional standard of statistical significance (P<.05), the observed result in Scenario 1 is significant (P=.0446), while the one in Scenario 2 is not (P=.0892). As you first encounter this logical distinction between directional and non-directional hypotheses, you are likely to find it a bit bizarre; for, after all, the observed fact is what it is, irrespective of whatever expectations we might have had about it prior to its occurrence. The 100 tosses and 59 heads that occur in Scenarios 1 and 2, respectively, do not materially differ from one another in any discernible way. So why should the probability of one 59-heads outcome be .0446 and the probability of the other be .0892? The answer is actually a rather simple one, though it will need to move around in your mind a bit before it comfortably settles down. It is simply that, when we perform a test of statistical significance, what we are testing is not the observed fact in and of itself, but rather the hypothesis to which the fact is connected. In general, the assessment of probability in connection with a directional hypothesis is spoken of as a one-way test, while the assessment for a non-directional (i.e., bi-directional) hypothesis is called a two-way test. Because such tests often refer to the left- and right-side "tails" of a symmetrical sampling distribution, they are also often described as one-tail and two-tail tests. As a practical rule of thumb, the probability of any particular observed result assessed by a two-way test is either exactly or approximately twice the probability that would be assessed for a one-way test.

The logic of the non-directional hypothesis is also the default position when there is no prior hypothesis at all. For example, imagine that you cut class on the day that the experiments of Scenarios 1 and 2 take place and thus know nothing about them. To amuse yourself, as well as to take the edge off the shame you rightly feel for cutting class, you toss a coin 16 times to see how closely the outcomes approximate the 8 heads and 8 tails that you now know would be calculated as MCE. Counting up a total of 12 heads among the observed outcomes, and only 4 tails, you excitedly say to yourself: "Hey, look at all those heads! P=.04. I must have the power of mind over matter!"

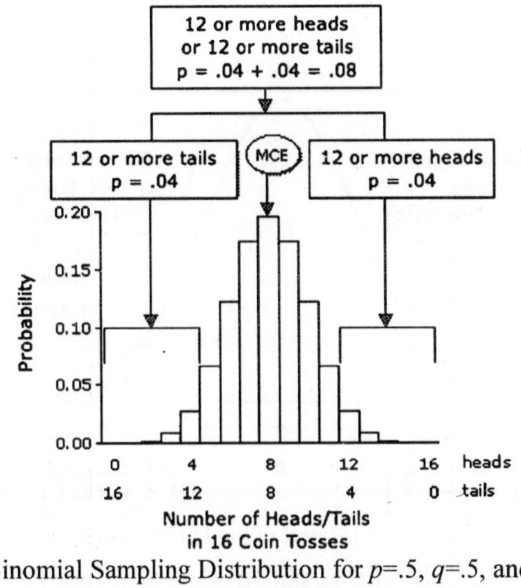

Fig. Binomial Sampling Distribution for p=.5, q=.5, and N=16

But the fact is that you were not expecting or aiming for heads. Essentially, you were saying "Let's do this and see what happens"; and your hypothesis, at most, was that the obtained result might prove different from MCE, in one direction or the other. The proof of this logical pudding is that, if the outcome were 12 tails instead of 12 heads, your reaction would be exactly the same, except now it would be occasioned by the obverse side of the coin: "Hey, look at all those tails! P=.04. I must have the power of mind over matter!" Given the bi-directional nature of your implicit hypothesis, the probability that would be assessed for either of these outcomes would not be .04, but rather .04+.04=.08.

Chapter 5

Basic Concepts of Probability

Imagine you are sitting in your statistics class one day when a man walks in and proclaims:

And so your class decides to put him to the test. One student stands in the front left corner of the room to toss the pennies one-by-one; another is positioned nearby to record the head or tail outcome of each toss; and the man who has made the claim stands in the right rear corner of the room, attempting to exert the sheer power of his thought upon the outcomes. The rest of the class watches closely to make sure that there is no hanky-panky or collusion.

You can begin eliciting these intuitions by asking yourself the question: How many heads would have to turn up in the 100 penny tosses before I would find the results "impressive" enough to take the man's claim seriously? When I ask the students in my own statistics classes to reflect on this question, an occasional ardent skeptic will proclaim that he or she would not be impressed even if 100% of the tosses turned up as heads. Most, however, say that their threshold for being sufficiently impressed falls somewhere between 70% and 80%, that is, between 70 and 80 heads out of 100 tosses. Never in all my years before the blackboard have I found a student who would be impressed with only 50% heads, nor with anything less than 50% heads.

But of course not, you will say–obviously! And indeed it is obvious–but do pause for a moment to consider why you find it so. I expect it comes down to something like this. Somewhere within you is the idea that a standard minted coin, on any particular toss, has a 50% chance of coming up heads and a 50% chance of coming up tails. And then there is a process of inference. For some students it might take place step-by-step as a deliberate conscious process; for others it is perhaps a kind of global intuitive leap. Either way, here is the underlying logic of it. If any particular toss of a coin has a 50% chance of coming up heads and a 50% chance of coming up tails, then in any multiplicity of coin tosses we would expect the total numbers of heads and tails outcomes to be about evenly divided, about half-and-half, 50/50. Thus you would not be impressed if our mind-over-matter claimant came up with only 50 heads out of 100 tosses, because that is just about what you would expect of any set of coin tosses, irrespective of whether anyone was there trying to zap the tosses with the sheer power of thought.

I expect you would also not be impressed by 51 heads out of 100, nor 52, nor 53. But what about 63, or 73, or 83? On the other hand, all but the most fervent skeptic would be impressed with 100 heads out of 100.

But what about 95, or 85, or 75? In brief, where do you draw the line? Somewhere between 50 heads and 100 heads is a point where a rational and open-minded person could reasonably say "I am impressed." Intuitively, you know the line is there, and you probably have an intuitive hunch of its general vicinity. Let us suppose that after reflecting on this question you decide that the line falls somewhere in the vicinity of 70%. Anything less than 70 heads out of 100 will not impress you; an even 70 heads out of 100 will begin to impress you; 71 out of 100 will impress you a bit more; 72 out of 100 still more; and so on.

Now for another question, also aimed at eliciting some intuitions that you already have about the subject. Suppose that the test of the man's claim had involved only 10 penny tosses. The question is, would you in this case still draw the line at 70%? That is, would you be anywhere near as impressed with 7 heads out of 10 tosses as you would be with 70 heads out of 100 tosses? I expect your answer to this question will be an immediate "no," based on a strong underlying intuition to the effect that a 70% line could be much more easily reached or exceeded with only 10 tosses than it could be with as many as 100 tosses. Later we will examine inferential procedures by which you can determine that the respective likelihoods for these two outcomes occurring by mere chance are

Outcome	Likelihood
>70% heads in 10 tosses:	17%
>70% heads in 100 tosses:	0.005%

The first of these would fall far short of the standard 5% criterion of statistical significance (i.e., equal to or less than 5%), while the second would far surpass it. So if our mind-over-matter claimant actually were to get as many as 70% heads in 100 tosses, you could allow yourself to be very impressed indeed–assuming of course that the pennies were not somehow mechanically biased in favor of heads, that there was no collusion, that the outcomes of the tosses were accurately recorded, and so on. For it is very highly unlikely (5/1000ths of 1%) that he or anyone else would get as many as 70% heads in 100 tosses by mere chance coincidence.

But more of such details later. First we must lay some foundations. A couple of centuries ago the great mathematician Laplace observed that the theory of probability is "at bottom only common sense reduced to calculation." I think a better way of putting it would be to say that the theory of probability is common sense expanded by calculation–for in fact there are many aspects of the subject that go well beyond common sense, and some that are even flatly contradictory of common sense.

Still, the conceptual and computational apparatus of probability does have its roots in common sense, and for that reason you will find you have a substantial head start in the task of studying it. As we begin developing the subject, please do not be lulled if you find some of the illustrative examples—coin tosses and the like—to be rather trivial. Keep your eye not on the examples, but on the general concepts that lie behind them. And bear in mind that these concepts are the foundation of all statistical inference, and thus of virtually everything else in this text that will follow.

ELEMENTARY PROBABILITIES: *"COMMON SENSE REDUCED TO CALCULATION"*

The basic concept is an idea of utter simplicity. Imagine you have four small balls, similar in every respect except that one is red and the other three are blue. Place the balls in a box, close the lid, and shake the box so as to jumble the order of its contents. Now reach into the box and blindly withdraw one ball–but before you do so, place a bet on whether the ball you draw will be red or blue. Assuming that you would want to place your bet rationally rather than just arbitrarily, the task is to determine which of the two colors has the greater chance of being drawn. And that is tantamount to asking which color has the greater probability of being drawn. The common sense of it is this. If you are blindly drawing 1 ball out of 4, and if these 4 balls include 3 blue and 1 red, then you have 3 chances out of 4 of drawing a blue ball, but only 1 chance out of 4 of drawing a red one. Thus the rational choice would be to bet on blue, for while you would still have 1 chance out of 4 of losing your bet, you would have 3 chances out of 4 of winning it. A bet on red, on the other hand, would have only 1 chance out of 4 of winning and 3 chances out of 4 of losing.

Although the calculations to which this common sense is "reduced" can sometimes grow quite complex, the basic operation is just elementary arithmetic. It amounts to taking common-sense concepts such as "3 chances out of 4" and converting them into a meaningful and useful numerical form. In general, for any common-sense concept of the form "this particular evext has x chances out of y of occurring," the probability of that event can be defined numerically as the ratio of x to y. Thus the probability, P, of drawing a blue ball (with 3 chances out of 4) is

$$P_{(blue)} = 3/4 = .75$$

and the probability of drawing a red ball (with 1 chance out of 4) is

$$P_{(red)} = 1/4 = .25$$

So far, so good. Calculating that the respective probabilities are $P_{(blue)} = .75$ and $P_{(red)} = .25$, you rationally place your bet on drawing a blue ball. The bet is down, you blindly reach into the box and draw a ball—and it is red! The moral of this scenario is that a probability value such as $P_{(blue)} = .75$ or $P_{(red)} = .25$ is not a reliable predictor of the outcome of a singular event. It is a reliable predictor only in reference to a multiplicity of events; and the greater the number of such events, the more reliable the prediction. For the present example, the practical predictive meaning of the two probability values is this: If you were to perform the ball-drawing operation an indefinitely large number of times (returning the drawn ball to the box and shaking the box prior to each new draw), you would draw one or another of the blue balls in 75% of the cases and the red ball in 25% of the cases. If you were to do it only a few times, your observed percentages of blue and red would perhaps differ markedly from these theoretical values of 75% and 25%; however, the more you repeated the operation, the closer they would come. To take another example, imagine a statistics class containing 30 students, of whom 7 are freshmen, 12 are sophomores, 10 are juniors, and one is a senior. If you were to select one student at random from this class, what is the probability that the student you select will be a freshman? As there is a total of 30 students in the class, of whom exactly 7 are freshmen, that probability is

$$P_{(freshman)} = 7/30 = .2333$$

By the same reasoning, the probability of selecting a sophomore is

$$P_{(sophomore)} = 12/30 = .40$$

and so on for the categories junior and senior:

$$P_{(junior)} = 10/30 = .3333$$

$$P_{(senior)} = 1/30 = .0333$$

These examples illustrate what is known as the relative frequency concept of probability, so named because it defines the probability of an event in terms of the number (or frequency) of possibilities favorable to the occurrence of that event, relative to the total number of possibilities. Thus, in its general conceptual structure, the probability of the occurrence of a certain event, x, is framed as

$$P_{(x)} = \frac{\text{number of possibilities favorable to the occurrence of } x}{\text{total number of pertinent possibilities}}.$$

So the probability of randomly drawing a blue ball from a box is

$$P_{(blue)} = \frac{\text{the total number of blue balls in the box}}{\text{the total number of balls in the box}}$$

and the probability of randomly selecting a sophomore from a statistics class is

$$P_{(sophomore)} = \frac{\text{the number of sophomores in in the class}}{\text{the total number of students in the class}}.$$

Similarly, if we were to select at random one person from a room full of persons, the probability of selecting a woman would be

$$P_{(woman)} = \frac{\text{the number of women in in the room}}{\text{the total number of persons in the room}}.$$

It will be fairly evident that probability values determined in this fashion will always fall somewhere between P=0 and P=1.0, inclusive; for the operation always involves the division of one number by another number, and the divisor is in every case equal to or larger than the number it divides. Abstract though this point might seem, it makes direct contact with common sense. If there are no women in a room of 100 persons, then the probability of randomly selecting a woman from that room is nil. Hence

$$P_{(woman)} = 0/100 = 0$$

Conversely, if the room with 100 persons contains *only* women, the probability of selecting a woman is what common sense would call one-hundred percent:

$$P_{(woman)} = 100/100 = 1.0$$

If the number of women in the room is greater than zero but less than 100, then the probability of selecting a woman will fall somewhere in between P=0 and P=1.0; and the greater the proportion of women, the greater the probability. That, indeed, is exactly what a probability value is–a statement of proportion. Multiply any probability value by 100, and you convert it into a common-sense statement of percentage.

There are certain kinds of situations where the particular values for "number of possibilities favorable to the occurrence of *x*" and "total number of pertinent possibilities" can be precisely known in advance, either by physically counting all the relevant possibilities, or else by enumerating them through logical analysis. In a case of this sort, the resulting probability value is said to be determined a priori. With our example of balls in a box, for instance, we know in advance that the box contains exactly 4 balls, of which exactly 3 are blue. Nothing else is needed. Given these prior facts, we can then proceed by "pure logic" to conclude that the probability of blindly drawing a blue ball from the box is exactly $P_{(blue)} = .75$. A variation on the theme of a priori probability reasoning is illustrated by our introductory example of tossing pennies. From everything we know about the physical properties of pennies and the physical principles involved in flipping them, it is reasonable to assume that the two possible outcomes, heads and tails, are equally likely. Thus, $P_{(head)}$=1/2=.5 and $P_{(tail)}$=1/2=.5.

There are many other kinds of situations, however, where the appropriate probability value cannot be known precisely in advance, but rather must be estimated on the basis of observing a large number of actual instances. Here the determination is said to be a posteriori. This is what lurks behind the scenes when a meteorologist tells us there is a 30-percent probability of rain today, or when a physician tells a patient there is an 87-percent chance that a certain surgical procedure will be successful. For the meteorologist it is an estimate based on observations of the frequency of rain in the past under similar combinations of temperature, humidity, atmospheric pressure, and other relevant factors:

$$P_{(rain)} = \frac{\text{number of previously observed similar days that produced rain}}{\text{total number of previously observed similar days}}$$

while for the physician it is an estimate based on the previously observed success rate of this particular surgical procedure for patients of this particular age, gender, physical condition, and so forth:

$$P_{(success)} = \frac{\text{number of similar patients for whom the surgery proved successful}}{\text{total number of similar patients on whom the surgery was performed}}.$$

COMPOUND PROBABILITIES

The usefulness of "reducing" common-sense concepts of probability to precise statements of proportionality is that they can then be subjected to the powerful analytical apparatus of mathematical calculation. Under certain circumstances they can be added, subtracted, multiplied, or divided. And while these operations themselves are elementary and commonplace, they end up telling us things about probability that go far beyond the raw intuitions of common sense.

There are basically two ways in which individual probability values can be linked together mathematically, and these in turn correspond to two basic kinds of logical linkage. The first is associated with the common-sense meaning of the word "and," and the second with the common-sense meaning of the word "or." In formal logic, the relationships denoted by these words are spoken of as conjunction and disjunction, respectively. Conjunctive probability questions take the general form, "What is the probability of having A and B occur?" and disjunctive questions take the form, "What is the probability of having A or B occur?" (Compound probabilities are sometimes described in the language of set theory. In this case, conjunction will be spoken of as "intersection" and disjunction will be described as "union.")

CONJUNCTIVE PROBABILITIES: 'A AND B', 'A AND B AND C', ETC.

Here again the basic concept is firmly rooted in common sense. Suppose you were to toss a penny twice. There is a 50% chance of getting a head on the first toss; and then, if you do get a head on the first toss, there is also a 50% chance of getting a head on the second toss. The probability that you will get a head on the first toss and on the second toss is therefore 50% of 50%. The rest is just elementary arithmetic. One-half of one-half is one-quarter; 50% of 50% is 25%; or in decimal form, .5x.5=.25. If you were to toss a penny three times, the probability that all three tosses would come up heads would be 50% of 50% of 50%, which is 12.5%; or again in decimal form, .5x.5x.5=.125. And so on for four tosses, five, ten, a hundred, or a thousand.

The general principle is that conjunctive probabilities are linked together by the mathematical act of multiplication. Thus, in the abstract, the probability of having A and B occur is the product of the two separate component probabilities for A and B:

$$P_{(A\ and\ B)} = P_{(A)} \times P_{(B)}$$

The probability of having A and B and C occur is the product of the three separate component probabilities for A and B and C:

$$P_{(A\ and\ B\ and\ C)} = P_{(A)} \times P_{(B)} \times P_{(C)}$$

and so on.

In the simple case of repeatedly tossing a coin, the probability of getting a head on any particular toss is completely independent of the outcome of any other toss, past, present, or future. If you get a head on the first toss, the probability of getting a head on the second toss is $P_{(H)} = .5$; and if you get a tail on the first toss, the probability of getting a head on the second toss is also $P_{(H)} = .5$. No matter how many times you have tossed the coin, no matter how many heads have already come up, or how many tails, the probability of getting a head on the next toss is still exactly $P_{(H)} = .5$. There are many other kinds of situations, however, where the probability of an event is not independent but dependent–that is, where the probability of one event *depends* on the outcome of some other event.

We will illustrate this distinction with an example that will also show how calculated probabilities sometimes end up telling a story quite different from what common-sense intuitions might have led you to expect. Imagine a room that contains 4 females and 6 males. Question 1: If you were to select 3 persons from this room at random, what is the probability that all 3 would be females? And question 2: If you were to select 3 persons from this room at random, what is the probability that all 3 would be males? Since the room contains more males than females, you will surely see intuitively that the probability for all-3-males will be greater than the probability for all-3-females. But I expect it will not be intuitively obvious to you at all that the probability for all-3-males is more than five times as great as the probability for all-3-females. Here is how it works. The critical point to take note of is that the probability of the outcome for each successive draw, after the first, depends on the outcome(*s*) of the preceding draw*s*.

First for the possibility that all 3 of the persons selected will be females. Clearly, the probability of selecting a female on the first draw is the ratio 4/10, since there are 10 persons in the room, of whom 4 are females. But then, once you have made your first selection, there remain only 9 persons in the room from whom to make your second selection; and if your first selection *is* a female, then only 3 of the remaining persons are females. Thus, if your first selection happens to be a female, the probability of selecting a female on the second draw is not 4/10, but rather 3/9.

After the second selection there are only 8 persons left in the room; and if both of the persons already drawn are females, then only 2 of these 8 remaining are females. Thus, the probability that the third draw will also be a female is not 4/10 nor 3/9, but rather 2/8. The probability of selecting females in all three draws in this situation is therefore

$$P_{(\text{all 3 females})} = (4/10) \times (3/9) \times (2/8) = .033$$

The logic is the same for the possibility of selecting males on all 3 draws. The probability of selecting a male on the first draw is 6/10, since the room contains 10 persons, of whom 6 are males. But then, once you have made your first selection, there remain only 9 persons in the room from whom to make your second selection; and if your first selection is a male, then only 5 of the persons remaining are males. Thus, if your first selection happens to be a male, the probability of selecting a male on the second draw is not 6/10, but rather 5/9. After the second selection there are only 8 persons left in the room; and if both of the persons already drawn are males, then only 4 of these 8 remaining are males. Thus, the probability that the third draw will also be a male is not 6/10 nor 5/9, but rather 4/8. The probability of selecting males on all three draws in this situation is therefore

$$P_{(\text{all 3 males})} = (6/10) \times (5/9) \times (4/8) = .167$$

which you will note is about 5 times larger than the P=.033 probability of selecting females on all three draws.

In both of these calculations the basic principle is the same as the one we outlined at the beginning of this section: the conjunctive probability of any two or more events is equal to the product of their separate component probabilities. It is simply a matter of making sure that the values assigned to the component probabilities are the *right*ones.

DISJUNCTIVE PROBABILITIES: 'A OR B', 'A OR B OR C', ETC.

The principle of disjunctive probabilities is even simpler. If you toss a coin, there is a 50% chance that it will come up heads, a 50% chance that it will come up tails, and thus a 100% chance that it will come up either heads or tails. With each lottery ticket you purchase you have a 1 in 100 million chance of winning. With two lottery tickets your chance of winning with either the first or the second is therefore 2 in 100 million. With three tickets, the chance of winning with either the first or the second or the third is 3 in 100 million; and so forth. When two or more chance events are disjunctively linked by the word "or," the corresponding mathematical linkage is just simple addition. Thus, the probability of having A or B occur is equal to the sum of the two component probabilities for A and B:

$$P_{(A \text{ or } B)} = P_{(A)} + P_{(B)}$$

The probability of having A or B or C occur is the sum of the three separate component probabilities for A and B and C:

$$P_{(A \text{ or } B \text{ or } C)} = P_{(A)} + P_{(B)} + P_{(C)}$$

and so on.

To put it concretely, imagine a class of 30 students composed of 4 freshmen, 12 sophomores, 10 juniors, and 4 seniors. The instructor announces that one of these students will be randomly selected to win the class lottery prize, which is an automatic A in the course. The probability that the lucky winner will be either a freshman or a sophomore is

$$\frac{4 \text{ freshmen}}{30 \text{ students}} + \frac{12 \text{ sophomores}}{30 \text{ students}} = \frac{16}{30} = .53$$

and the probability that it will be either a freshman or a sophomore or a junior is

$$\frac{4 \text{ freshmen}}{30 \text{ students}} + \frac{12 \text{ sophomores}}{30 \text{ students}} = \frac{10 \text{ juniors}}{30 \text{ students}} = \frac{26}{30} = .87.$$

The restriction on this additive operation is that component probabilities can be added together in this simple fashion only when they pertain to possibilities that are *mutually exclusive.* In the example that we have just examined, each of the four academic-class categories excludes all the others. If you are a freshman, you cannot at the same time be a sophomore or a junior or a senior. If you are a sophomore, you cannot at the same time be a freshman or a junior or a senior. And so forth.

But now consider the following variation on this classroom theme. Suppose the class has a total of 26 students, of whom 12 are sophomores and 14 are juniors. Of the sophomores, 7 are females and 5 are males; and of the 14 juniors, 8 are females and 6 are males. Question: In randomly selecting one of the members of this class to win the lottery, what is the probability that the student selected will be either a sophomore or a female?

Here is how not to answer the question. The probability of selecting a sophomore is 12/26 (right!), and the probability of selecting a female is 15/26. The probability of selecting either a sophomore or a female is therefore

$$P_{(soph.\ or\ female)} = P_{(soph.)} + P_{(female)}$$
$$= \frac{12}{26} + \frac{15}{26} = \frac{27}{26} = 1.038 \text{ [Wrong!].}$$

Applying the simple additive formula in this situation would lead you to conclude that the chance of selecting either a sophomore or a female is about 104%—which is patently absurd. For reasons described earlier, the probability that any particular event or outcome will occur must always fall within the range bounded at the bottom by P=0 (0%) and at the top by P=1.00 (100%).

The reason why simple addition does not work in this situation can be gleaned by going back and looking closely at some details. When you say that the component probability of selecting a sophomore is 12/26, what you are really saying is

$$P_{(soph.)} = \frac{\text{7 soph. females + 5 soph. males}}{\text{26 students}} = \frac{12}{26}.$$

And when you say that the component probability of selecting a female is 15/26, what you are really saying is

$$P_{(female)} = \frac{\text{7 soph. females + 8 jr. females}}{\text{26 students}} = \frac{15}{26}.$$

In and of themselves, these two component probabilities are quite correct—but now see what happens when you add them together.

$$\begin{matrix} P_{(soph.)} & & \dfrac{\text{7 soph. females + 5 soph. females}}{\text{26 students}} & & \dfrac{12}{26} & & \\ + & = & + & = & + & = & \dfrac{27}{26} = 1.038. \\ P_{(female)} & & \dfrac{\text{7 soph. females + 8 jr. females}}{\text{26 students}} & & \dfrac{15}{26} & & \end{matrix}$$

The complication here is that the two components, sophomore and female, are not mutually exclusive. Seven of the 26 students are both sophomores and females, and these 7 students are being counted twice: once because they are sophomores, and then again because they are females. If you are counting apples in a basket and mistakenly count some of them twice, you end up with an inflated measure of the number of apples. If you are counting the elements that enter into a probability calculation and count some of them twice, you end up with an inflated measure of probability.

The method commonly recommended for correcting this inflation is to calculate the disjunctive probability of "A or B" by simple addition, as we have just done, and then subtract from that inflated sum the conjunctive (multiplicative) probability of "A and B." That is, for cases where the component probabilities are not mutually exclusive

$$P_{(A \text{ or } B)} = P_{(A)} + P_{(B)} - P_{(A \text{ and } B)}$$

The reason why this method works is that in subtracting the conjunctive probability of "A and B," you are in effect removing from the inflated sum the amount that was counted twice.

There are, however, two limitations of this method. The first is that the elemental probabilities, $P_{(A)}$ and $P_{(B)}$, are sometimes not independent, in which case the conjunctive portion of the above formula, "$P_{(A \text{ and } B)}$," will involve some more or less complicated conditional probabilities. Here again is an illustration of how not to do something.

$$\begin{aligned} P_{(soph. \text{ or } female)} &= P_{(soph.)} + P_{(female)} - P_{(soph \text{ and } female)} \\ &= (12/26)+(15/26) - [(12/26)(15/26)] \\ &= 1.038 - .266 \\ &= .772 \quad \text{[Wrong!]} \end{aligned}$$

Although this result is not obviously preposterous, as was our earlier P=1.038, a moment's reflection will show that it clearly cannot be correct. Of the 26 students, 6 are junior males, which is to say, neither sophomores nor females. So the probability of not selecting a sophomore or a female is 6/26=.231; and the probability of selecting someone who is not a junior male, but rather either a sophomore or a female, is accordingly 1–.231=.769.

The first result, .772, is not merely wrong by a certain small amount. It is wrong fundamentally and in principle, because the probability of selecting someone who is both a sophomore and a female is not simply

$$\begin{array}{ccc} P_{(soph.)} & \times & P_{(female)} \\ \dfrac{12 \text{ soph.}}{26 \text{ students}} & \times & \dfrac{15 \text{ females}}{26 \text{ students}} \end{array} = .266 \ \text{[Wrong!]}.$$

Sure enough, the probability of selecting a sophomore is 12/26 (the number of sophomores divided by the total number of students). But then, if you do select a sophomore, the probability that this particular one of the 12 sophomores will also be a female is not 15/26 (the number of females divided by the total number of students), but rather 7/12 (the number of female sophomores divided by the total number of sophomores). Hence

$$\begin{array}{ccc} P_{(soph.)} & \times & P_{(female)} \\ \dfrac{12 \text{ soph.}}{26 \text{ students}} & \times & \dfrac{7 \text{ soph. females}}{12 \text{ soph}} \end{array} = \dfrac{7 \text{ soph. females}}{26 \text{ students}} = .269 \ \text{[Correct!]}.$$

So the true probability of selecting either a sophomore or a female, correcting for inflation and taking account of the overlap between "sophomore" and "female," is

$$P_{(soph.\text{ or }female)}$$
$$= P_{(soph.)} + P_{(female)} - P_{(soph\text{ and }female)}$$
$$= (12/26) + (15/26) - [(12/26)(7/12)]$$
$$= 1.038 - .269$$
$$= .769 \text{ [Correct!]}$$

which is of course identical with our earlier calculation of 1–.231 =.769.

The second limitation is that even when the elemental probabilities are independent, the method will not work when there are more than two items in the disjunction (A or B or C; A or B or C or D; etc.). For example, suppose you are tossing 3 coins, A, B, and C. What is the probability of getting a head (H) on either A or B or C? When you set it up in the manner just described, the result is

$$P_{(H[A]\text{ or }H[B]\text{ or }H[C])}$$
$$= P_{(H[A])} + P_{(H[B])} + P_{(H[C])} - P_{(H[A]\text{ and }H[B]\text{ and }H[C])}$$
$$= (.5 + .5 + .5) - (.5 \times .5 \times .5) = 1.375 \text{ [Wrong!]}$$

which, again, is patently absurd.

The method that *will* work in such multi-item disjunctions, and which I recommend for two-item disjunctions as well, is actually quite a simple one. In any particular situation, there is a certain probability that the event or outcome in question will occur, and a certain probability that it will not occur; and taken together, these two complementary probabilities must always add up to $P_{(Total)}$=1.0. Thus, one way of determining the probability that a disjunction ("A or B," "A or B or C," etc.) will occur is to figure out the probability that it will not occur, and then subtract that amount from 1.0. That is

$$P_{(\text{that} \times \text{will occur})}$$
$$= P_{(\text{Total})} - P_{(\text{that} \times \text{will nor occur})}$$
$$= 1 - P_{(\text{that} \times \text{will nor occur})}$$

For example, when tossing 3 coins, A, B, and C, the only way you could not get the disjunctive outcome of a head (H) on either A or B or C would be to get the conjunctive outcome of a tail (T) on A and B and C. The probability of getting a tail on all 3 of the tosses is

$$.5 \times .5 \times .5 = .125.$$

So the complementary probability of not getting a tail on all 3 tosses, which can happen only if you get a head on at least one of the tosses–A or B or C–is

$$1.0 - .125 = .875.$$

In principle, virtually any probability situation you might ever encounter within the context of inferential statistics can be seen as a more or less complex combination of elemental conjunctive and disjunctive probabilities. In actual practice you will not normally need to see it this way, because most inferential statistical procedures are streamlined to the point where these elemental constituents are no longer visible. Nonetheless, it is important that you have a sense of what is going on behind the scenes in these streamlined procedures. In the next section we will examine some of the basic ways in which conjunctive and disjunctive probabilities can be combined, we will see how these potentially very complex and unwieldy combinations can be transformed into structures of elegant simplicity.

CONJUNCTION AND DISJUNCTION

An imaginary example taken from a medical context. There is a certain disease that has an abrupt and unmistakable onset, and for which there is currently no effective treatment. Of all the persons who come down with this disease, 40% spontaneously recover within two months and the remaining 60% do not recover within two months. As there is no way of predicting which patient will recover and which will not, the two-month outcome for any particular patient is regarded as a matter of mere chance. Thus, for any particular patient who comes down with the disease, the probability of spontaneously recovering within two months is

$$P_{(recovery)} = .4$$

and the probability of not recovering within two months is

$$P_{(non\text{-}recovery)} = .6$$

Then comes a report from a field botanist working in the Brazilian rain forests of a certain plant that the indigenous peoples of the region have long used to treat the disease, apparently with some degree of success. A team of researchers reasons that perhaps the plant contains an ingredient that could be developed into an effective medical treatment for the disease. But as the isolation and refinement of that ingredient would surely prove quite costly, they first undertake to determine experimentally whether an extract from the plant, in its raw form, shows any effectiveness at all in treating the disease. In the next few paragraphs we will be describing their two-stage effort to put the raw plant extract to an experimental test—first, with a small sample of patients; and then, when the first experiment proves promising, with a much larger sample. We will stipulate that the researchers determine beforehand that the raw plant extract has no toxic effects. Also, please note that in both of these investigations the experimental hypothesis is directional. The researchers are expecting that the extract will produce a recovery rate significantly greater than the 40% rate attributable to mere chance spontaneous recovery.

In their preliminary test of this hypothesis, the researchers administer the plant extract to 10 randomly selected patients, beginning for each patient immediately after the onset of the disease and continuing for two months. As the experiment runs its course, the investigators observe that 7 of the 10 patients have recovered within their respective two-month periods, while 3 have not recovered.

Now clearly, the 70% recovery rate observed in this experiment is greater than the 40% baseline rate for spontaneous recovery. But the question is, is it significantly greater? This is the question of statistical significance, which you will recall is essentially a question of how likely or unlikely it is that the observed result could have been produced by nothing more than mere chance coincidence. Specifically for the present example, how likely or unlikely it is that as many as 7 of the 10 patients could have recovered by mere chance ("spontaneously"), if the plant extract was completely ineffective?

The first thing to notice about a question of this general type is that it has several layers. The first layer is indicated by the phrase "as many as," which reveals itself upon analysis to be simply another way of saying "this many or more." The logic of this point applies to scientific research in general. In most cases, the pertinent probability question is not, What is the probability of getting *exactly* this result? Rather it is, How likely is it that mere chance coincidence might have produced a result "as large as this," which is to say, "this large or larger?" (Sometimes the question is in the opposite direction: "... this small or smaller?") So the first layer of the question takes the form of a disjunction: How likely is it that mere chance coincidence could have produced as many as 7 recoveries out of 10, which is to say, either exactly 7 recoveries out of 10, or exactly 8 out of 10, or 9 out of 10, or 10 out of 10? Abbreviating "recoveries" as "*R*," the formal expression would be

$$
\begin{aligned}
&P_{(7 \text{ or more } R \text{ out of } 10)} \\
&\quad = P_{(7R \text{ or } 8R \text{ or } 9R \text{ or } 10R)} \\
&\quad = P_{(7R)} + P_{(8R)} + P_{(9R)} + P_{(10R)}
\end{aligned}
$$

So, essentially, it is just a task of adding up several component probabilities. The only complication is that, in order to add them, you first have to get them. We will introduce the logic of this type of task by examining a much simpler analogy involving coin tosses. If you toss 3 coins, A, B, and C, what is the mere-chance probability of getting as many as 2 heads, that is, either exactly 2 heads or exactly 3 heads? Here again (abbreviating "heads" as "H"), the first statement of the question takes the form of a disjunction:

$$
\begin{aligned}
&P_{(2 \text{ or more } H \text{ in } 3 \text{ tosses})} \\
&\quad = P_{(2H \text{ or } 3H)} \\
&\quad = P_{(2H)} + P_{(3H)}
\end{aligned}
$$

But once you start getting into its details, you find that even this relatively simple situation has a rather complex structure. The source of the complexity is that, while there is only one combination of heads (H) and tails (T) that will yield exactly 3 heads in 3 tosses:

A B C
H H H

there are several different ways of getting exactly two heads in 3 tosses, namely:

A B C
H H T
H T H
T H H

As illustrated in the following table, you can think of this complex situation as a network of probability pathways. The network has two main branches, labeled 2H and 3H, of which the first has three separate sub-branches, one for each of the three ways in which it is possible to get exactly 2 heads in 3 tosses. Each separate branch or sub-branch represents a conjunctive (multiplicative) probability, and where branches or sub-branches converge there is a disjunctive (additive) probability. Thus, there are 3 separate sub-pathways for reaching the outcome of exactly 2 heads, each with the conjunctive probability .5x.5x.5=.125. (Recall that $P_{(H)} = .5$ and $P_{(T)} = .5$.) The disjunctive probability of reaching this outcome one way or the other is therefore .125+.125+.125=.375. There is only one pathway for reaching the outcome of exactly 3 heads; it also has a conjunctive probability of .5x.5x.5=.125. Thus, the overall disjunctive probability of getting as many as 2 heads in 3 tosses (i.e., exactly 2 heads or exactly 3 heads) is .375+.125=.5.

Outcome		*Probability*	
	HHT	.5x.5x.5=.125	
2H	HTH	.5x.5x.5=.125	.375
	THH	.5x.5x.5=.125	
3H	HHH	.5x.5x.5=.125	.125
		Total =	.5

Let us now apply the same logic to the more complex example of our medical experiment. The task, recall, is to figure out the disjunctive probability of getting by mere chance, in the absence of any effective treatment, one or the other of the following potential outcomes:

7 patients recover and 3 do not recover; or

8 patients recover and 2 do not recover; or

9 patients recover and 1 does not recover; or
all 10 patients recover

We will begin with the most extreme outcome, because that is the simplest. Just as there is only one way of getting 3 heads in 3 coin tosses, there is only one way of getting 10 recoveries in 10 patients. Obviously that could occur only if patient 1 recovers, and patient 2 recovers, and patient 3 recovers, and so on through patient 10. (Recall that $P_{(recovery)} = .4$ and $P_{(non\text{-}recovery)} = .6$.) The probability for the extreme outcome of 10 recoveries out of 10 is therefore just the single conjunctive pathway

.1×.2×.3×.4×.Patient

1×.2×.3×.4×.5×.6×.7×.8×.9×10

.4×.4×.4×.4×.4×.4×.4×.4×.4×.4 = .000,105

As we move on to the less extreme outcomes, however, we begin to find the pathways branching into a multiplicity of sub-pathways. The extreme outcome of 10 recoveries out of 10 patients can be reached by only one route. The nearest less extreme outcome, 9 recoveries out of 10, on the other hand, can be reached by any one or another of 10 possible sub-pathways. That is, this outcome would be produced by any one or another of 10 different combinations of recoveries and non-recoveries. The first of these would be for the case where all patients except patient 1 recover; the second would be for the case where all except patient 2 recover; the third for the case where all except patient 3 recover; and so on, up through the case where all patients except patient 10 recover. As shown by the following calculations, this would constitute a total of 10 conjunctive sub-pathways for the potential outcome of 9 recoveries out of 10 patients, each with a probability equal to $.4^9$x.6=.000,157.

$.1\times.2\times.3\times.4\times$.Patient

$1\times.2\times.3\times.4\times.5\times.6\times.7\times.8\times.9\times10$

$$.6\times.4\times.4\times.4\times.4\times.4\times.4\times.4\times.4\times.4 = .4^9\times.6 = .000{,}157$$

$$.4\times.6\times.4\times.4\times.4\times.4\times.4\times.4\times.4\times.4 = .4^9\times.6 = .000{,}157$$

$$.4\times.4\times.6\times.4\times.4\times.4\times.4\times.4\times.4\times.4 = .4^9\times.6 = .000{,}157$$

$$.4\times.4\times.4\times.6\times.4\times.4\times.4\times.4\times.4\times.4 = .4^9\times.6 = .000{,}157$$

$$.4\times.4\times.4\times.4\times.6\times.4\times.4\times.4\times.4\times.4 = .4^9\times.6 = .000{,}157$$

$$.4\times.4\times.4\times.4\times.4\times.6\times.4\times.4\times.4\times.4 = .4^9\times.6 = .000{,}157$$

$$.4\times.4\times.4\times.4\times.4\times.4\times.6\times.4\times.4\times.4 = .4^9\times.6 = .000{,}157$$

$$.4\times.4\times.4\times.4\times.4\times.4\times.4\times.6\times.4\times.4 = .4^9\times.6 = .000{,}157$$

$$.4\times.4\times.4\times.4\times.4\times.4\times.4\times.4\times.6\times.4 = .4^9\times.6 = .000{,}157$$

$$.4\times.4\times.4\times.4\times.4\times.4\times.4\times.4\times.4\times.6 = .4^9\times.6 = .000{,}157$$

The disjunctive probability of reaching the outcome one way or the other is therefore the sum of these 10 sub-pathway probabilities, which (since the 10 separate probabilities are all the same) can be calculated multiplicatively as

$$10 \times (.4^9 \times .6) = .00157$$

The principle illustrated by this calculation applies to all such cases where we are interested in the probability of getting "10 out of 10," "9 out of 10," "22 out of 30," and so forth. Abstractly, we can speak of it as the probability of getting "k out of N," where k represents the first number in such an expression and N represents the second. Two other abstract terms that we will need for this formulation are p, which is the probability that the event in question (e.g., the recovery of a patient) will occur in any particular instance; and q, which is the complementary probability that the event in question will not occur. For all situations of this general type, the number of ways in which it is possible to get the result of "k out of N" (10 out of 10, 9 out of 10, 8 out of 10, and so on) is given by the formula

$$\text{number of ways}_{(k \text{ out of } N)} = \frac{N!}{k!(N-k)!}$$

and the probability of getting the result in any particular one of these ways is

$$p^k \times q^{N-k}$$

Put it all together and you have

$$P_{(k \text{ out of } N)} = \frac{N!}{k!(N-k)!} \times p^k \times q^{N-k}.$$

(A brief refresher course on the factorial and exponential operations required to apply this formula is provided in SideTrip 5.1).

Listed below are the full calculations, using this formula, for the four pathway probabilities that are needed to answer our original question concerning the probability of getting as many as 7 recoveries out of 10 patients, by mere chance, in the absence of any effective treatment. The results of the intermediate exponential operations are rounded to six decimal places, and the final result of each calculation is rounded to four decimal places.

Probability that Exactly 7 out of 10 patients will recover:

[N=10, k=7, p=.4, q=.6]

$$\frac{10!}{7!\times 3!} \times\times .4^7 \times .6^3 =$$

$$\frac{3{,}628{,}800}{5{,}040\times 6} \times .001{,}638 \times .216 =$$

$$120 \times .000{,}354 = .0425.$$

Translation: There are 120 different ways of reaching the result of 7 recoveries out of 10 patients, and each of those ways has a mere-chance probability of .000,354. The probability of reaching the result one way or the other is therefore 120x.000,345=.0425.

Probability that Exactly 8 out of 10 Patients will Recover:

[N=10, k=8, p=.4, q=.6]

$$\frac{10!}{8!\times 2!} \times .4^8 \times .6^2 =$$

$$\frac{3{,}628{,}800}{40{,}320\times 2} \times .000{,}655 \times .36 =$$

$$45 \times 000{,}236 = .0106.$$

Translation: There are 45 different ways of reaching the result of 8 recoveries out of 10 patients, and each of those ways has a mere-chance probability of .000,236. The probability of reaching the result one way or the other is therefore 45x.000,236=.0106.

Probability that Exactly 9 out of 10 Patients will Recover:

[N=10, k=9, p=.4, q=.6]

$$\frac{10!}{9!\times 1!} \times .4^9 \times .6^1 =$$

$$\frac{3{,}628{,}800}{362{,}880\times 1}\times .000{,}262\times .6 =$$

$$10\times .000{,}157 = .0016.$$

Translation: There are 10 different ways of reaching the result of 9 recoveries out of 10 patients, and each of those ways has a mere-chance probability of .000,157. The probability of reaching the result one way or the other is therefore 10x.000,157=.0016.

Probability that all 10 Patients will Recover:

$[N=10, k=10, p=.4, q=.6]$

$$\frac{10!}{10!\times 0!}\times .4^{10}\times .6^{0} = 3$$

$$\frac{3{,}628{,}800}{3{,}628{,}800}\times .000{,}105\times 1 =$$

$$1\times .000{,}105 = .0001.$$

Translation: There is only one way of reaching the result of 10 recoveries out of 10 patients, and that way has a mere-chance probability of .0001.

The full answer to the question is then simply the sum of these separate pathway probabilities:

$$P_{(7\text{ or more }R\text{ out of }10)}$$
$$= P_{(7R)} + P_{(8R)} + P_{(9R)} + \mathrm{P}_{(10R)}$$
$$= .0425 + .0106 + .0016 + .0001 = .0548$$

Translation: There is a 5.48% likelihood that as many as 7 recoveries out of 10 patients would occur by mere chance coincidence, in the absence of any effective treatment.

The standard criterion for statistical significance is the 5% level. Any observed result that has a mere-chance likelihood equal to or less than 5% (P<.05) is regarded as significant, and any observed result that has a mere-chance likelihood greater than 5% (P>.05) is regarded as non-significant. By this criterion the present result (P=.0548) falls short of significance, though only by a very slight distance. Most investigators would be inclined to describe such a narrow miss as marginally significant, which is a compact way of saying "not quite significant, but close enough to warrant further investigation."

Following this custom, our medical researchers regard their results as promising and proceed to put the plant extract to a much fuller test. Making use of the contacts they have with a number of clinical facilities throughout the land, they arrange to have the extract administered to a total of 1,000 patients, again beginning for each patient immediately after the onset of the disease and continuing for two months. As the new experiment runs its course, the investigators observe that 430 of the 1,000 patients have recovered within their respective two-month periods, while 570 have not recovered. This 43% recovery rate is of course much lower than the non-significant 70% rate observed in the first experiment—and, indeed, only slightly above the 40% baseline rate of mere chance spontaneous recovery. But keep in mind that we are now dealing with a much larger sample. It is the same point made earlier in our introductory mind-over-matter example, concerning the difference between a 10-toss test and a 100-toss test. Clearly, it would be much easier for mere chance to produce a 70% recovery rate in a sample of 10 patients than in a sample of 1,000 patients. What you are now being asked to consider is the possibility that for a sample of 1,000 patients even a recovery rate of 43% would be very unlikely to result from mere chance.

From the fact that I am bothering to use this "430 out of 1,000" scenario as an example, you can safely assume that the result *will* prove significant, once we figure out the details. Except for the difference in the numbers, the question of statistical significance here is exactly the same as before: given that any particular patient has a 40% chance of spontaneous recovery within two months, and a 60% chance of non-recovery, how likely or unlikely it is that as many as 430 out of any particular set of 1000 patients could recover by mere chance, if the plant extract has no effect whatsoever? Essentially, it is the disjunctive probability of getting either 430 recoveries out of 1000, or 431 out of 1000, or 432 out of 1000, and so on, up through 1000 out of 1000.

The logic for this question is exactly the same as for the one we have just examined. The only difference is in the complexity of the details. For the probability of 7 or more recoveries out of 10 you only need to perform four main pathway calculations, whereas for 430 or more out of 1000 you need to perform a total of 571—and most of these would involve factorial and exponential operations of rather staggering proportions. Take a few minutes to try to work out just the first of these calculations

$$P_{(430\,R \text{ out of } 1000)} = \frac{1000!}{430! \times 570!} \times .4430 \times .6570$$

and you will find yourself hoping there might be an easier way.

Chapter 6

Measures of Central Tendency

There are two types of data that can be manipulated with statistical tools: cross-sectional data and time-series data. *Cross-sectional data* is a snapshot that is collected at a single point in time. It has no time dimension. A typical example of cross-sectional data would be a set of height measurements of children of different ages. Computing the average of all the measurements for a given age gives you the average height of a child of that age. The specific date and time on which any given measurement was taken is irrelevant. All that matters is the height of each child and the child's age when that height was measured.

The second type of data that you'll encounter is time-series data. With *time-series data*, every measurement has a time dimension associated with it. For example, you can look at stock-market prices over some period of time, and you'll find that the share price for any given company differs from day to day. And, not only from day to day, but also typically from hour to hour and minute to minute. A share price by itself means nothing. It's the share price coupled with the time at which that price was effective that is meaningful. A list of such prices and times over a given period is referred to as a time series.

We need to define five more terms before proceeding into the sections that describe the calculations behind the various statistics discussed in this chapter. These terms are population, sample, sample size, case, and value.

Consider the problem of having a large box of oranges for which you need to determine the sweetness level so that you can sell the box at the best possible price. That box of oranges represents your *population*. Now you could taste each orange in the box to check the sweetness level, but then you wouldn't have any oranges left to sell. Instead, you might choose to test only a few oranges—say, three.

Those oranges that you test represent a *sample*, and the number of oranges in the sample represents the *sample size*. From a statistical point of view, you are testing a representative sample and then extrapolating the results to the oranges remaining in the box.

Continuing with our example, each orange in a sample represents a specific *case*. For each case—or orange—you must test the sweetness level. The result of each test will be a *value*—usually a number—indicating how sweet the orange really is. We use the term *case* to refer to a specific measurement in the sample. We use the term *value* to refer to a distinct numeric result. A value may occur more than once in a sample. For example, three oranges (three cases) may all have the same sweetness value.

Now that you have a good grasp of the statistical terms used in this chapter, you can proceed to the following sections describing various statistical measurements and calculations. Whenever you collect data, you end up with a group of scores on one or more variables. If you take the scores on one variable and arrange them in order from lowest to highest, what you get is a distribution of scores.

Researchers often want to know about the characteristics of these distributions of scores, such as the shape of the distribution, how spread out the scores are, what the most common score is, and so on. One set of distribution characteristics that researchers are usually interested in is central tendency. This set consists of the mean, median, and mode.

STRANGE SYMBOLS USED

The Σ symbol is the summation operator. It's used as a shorthand way of telling you to add up all the values in a sample set. When you see ΣX in a formula, you know that you should sum all the values in the sample set labeled *X*. Some formulas are more complex than that. For example, you might see ΣXY in a formula. The *X* and *Y* refer to two sample sets, and this notation is telling you to multiply corresponding elements of each set together and sum the results. So, multiply the first case in *X* by the first case in *Y*. Then, add to that the result of multiplying the second case in *X* by the second case in *Y*. Continue this pattern for all cases in the samples.

Whenever you see two elements run together, you should multiply. Thus, *XY* means to multiply the values *X* and *Y* together. $n\Sigma XY$ means to multiple the value n by the value of ΣXY.

The $\sqrt{\ }$ symbol is the square-root operator. When you see $\sqrt{n}$, you need to find the value that, when multiplied by itself, yields *n*. For example, $\sqrt{9}$ is 3, because 3×3 = 9. Most calculators allow you to easily compute square roots. Another way to express 3×3, by the way, is to write it as 32. The small, raised 2 indicates that you should multiply the value by itself.

Finally, in our notation, we use a horizontal bar over a value to indicate the mean of a sample. For example: indicates that you should compute the mean of the values for all cases in the sample labeled *X*.

SUMMATION NOTATION

The Greek letter Σ (a capital sigma) is used to designate summation. For example, suppose an experimenter measured the performance of four subjects on a memory task. Subject 1's score will be referred to as X_1, Subject 2's as X_2, and so on. The scores are shown below:

Subject		*Score*
1	X_1	7
2	X_2	6
3	X_3	5
4	X_4	8

The way to use the summation sign to indicate the sum of all four X's is:

$$\sum_{i=1}^{4} X_i$$

This notation is read as follows:

Sum the values of X from X_1 through X_4.

The index i (shown just under the Σ sign) indicates which values of X are to be summed. The index i takes on values beginning with the value to the right of the "=" sign (1 in this case) and continues sequentially until it reaches the value above the Σ sign (4 in this case). Therefore i takes on the values 1, 2, 3, and 4 and the values of X_1, X_2, X_3, and X_4 are summed (7 + 6 + 5 + 8 = 26).

In order to make formulas more general, variables can be used with the summation notation. For example,

$$\sum_{i=1}^{N} X_i$$

means to sum up values of X from 1 to N where N can be any number but usually indicates the sample size.

Often an abbreviated form of the summation notation is used. For example, ΣX means to sum all the values of X. When only a subset of the values of X are to be summed then the full version is required. Thus, the sum of all elements of X except the first and the last (the N'th) would be indicated as:

$$\sum_{i=2}^{N-1} X_i$$

which would be read as the sum of X with i going from 2 to N-1. Some formulas require that each number be squared before the numbers are summed. This is indicated by:

$$\sum_{i=2}^{N} X_i^2$$

and is equal to $7^2 + 6^2 + 5^2 + 8^2 = 174$.

The abbreviated version is simply: ΣX^2. It is very important to note that it makes a big difference whether the numbers are squared first and then summed or summed first and then squared. The symbol $(\Sigma X)^2$ indicates that the numbers should be summed first and then squared. For the present example, this equals:

$(7 + 6 + 5 + 8)^2 = 26^2 = 676$. This, of course, is quite different from 174.

Sometimes a formula requires that the sum of cross products be computed. For instance, if 3 subjects were each tested twice, they might each have a score on X and on Y.

Subject	X	Y
1	2	3
2	1	6
3	4	5

The sum of cross products (2 x 3) + (1 x 6) + (4 x 5) = 32 can be represented in summation notation simply as: ΣXY.

Basic Theorems:

The following data will be used to illustrate the theorems:

X	Y
3	8
2	3
4	1

$\Sigma(X + Y) = \Sigma X + \Sigma Y$

$\Sigma(X + Y) = 11 + 5 + 5 = 21$

$\Sigma X = 3 + 2 + 4 = 9$

$\Sigma Y = 8 + 3 + 1 = 12$

$\Sigma X + \Sigma Y = 9 + 12 = 21$

$\Sigma aX = a\Sigma X$ (a is a constant)

For an example, let a = 2.

$\Sigma aX = (2)(3) + (2)(2) + (2)(4) = 18$

$a\ \Sigma X = (2)(9) = 18$

$\Sigma(X\text{-}M)^2 = \Sigma X^2 - (\Sigma X)^2/N$

(N is the number of numbers, 3 in this case, and M is the mean which is also equal to 3 in this case.

$\Sigma(X\text{-}M)^2 = (3\text{-}3)^2 + (2\text{-}3)^2 + (4\text{-}3)^2 = 2$

$\Sigma X^2 = 3^2 + 2^2 + 4^2 = 29$

$(\Sigma X)^2/N = 9^2/3 = 27$

$\Sigma X^2 - (\Sigma X)^2/N = 29 - 27 = 2$

MEAN

A *mean* is a type of average, often referred to as an *arithmetic average*. Given a sample set of values, you calculate the mean by summing all the values and dividing the result by the sample size. Consider the following set of height values for seven kids in a hypothetical fifth-grade class:

100 cm
100 cm
100 cm
110 cm
115 cm
120 cm
180 cm

Given these values, you compute the mean height by summing the values and dividing by the sample size of seven. For example:

$$100+100+100+110+115+120+180 = 825 \text{ cm}$$

$$825 \text{ cm} / 7 = 117.86 \text{ cm}$$

In this case, the mean height is 117.86 cm. As you can see, the mean height may not correspond to any student's actual height.

Arithmetic Mean

The arithmetic mean is probably the most commonly used statistic in all social science research. The mean is simply the arithmetic average of a distribution of scores, and researchers like it because it provides a single, simple number that gives a rough summary of the distribution. It is important to remember that although the mean provides a useful piece of information, it does not tell you anything about how spread out the scores are (i.e., variance) or how many scores in the distribution are close to the mean. It is possible for a distribution to have very few scores at or near the mean.

The arithmetic mean is what is commonly called the average: When the word "Mean" is used without a modifier, it can be assumed that it refers to the arithmetic mean. The mean is the sum of all the scores divided by the number of scores. The formula in summation notation is:

$$\mu = \Sigma X/N$$

where μ is the population mean and N is the number of scores.

If the scores are from a sample, then the symbol M refers to the mean and N refers to the sample size. The formula for M is the same as the formula for μ.

$$M = \Sigma X/N$$

The mean is a good measure of central tendency for roughly symmetric distributions but can be misleading in skewed distributions since it can be greatly influenced by scores in the tail. Therefore, other statistics such as the median may be more informative for distributions such as reaction time or family income that are frequently very skewed

The sum of squared deviations of scores from their mean is lower than their squared deviations from any other number.

For normal distributions, the mean is the most efficient and therefore the least subject to sample fluctuations of all measures of central tendency.

The formal definition of the arithmetic mean is $\mu = E[X]$ where μ is the population mean of the variable X and E[X] is the expected value of X.

MEDIAN

A *median* is yet another type of statistic that you could loosely say refers to the middle value in a sample set. To be more precise, the median is the value in a sample set that has the same number of cases below it as above it. Consider again the following height data, which is repeated from the earlier section on means:

100 cm
100 cm
100 cm
110 cm
115 cm
120 cm
180 cm

There are seven cases total. Look at the 110 cm value. There are three cases with values above 110 cm and three cases with values below 110 cm. The median height, therefore, is 110 cm.

Computing a median with an odd number of cases is relatively straightforward. Having an even number of cases, though, makes the task of computing a median more complex. With an even number of cases, you simply can't pick just one value that has an equal number of cases above and below it. The following sample amply illustrates this point:

100 cm
100 cm
100 cm
110 cm
115 cm
120 cm
125 cm
180 cm

There is no "middle" value in this sample. 110 cm won't work because there are four cases having higher values and only three having lower values. 115 cm won't work for the opposite reason—there are three cases having values above and four cases having values below. In a situation like this, there are two approaches you can take. One solution is to take the lower of the two middle values as the median value. In this case, it would be 110 cm, and it would be referred to as a *statistical median*. Another solution is to take the mean of the two middle values. In this case, it would be 112.5 cm [(115 cm+110 cm)/2]. A median value computed by taking the mean of the two middle cases is often referred to as a *financial median*.

It is actually possible to find an example where there is a median with an even number of cases. Such an example is: 1,2,3,3,4,5. There is an even number of cases in the sample; however, 3 has an equal number of cases above and below it, so that is the median. This example is exceptional, though, because the two middle cases have the same value. You can't depend on that occurring all the time.

The median is the score in the distribution that marks the 50th percentile. That is, 50% percent of the scores in the distribution fall above the median and 50% fall below it. Researchers often use the median when they want to divide their distribution scores into two equal groups (called a median split). The median is also a useful statistic to examine when the scores in a distribution are skewed or when there are a few extreme scores at the high end or the low end of the distribution.

The median is the middle of a distribution: half the scores are above the median and half are below the median. The median is less sensitive to extreme scores than the mean and this makes it a better measure than the mean for highly skewed distributions. The median income is usually more informative than the mean income, for example.

The sum of the absolute deviations of each number from the median is lower than is the sum of absolute deviations from any other number.

The mean, median, and mode are equal in symmetric distributions. The mean is typically higher than the median in positively skewed distributions and lower than the median in negatively skewed distributions, although this may not be the case in bimodal distributions.

Computation of Median

When there is an odd number of numbers, the median is simply the middle number. For example, the median of 2, 4, and 7 is 4.

When there is an even number of numbers, the median is the mean of the two middle numbers. Thus, the median of the numbers 2, 4, 7, 12 is (4+7)/2 = 5.5.

MODE

A *mode* is another type of statistic that refers to the most frequently occurring value in a sample set. Look again at the numbers in the previous section on means. You'll see that the number of occurrences is as follows:

100 cm three
110 cm one
115 cm one
120 cm one
180 cm one

It's obvious that the most frequently occurring value is 100 cm. That value represents the mode for this sample set.

The mode is the least used of the measures of central tendency because it provides the least amount of information. The mode simply indicates which score in the distribution occurs most often, or has the highest frequency.

The mode is the most frequently occurring score in a distribution and is used as a measure of central tendency. The advantage of the mode as a measure of central tendency is that its meaning is obvious. Further, it is the only measure of central tendency that can be used with nominal data.

The mode is greatly subject to sample fluctuations and is therefore not recommended to be used as the only measure of central tendency. A further disadvantage of the mode is that many distributions have more than one mode. These distributions are called "multi modal."

In a normal distribution, the mean, median, and mode are identical.

STANDARD ERROR

A *standard error* is a measurement of a statistical projection's accuracy based on the sample size used to obtain that projection. Suppose that you were trying to calculate the average height of students in the fifth grade. You could randomly select seven children and calculate their average height. You could then use that average height as an approximation of the average height for the entire population of fifth graders. If you select a different set of seven children, your results will probably be slightly different. If you increase your sample size, the variation between samples should decrease, and your approximations should become more accurate. In general, the greater your sample size, the more accurate your approximation will be.

You can use the sample size and the standard deviation to estimate the error of approximation in a statistical projection by computing the *standard error of the mean*. The formula to use is:

$$\frac{s}{\sqrt{n}}$$

where:

s

Is the standard deviation in the sample.

n

Is the number of cases in the sample.

We can use the data shown in the previous section on standard deviation in an example here showing how to compute the standard error. Recall that the sample size was seven elements and that the standard deviation was 28.56. The standard error, then, is calculated as follows:

$$\frac{28.56}{\sqrt{7}} = \frac{28.56}{2.65} = 10.7$$

Double the sample size to 14 elements, and you have the following:

$$\frac{28.56}{\sqrt{14}} = \frac{28.56}{3.74} = 7.6$$

A standard-error value by itself doesn't mean much. It only has meaning relative to other standard-error values computed for different sample sizes for the same population. In the examples shown here, the standard error is 10.78 for a sample size of 7 and 7.64 for a sample size of 14. A sample size of 14, then, yields a more accurate approximation than a sample size of 7.

A WORD ABOUT POPULATIONS AND SAMPLES

There are two different symbols used for the mean, X and μ. Two different symbols are needed because it is important to distinguish between a statistic that applies to a sample and a parameter that applies to a population. A population is an individual or group that represents all the members of a certain group or category of interest. A sample is a subset drawn from the larger population. For example, let us suppose that wanted to know the average income of the current full-time, tenured faculty at Harvard. There are two ways that we can find this average. First, get a list of every full-time, tenured faculty member at Harvard and find out the annual income of each member on this list. Because this list contains every member of the group is interested in, it can be considered a population. If we have to collect these data and calculate the mean, we would have generated a parameter, because a parameter is a value generated from, or applied to, a population. The symbol used to represent the population mean is u. Another way to generate the mean income of the tenured faculty at Harvard would be to randomly select a sub-set of faculty names from our list and calculate the average income of this sub-set. The sub-set is known as a sample (in this case it is a random sample), and the mean that is generated from this sample is a type of statistic. Statistics are values derived from sample data, whereas parameters are values that are either derived from, or applied to, population data. It is important to note that all samples are representative of some population and that all sample statistics can be used as estimates of population parameters. In the case of the mean, the sample statistic is represented with the symbol.

MEASURES OF CENTRAL TENDENCY IN DEPT

The calculations for each measure of central tendency are mercifully straightforward. With the aid of a calculator or statistics software programme, you will probably never need to calculate any of these statistics. But for the sake of knowledge and in the event you find yourself without a calculator and in need of these statistics, here is the information you will need.

Because the mean is an average, calculating the mean involves adding, or summing, all of the scores in a distribution and dividing by the number of scores. So, if you have 10 scores in a distribution, you would add all of the scores together to find the sum and then divide the sum by 10, which is the number of scores in the distribution. The formula for calculating the mean is presented in Table,

Formula for calculating the mean of a distribution

$$\overline{X}, \mu = \frac{\Sigma X}{n, N}$$

where $\overline{X}$ is the sample mean

μ is the population mean

Σ means "the sum of"

X is an individual score in the distribution

n is the number of scores in the sample

N is the number of scores in the population

The calculation of the median for a simple distribution of scores is even simpler than the calculation of the mean. To find the median of a distribution, you need to first arrange all of the scores in the distribution in order, from smallest to largest. Once this is done, you simply need to find the middle score in the distribution. If there are an odd number of scores in the distribution, there will be a single score that marks the middle of the distribution. For example, if there are 11 scores in the distribution arranged in descending order from smallest to largest, the 6th score will be the median because there will be 5 scores below it and 5 scores above it. However, if there are an even number of scores in the distribution, there is no single middle score. In this case, the median is the average of the two scores in the middle of the distribution (as long as the scores are arranged in order, from largest to smallest). For example, if there are 10 scores in a distribution, to find the median you will need to find the average of the 5th and 6th scores. To find this average, you simply add the two scores together and divide by two.

To find the mode, there is no need to calculate anything. The mode is simply the category in the distribution that has the highest number of scores, or the highest frequency. For example, suppose you have the following distribution of IQ test scores from 10 students: 86 90 95 100 100 100 110 110 115 120

In this distribution, the score that occurs most frequently is 100, making it the mode of the distribution. If a distribution has more than one category with the most common score, the distribution has multiple modes and is called multimodal. One common example of a multimodal distribution is the bimodal distribution. Researchers often get bimodal distributions when they ask people to respond to controversial questions that tend to polarize the public. For example, if we were to ask a sample of 100 people how they feel about capital punishment, we might get the results presented in Table.

On the following scale, please indicate how you feel about capital punishment.

1———2———3———4———5

. Strongly opposed — Strongly in Favour

In this example, because most people either strongly opposed or strongly supported capital punishment, we ended up with a bimodal distribution of scores.

EXAMPLE: THE MEAN, MEDIAN, AND MODE OF A SKEWED DISTRIBUTIO

When scores in a distribution are normally distributed, the mean, median, and mode are all at the same point: the centre of the distribution. In the messy world of social science, however, the scores from a sample on a given variable are often not normally distributed. When the scores in a distribution tend to bunch up at one end of the distribution and there are a few scores at the other end, the distribution is said to be skewed. When working with a skewed distribution, the mean, median, and mode are usually all at different points.

It is important to note that the procedures used to calculate a mean, median, and mode are the same whether you are dealing with a skewed or a normal distribution. All that changes are where these three measures of central tendency are in relation to each other. To illustrate, this a fictional distribution of scores is created based on a sample size of 30. Suppose that we were to ask a sample of 30 randomly selected fifth graders whether they think it is important to do well in school. Suppose, further it is asked them to rate how important they think it is to do well in school using a 5-point scale, with 1= "not at all important" and 5 = "very important" Because most fifth graders tend to believe it is very important to do well in school, most of the scores in this distribution are at the high end of the scale, with a few scores at the low end. We have arranged our fictitious scores in order from smallest to largest and get the following distribution:

1	1	1	2	2	2	3	3	3	3
4	4	4	4	4	4	4	4	5	5
5	5	5	5	5	5	5	5	5	5

As you can see, there are only a few scores near the low end of the distribution (1's and 2's) and more at the high end of the distribution (4's and 5's). To get a clear picture of what this skewed distribution looks like, the graph in Figure is created.

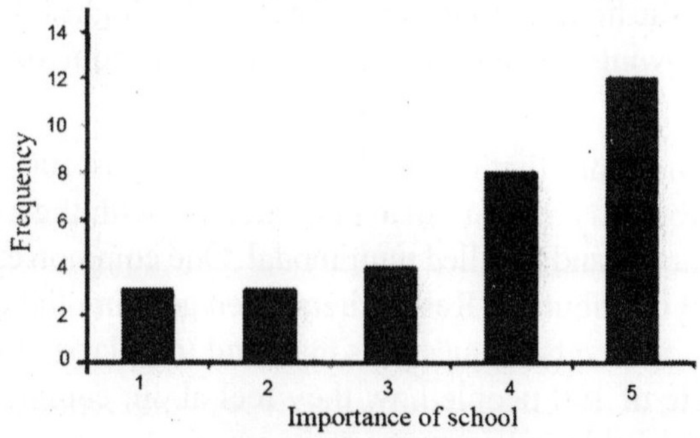

A skewed distribution

This graph provides a picture of what some skewed distributions look like. Notice how most of the scores are clustered at the higher end of the distribution and there are a few scores creating a tail toward the lower end. This is known as a negatively skewed distribution, because the tail goes toward the lower end. If the tail of the distribution were pulled out toward the higher end, this would have been a positively skewed distribution.

A quick glance at the scores in the distribution, or at the graph, reveals that the mode is 5 because there were more scores of 5 than any other number in the distribution.

To calculate the mean, we simply apply the formula mentioned earlier. That is, we add up all of the scores (ΣX) and then divide this sum by the number of scores in the distribution (n). This gives us a fraction of 113/30, which reduces to 3.7666. When we round to the second place after the decimal, we end up with a mean of 3.77.

To find the median of this distribution, we arrange the scores in order from smallest to largest and find the middle score. In this distribution, there are 30 scores, so there will be 2 in the middle. When arranged in order, the 2 scores in the middle (the 15th and 16th scores) are both 4. When we add these two scores together and divide by 2, we end up with 4, making our median 4.

As mentioned earlier, the mean of a distribution can be affected by scores that are unusually large or small for a distribution, sometimes called outliers, whereas the median is not affected by such scores. In the case of a skewed distribution, the mean is usually pulled in the direction of the tail, because the tail is where the outliers are. In a negatively skewed distribution, such as the one presented previously, we would expect the mean to be smaller than the median, because the mean is pulled toward the tail whereas the median is not. In our example, the mean (3.77) is somewhat lower than the median (4). In positively skewed distributions, the mean is usually somewhat higher than the median.

Measures of central tendency, particularly the mean and the median, are some of the most used and useful statistics for researchers. They each provide important information about an entire distribution of scores in a single number. For example, we know that the average height of a man in the United States is five feet nine inches tall. This single number is used to summarize information about millions of men in this country. But for the same reason that the mean and median are useful, they can often be dangerous if we forget that a statistic such as the mean *ignores* a lot of information about a distribution, including the great amount of variety that exists in many distributions. Without considering the variety as well as the average, it becomes easy to make sweeping generalizations, or stereotypes, based on the mean.

MEASURES OF VARIABILITY

Measures of central tendency, such as the mean and the median, provide useful information. But it is important to recognize that these measures are limited and, by themselves, do not provide a great deal of information. To illustrate, consider this example: Suppose we gave a sample of 100 fifth-grade children a survey to assess their level of depression. Suppose further that this sample had a mean of 10.0 on our depression survey and a median of 10.0 as well.

All we know from this information is that the mean and median are in the same place in our distribution, and this place is 10.0. Now consider what we do not know. We do not know if this is a high score or a low score. We do not know if all of the students in our sample have about the same level of depression or if they differ from each other. We do not know the highest depression score in our distribution or the lowest score. Simply put, we do not yet know anything about the dispersion of scores hi the distribution. In other words, we do not yet know anything about the variety of the scores in the distribution.

There are three measures of dispersion that researchers typically examine: the range, the variance, and the standard deviation. Of these, the standard deviation is perhaps the most informative and certainly the most widely used.

RANGE

The range is simply the difference between the largest score (the maximum value) and the smallest (the minimum value) score of a distribution. This statistic gives researchers a quick sense of how spread out the scores of a distribution are, but it is not a particularly useful statistic because it can be quite misleading. For example, in our depression survey described earlier, we may have 1 student score a 1 and another score a 20, but the other 98 may all score 10. In this example, the range will be 19 (20 - 1 = 19), but the scores really are not as spread out as the range might suggest. Researchers often take a quick look at the range to see whether all or most of the points on a scale, such as a survey, were covered in the sample.

Another common measure of the range of scores in a distribution is the interquartile range (IQR). Unlike the range, which is the difference between the largest and smallest score in the distribution, the IQR is the difference between the score that marks the 75th percentile (the third quartile) and the score that marks the 25th percentile (the first quartile). If the scores in a distribution were arranged in order from largest to smallest and then divided into groups of equal size, the IQR would contain the scores in the two middle quartiles.

VARIANCE

The variance provides a statistical average of the amount of dispersion in a distribution of scores. Because of the mathematical manipulation needed to produce a variance statistic (more about this in the next section), variance, by itself, it not often used by researchers to gain a sense of a distribution. In general, variance is used more as a step in the calculation of other statistics (e.g., analysis of variance) than as a stand-alone statistic. But with a simple manipulation, the variance can be transformed into the standard deviation, which is one of the statistician's favourite tools.

The variance and the closely-related standard deviation are measures of how spread out a distribution is. In other words, they are measures of variability.

The variance is computed as the average squared deviation of each number from its mean. For example, for the numbers 1, 2, and 3, the mean is 2 and the variance is:

$$\sigma^2 = \frac{(1-2)^2 + (2-2)^2 + (3-2)^2}{3} = 0.6$$

The formula (in summation notation) for the variance in a population is

$$\sigma^2 = \frac{\sum(X-\mu)^2}{N}$$

Where μ is the mean and N is the number of scores.

When the variance is computed in a sample, the statistic

$$S^2 = \frac{\sum(X-M)^2}{N}$$

(Where M is the mean of the sample) can be used. S^2 is a biased estimate of σ^2, however. By far the most common formula for computing variance in a sample is:

$$S^2 = \frac{\sum(X-M)^2}{N-1}$$

Which gives an unbiased estimate of σ^2. Since samples are usually used to estimate parameters, s^2 is the most commonly used measure of variance. Calculating the variance is an important part of many statistical applications and analyses. It is the first step in calculating the standard deviation.

STANDARD DEVIATION

The best way to understand what a standard deviation is to consider what the two words mean. *Deviation,* in this case, refers to the difference between an individual score in a distribution and the average score for the distribution. So if the average score for a distribution is 10 (as in our previous example), and an individual child has a score of 12, the deviation is 2. The other word in the term standard deviation is *standard.* In this case, standard means typical, or average. So a standard deviation is the typical, or average, deviation between individual scores in a distribution and the mean for the distribution. [2] This is a very useful statistic because it provides a handy measure of how spread out the scores are in the distribution. When combined, the mean and standard deviation provide a pretty good picture of what the distribution of scores is like.

In a sense, the range provides a measure of the total spread in a distribution (i.e., from the lowest to the highest scores), whereas the variance and standard deviation are measures of the average amount of spread within the distribution. Researchers tend to look at the range when they want a quick snapshot of a distribution, such as when they want to know whether all of the response categories on a survey question have been used (i.e., did people use all 5 points on the 5-point Likert scale?) or they want a sense of the overall balance of scores in the distribution. Researchers rarely look at the variance alone, because it does not use the same scales as the original measure of a variable, although the variance statistic is very useful for the calculation of other statistics. The standard deviation is a very useful statistic that researchers constantly examine to provide the most easily interpretable and meaningful measure of the average dispersion of scores in a distribution.

The standard deviation formula is very simple: it is the square root of the variance. It is the most commonly used measure of spread.

An important attribute of the standard deviation as a measure of spread is that if the mean and standard deviation of a normal distribution are known, it is possible to compute the percentile rank associated with any given score. In a normal distribution, about 68% of the scores are within one standard deviation of the mean and about 95% of the scores are within two standard deviations of the mean.

The standard deviation has proven to be an extremely useful measure of spread in part because it is mathematically tractable. Many formulas in inferential statistics use the standard deviation.

Although less sensitive to extreme scores than the range, the standard deviation is more sensitive than the semi-interquartile range. Thus, the standard deviation should be supplemented by the semi-interquartile range when the possibility of extreme scores is present.

If variable Y is a linear transformation of X such that:

Y = bX + A,

then the variance of Y is:

$$\sigma_y^2 = b^2\sigma_x^2$$

where σ_x^2 is the variance of X.

The standard deviation of Y is $b\sigma_X$ where σ_X is the standard deviation of X.

MEASURES OF VARIABILITY

Standard Deviation and Variance

Standard deviations and variance are very closely related. A *standard deviation* will tell you how evenly the values in a sample are distributed around the mean. The smaller the standard deviation, the more closely the values are condensed around the mean. The greater the standard deviation, the more values you will find at a distance from the mean. Consider again, the following values:

100 cm

100 cm

100 cm

110 cm

115 cm

120 cm

180 cm

The standard deviation for the sample is calculated as follows:

$$\sqrt{\frac{\sum(X x)}{n\ 1}}$$

Where:

X

Is the value for a specific case in the sample.

Is the mean value for all the cases in the sample.

n

Is the number of cases in the sample.

Take the formula shown here for standard deviation, expand the summation for each of the cases in our sample, and you have the following calculation:

$$\sqrt{\frac{(100\ 117.86)^2+(100\quad 117.86)^2+(100\ \ 117.86)^2+ \\ (115\ \ 117.86)^2+(120\ 117.86)^2+(180\ 117.86)^2}{7\ \ 1}}$$

Work through the math, and you'll find that the calculation returns 28.56 as the sample's standard deviation.

A *variance* is nothing more than the square of the standard deviation. In our example, the standard deviation is 28.56, so the variance is calculated as follows:

$28.56 \times 28.56 = 815.6736$

If you look more closely at the formula for calculating standard deviation.

Calculating the Variance and Standard Deviation

There are two central issues that we need to address when considering the formulas for calculating the variance and standard deviation of a distribution:

1. Whether to use the formula for the sample or the population,
2. How to make sense of these formulas.

Sample Statistics as Estimates of Population Parameters

Although researchers are sometimes interested in simply describing the characteristics (e.g., mean, median, range, etc.) of a sample, for the most part we are much more concerned with what our sample tells us about the population from which the sample was drawn. If we are conducting a study about the study habits of eighth graders, we are more concerned about eighth graders *in general* than we are about the particular group of eighth graders in our sample. It is important to keep this in mind, because most of our statistics, although generated from sample data, are used to make estimations about the population.

This point is discussed in more detail later in the book when examining inferential statistics. But it is important to keep this in mind as you read about these measures of variation. The formulas for calculating the variance and standard deviation of sample data are actually designed to make these sample statistics better *estimates* of the population parameters (i.e., the population variance and standard deviation).

It is important to note that the formulas for calculating the variance and the standard deviation differ depending on whether you are working with a distribution of scores taken from a sample or from a population. The reason these two formulas are different is quite complex, and requires more space than allowed in a short book like this. We provide an overly brief explanation here and then encourage you to find a more thorough explanation in a traditional statistics textbook.

Briefly, when we do not know the population mean, we must use the sample mean as an estimate. But the sample mean will probably differ from the population mean. Whenever we use a number *other than* the actual mean to calculate the variance, we will end up with a *larger* variance, and therefore a larger standard deviation, than if we had used the actual mean. This will be true regardless of whether the number we use in our formula is smaller or larger than our actual mean.

Because the sample mean usually differs from the population mean, the variance and standard deviation that we calculate using the sample mean will probably be smaller than it would have been had we used the population mean.

Therefore, when we use the sample mean to generate an *estimate* of the population variance or standard deviation, we will actually *underestimate* the size of the true variance in the population because if we had used the population mean in place of the sample mean, we would have created a larger sum of squared deviations, and a larger variance and standard deviation.

To adjust for this underestimation, we use n - 1 in the denominator of our sample formulas. Smaller denominators produce larger overall variance and standard deviation statistics, which will be more accurate estimates of the population parameters.

The formulas for calculating the variance and standard deviation of a population and the estimates of the population variance and standard deviation based on a sample are presented in Table. As we can see, the formulas for calculating the variance and the standard deviation are virtually identical. Because both require that you calculate the variance first, we begin with the formulas for calculating the variance. This formula is known as the *deviation score formula.*

When working with a population distribution, the formulas for the variance and for the standard deviation both have a denominator of N, which is the size of the population. In the real world of research, particularly social science research, we usually assume that we are working with a sample that represents a larger population. For example, if we study the effectiveness of our new reading programme with a class of second graders, as a researcher we assume that these particular second graders represent a larger population of second graders, or students more generally.

Because of this type of inference, researchers generally think of their research participants as a sample rather than a population, and the formula for calculating the variance of a sample is the more often used. Notice that the formula for calculating the variance of a sample is identical to that used for the population except the denominator for the sample formula is n - 1.

How much of a difference does it make if we use *N or n-1* in our denominator? Well, that depends on the size of the sample. If we have a sample of 500 people, there is virtually no difference between the variance formula for the population and for the estimate based on the sample. After all, dividing a numerator by 500 is almost the same as dividing it by 499. But when we have a small sample, such as a sample of 10, then there is a relatively large difference between the results produced by the population and sample formulas.

The second issue to address involves making sense of the formulas for calculating the variance. In all honesty, there will be very few times that you will need to use this formula. Outside of our teaching duties, we haven't calculated a standard deviation by hand since our first statistics course. Thankfully, all computer statistics and spreadsheet programmes, and many calculators, compute the variance and standard deviation for us. Nevertheless, it is mildly interesting and quite informative to examine how these variance formulas work.

To begin this examination, let me remind you that the variance is simply an average of a distribution. To get an average, we need to add up all of the scores in a distribution and divide this sum by the number of scores in the distribution, which is n (remember the formula for calculating the mean.

With the variance, however, we need to remember that we are not interested in the average *score* of the distribution. Rather, we are interested in the average *difference,* or *deviation,* between each score in the distribution and the mean of the distribution.

To get this information, we have to calculate a *deviation score* for each individual score in the distribution. This score is calculated by taking an individual score and subtracting the mean from that score. If we compute a deviation score for each individual score in the distribution, then we can sum the deviation scores and divide by n to get the average, or standard, deviation, right? Not quite.

The problem here is that, by definition, the mean of a distribution is the mathematical middle of the distribution. Therefore, some of the scores in the distribution will fall above the mean (producing positive deviation scores), and some will fall below the mean (producing negative deviation scores). When we add these positive and negative deviation scores together, the sum will be zero.

Because the mean is the mathematical middle of the distribution, we will get zero when we add up these deviation scores no matter how big or small our sample, or how skewed or normal our distribution. And because we cannot find an average of zero (i.e., zero divided by n is zero, no matter what n is), we need to do something to get rid of this zero.

Variance and standard deviation formulas.

	Population	**Estimate Based on a sample**
Variance	$\sigma^2 = \frac{\Sigma(x-\mu)^2}{N}$	$r^2 = \frac{\Sigma(X-\overline{X})^2}{n-1}$
	Where $\sum$ = sum	Where $\sum$ = sum
	X = a score in the distribution	X = a score in the distribution
	μ = the population mean	$\overline{X}$ = the sample mean
	N = the number of cases in the population	n = the number of cases in the sample
Standard Deviation	$\sigma = \frac{\Sigma(x-\mu)^2}{N}$	$r = \frac{\Sigma(x-\overline{x})^2}{n-1}$
	Where $\sum$ = sum	Where $\sum$ = sum
	X = a score in the distribution	X = a score in the distribution
	μ = the population mean	$\overline{X}$ = the sample mean
	N = the number of cases in the population	n = the number of cases in the sample

The solution statisticians came up with is to make each deviation score positive by squaring it. So, for each score in a distribution, we subtract the mean of the distribution and then square the deviation.

If you look at the deviation score formulas in Table, we will see that all that the formula is doing with $(X - \mu,)^2$ is to take each score, subtract the mean, and square the resulting deviation score. What you get when you do this is the all-important squared deviation, which is used all the time in statistics. If we then put a summation sign in front such that we have $\Sigma(X-\mu)^2$.

What this tells us is that after we have produced a squared deviation score for each case in our distribution, we then need to add up all of these squared deviations, giving us the sum of squared deviations, or the sum of squares *(SS)*. Once this is done, we divide by the number of cases in our distribution, and we get an average, or mean, of the squared deviations. This is our variance. The final step in this process is converting the variance into a standard deviation. Remember that in order to calculate the variance, we had to square each deviation score.

We did this to avoid getting a sum of zero in our numerator. When we squared these scores, we changed our statistic from our original scale of measurement (i.e., whatever units of measurement were used to generate our distribution of scores) to a squared score.

To reverse this process and give us a statistic that is back to our original unit of measurement, we merely need to take the square root of our variance. When we do this, we switch from the variance to the standard deviation. Therefore, the formula for calculating the standard deviation is exactly the same as the formula for calculating the variance, except we put a big square root symbol over the whole formula. Notice that because of the squaring and square rooting process, the standard deviation and the variance are always positive numbers.

WHY HAVE VARIANCE?

If the variance is a difficult statistic to understand, and rarely examined by researchers, why not just eliminate this statistic and jump straight to the standard deviation? There are two reasons. First, we need to calculate the variance before we can find the standard deviation anyway, so it is not more work. Second, the fundamental piece of the variance formula, which is the sum of the squared deviations, is used in a number of other statistics, most notably analysis of variance (ANOVA).

When you learn about more advanced statistics such as ANOVA, factorial ANOVA, and even regression, you will see that each of these uses the *sum of squares,* which is just another way of saying the sum of the squared deviations. Because the sum of squares is such an important piece of so many statistics, the variance statistic has maintained a place in the teaching of basic statistics.

EXAMPLE: EXAMINING THE RANGE, VARIANCE, AND STANDARD DEVIATION

We recently conducted a study in which we gave questionnaires to approximately 500 high school students in the 9th and 11th grades. In the examples that follow, we examine the mean, range, variance, and standard deviation of the distribution of responses to two of these questions.

To make sense of these (and all) statistics, you need to know exactly what the wording of the survey items was and the response scale that the students in the sample used to answer the survey items. Although this may sound obvious, we mention it here because, if you notice, much of the statistical information reported in the news (e.g., the results of polls) does not provide the exact wording of the questions or the choices the sample had to respond to the questions. Without this information, it is difficult to know exactly what the responses mean, and "lying with statistics" becomes easier.

The first survey item we examine reads "If we have enough time, we can do even the most difficult work in this class." This item is designed to measure students' confidence in their abilities to succeed in their classwork. Students were asked to respond to this question by circling a number on a scale from 1 to 5. On this scale, circling the 1 means that the statement is "not at all true" and the 5 means "very true." So students were basically asked to indicate how true they felt the statement was on a scale from 1 to 5, with higher numbers indicating a stronger belief that the statement was true.

The distribution of responses produced the following statistics:

Sample Size = 491 Mean = 4.21 Standard Deviation = .98 Variance = $(.98)^2$ = .96 Range = 5-1=4

A graph of the frequency distribution for the responses on this item appears in Figure. As you can see in this graph, most of the students in the sample circled number 4 or number 5 on the response scale, indicating that they felt the item was quite true (i.e., that they were confident in their ability to do their classwork if they were given enough time). Because most students circled a 4 or a 5, the average score on this item is quite high (4.21 out of a possible 5). This is a negatively skewed distribution.

The graph in Figure. also provides information about the variety of scores in this distribution. Although our range statistic is 4, indicating that students in the sample circled both the highest and the lowest number on the response scale, we can see that the range does not really provide much useful information. For example, the range does not tell us that most of the students in our sample scored at the high end of the scale. By combining the information from the range statistic with the mean statistic, we can reach the following conclusion: "Although the distribution of scores on this item covers the full range, it appears that most scores are at the higher end of the response scale."

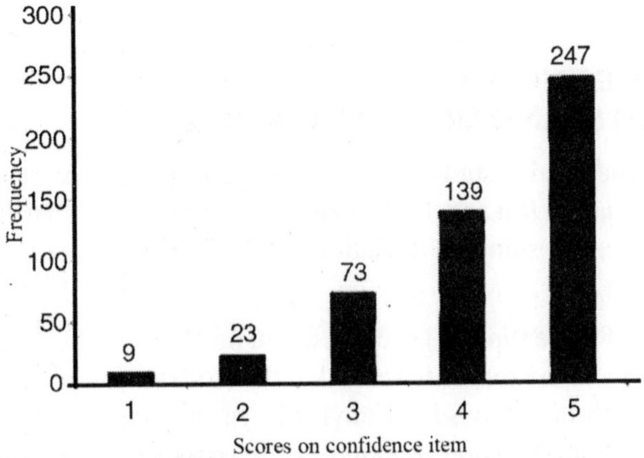

Fig. Frequency Distribution of Scores on Confidence Item

Now that we've determined that (a) the distribution of scores covers the full range of possible scores (i.e., from 1 to 5), and (b) most of the responses were at the high end of the scale (because the mean is 4.21 out of a possible 5), we may want a more precise measure of the average amount of variety among the scores in the distribution. For this we turn to the variance and standard deviation statistics. In this example, the variance (.96) is almost exactly the same as the standard deviation (.98).

This is something of a fluke. Do not be fooled. It is quite rare for the variance and standard deviation to be so similar. In fact, this only happens if the standard deviation is about 1.0, because 1.0 squared is 1.0. So in this rare case, the variance and standard deviation provide almost the same information. Namely, they indicate that the average difference between an individual score in the distribution and the mean for the distribution is about 1 point on the 5-point scale.

Taken together, these statistics tell us the same things that the graph tells us, but more precisely. Namely, we now know that (a) students in the study answered this item covering the whole range of response choices (i.e., 1 - 5); (b) most of the students answered at or near the top of the range, because the mean is quite high; and (c) the scores in this distribution generally packed fairly closely together with most students circling a number within 1 point of the mean, because the standard deviation was .98. The variance tells us that the average *squared deviation* was .96, and we scratch our heads, wonder what good it does us to know the average squared deviation, and move on.

In our second example, we examine students' responses to the item "We would feel really good if we were the only one who could answer the teacher's question in class." This item is one of several on the survey designed to measure students' desires to demonstrate to others that they are smart, or academically able.

We received responses from 491 students on this item and the distribution produced the following statistics:

Sample Size = 491 Mean = 2.92 Standard Deviation = 1.43 Variance = $(1.43)^2$ = 2.04 Range = 5-1= 4

In Figure, a graph is presented that illustrates the distribution of students' responses to this item across each of the five response categories. It is obvious, when looking at this graph, how the distribution of scores on this item differs from the distribution of scores on the confidence item presented earlier. But if we didn't have this graph, how could we use the statistics to discover these differences between the distributions of scores on these two items?

Notice that, as with the previous item, the range is 4, indicating that some students circled the number 1 on the response scale and some circled the number 5. Because the ranges for both the confidence and the wanting to appear able items are equal (i.e., 4), they do nothing to indicate the differences in the distributions of the responses to these two items. That is why the range is not a particularly useful statistic-it simply does not provide very much information.

Our first real indication that the distributions differ substantially comes from a comparison of the means. In the previous example, the mean of 4.21 indicated that most of the students must have circled either a 4 or a 5 on the response scale. For this second item, the mean of 2.92 is a bit less informative. Although it provides an average score, it is impossible from just examining the mean to determine whether most students circled a 2 or 3 on the scale, or whether roughly equal numbers of students circled each of the five numbers on the response scale, or whether almost half of the students circled 1 whereas the other half circled 5. All three scenarios would produce a mean of about 2.92, because that is roughly the middle of the response scale.

To get a better picture of this distribution, we need to consider the standard deviation in conjunction with the mean. Before discussing the actual standard deviation for this distribution of scores, let us briefly consider what we would expect the standard deviation to be for each of the three scenarios just described. First, if almost all of the students circled a 2 or a 3 on the response scale, we would expect a fairly small standard deviation, as we saw in the previous example using the confidence item.

The more similar the responses are to an item, the smaller the standard deviation. However, if half of the students circled 1 whereas the other half circled 5, we would expect a large standard deviation (about 2.0) because each score would be about two units away from the mean (if the mean is about 3.0 and each response is either 1 or 5, each response is about two units away from the mean). Finally, if the responses are fairly evenly spread out across the five response categories, we would expect a moderately sized standard deviation (about 1.50).

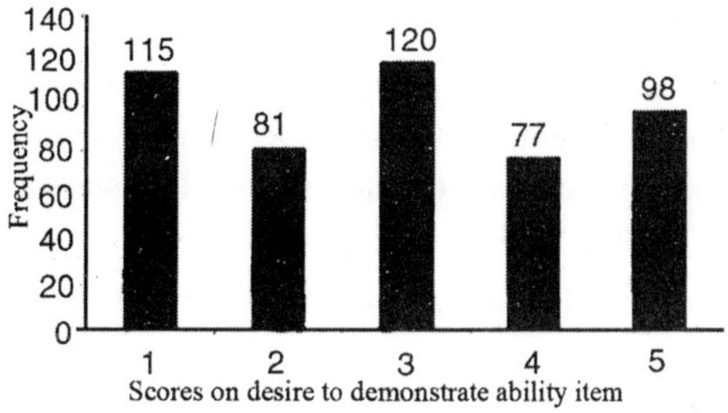

Fig. Frequency Distribution for Scores on Desire to Demonstrate Ability Item

Now, when we look at the actual mean for this distribution (2.92) and the actual standard deviation (1.43), we can develop a rough picture of the distribution in our minds. Because we know that on a scale from 1 to 5, a mean *of 2.92* is about in the middle, we can guess that the distribution looks somewhat symmetrical (i.e., that there will be roughly the same number of responses in the 4 and 5 categories as there are in the 1 and 2 categories.

Furthermore, because we've got a moderately sized standard deviation of 1.43, we know that the scores are pretty well spread out, with a healthy number of students in each of the five response categories. So we know that we didn't get an overwhelming number of students circling 3 and we didn't get students circling only 1 or 5. At this point, that is about all we can say about this distribution: The mean is near the middle of the scale and the responses are pretty well spread out across the five response categories. To say any more, we would need to look at the number of responses in each category, such as that presented in Figure.

As we look at the actual distribution of scores presented in the graph in Figure, we can see that our predictions that we generated from our statistics about the shape of the distribution were pretty accurate. Notice that we did not need to consider the variance at all, because the variance in this example (2.04) is on a different scale of measurement than our original 5-point response scale, and therefore is very difficult to interpret. Variance is an important statistic for many techniques (e.g., ANOVA, regression), but it does little to help us understand the shape of a distribution of scores. The mean, standard deviation, and to a lesser extent the range, when considered together, can provide a rough picture of a distribution of scores. Often, a rough picture is all a researcher needs or wants. Sometimes, however, researchers need to know more precisely the characteristics of a distribution of scores. In that case, a picture, such as a graph, may be worth a thousand words.

Another useful way to examine a distribution of scores is to create a boxplot. In Figure, a boxplot is presented for the same variable that is represented in Figure, wanting to demonstrate ability. This boxplot was produced in the SPSS statistical software programme. The box in this graph contains some very useful information. First, the thick line in the middle of the box represents the median of this distribution of scores. The top line of the box represents the 75th percentile of the distribution and the bottom line represents the 25th percentile.

Therefore, the top and bottom lines of the box reveal the interquartile range (IQR) for this distribution. In other words, 50 percent of the scores on this variable in this distribution are contained within the upper and lower lines of this box (i.e., 50% of the scores are between just above a score of 2 and just below a score of 4). The vertical lines coming out of the top and bottom of the box and culminating in horizontal lines reveal the largest and smallest scores in the distribution, or the range. These scores are 5 and 1, producing a range of 5 - 1 = 4. As you can see, the boxplot in figure contains a lot of useful information about the spread of scores on this variable in a single picture.

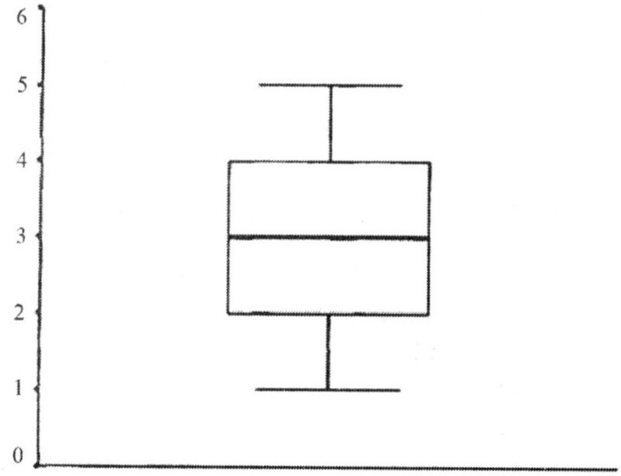

Fig. Boxplot for the Desire to Appear able Variable

Measures of variation, such as the variance, standard deviation, and range, are important descriptive statistics. They provide useful information about how spread out the scores of a distribution are, and the shape of the distribution. Perhaps even more important than their utility as descriptors of a single distribution of scores is their role in more advanced statistics such as those coming in later chapters. We examine the properties of a theoretical distribution, the normal distribution, that has a specific shape and characteristics. We can see how the normal distribution can be used make inferences about the population based on sample data.

ANALYSIS OF VARIANCE

The purpose of a one-way analysis of variance (one-way ANOVA) is to compare the means of two or more groups (the independent variable) on one dependent variable to see if the group means are significantly different from each other. In fact, if you want to compare the means of two independent groups on a single variable, you can use either an independent samples *t* test or a oneway ANOVA. The results will be identical, except instead of producing a *t* value, the ANOVA will produce an Fratio, which is simply the *t* value squared. Because the *t* test and the one-way ANOVA produce identical results when there are only two groups being compared, most researchers use the one-way ANOVA only when they are comparing three or more groups. To conduct a one-way ANOVA, you need to have a categorical (or nominal) variable that has at least two independent groups as the independent variable and a continuous variable as the dependent variable.

Because the independent *t* test and the one-way ANOVA are so similar, people often wonder, Why don't we just use *t* tests instead of one-way ANOVAs? Perhaps the best way to answer this question is by using an example. Suppose that we want to go into the potato chip business. We've got three different recipes, but because I'm new to the business and don't have a lot of money, we can produce only one flavour. We want to see which flavour people like best and produce that one. randomly select 90 adults and randomly divide them into three groups. One group tries our BBQ-flavoured chips, the second group tries our ranch-flavoured chips, and the third group tastes our cheese-flavoured chips. All participants in each group fill out a rating form after tasting the chips to indicate how much they liked the taste of the chips. The rating scale goes from a score of 1 ("Hated it") to 7 ("Loved it"). We then compare the average ratings of the three groups to see which group liked the taste of their chips the most. In this example, the chip flavour (BBQ, Ranch, Cheese) is our categorical, independent variable and the rating of the taste of the chips is our continuous, dependent variable.

To see which flavour received the highest average rating, we could run three separate independent *t* tests comparing (a) BBQ with Ranch, (b) BBQ with Cheese, and (c) Ranch with Cheese. The problem with running three separate *t* tests is that each time we run a *t* test, we must make a decision about whether the difference between the two means is meaningful, or statistically significant. This decision is based on probability, and every time we make such a decision, there is a slight chance we might be wrong. The more times we make decisions about the significance of *t* tests, the greater the chances are that we will be wrong. In other words, the more *t* tests we run, the greater the chances become of deciding that a *t* test is significant (i.e., that the means being compared are really different) when it really is not. A one-way ANOVA fixes this problem by adjusting for the number of groups being compared. To see how it does this, let's take a look at one-way ANOVA in more detail.

ONE-WAY ANOVA IN DEPTH

The purpose of a one-way ANOVA is to divide up the variance in some dependent variable into two components: the variance attributable to between-group differences, and the variance attributable to within-group differences, also known as *error.* When we select a sample from a population and calculate the mean for that sample on some variable, that sample mean is our best predictor of the population mean. In other words, if we do not know the mean of the population, our best guess about what the population mean is would have to come from the mean of a sample drawn randomly from that population. Any scores in the sample that differ from the sample mean are believed to include what statisticians call error. For example, suppose we have a sample of 20 randomly selected fifth graders. We give them a test of basic skills in math and find out that, in our sample, the average number of items answered correctly on our test is 12. If we were to select one student in our sample and find that she had a score of 10 on the test, the difference between her score and the sample mean would be considered error.

The variation that we find among the scores in a sample is not just considered error. In fact, it is thought to represent a specific kind of error: *random error.* When we select a sample at random from a population, we expect that the members of that sample will not all have identical scores on our variable of interest (e.g., test scores). That is, we expect that there will be some variability in the scores of the members of the sample. That's just what happens when you select members of a sample randomly from a population. Therefore, the variation in scores that we see among the members of our sample is just considered random error.

The question that we can address using ANOVA is this: Is the average amount of difference, or variation, between the scores of members *of different* samples large or small compared to the average amount of variation *within* each sample, otherwise known as random error (a.k.a. error)? To answer this question, we have to determine three things. First, we have to calculate the average amount of variation within each of our samples. This is called the mean square within (MS_{W}) or the mean square error (MS_{e}). Second, we have to find the average amount of variation *between* the groups. This is called the mean square between (MS_{b}). Once we've found these two statistics, we must find their ratio by dividing the mean square between by the mean square error. This ratio provides our *F* value, and when we have our *F* value we can look at our family of *F* distributions to see if the differences between the groups are statistically significant.

Formula for the F value

$$F = \frac{mean\ square\ between}{mean\ square\ error}$$

or

$$F = \frac{MS_b}{MSe}$$

where *F is the F value*

MS_b is mean square between group

MS_e is mean square error, or within groups

Recall that the formula for calculating an independent *t* test also involves finding a ratio. The top portion of the fraction is the difference between two sample means, which is analogous to the mean square between *(MSb)* just presented. The only differences between the two are (a) rather than finding a simple difference between two means as in a *t* test, in ANOVA we are finding the *average* difference between means, because we often are comparing more than two means; and (b) we are using the squared value of the difference between the means. The bottom portion of the fraction for the *t* test is the standard *error* of the difference between two sample means. This is exactly the same as the *average,* or standard, error within groups. In the formula used to calculate the *F* value in ANOVA, we must square this average within-group error. So, just as in the *t* test, in ANOVA we are trying to find the average difference *between* group means relative to the average amount of variation *within* each group.

To find the MS_e and MS_b, we must begin by finding the sum of squares error (SS_e) and the sum of squares between (SS_b). This sum of squares idea is not new. It is the same old sum of squares introduced back when talking about variance and standard deviation. Sum of squares is actually short for *sum of squared deviations.* In the case of ANOVA, we have two types of deviations. The first is deviations between each score in a sample and the mean for that sample (i.e., error). The second type of deviation is between each sample mean and the mean for all of the groups combined, called the grand mean (i.e., between groups). To find the sum of squares error

1. Subtract the group mean from each individual score in each group: $(X–X)$.
2. Square each of these deviation scores: $(X–X)^2$.
3. Add them all up for each group: $\Sigma(X–X)^2$.
4. Then add up all of the sums of squares for all of the groups:

$$\Sigma(X_1 - \overline{X}_1)^2 + \Sigma(X_2 - \overline{X}_2)^2 + \ldots + \Sigma(X_k - \overline{X}_k)^2$$

Note: The subscripts indicate the individual groups, through the last group, which is indicated with the subscript k.The method used to calculate the sum of squares between groups *(SSb)* is just slightly more complicated than the SS_e formula. To find the SS_b we

1. Subtract the grand mean from the group mean: $(X–X_T)$; $_T$ indicates total, or the mean for the total group.
2. Square each of these deviation scores: $(X–X_T)^2$.
3. Multiply each squared deviation by the number of cases in the group: $[n(X – X_T)^2]$.
4. Add these squared deviations from each group together: $\Sigma[n(X – X_T)^2]$.

The only real differences between the formula for calculating the SS_e and the SS_b are:

1. In the SS_e we subtract the group mean from the individual scores in each group whereas in the SS_b we subtract the grand mean from each group mean.
2. In the SS_b we multiply each squared deviation by the number of cases in each group. We must do this to get an approximate deviation between the group mean and the grand mean for *each case in every group.*

If we were to add the SS_e to the SS_b, the resulting sum would be called the sum of squares total (SS_T). A brief word about the SS_T is in order. Suppose that we have three randomly selected samples of children. One is a sample of 5th graders, another is a sample of 8th graders, and the third is a sample of 11th graders. If we were to give each student in each sample a spelling test, we could add up the scores for all of the children in the three samples combined and divide by the total number of scores to produce one average score. Because we have combined the scores from all three samples, this overall average score would be called the grand mean, or total mean, which would have the symbol X_T. Using this grand mean, we could calculate a squared deviation score for each child in all three of our samples combined using the familiar formula $(X - X_T)^2$. The interesting thing about these squared deviations is that, for each child, the difference between each child's score and the grand mean is the sum of that child's deviation from the mean of his or her own group plus the deviation of that group mean from the grand mean. So, suppose Jimmy is in the fifth-grade sample. Jimmy gets a score of 25 on the spelling test. The average score for the fifth-grade sample is 30, and the average score for all of the samples combined (i.e., the grand mean) is 35. The difference between Jimmy's score (25) and the grand mean (35) is just the difference between Jimmy's score and the mean for his group (25 – 30=–5) plus the difference between his group's mean and the grand mean (30 – 35=–5). Jimmy's deviation from the grand mean is –10. If we square that deviation score, we end up with a squared deviation of 100 for Jimmy.

Now, if we calculated a deviation score for each child in all three samples and added up all of these deviation scores using the old E $(X - X_T)^2$ formula, the result would be the sum of squares total, or the SS_T. The interesting thing about this SS_T is that it is really just the sum of the SS_b and the SS_e. $SS_T = SSb + SS_e$. This makes sense, because, as we saw with Jimmy, the difference between any individual score and the grand mean is just the sum of the difference between the individual score and the mean of the group that the individual is from plus the difference between that group mean and the grand mean. This is the crux of ANOVA.

DECIDING IF THE GROUP MEANS ARE SIGNIFICANTLY DIFFERENT

Once we have calculated the SS_b and the SS_e, we have to convert them to average squared deviation scores, or *MSb* and MS_e. This is necessary because there are far more deviation scores in the SS_e than there are in the SS_b, so the sums of squares can be a bit misleading. What we want to know in an ANOVA is whether the *average* difference between the group means is large or small relative to the *average* difference between the individual scores and their respective group means, or the average amount of error within each group. To convert these sums of squares into mean squares, we must divide the sums of squares by their appropriate degrees of freedom.

For the SS_b, remember that we are only making comparisons between each of the groups. The degrees of freedom for the SS_b is always the number of groups minus 1. If we use K to represent the number of groups, and *df* to represent degrees of freedom, then the formula for the between groups degrees of freedom is $df=K-1$. So, to convert an *SSb* to an *MSb, we* divide SS_b *by* $K-1$. The degrees of freedom for SS_e is found by taking the number of scores in each group and subtracting 1 from each group. So, if we have three groups, our *df for* SS_e will be $(n_1 - 1) + (n_2 - 1) + (n_3 - 1)$. Notice that this is the same formula for the degrees of freedom that was used for the independent samples *t* test. The only difference is that we have one more group here. A simpler way to write this *df* formula is *N-K,* where N is the total number of cases for *all* groups combined and *K* is the number of groups. Once we have this *df,* we can convert the SS_e into an MS_e by simply dividing SS_e *by* $N-K$. Table contains a summary of the formulas for converting the sums of squares into mean squares.

Converting Sums of Squares into mean squares

$MS_e = \frac{SS_e}{K-1}$	$MS_e = \frac{SS_e}{N-K}$
MS_e =Mean squres between groups	MS_e =Mean squres error
SS_e =Sum of squres between groups	SS_e =Sum of squres error
N=The number of cases combined	K=The number of groups
across all groups	N=The number of cases combined across all groups

Once we have found our *MSb and* our MS_e, all we have to do is divide *MSb by* MS_e to find our *F* value. Once we've found our *F* value, we need to look in our table of *F* values to see whether it is statistically significant. This table of *F* values is similar to the table of *t* values we used, with one important difference. Unlike *t* values, the significance of *F* values depends on both the number of cases in the samples (i.e., the *df for* $MS_{e)}$ *and* the number of groups being compared (i.e., the *df for* $MS_{b)}$. This second *df is* critical, because it is what is used to control for the fact that we are comparing more than two groups. Without it, we might as well conduct multiple *t* tests, and this is problematic for the reasons discussed at the beginning of the chapter. Critical values for *F* associated with different alpha levels. If our observed value of *F* is larger than our critical value of *F,* we must conclude that there are statistically significant differences between the group means.

FACTORIAL ANALYSIS OF VARIANCE

We examined one-way ANOVA. In this chapter and the two that follow, we explore the wonders of two more advanced methods of analyzing variance: factorial ANOVA and repeated-measures ANOVA. These techniques are based on the same general principles as one-way ANOVA. Namely, they all involve the partitioning of the variance of a dependent variable into its component parts (e.g., the part attributable to between-group differences, the part attributable to within-group variance, or error). In addition, these techniques allow us to examine more complex, and often more interesting questions than is allowed by simple one-way ANOVA. These more advanced statistical techniques involve much more complex formulas than those we have seen previously. Therefore, in this chapter and those that follow only a basic introduction to the techniques is offered. You should keep in mind that there is much more to these statistics than described in these pages, and should consider reading more about them in the suggested readings at the end of each chapter.

WHEN TO USE FACTORIAL ANOVA

Factorial ANOVA is the technique to use when you have one continuous (i.e., interval or ratio scaled) dependent variable and two or more categorical (i.e., nominally scaled) independent variables. For example, suppose we want to know whether boys and girls differ in the amount of television they watch per week, on average. Suppose we also want to know whether children in different regions of the United States (i.e., East, West, North, and South) differ in their average amount of television watched per week. In this example, average amount of television watched per week is our dependent variable, and gender and region of the country are our two independent variables. This is known as a 2 X 4 factorial analysis, because one of our independent variables has two levels (gender) and one has four levels (region). If we were writing about this analysis in an academic paper, we would write "we conducted a 2 (gender) x 4 (region) factorial ANOVA."

Now when we run our factorial ANOVA, we will get three interesting results. First, we will get two main effects, one for our comparison of boys and girls and one for our comparison of children from different regions of the country. These results are similar to the results we would get if we simply were to run two one-way ANOVAs, with one important difference, which we describe in the next section. In addition to these main effects, our factorial ANOVA also produces an interaction effect, or simply an interaction. An interaction is present when the differences between the groups of one independent on the dependent variable vary according to the level of a second independent variable. Interaction effects are also known as moderator effects. We discuss interactions in greater detail in the next section as well. For now, suffice it to say that interaction effects are often very interesting and important pieces of information for social scientists.

SOME CAUTIONS

Just as with one-way ANOVA, when conducting a factorial ANOVA it is important to determine whether there are roughly an equal number of cases in each group and whether the amount of variance within each group is roughly equal (known as homogeneity of variance). The ideal situation in ANOVA is to have roughly equal sample sizes in each group and a roughly equal amount of variation (e.g., the standard deviation) in each group. If the sample sizes are equal but the standard deviations are significantly different, this is not a big problem. But if the sample sizes *and* the standard deviations are quite different in the various groups, there is a problem. This is particularly likely to be a problem in factorial ANOVA because the combination of independent variables often produces a small number of cases in each group.

In our previous example, suppose we have 20 boys and 20 girls in our entire sample. In addition, suppose that we have 20 children from each of the four regions in our sample. To test the main effects, these numbers are acceptable. That is, it is reasonable to compare 20 boys to 20 girls if we want to know whether boys and girls differ in their average amount of television viewing. Similarly, it is reasonable to compare 20 children from each of the four different regions of the country. But suppose that in the West, our sample of 20 children includes only 5 girls and 15 boys, whereas in the North our sample includes 15 girls and only 5 boys. When we divide up our sample by *two* independent variables, it is easy to wind up with cell sizes that are too small to conduct meaningful ANOVAs. A "cell" is a subset of cases representing one unique point of intersection between the independent variables. In the aforementioned example, there would be eight cells: girls from the West, boys from the West, girls from the South, boys from the South, and so on. When you consider that factorial ANOVAs can have more than two independent variables, the sample can be subdivided a number of times. Without a large initial sample, it is easy to wind up with cells that contain too few cases. As a general rule, cells that have fewer than 10 cases are too small to include in ANOVAs, and cell sizes of at least 20 are preferred.

FACTORIAL ANOVA IN DEPTH

When dividing up the variance of a dependent variable, such as hours of television watched per week, into its component parts, there are a number of components that we can examine. In this section, we examine three of these components: The main effects, interaction effects, and simple effects. In addition, we also present an introduction to the idea of partial and controlled effects.

MAIN EFFECTS AND CONTROLLED OR PARTIAL EFFECTS

As mentioned earlier, a factorial ANOVA will produce main effects for each independent variable in the analysis. These main effects will each have their own F value, and are very similar to the results that would be produced if you just conducted a one-way ANOVA for each independent variable on the dependent variable. However, there is one glorious benefit of looking at the main effects in a factorial ANOVA rather than separate one-way ANOVAs: When looking at the main effects from a factorial ANOVA, it is possible to test whether there are significant differences between the groups of one independent variable on the dependent variable while controlling for, or partialing out the effects of the other independent variable(s) on the dependent variable. Let me clarify this confusing sentence by returning to our example of television viewing.

Suppose that when we examine whether boys and girls differ in the average amount of television they watch per week, we find that there is a significant difference: Boys watch significantly more television than girls. In addition, suppose that children in the North watch, on average, more television than children in the South. Now, suppose that, in our sample of children from the Northern region of the country, there are twice as many boys as girls, whereas in our sample from the South there are twice as many girls as boys. Now I've got a potential problem. How do we know whether our finding that children in the North watch more television than children in the South is not just some *artifact* caused by the greater proportion of boys in our Northern sample? By "artifact" we mean that the North-South difference is merely a *by-product* of the difference between boys and girls; region of the country is not an important factor in and of itself. Think about it: If we already know that boys watch more television, on average, than girls, then we would *expect* our Northern sample to watch more television than our Southern sample because there is a greater proportion of boys in our Northern sample than in the Southern sample. So our question is this: How can we determine whether there is a difference in the average amount of television watched by children in the North and South *beyond* the difference caused by the unequal proportions of boys and girls in the samples from these two regions. Phrased another way, is there an effect of region on television viewing *beyond*, or *in addition to* the effect of gender?

To answer this intriguing question, we must examine the main effect of region on television viewing after *controlling for*, or *partialing out* the effect of gender. We can do this in a factorial ANOVA. To understand how this is accomplished, keep in mind that what we are trying to do with an ANOVA is to *explain* the variance in our dependent variable (amount of television children watch per week) by dividing that variance up into its component parts. If boys and girls differ in how much they watch television, then part of the variance is explained, or accounted for, by gender. In other words, we can understand a bit of the differences among children in their weekly television viewing if we know their gender. Now, once we remove that portion of the total variance that is explained by gender, we can test whether any *additional* part of the variance can be explained by knowing what region of the country children are from. If children from the North and South *still* differ in the amount of television they watch, after *partialing out* or *controlling for* the chunk of variance explained by gender, then we know that there is a main effect of region *independent* o/the effect of gender. In statistical jargon, we would say "There is a main effect of region on amount of television watched *after controlling for the effect of gender.*" This is powerful information. In factorial ANOVA, it is possible to examine each main effect and each interaction effect when controlling for *all* other effects in the analysis.

INTERACTIONS

A second benefit of factorial ANOVA is that it allows researchers to test whether there are any statistical interactions present. Interactions can be a complex concept to grasp. Making the whole issue even more confusing is that the level of possible interactions increases as the number of variables increases. For example, when there are two independent variables in the analysis, there are two possible main effects and one possible two-way interaction effect (i.e., the interaction between the two independent variables). If there are three independent variables in the analysis, there are three possible main effects, three possible two-way interaction effects, and one possible three-way interaction effect. The whole analysis can get very complicated very quickly. To keep things simple, let's take a look at two-way interactions first.

In our television-viewing example, suppose that we randomly selected 25 boys and 25 girls from each of the four regions of the country, measured the number of hours each child spent watching television, and calculated the averages for each group. (Note: Unlike the example provided earlier, in this example there are equal numbers of boys and girls from each region in the sample.) These averages are presented in Table.

Television Viewed by Gender and Region

	North	East	West	South	Overall average by Gender
Girls	20	15	15	10	15
Boys	25	20	20	25	22.5
Overall average by region	22.5	17.5	17.5	17.5	

As we see when examining the means in Table, boys in each region watch more television, on average, than girls. The overall averages by gender presented in the last column indicate that there appears to be a main effect for gender, with boys watching an average of 22.5 hours of television per week and girls watching an average of only 15 hours per week.

When we look at the overall averages presented for each region (bottom row), we can see that children in the North watch more television, on average, than do children in the other three regions. Therefore, we can tell that there appear to be main effects for gender and region on amount of television watched.

Notice that it is said *"appear* to be main effects." To determine whether these main effects of statistically significant, we would have to determine the probability of obtaining differences of this size between randomly selected groups of this size.

Once we have examined the main effects, we can turn our attention to the possible interaction effects. To do this, we need to examine the means in each of the eight cells presented in Table (i.e., Northern boys, Northern girls, Eastern boys, Eastern girls, etc.). When we examine these means, we can see that in the North, East, and West, boys watch an average of 5 hours more television per week than girls.

But in the South, boys watch an average of 15 more hours of television than girls. Therefore, it appears that the differences in the amount of television watched by girls and boys are not uniform across the four regions of the country. In other words, the relationship between gender and amount of television watched *depends on,* or *is moderated by,* the region of the country. Because the definition of a two-way interaction is that the relationship between an independent variable and a dependent variable is moderated by a second independent variable, we appear to have a two-way interaction here.

When we find a statistically significant interaction (again, we must examine the p value of the F ratio for the interaction term to determine if it is statistically significant), we must determine the *nature* of the interaction and then *describe* the interaction.

One excellent method for getting a handle on the nature of the interaction is to depict it graphically. To do this, all we need to do is to graph the means. Line graphs and bar graphs work best. We have produced a line graph that represents the data presented in Table. This graph is presented in Figure.

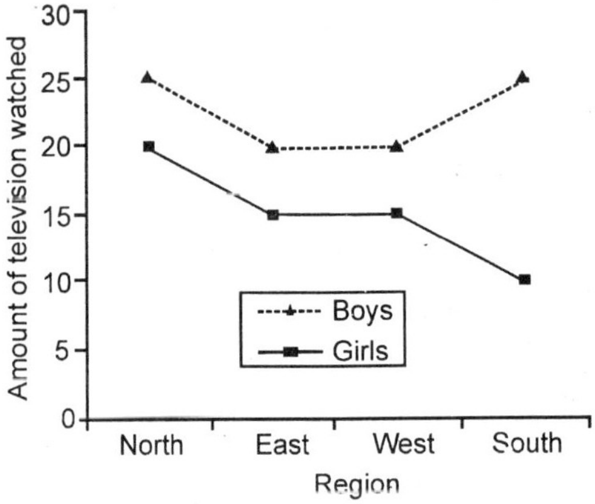

Fig. Interaction of Gender and Region

When we look at this graph, the nature of the interaction becomes readily apparent. Specifically, what we can see is that there is a consistent pattern for the relationship between gender and amount of television viewed in three of the regions (North, East, and West), but in the fourth region (South) the pattern changes somewhat. Specifically, the gap between boys and girls in the average amount of television watched per week is much wider in the South than in the other three regions. In Figure, we can see that the means for boys and girls on the dependent variable are further apart in one region than in the other, but the lines never cross. That is, there is no region *in* which girls have higher average scores than boys on the dependent variable.

As you look at the graph presented in Figure, notice that you can see both the main effects and the interaction effects. Recall that the main effect for gender indicates that, when we combine the scores from all four regions, boys appear to have higher average scores than girls on our dependent variable (i.e., amount of television watched per week). In Figure this effect is clear, as we can see that the line for boys is higher than the line for girls in all four regions. We can also see evidence of a main effect for region, although this effect is somewhat less clear than the main effect for boys. We see the region effect by noting that for both boys and girls, the average amount of television viewing is higher in the North than in either the East or West regions. This main effect is complicated a bit, however, by the presence of the interaction. Notice that whereas the mean is lower in the South than in the North for girls (supporting our main effect for region), the mean for boys in the South is equal to the mean for boys in the North. This raises a difficult question: When we say there is a main effect for region, with children in the North watching more television, on average, than children in the other three regions, are we being accurate?

Interpreting Main Effects in the Presence of an Interaction Effect

Researchers do not always agree on the best way to interpret main effects when there is a significant interaction effect. Some argue that it makes little sense to interpret main effects at all when there is an interaction effect present, because the interaction effect essentially modifies the meaning of the main effect. In the preceding example, the main effect for region that shows children in the North watch more television than children elsewhere is really only true within the girls sample. In fact, boys in the South watch as much television as boys in the North, and girls in the North do not watch more television than boys in any region. Therefore, some would argue that we should just describe the nature of the interaction, and not interpret the main effects. The logic of this argument is as follows: If we say that children in the North watch more television than children in other regions, the statement is misleading because it is not true for boys. To be accurate, we should just say that *girls* in the North watch more television that *girls* in other regions.

Others, ourself included, think it makes sense to interpret all of the effects and to consider them in relation to each other. Returning to our earlier example, we can see that there is a main effect for gender, with boys watching more television, on average, than girls. We can also see that this effect is especially pronounced in the South. In addition, we can say that overall, when we *combine* the samples of boys and girls together, there is a main effect for region such that Northern children watch more television than children in other regions, on average. When we add the consideration of the interaction effect, we can further argue that this overall effect is due *primarily* to differences within the sample of girls, and less to variation within the sample of boys. It is possible to get an interaction effect without a main effect. In this example, boys and girls have equal means, as do children in each of the four geographic regions). Therefore, it makes sense to report and interpret significant main effects, even in the presence of an interaction effect. The key is to provide enough information so that readers of your results can makes sense of them. To do this, it may be necessary to discuss your interaction and main effects in relation to each other.

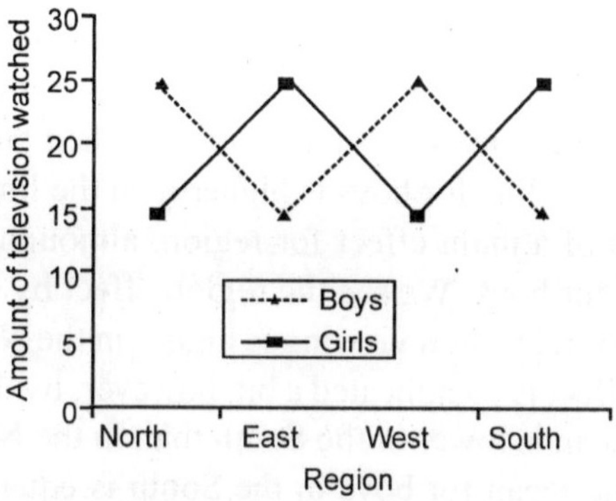

Fig. Interaction with Equal Means

Here is another example to more clearly illustrate the problems of interpreting main effects in the presence of significant interactions. Suppose that we were to examine the math skills of boys and girls in two different types of mathematics programmes. Students in the "Traditional" programme study math in the usual way, reading a textbook and working out math problems in class. Students in the "Experimental" programme work in groups to solve problems collaboratively and work with more real-world, applied problems. After one year, we give a math test to 25 randomly selected boys and 25 randomly selected girls from each math programme. We calculate the averages for these four groups, which are presented in Figure.

The means presented in the figure clearly show that although boys and girls in the Traditional math programme had similar average scores on the math test, girls did much better than boys in the Experimental math programme. This is an interaction. In addition, because girls in the Experimental programme did so well on their math test, the overall mean for the Experimental group was significantly higher than the overall mean for the Traditional group, thereby creating a main effect for math programme. But does it make sense to say that students in the Experimental math programme did better on the test than students in the Traditional programme? Clearly, this was not the case for boys, and some would argue that it would be misleading to point out the main effect for math programme because the effect is only present for the girls, not the boys.

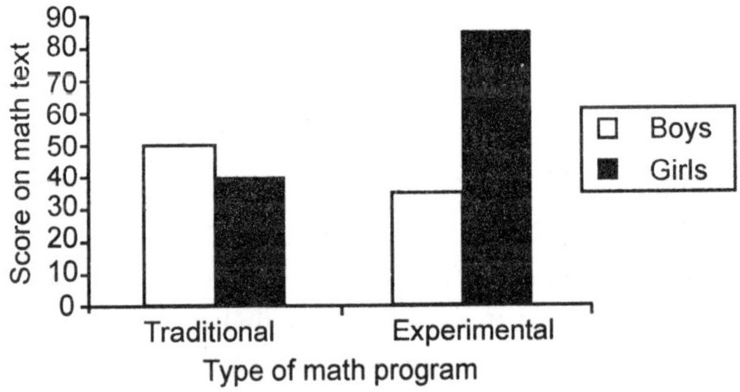

Fig. Interaction of Gender by math Programme

TESTING SIMPLE EFFECTS

Once we have found our main and interaction effects in factorial ANOVA, we can conduct one final set of analyses to examine the simple effects. The methods used to calculate the simple effects and determine whether they are statistically significant are analogous to the post-hoc tests described in Chapter. What simple effects analysis allows us to do is to test whether there are significant differences in the average scores of any particular cells. One of the benefits of simple effects analysis is that is allows us to better understand some of the complexities in our data, particularly how to make sense of significant interaction effects.

Returning to our sample data presented in Figure we can see that we have four cells: girls in the Traditional math programme, Traditional boys, Experimental girls, and Experimental boys. With a simple effects analysis, we can test whether boys in the Traditional math programme *(X* = 50) had significantly higher average math test scores than did boys in the Experimental programme *(X* " 35). We could also test whether boys and girls in the Traditional programme differed significantly. Perhaps most important for helping us understand the interaction effect, we can test whether girls in the Experimental programme had higher average math test scores than students in each of the three other groups.

Analysis of Covariance

Earlier in this chapter, we suggested that one of the benefits of conducting factorial ANOVAs is that it allows us to determine whether groups differ on some dependent variable while controlling for, or partialing out, the effects of other independent variables. A closely related concept that applies to all types of ANOVA, including one-way, factorial, and repeated-measures, is the use of covariates in these analyses. In analysis of covariance (ANCOVA), the idea is to test whether there are differences between groups on a dependent variable after controlling for the effects of a different variable, or set of variables. The difference between an ANCOVA and the types of controlled variance we described earlier is that with an ANCOVA, the variable(s) that we are controlling, or partialing out, the effects of is not necessarily an independent variable. Let me explain.

In our earlier example, we was able to test whether boys and girls differed in the amount of television they watched while controlling for the effects of which region of the country they lived in (the second independent variable), as well as the interaction between the two independent variables.

But in an ANCOVA analysis, we can control the effects of variables *besides* independent variables. For example, we could use socioeconomic status (SES) as a covariate and test whether children in different regions of the country differ in the amount of television they watch *after* partialing out the effects of their SES. Suppose that our sample of children from the North is less wealthy than our samples from the three other regions. Suppose further than children from poorer families tend to watch more television than children from wealthier families. Because of this, our earlier results that found greater television watching among children in the Northern region may simply be due to the fact that these children are less wealthy than children in the other regions. With ANCOVA, we can test whether the difference in the viewing habits of children from different regions is due strictly to differences in SES, or whether there are regional differences *independent* of the effects of SES. This is particularly handy because even though factorial ANOVA only allows us to use categorical (i.e., nominally scaled) independent variables, with ANCOVA we can also control the effects of continuous (i.e., intervally scaled) variables.

Effect Size

It is illustrated effect size in factorial ANOVA, along with some of the particulars about sums of squares, mean squares, and *F* values, using output from an analysis of our own data using the SPSS computer software programme. In this example, students' confidence in their ability to understand and successfully complete their English classwork, referred to here as self-efficacy, was the dependent variable.

We wanted to see whether high school boys and girls differed in their self-efficacy (i.e., a main effect for gender), whether students with relatively high grade point averages (GPA) differed from those with relatively low GPAs in their self-efficacy (i.e., a main effect for GPA), and whether there was an interaction between gender and GPA on self-efficacy. Gender, of course, is a two-category independent variable. To make GPA a two-category variable, we divided students into high- and low-GPA groups by splitting the sample in two using the median GPA. This allowed me to perform a 2 (gender) x 2 (GPA) factorial ANOVA. Self-efficacy was measured using a survey with a 5-point scale (1="not at all confident" and 5="very confident"). Our sample consisted of 468 high school students.

The results presented in Table begin with descriptive statistics. These statistics are presented separately by subgroups (e.g., low-achieving girls, high-achieving girls, all girls combined, low-achieving boys, etc.). The means and standard deviations presented are for the dependent variable, self-efficacy. By glancing over the means we can see that the boys in our sample had slightly higher average feelings of self-efficacy than did the girls, and this difference appears to be largest among the boys and girls in the low-GPA group.

Turning our attention to the ANOVA results, there are a number of important features to notice. In the far left column titled "Source, " there are the various *sources* of variation in self-efficacy. These are the different ways that the variance of the dependent variable, self-efficacy, is sliced up by the independent variables. The first source is called the "Corrected Model." This is the combination of all of the main and interaction effects. If covariates were used, these effects would be included in the "Corrected Model" statistics. Reading from left to right, we can see that the full model has a sum of squares (11.402), which when divided by three degrees of freedom *("df")* produces a "Mean Square" of 3.801. When we divide this by the mean square error a few rows down (MS_e = .649) we get an F value of 5.854. This has a "Sig." of .001 (in other words, p = .001). Because this value is less than .05, the overall model is statistically significant. But is it *practically* significant? In the final column labeled "Eta Squared" we can see that the overall model accounts for only 3.6 Percent of the variance in self-handicapping scores. In other words, gender, GPA level, and the interaction of these two *combined* only explain 3.6 Percent of the variance. Although this is *statistically* significant, this may not be a big enough effect size to be considered *practically* significant. Remember that statistical significance is influenced by sample size, and 468 cases is a pretty large sample. An effect size of .036, in contrast, is not affected by sample size and therefore may be a better indicator of practical significance.

SPSS results for gender GPA factorial ANOVA

Gender	**GPA**	**Mean**	**Std.**	**N derivation**	
Girl		1.00	3.6667	.7758	121
		2.00	4.0050	.7599	133
		Total	3.8438	.7845	254
Boy		1.00	3.9309	.8494	111
		2.00	4.0031	.8503	214
		Total	4.0031	.8503	214
Total	1.00	3.7931	.8208	232	
		2.00	4.0381	.7989	236
		Total	3.9167	.8182	468.

ANOVA Results

Source	Type III	off	Mean Sum of	F	Sig. Square	Etc. Squard
Corrected Model	11.402	3	6.801	5.854	.001	.036
Intercept	7129.435	1	7129.435	10981.566	.000	.959
Cendet	3.354	1	3.354	5.166	.003	.011
GPA	6.912	1	6.912	10.646	.001	.022
Gender+ GPA	1.028	1	1.028	1.584	.209	.003
Error	301.237	464	.649			
Total	7491.889	468				
Cunnected Total	312.639	467				

In addition to the F value, p value, and effect size for the entire model, SPSS prints out statistics for each of the main effects and the interaction effect as well. Here we can see that gender is a statistically significant predictor of self-efficacy ("Sig."=.023), but the effect size of .011 is very small. By looking at the overall means for girls and boys in the top portion of Table, we can see that boys (X_{boys} = 4.0031) have the slightly higher average feelings of self-efficacy in English than girls ($Xgirls$ = 3.8438). GPA has a larger F value (F = 10.646) and is statistically significant, but it also has a small eta squared value (.022). Students in the high-GPA group had slightly higher average feelings of self-efficacy (X_{high} = 4.0381) than did students in the low-GPA group ($X_{lo\ w}$ = 3.793). The gender by GPA interaction was not statistically significant and has a tiny effect size. Overall, then, the statistics presented in Table reveal that although there are statistically significant main effects for gender and GPA on self-efficacy, and the overall model is statistically significant, these effect sizes are quite small and suggest that there is not a strong association between either gender nor GPA with self-efficacy among this sample of high school students.

There are two other features of the SPSS output presented in Table worth noting. First, the sum of squares that SPSS uses by default in a factorial ANOVA is called "Type III" sum of squares. This means that when SPSS calculates the sum of squares for a particular effect, it does so by accounting for the other effects in the model. So when the sum of squares for the gender effect is calculated, for example, the effect of GPA and the gender by GPA interaction effects have been partialed out already. This allows us to determine the *unique* effect of each main effect and interaction effect. Second, notice that the F value for each effect is obtained by dividing the mean square for that effect by the mean square error. This is the same way F values were calculated in one-way ANOVA discussed.

REPEATED-MEASURES

One type of *t* test discussed the paired *t* test. One type of study in which a paired *t* test would be used is when we have two scores for a single group on a single measure. For example, if we had a group of third-grade students and we gave them a test of their math abilities at the beginning of the school year and again at the end of the school year, we would have one group (third graders) with two scores on one measure (the math test). In this situation, we could also use a repeated-measures analysis of variance (ANOVA) to test whether students scores on the math test were different at the beginning and end of the academic year.

Repeated-measures ANOVA has a number of advantages over paired *t* tests, however. First, with repeated-measures ANOVA, we can examine differences on a dependent variable that has been measured at more than two time points, whereas with an independent *t* test we can only compare scores on a dependent variable from two time points. Second, as discussed on factorial ANOVA, with a repeated-measures ANOVA we can control the for effects of one or more covariates, thereby conducting a repeated-measures analysis of covariance (ANCOVA). Third, in a repeated-measures ANOVA, we can also include one or more independent categorical, or group variables. This type of mixed model is a particularly useful technique and is discussed in some detail.

WHEN TO USE EACH TYPE OF REPEATED-MEASURES TECHNIQUE

The most basic form of a repeated-measures ANOVA occurs when there is a single group (e.g., third graders) with two scores (e.g., beginning of the year, end of the year) on a single dependent variable (e.g., a mathematics test). This is a very common model that is often used in simple laboratory experiments. For example, suppose we wanted to know whether drinking alcohol affects the reaction time of adults when driving. We could take a group of 50 adults and test their reaction time when driving by flashing a red light at each one of them when they are driving and measuring how long it takes for each one to apply the brakes. After calculating the average amount of time it takes this group to apply the brakes when sobre, we could then ask each member of our group to consume two alcoholic drinks and then again test their reaction time when driving, using the same methods. In this example, I've got one group (50 adults) with two scores on one dependent variable (reaction time when driving). After the second measure of reaction time, we could ask each of our participants to consume two more alcoholic drinks and again test their reaction time. Now we've got three measures of reaction time that we can use in our repeated-measures ANOVA. Notice that our dependent variable is always the same measure (reaction time), and our group is always the same (sample of 50 adults). The results of our repeated-measures ANOVA will tell me whether, on average, there are differences in reaction time across our three trials. If there are, we might logically conclude that drinking alcohol affects reaction time, although there may be other explanations for our results (e.g., our participants may be getting tired or bored with the experiment, they may be getting used to the test situation, etc.).

In a slightly more advanced form of the test of reaction time, we could include a covariate. In the previous example, suppose that we found the reaction time was fastest when our participants were sobre, a bit slower after two drinks, and a lot slower after four drinks. Suppose that we publish these results and the national beer, wine, and liquor companies become worried that, because of our study, people will stop drinking their products for fear of getting in automobile accidents. These producers of alcoholic drinks begin to criticize our study. They suggest that because equal amounts of alcohol generally have greater effects on those who weigh less than on heavier people, our results may have been skewed by the effects of alcohol on the lighter people in our study. "Although the effects of two alcoholic drinks may impair the reaction time of lighter people, even four alcoholic drinks will not impair the reaction time of heavier people" said the United Alcohol Makers of America (a fictitious group).

Stung by the criticism of the UAMA, we decide to replicate our study, but this time we use weight as a covariate. Again, we measure participants' reaction time when driving completely sobre, after two alcoholic drinks, and after four alcoholic drinks. In addition, this time we weigh each of our participants. Now when we analyse our data, we include our weight covariate. We find that, after controlling, or partialing out the effects of weight, there is no difference in the reaction time of participants before they have any drinks and after they have two drinks, but after four drinks our participants react more slowly, on average, than they did after zero or two drinks. These results suggest that drinking may increase the reaction time of lighter people after only two drinks, but it seems to impair the reaction time of people, regardless of weight, after four drinks.

Still bothered by our results, the UAMA suggests that our results are skewed because we did not look at the effects of drinking on reaction time separately for men and women. "Women are more dramatically affected by alcohol than men, regardless of weight" claims the UAMA. They argue that although consuming four alcoholic drinks may slow the reaction time of women, it will not have an effect on heavy men. Though we are dubious of the argument that heavy men should have their rights to drink and drive protected, in the name of science we decide to conduct one final study. In this study, again with 50 adults (25 women and 25 men) of various weights, we again test their reaction time while driving after zero, two, and four alcoholic drinks. Now I've got one dependent variable (reaction time) measured at three time points, one covariate (weight), and one independent group variable (gender of participant). Notice that although number of drinks is technically an independent variable, it is not a categorical, or group, variable. In other words, we do not have three independent groups (the zero-drink group, the two-drink group, and the four-drink group). Rather, we have three *dependent,* or repeated, measures of the same dependent variable, reaction time.

When we examine the results of our study, we find that, after controlling for the effects of our covariate (weight), there is still no difference in reaction time measured after zero and two drinks, but still slower reaction time, on average, after four drinks. In addition, we find no interaction between gender and number of drinks on reaction time. This tells us that *both* men and women have slower reaction time after four drinks, regardless of their weight.

To summarize, our three different repeated-measures ANOVAs produced the following results. The first one found that adults' reaction times while driving were slower, on average, after two drinks and slower still after four drinks. Our second test included the covariate of weight, and we found that when we control for the effects of weight, reaction time is not slower after two drinks but is slower after four drinks. Finally, in our third analysis, we examined whether changes in reaction time after two and four drinks, when controlling for weight, was different for men and women. found that it was not. These three analysis provide a snapshot of how repeated-measures ANOVA works and what information it can provide. Now let's take a closer look at how it works.

REPEATED-MEASURES ANOVA

Repeated-measures ANOVA is governed by the same general principles as all ANOVA techniques. As with one-way ANOVA and factorial ANOVA, in repeated-measures ANOVA we are concerned with dividing up the variance in the dependent variable. Recall that in a one-way ANOVA, we separated the total variance in the dependent variable into two parts: that attributable to differences between the groups, and that attributable to differences among individuals in the same group (a.k.a., the error variance). In a repeated-measures ANOVA with no independent group variable, we are still interested in the error variance. However, we also want to find out how much of the total variance can be attributed to time, or trial. That is, how much of the total variance in the dependent variable is attributable to differences *within individuals* across the tunes they were measured on the dependent variable.

Consider an example. Suppose that we are interested in examining whether a group of students increase their knowledge and skills from one academic year to the next. To do this, we give our sample a standardized test of vocabulary (with a possible range of 1"100), once when they are finishing third grade and again when they are finishing fourth grade. When we do this, suppose we get the data presented in Table.

Vocabulary test scores at two time points

Case Number	Text Scene, Time I (Third Grade)	Text Scene, Time 2 (Fourth Grade)
1	40	60
2	55	55
3	60	70
4	40	45
5	75	70
6	80	85
7	65	75
8	40	60
9	20	35
10	45	60

For each of the 10 cases in Table, we have two test scores, giving us a total of 20 scores in the table. We could find an average for these 20 scores, and a standard deviation, and a variance. In a repeated-measures ANOVA, we want to try to partition this total variance into different pieces. In the most basic form of repeated-measures ANOVA, there are three ways that we can slice up this variance. First, there is the portion of variance attributable to deviations in scores between the individual cases in the sample. For each case in our sample, we have two scores (one for Time 1 and one for Time 2). We can find an average of these two scores, for each individual, and then see how much this individual average differs from the overall average. In Table, for example, the first case has an average score of 50 across the two trials (40 + 60 2=50). The overall average for the scores in the table if 56.75. So there is some variation in the average scores of the 10 individuals in the sample. This is one source of variation.

The second source of variation in the scores involves the within-subject variance, or differences, between Time 1 and Time 2 scores. As we can see by looking at the scores in Table and in Figure, it appears that students generally had different scores on the test at Time 1 than they did at Time 2. These infra-individual, or within-subject, differences between Time 1 and Time 2 scores can be seen more easily in the graph presented in Figure. These intra-individual changes reflect differences, or variance, *within each individual,* and therefore are called within-subject effects. What we are interested in is whether, on average, individuals' scores were different at Time 1 (in third grade) than they were at Time 2 (in fourth grade). Notice that we are asking whether there were differences in the scores between Time 1 and Time 2 *on average.* If the scores of some of the cases went up from Time 1 to Time 2, but the scores of other cases went down by the same amount, then these changes would cancel each other out, and there would be no *average* difference between the Time 1 and Time 2 scores. But if the scores either went up or down on average between Time 1 and Time 2, then we could say that some of the total variation can be attributable to within-subject differences across time. A look at the scores in Table and Figure reveals that scores *appeared* to increase from Time 1 to Time 2. To examine whether there are differences in the average scores across time, all we need to do is calculate the average score at each time and find the difference between these average scores and the overall average. In the preceding paragraph, we found that the overall average score was 56.75. In Table, we can see that the average for Time 1 is 51.5, and the average score for Time 2 is 61.5. So we can see that there is some variance in the average scores at the two times (i.e., third and fourth grade), suggesting that there may be a within-subjects effect.

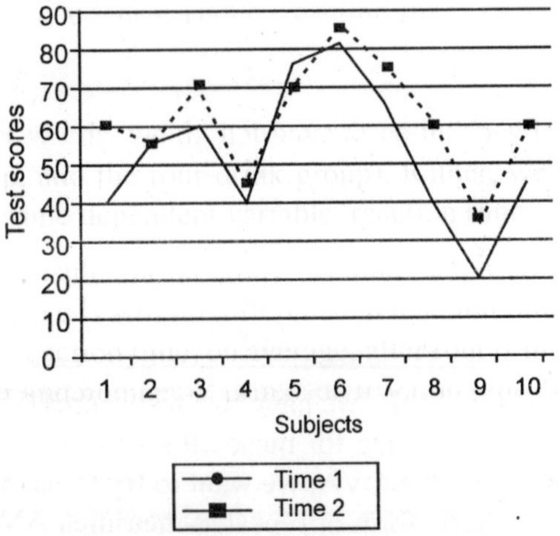

Fig. Time 1 and Time 2 Test Scores

The third source of variation in the scores comes from the interaction between the within-subject scores and the variance in scores across the subjects. Although it appears that the scores of the members in our study increased, *on average,* from Time 1 to Time 2, it does not appear that these within-subject changes over time were the same across all of the subjects in the sample. As Figure clearly shows, some subjects had large increases in their test scores from Time 1 to Time 2 (e.g., Subjects 1, 8, 9, and 10), whereas others had more modest increases, one had no change (Subject 2) and one actually had a lower score at Time 2 (Subject 5). So there appears to be a case, or subject, by time interaction. In other words, the size of the increase in test score from third to fourth grade depended on the which subject we were looking at. This difference among the subjects in the magnitude of change from Time 1 to Time 2 represents the third source of variance.

Using these three sources of variance, we can then calculate an *F* ratio and determine whether there are statistically significant differences in the average scores at Time 1 and the average scores at Time 2. To do this, we divide the mean square for the differences between the trials, or time, averages *(*MS_{T}*)* by the mean square for the subject by trial interaction *(*$MS_{S \times r}$*)*. The degrees of freedom *F* ratio is the number of trials minus 1 *(T*- 1) and *(T* “ *l)(S*- 1), where *S* represents the number of subjects in the sample. What we get when we calculate this *F* ratio is an answer to the following question: How large is the difference between the average scores at Time 1 and Time 2 relative to (i.e., ivided by) the average amount of variation among subjects in their change from Time 1 to Time 2? Because differences in the rate of change across time are just considered random fluctuations among individuals, this *F* ratio, like all *F* ratios, is a measure of *systematic* variance in scores divided by *random* variance in scores

In this most basic form of repeated-measures ANOVA, notice that what we are primarily concerned with is whether there is a systematic pattern of differences *within individuals, or subjects,* in the scores on the dependent variable measured at two time points. Also notice that if we had three points of data (e.g., test scores from third, fourth, and fifth grades), our question would remain the same: Is there a pattern of differences in the scores *within subjects* over time? Keep in mind that when we say “a pattern” or “a systematic pattern, “ we mean *on average.* So, to rephrase the question, a simple repeated-measures ANOVA can help us detect whether, *on average,* scores differ *within subjects* across multiple points of data collection on the dependent variable. This type of simple repeated-measures ANOVA is sometimes referred to as a within-subjects design.

REPEATED-MEASURES ANALYSIS OF COVARIANCE (ANCOVA)

A slightly more complicated form of repeated-measures ANOVA can be produced by adding one or more covariates to the model. As well as in factorial ANOVA, covariates can be used to partial out some portion of the variance in the dependent variable. We illustrate how this works by returning to the example data presented in Table.

One could argue that the results of our repeated-measures ANOVA were skewed by the scores of the more intelligent students in our sample. Although the students in our sample scored higher on the vocabulary test at the end of fourth grade than they did at the end of third grade, we must keep in mind that the change in scores over time represents an *average* change. Some students in our sample improved quite a bit over time, whereas others did not increase at all, and one actually declined. So it is possible that this overall average improvement over time was caused by large increases among the brightest students. To explore this hypothesis, we conduct a new repeated-measures ANOVA, but this time we include a covariate: IQ test scores.

When we conduct our repeated-measures ANOVA, we now have three ways of partitioning the total variance in our vocabulary test scores. First, there is the portion of variance that is accounted for by our covariate, IQ test scores. If students' IQ test scores are related to (i.e., correlated with) their vocabulary test scores, then the IQ test scores will explain, or account for, some percentage of the variance in students' vocabulary test scores. Second, after partialing out the portion of variance attributable to IQ test scores, we can see whether any of the remaining variance in vocabulary test scores is accounted for by *changes* in vocabulary test scores from third to fourth grade.

In other words, once we control for the effects of IQ test scores, do the scores of our sample change significantly from Time 1 (third grade) to Time 2 (fourth grade), on average? Is there still a *within-subjects* effect after controlling for IQ test scores? Finally, after accounting for the variance in vocabulary test scores that is attributable to the covariate (i.e., IQ test scores) and the within-subjects effect (i.e., changes from third to fourth grade), there will still be some variance in vocabulary test scores that is not explained. This is error variance, which is the same as the random variance that we normally find between different members of the same sample.

To reiterate, when a covariate (or several covariates) are added to the repeated-measures ANOVA model, they are simply included to "soak up" a portion of the variance in the dependent variable. Then, we can see whether there are any within-subject differences in the scores on the dependent variable, when *controlling for, or partialing out* that portion of the variance accounted for by the covariate(s). In the example we have been using, the addition of the IQ score covariate allows us to answer this question: Do students' vocabulary test scores change, on average, from third to fourth grade *independently* of their IQ scores? Phrased another way, we can ask whether, when controlling for IQ, students' vocabulary test scores change from third to fourth grade.

ADDING AN INDEPENDENT GROUP VARIABLE

Now that we have complicated matters a bit by adding a covariate to the model, let's finish the job by adding an independent categorical, or *group* variable. Suppose, for example, that our 10 cases listed in Table included an equal number of boys and girls. This two-level independent variable may allow us to divvy up the variance in our dependent variable even more, but only if there are differences in the scores of boys and girls..

Table. Vocabulary Text scores at the time point

Case Number	Text Source, Time I (Third Grade)	Text Source, Time 2 (Fourth Grade)
Boys		
9	20	35
4	30	45
8	40	60
1	40	60
10	45	60
Girls		
2	55	55
3	60	70
7	65	75
5	75	70
6	80	85

There are two ways that this independent group variable may explain variance in vocabulary test scores. First, boys and girls may simply differ in their average scores on the vocabulary test. Suppose that when we divide the scores in Table by gender, we get the results presented in Table. If the data were aligned in this way, we would find that at Time 1, the average score on the vocabulary test was 35 for boys and 65 for girls. Similarly, at Time 2, the average score for boys was 52 whereas for girls it was 71,

At both Time 1 and Time 2, boys appear to have lower average scores on the vocabulary test than girls. Therefore, there appears to be a *main effect* for gender. Because this main effect represents a difference between groups of cases in the study, this type of effect is called a between-groups or between-subjects main effect. In other words, some of the variance in vocabulary test scores can be explained by knowing the group (i.e., gender) to which the student belongs.

The second way that our independent group variable can explain some of the variance in our dependent variable is through an interaction effect. If we were to graph the means for boys and girls at both time points, we would get an interesting picture. The main effect for gender is clear. In addition, it is also clear that there is a within-subject effect, because both boys and girls have higher scores at Time 2 than they did at Time 1.

But what also becomes clear in Figure is that the *amount* of change from Time 1 to Time 2 appears to be greater for boys than for girls. Whereas the average score for girls increased 6 points from third to fourth grade, it grew 17 points for boys. These different amounts of change represent another source of explained variance in vocabulary test scores: the interaction of the within-subjects effect with the *between-subjects* effect. In other words, there appears to be a gender (i.e., between-subjects) by time (i.e., within-subjects) interaction on vocabulary test scores.

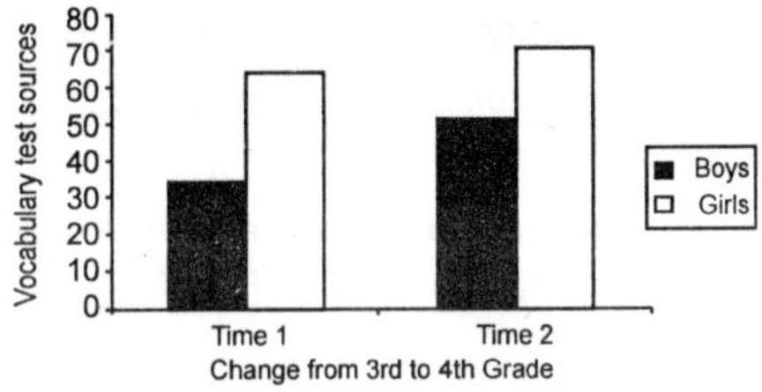

Fig. Gender by Time Interaction

To summarize, our final model has a number of effects, each of which can explain some of the variance in the vocabulary test scores of the cases in our sample. First, some of the variance in vocabulary test scores can be explained by students' IQ test scores. On average, students with higher IQ test scores had higher vocabulary test scores. Second, even after controlling for IQ, there was a within-subjects main effect. That is, we can know something about students' scores on the vocabulary test by knowing whether we are talking about Time 1 or Time 2 test scores because, on average, students had higher scores at Time 2. Third, there was a between-subjects effect for gender, so we can explain some of the variance in vocabulary test scores by knowing the gender of the student. Girls had higher scores, on average, than did boys.

Fourth, our time by gender interaction explains some additional variance in vocabulary test scores. Although both boys' and girls' scores improved over time, this improvement was more dramatic among boys, on average. Finally, there is some variance in vocabulary test scores that we cannot explain with our covariate, time, gender, or interaction effects. This is error variance.

Please keep in mind that our different effects (time, gender, interaction, covariate) will only explain variation in our dependent variable if the relations between our dependent variable and these effects are statistically significant. With only 10 cases in our sample, many of these effects may not be statistically significant.

Chapter 7

The Histogram

A histogram is a graphical method of representing a probability distribution over an interval of the real number line. First, we discuss the formal representation of a histogram and follow this up with an informal discussion. We assume that one has n data points from a particular probability distribution, $x_1.....x_n$ over an interval [a, b]. To obtain histograms, we partition the interval [a, b] into m equally sized intervals called bins. Bin j is then:

$$B_j = [a + \frac{(b-a).(j-1)}{m},,, a + \frac{(b-a).j}{m} \textit{ for } \text{j=1,....m}$$

The endpoints a and b are usually taken to be the minimum and maximum of the data set. The number of data points in bin j is n_j. Of course it will turn out that

$$\sum_{1}^{j} n_j = n$$

Essentially, a histogram is the simplest non-parametric density estimator and is the that is used by most researchers. The figure below depicts a histogram of 100 pseudo random numbers from a normal distribution with a mean of zero and standard deviation of one:

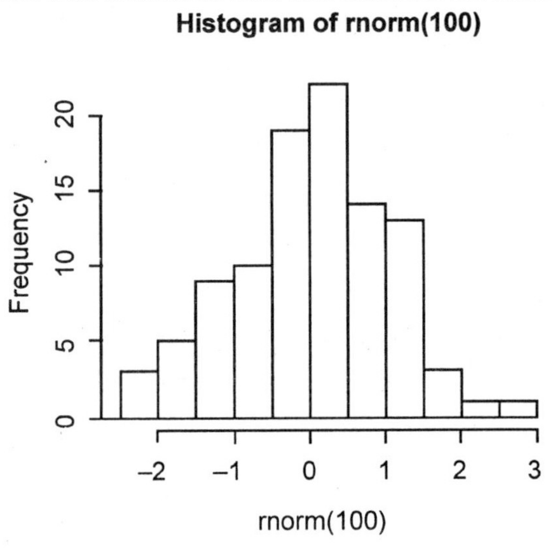

This histogram is constructed by dividing the interval covered by the data values into equal sub-intervals called bins. Each time a data point falls into a particular bin, then the bin is incremented by 1. The choice of endpoints and the choice of the number of sub-intervals can have marked effects on the shape of the histogram. Data can look bimodal when represented with a particular number of bins and bin width, but can appear uni-modal when represented with, for example, less bins and a wider bin width:

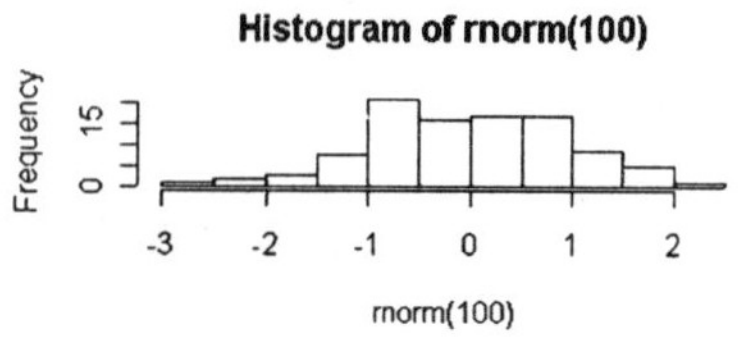

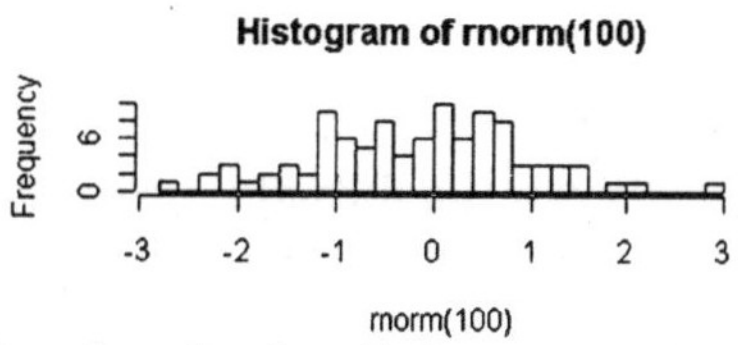

Thus, histograms have a few drawbacks: 1) they are not smooth, 2) the depend greatly on the end points of the bins, and 3) they depend on the width of the bins. These first two problems can be addressed by a histogram smoothing technique called "kernel density estimation". Before we look at kernel density estimation, we want to illustrate a method for simulating correlated bivariate and multivariate data sets. We will use a simulated data set with a known population correlation to illustrate kernel density estimation in two dimensions.

The effect is a rough approximation of the frequency distribution of the data. The groups of data are called *classes*, and in the context of a histogram they are known as *bins*, because one can think of them as containers that accumulate data and "fill up" at a rate equal to the frequency of that data class. Consider the exam scores of a group of students. By defining data classes each spanning an interval of 10 points and counting the number of scores in each data class, a frequency table can be constructed as in the following example:

Table: Frequency Table

Group	Count
0 - 9	1
10 - 19	2
20 - 29	3
30 - 39	4
40 - 49	5
50 - 59	4
60 - 69	3
70 - 79	2
80 - 89	2
90 - 99	1

To construct the histogram, groups are plotted on the *x* axis and their frequencies on the *y* axis. The following is a histogram of the data in the above frequency table.

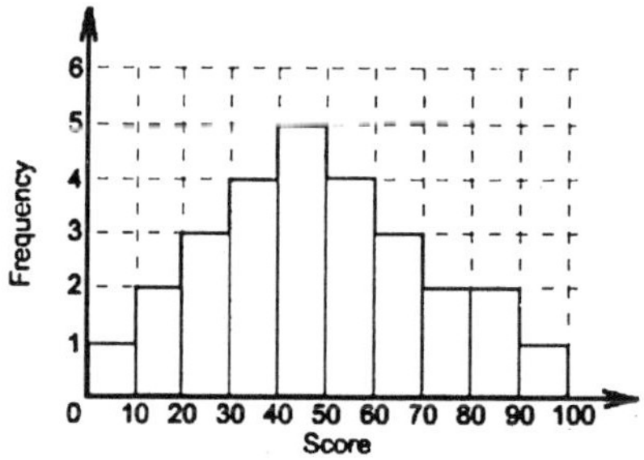

Fig. Histogram

Information Conveyed by Histograms

Histograms are useful data summaries that convey the following information:

1. The general shape of the frequency distribution (normal, chi-square, etc.)
2. Symmetry of the distribution and whether it is skewed
3. Modality-unimodal, bimodal, or multimodal

The histogram of the frequency distribution can be converted to a probability distribution by dividing the tally in each group by the total number of data points to give the relative frequency.

The shape of the distribution conveys important information such as the probability distribution of the data. In cases in which the distribution is known, a histogram that does not fit the distribution may provide clues about a process and measurement problem. For example, a histogram that shows a higher than normal frequency in bins near one end and then a sharp drop-off may indicate that the observer is "helping" the results by classifying extreme data in the less extreme group.

BIN WIDTH

The shape of the histogram sometimes is particularly sensitive to the number of bins. If the bins are too wide, important information might get omitted. For example, the data may be bimodal but this characteristic may not be evident if the bins are too wide. On the other hand, if the bins are too narrow, what may appear to be meaningful information really may be due to random variations that show up because of the small number of data points in a bin. To determine whether the bin width is set to an appropriate size, different bin widths should be used and the results compared to determine the sensitivity of the histogram shape with respect to bin size. Bin widths typically are selected so that there are between 5 and 20 groups of data, but the appropriate number depends on the situation.

HISTOGRAMS AND BOXPLOTS

The histogram provides a graphical summary of the shape of the data's distribution. It often is used in combination with other statistical summaries such as the boxplot, which conveys the median, quartiles, and range of the data.

Purpose of A Histogram

A histogram is used to graphically summarize and display the distribution of a process data set.

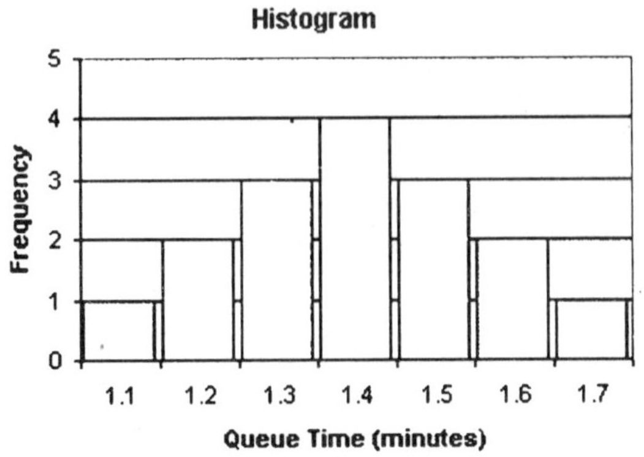

Fig. Sample Bar Chart Depiction

How To Construct A Histogram

A histogram can be constructed by segmenting the range of the data into equal sized bins (also called segments, groups or classes). For example, if your data ranges from 1.1 to 1.8, you could have equal bins of 0.1 consisting of 1 to 1.1, 1.2 to 1.3, 1.3 to 1.4, and so on.

The vertical axis of the histogram is labeled Frequency (the number of counts for each bin), and the horizontal axis of the histogram is labeled with the range of your response variable.

You then determine the number of data points that reside within each bin and construct the histogram. The bins size can be defined by the user, by some common rule, or by software methods (such as Minitab).

What Questions The Histogram Answers

1. What is the most common system response?
2. What distribution (centre, variation and shape) does the data have?
3. Does the data look symmetric or is it skewed to the left or right?
4. Does the data contain outliers?

PROBABILITY

A Short History of Probability

"A gambler's dispute in 1654 led to the creation of a mathematical theory of probability by two famous French mathematicians, Blaise Pascal and Pierre de Fermat. Antoine Gombaud, Chevalier de Mırı, a French nobleman with an interest in gaming and gambling questions, called Pascal's attention to an apparent contradiction concerning a popular dice game. The game consisted in throwing a pair of dice 24 times; the problem was to decide whether or not to bet even money on the occurrence of at least one "double six" during the 24 throws. A seemingly well-established gambling rule led de Mırı to believe that betting on a double six in 24 throws would be profitable, but his own calculations indicated just the opposite.

This problem and others posed by de Mıri led to an exchange of letters between Pascal and Fermat in which the fundamental principles of probability theory were formulated for the first time. Although a few special problems on games of chance had been solved by some Italian mathematicians in the 15th and 16th centuries, no general theory was developed before this famous correspondence.

The Dutch scientist Christian Huygens, a teacher of Leibniz, learned of this correspondence and shortly thereafter published the first book on probability; entitled *De Ratiociniis in Ludo Aleae*, it was a treatise on problems associated with gambling. Because of the inherent appeal of games of chance, probability theory soon became popular, and the subject developed rapidly during the 18th century. The major contributors during this period were Jakob Bernoulli and Abraham de Moivre.

In 1812 Pierre de Laplace introduced a host of new ideas and mathematical techniques in his book, *Thıorie Analytique des Probabilitıs*. Before Laplace, probability theory was solely concerned with developing a mathematical analysis of games of chance. Laplace applied probabilistic ideas to many scientific and practical problems. The theory of errors, actuarial mathematics, and statistical mechanics are examples of some of the important applications of probability theory developed in the l9th century.

Like so many other branches of mathematics, the development of probability theory has been stimulated by the variety of its applications. Conversely, each advance in the theory has enlarged the scope of its influence. Mathematical statistics is one important branch of applied probability; other applications occur in such widely different fields as genetics, psychology, economics, and engineering. Many workers have contributed to the theory since Laplace's time; among the most important are Chebyshev, Markov, von Mises, and Kolmogorov.

One of the difficulties in developing a mathematical theory of probability has been to arrive at a definition of probability that is precise enough for use in mathematics, yet comprehensive enough to be applicable to a wide range of phenomena. The search for a widely acceptable definition took nearly three centuries and was marked by much controversy. The matter was finally resolved in the 20th century by treating probability theory on an axiomatic basis. In 1933 a monograph by a Russian mathematician A. Kolmogorov outlined an axiomatic approach that forms the basis for the modern theory. Since then the ideas have been refined somewhat and probability theory is now part of a more general discipline known as measure theory."

Introductory Note

The term Probability, as the name suggests, refers to a calculation, a mathematical estimate concerning the possibility of a particular outcome in a situation where there are a number of different possible outcomes. If, for example, we know that in a certain situation (e.g., a roll of the dice) there is a determined number of possible outcomes (e.g., six possible faces of the die), we can calculate the probability of a particular outcome or of a particular sequence of outcomes if the event is repeated.

For example, with a normal die in a fair roll, there is a 1 in 6 chance of any particular face of the die showing on the top (since there are six surfaces on the cube); in a fair toss of a normal coin there is a 1 in 2 chance of a head appearing on top, a 1 in 2 chance of a tail appearing on top, and a 2 in 2 chance of a head or a tail appearing on top.

It is important to stress at the outset that probability is a *theoretical* calculation of the likelihood of a particular outcome. Thus, it is based, not upon observation of actual events (e.g., listing the results from several tosses of the coin) but on calculations which are based on certain assumptions (e.g., that the coin or die toss is fair, that there are no hidden factors, like cheating, influencing the outcome).

Because the probability of an outcome is a theoretical calculation, knowing the probability does not mean that we can always predict a particular outcome with certainty (please understand this very important point). Often the result of a particular event or series of events will not match the theoretical probabilities. For example, in a sequence of ten coin tosses, theory tells us that the most probable result will be 5 outcomes with heads on top and 5 outcomes with tails on top. But that calculation does not mean that, if we observe an actual 10-toss sequence, we will always get that most probable result.

Checking probabilities (that is, the theoretical calculations) against the observed outcomes of a real event is, as we shall see, an essential part of the study of probability and statistics (and a central exercise in this module). But always keep in mind this opening point: the probability of an event is a theoretical calculation; the observed frequency with which an event actually occurs is something different: the first is based upon mathematical analysis of an event (independent of observation); the second is based upon an actual experiment.

Probability theory has a number of extremely important uses. A simple but powerful application is its ability to assist us in establishing the mathematical likelihood of a particular outcome and thus to advise ourselves of the prudence of a certain action (e.g., betting). For instance, to take a very simple example, in a fair toss of a coin, theory indicates that the probability of the coin landing with heads on top is 50 percent, with tails on top 50 percent, and with either heads of tails on top 100 percent. In a bet on the result of a coin toss, therefore, one would be unwise to wager, say, ten dollars that a head would appear against your opponent's bet of two dollars that a tail would appear.

In the course of this module, we will become more familiar with a number of important applications of probability theory.

EXPRESSING PROBABILITY MATHEMATICALLY

We can express the probability of an event mathematically in a number of ways, but the most common is as a decimal between 1 and 0. Thus, a probability of 1 ($p = 1$) is certainty that the event will occur. A probability of 0 ($p = 0$) is certainty that event will not occur. Thus, all probabilities for an event are expressed as a decimal between 0 and 1.

The probability that a single fair coin toss will produce a head on top is expressed as follows: $p = .5$. This we might also express as meaning that the probability is 5 out of 10 or 50 percent. The decimal notation is the most common and will be standard throughout the rest of this workbook.

In the same way, if $p = .75$, that means that the outcome will theoretically occur 75 times out of 100; if $p = .234$, that means that the outcome will theoretically occur 234 times out of 1000; if $p = .01345$, that means the outcome will theoretically occur in 1345 cases out of 100,000.

It will be clear that any value for p greater than .5 means that the result is theoretically more likely to occur than not; a value for p less than .5 means that the result is more likely not to occur than it is to occur. The smaller the value of p, the less likely that outcome.

Probability Theory and Observation

One of the major applications of probability is to check the theoretical possibility of a particular outcome (calculated mathematically) with the observed result of an event. If we find that the observed frequency of a particular occurrence departs widely from theoretical probabilities (i.e., the result is occurring more frequently than we would expect from our calculations), sometimes we have good reason to investigate the event more carefully.

For example, in a fair toss of a coin, we theoretically calculate p as .5 for a head to appear on any single toss. But what is the probability that someone will throw the coin four times in a row and have the outcome heads each time? We can calculate the probability by multiplying the probabilities together. Since there are four tosses and the probability of a head is .5 on each of these, the probability of a sequence of four heads is as follows:

$$p = .5 \times .5 \times .5 \times .5 = .0625$$

In other words, in a sequence of four consecutive coin tosses, probability theory indicates that in 625 cases out of 10,000 (or 1 outcome in 16) four heads will come up in a row . Thus, if you were making a bet in a number of different trials of four consecutive coin tosses that four heads would appear in a row, you could theoretically expect to win 1 out of every 16 trials and lose 15 out of 16 trials.

Now, this figure of 1 in 16 ($p = .0625$) does not claim that in every 16 trials of a four-coin-toss sequence, one result will *always* be four heads. The theory states that, on average, in a sequence of four-toss trials, the most likely outcome is that there will be four heads appearing in 1 out of every 16 trials. But in any particular series of four-toss trials, it is quite possible for more than sixteen four-toss trials to happen before the first four-head result appears or for a four-head result to occur before sixteen trials are completed or more than once in one sixteen four-toss sequence. As we shall see, the more times one repeats the four-toss sequence, the closer the observed result will come to the theoretically calculated result.

Thus, given that one four-head sequence every sixteen throws is the most likely (but not the inevitable) outcome of a series of four-toss sequences, if you were betting on the outcome of many four-toss sequences (that there would be four consecutive heads appearing), for the game to be truly fair, you should put up \$1 against your opponent's \$15 for each four-coin sequence (i.e., if four heads appear you win \$15; if they do not your opponent wins \$1). With a great many four-toss trials, you and your opponent should come out more or less equal.

Hence, if you are betting \$1 that four heads will appear in a four-toss sequence of coins and if your opponent is unwilling to put up \$15, you should not undertake the bet (the probabilities are against you). If she is willing to put up more than \$15, then you should take the bet (the probabilities are in your favour).

An Introductory Note about Significance

When the probability of an event (theoretically calculated) is very low and yet the event occurs, sometimes we are entitled to be suspicious or alarmed at the result. Such moments are an invitation to investigate the event to see if there are hidden factors at work (e.g., loaded dice, cheating, trick coins, some hitherto unknown cause). When do such moments occur? At what point is the probability so unlikely (i.e., so low) that we should investigate further?

Well, let us first, as an example, consider the probability of someone's tossing a coin with ten consecutive heads appearing. Using the same way of calculating the probability as we did above (multiplying together the probabilities for a single event), we can calculate the theoretical probability of ten consecutive heads appearing as follows:

$$p = .5 \times .5 \times .5 \times .5 \times .5 \times .5 \times .5 \times .5 \times .5 \times .5 = .000977$$

This rounded off result indicates that the probability for ten consecutive heads in a series of ten fair coin tosses is 977 out of 1,000,000. Obviously the probability here is much smaller than for the four-head sequence ($p = .0625$), and we would be right to be much more surprised at such a result. But do we have the right to be suspicious? In other words, how improbable does an event have to get before we begin to think that there is something strange going on?

Scientists and social scientists have set conventional standards for how much an observed result has to differ from the theoretical expectation in order for the result to be considered significant (i.e., worthy of notice or of further investigation, because it is not likely to have occurred by chance). That level is $p = .05$ (or 5 results in 100). If there is more than a probability of .05 (i.e., if p is greater than .05) that an event could have occurred by chance, scientists do not usually pay much attention to it. If p is less than .05, however, then the result is considered significant and warrants analysis, because something other than chance may be affecting the outcome.

Thus, for example, someone's tossing a coin and getting four consecutive heads (where $p = .0625$) would not prompt any enquiry; the probability is low, but not so low that we need to start wondering about the fairness of the toss (remember the cut off point is $p = .05$). However, someone's tossing a coin and getting ten heads in a row ($p = .000977$ or, rounded off, .001) is significant. The probability is so low (1 chance in 1000) that we have some grounds for thinking the result might not be produced by chance and that it is time to check the coin.

The justification for this convention of a probability of .05 as the dividing line between significant and non-significant results is not mathematical but methodological and practical. Given the constraints, economic for example, upon investigation, it has been found to pay, in terms of success in explaining the world, to concentrate research in areas where such a degree of significance is obtainable. In some tests, as we shall see, where the procedure requires a much higher level of certainty, the cut-off line is sometimes reduced to .01 (or 1 chance in 100). In general, however, the figure of .05 probability is the accepted level below which results become significant and above which we can still ascribe the results to chance.

Note the very important point that a low probability (e.g., p = .001) does not indicate impossibility. It may well be that the ten-head coin-toss sequence, although very very improbable, is fair (i.e., produced by chance). The very low figure for p is a warning that something may be (and probably is) interfering with the result, but the figure does not *prove* such interference . These basic principles indicate one important aspect of probability, its use as a diagnostic tool, a means of recognizing that in a natural event something may be interfering with the results (e.g., cheating or some hidden factor). Significant discrepancies between theoretical calculations of probable outcomes and the observed results of actual outcomes are often the first clue that something needs investigating. And from this initial observation, the investigators often discover something they had not known about before (e.g., a new disease, like Legionnaire's Disease or AIDS, or a new natural phenomenon).

The Fundamental Basis of Probability Theory

The basis upon which the mathematical treatment of probability rests is as follows: we assume that the phenomenon we wish to investigate can be described in terms of a given number of different possible but equally probable outcomes. For instance, in a simple coin toss there are two equally probable outcomes, heads or tails. In one roll of a die, there are six equally probable outcomes, represented by the six sides of the die, each with a number of dots from 1 to 6. In drawing a card from a full and well shuffled normal deck with no jokers, we know that there are four equally probable outcomes for the suit of the drawn card (hearts, diamonds, spades, or clubs), there are two possible outcomes for the colour of the card (red or black), and there are thirteen possible outcomes for the value of the card. Naturally we are assuming that in all these trials the procedure is fair (i.e., no cheating).

Multiple outcomes (e.g., so many heads in consecutive coin tosses or so many sixes in consecutive rolls of a single dice) can be calculated from the probability of a single event (as we did above by multiplying the frequency of repeated events by the probability of the single outcome in one event). Events which cannot be described in terms of equally probable outcomes are beyond the scope of probability theory discussed here. The probability of any specific outcome is given by the number which is equal to the number of possibilities which involve that outcome divided by the total number of possibilities. For example, in a simple coin toss, there is only one possibility that a head will appear, and the total number of possible outcomes is two. Thus, the probability that a head will appear is 1 divided by 2 or .5.

In a roll of a single die, there is one chance that a six will appear, and there are six possible outcomes. Therefore the probability of a six appearing on a single roll of a normal die is 1 divided by 6 (p = .1667). If we draw a card from a normal full deck, there is only one chance that a heart will be on the card, and there are four possibilities for different suits. Therefore the probability of drawing a heart from a full deck is 1 divided by 4 (p = .25).

What is the probability that on a single roll of a die, the result will be an even number? Well, there are three possible even outcomes (the numbers 2, 4, and 6) and there are 6 possible outcomes (1, 2, 3, 4, 5, 6). Thus, the probability that on a single roll of a die the result will be an even number of dots is 3 divided by 6: $p = .5$. More complicated outcomes (e.g., four heads in a row in consecutive coin tosses, three sixes in a row in consecutive rolls of the die, four number 12 results in four consecutive spins of the roulette wheel, and so on) are more complicated, but we can work out the possible outcomes for such events and calculate the probabilities involved (as we did above for sequences of four and ten coin tosses).

Randomness

Random phenomenon...outcome cannot be predicted, but that has a regulardistribution over many repetitions.

Probability...proportion of times an event occurs in many repeatedevents of a random phenomenon.

Independent events (trails)...outcome of an event (trial) does not influence theoutcome of any other event (trial).

Example of independent events: Rollinga single die twice. The result on the second roll is independent ofthe first.

Example of events not independent: Picking twocards from a deck, one at a time, without replacement. (If you pick acard, and don't put it back in the deck and reshuffle, then theprobability of a red card on the second pick is dependent upon whatthe first card happened to be.)

Probability Models

Overview: It is often important andnecessary to provide a mathematical description or model forrandomness. This section provides insights into how this can bedone.

Sample space: Theset of all possible outcomes of a random phenomenon.

Example:

Consider the sum obtained when two dice are rolled.

The following table displays the sums that can be obtained from thisevent.

		RED DIE					
		1	2	3	4	5	6
GREEN	1	**2**	**3**	**4**	**5**	**6**	**7**
DIE	2	**3**	**4**	**5**	**6**	**7**	**8**
	3	**4**	**5**	**6**	**7**	**8**	**9**
	4	**5**	**6**	**7**	**8**	**9**	**10**
	5	**6**	**7**	**8**	**9**	**10**	**11**
	6	**7**	**8**	**9**	**10**	**11**	**12**

The sample space contains 11 outcomes: S ={2,3,4,5,6,7,8,9,10,11,12}

Probability Rules

o If A is an event, then the probability P(A) is anumber between 0 and 1, inclusive.

o Prob(A does not happen) is 1 - P(A).

o Two events are disjoint if they have nooutcomes in common. If A and B are disjoint, then P(A or B) = P(A) +P(B). Some sources use the phrase mutuallyexclusive to describe events that aredisjoint.

- In a roll of two dice, if A = "roll asum of 7" and B = "roll a sum of 2", then P(A or B) = 6/36 + 1/36 =7/36, since A and B are disjoint.
- If C = "roll a sum of 7" and D = "roll an oddsum" then C and D are not disjoint since 7 is an odd sum. In thiscase, P(C or D) = P(C) + P(D) - P(C and D) = 6/36 + 18/36 - 6/36 =18/36 = 1/2.

Two events are independent if knowing that oneoccurs does not change the probability of the other. If events A andB are independent, then prob(A and B) = prob(A)xprob(B).

- If A is the event "getting a head onthe first coin toss" and B is the event "getting a head on a secondtoss of the coin." then A and B are independent, and prob(A and B) =(1/2)(1/2) = 1/4.
- If C is the event "getting a red card by pickinga card from a randomly shuffled deck," and D is the event "getting ared card by picking a second card from the deck" (when the first cardhas not been replaced), then C and D are not independent. Prob(C) =26/52 = 1/2. If the first card is red, then prob(C and D) =(26/52)(25//51), which is not equal to (26/52)(26/52). If the firstcard was replaced in the deck, and the deck reshuffled before eventD, then C and D would be independent.

More About Probability

Overview: There are basic laws thatgovern wise and efficient uses of probability. An understanding ofthese laws is important for those who wish to utilize statistics inpractical and meaningful ways..

For any two events A and B,

Prob(A or B) = Prob(A) + Prob(B) -Prob(A and B).

If A and B are disjoint, then Prob(A and B) =0.

Prob(B|A) = prob(A and B)/prob(A).

Consider a family with two children. There arefour possible boy/girl combinations, all equally likely.

Oldest	Youngest
G	G
G	B
B	G
B	B

Prob(at least one child is a girl) = 3/4.

Prob(two girls) = 1/4.

Prob(two girls|one child is a girl) = 1/3.Sample space is {(G,G), (G,B), (B,G)}.

Prob(two girls|oldest child is a girl) = 1/2.Sample space is {(G,G), (G,B)}.

Prob(exactly one boy|one child is a girl) =2/3.

Prob(exactly one boy|youngest child is a girl) =1/2.

OK, now suppose that a family unknown to you hastwo children. Here are two questions relating to this family.

1. If one of the children is a boy,what is the probability that the other child is aboy?
2. If the oldest child is a boy, whatis the probability that the other child is a boy?

Many intelligent people think that 50% is the answer to both (1) and(2). You should be able to show that this is not the case.

BINOMIAL DISTRIBUTION

The probability distribution of a random variable X which assumes integral values x=0,..., n with the probabilities

$$\mathrm{P}\{X = x\} = b_x(n, p) = \binom{n}{x} P^x (1-p)^{n-x}$$

where p is a parameter of the binomial distribution, called the probability of a positive outcome, which can take values in the interval $0 \le p \le 1$. The binomial distribution is one of the fundamental probability distributions connected with a sequence of independent trials. Let $Y_{1,}$ $Y_{1\ldots,}$ be a sequence of independent random variables, each one of which may assume only one of the values 1 and 0 with respective probabilities and (i.e. all are binomially distributed with). The values of may be treated as the results of independent trials, with if the result of the -th trial is "positive" and if it is "negative". If the total number of independent trials is fixed, such a scheme is known as Bernoulli trials, and the total number of positive results,

$$X = Y_1 + \ldots + Y_n, \quad n \ge 1,$$

is then binomially distributed with parameter P.

The mathematical expectation $\mathrm{E}z^X$ (the generating function of the binomial distribution) for any value of is the polynomial $[pz + (1-p)]^n$, the representation of which by Newton's binomial series has the form

$$b_0 + b_1 z + \ldots + b_n z^n$$

(Hence the very name "binomial distribution".) The moments (cf. Moment) of a binomial distribution are given by the formulas

$$\mathrm{EX} = np,$$

$$DX = E(X - np)^2 = np(1-p),$$

$$E(X - np)^3 = np(1-p)(1-2p)$$

The binomial distribution function is defined, for any real y,0<y<n, by the formula

$$F(y) = \mathrm{P}\{X \le u\} = \sum_{x=0}^{[y]} \binom{n}{x} p^x (1-p)^{n-x},$$

where is the integer part of, and

$$F(y) = \frac{1}{B([y]+1, n-[y])} \int_p^1 t^{[y]} (1-t)^{n-[y]-1}\, dt,$$

B(a, b) is Euler's beta-function, and the integral on the right-hand side is known as the incomplete beta-function.

As, $n \to \infty$ the binomial distribution function is expressed in terms of the standard normal distribution function by the asymptotic formula (the de Moivre–Laplace theorem):

$$F(y) = \Phi\left[\frac{y - np + 0.5}{\sqrt{np(1-p)}}\right] + R_n(y, p),$$

where,

$$R_n(y, p) = O(n^{-1/2})$$

uniformly for all real . There also exist other, higher order, normal approximations of the binomial distribution.

If the number of independent trials *n* is large, while the probability *P* is small, the individual probabilities $b_x(n, p)$can be approximately expressed in terms of the Poisson distribution:

$$b_x(n, p) = \binom{n}{x} p^x (1-p)^{n-x} \approx \frac{(np)}{x!} e^{-np}$$

If $n \to \infty$ and $0 < c \le y \le C$ (where and are constants), the asymptotic formula

$$F(y) = \sum_{x=0}^{[y]} \frac{\lambda^x}{x!} e^{-\lambda} + 0(n^{-2}),$$

where, is uniformly valid with respect to all in the interval .

$$\lambda = (2n - [y])p/(2 - P)$$

THE BINOMIAL DISTRIBUTION

Many events have only two possible results: success or failure. In tossing a coin, there are only two possible results—heads and tails—each with a probability of .5. Success in a coin toss is calling the correct result. In answering multiple choice tests where there are four possible answers for each question, there are again only two possible results for each question, a correct answer or an incorrect one. If we are trusting entirely to chance in answering such a test (i.e., guessing), we have a one in four chance of getting the answer for any question correct (i.e., on each question for success $p = .25$, for failure $p = .75$).

Sorting out the probabilities for a particular outcome requires us to figure out all the possible outcomes and to calculate the distribution for all those possible outcomes. And these calculations will vary depending on the number of possible outcomes. Thus, there are many possible probabilities for different events (e.g., the distribution of outcomes—heads or tails—for a four coin-toss sequence will be different from the distribution of outcomes—heads or tails—for a ten coin-toss sequence; the distribution for the results of a repeated event with a .5 probability will be different from a similar sequence with a .1667 probability, and so on).

Constructing such a mathematical calculation for a particular series of outcomes is called calculating the binomial distribution. Constructing a binomial distribution for a particular sequence of events is not very difficult, but it is rather time consuming. Fortunately we have tables of binomial distributions so that we do not have to go through the calculations here.

However, in this section we are concerned with being above to read and use a simple binomial distribution table, so that we can judge the probabilities of simple events with only two possible outcomes (success or failure). The simple table below provides an example of two binomial distributions. An explanation of the table follows immediately after it.

Table: Selected Binomial Distributions

$p = .5$					
$n = 10$			$n = 20$		
r	*ind*	*com*	*r*	*ind*	*com*
0	.001	1-	0	0+	1-
1	.010	.999	1	0+	1-
2	.044	.989	2	0+	1-
3	.117	.945	3	.001	1-
4	.205	.828	4	.005	.999
5	.246	.623	5	.015	.994
6	.205	.377	6	.037	.979
7	.117	.172	7	.074	.942
8	.044	.055	8	.120	.868
9	.010	.011	9	.160	.748
10	.001	.001	10	.176	.588
		11	.160	.412	
		12	.120	.252	
		13	.074	.132	
		14	.037	.058	
		15	.015	.021	
		16	.005	.006	
		17	.001	.001	
		18	0+	0+	
		19	0+	0+	
		20	0+	0+	

Note that the entry 1- means a probability larger than .9995 but less than 1 (about as certain as one can be); the entry 0+ means a probability less than .0005 but greater than 0 (extremely unlikely).

This table indicates two different binomial distributions. The one on the left side indicates the probabilities of success in a series of 10 coin tosses; the one on the right side indicates the probabilities of success in a series of 20 coin tosses; this distinction is indicated by the label n = 10 on the left and n = 20 on the right. For both tables the probability of success in any one toss is .5 (hence heading p = .5 below the title). The first column on each side has the heading r. This indicates the number of successes in the sequence. So, for example, on the left side, the values go from 0 to 10, indicating the possible outcomes of 0 heads in 10 tosses of the coin to 10 consecutive heads in 10 tosses. On the right side, the r values go from 0 to 20, indicating the possible outcomes from 0 heads in 20 tosses to 20 heads in 20 tosses.

The *ind* (= individual) column in the middle of each side of the table gives the probability of one's achieving that exact r value in a full sequence of tosses. So on the left side, for example, the probability of getting exactly 5 heads in a sequence of 10 tosses is .246 (the *ind* value on the same line as the r value of 5). In the right side of the table, the probability of getting exactly 5 heads in a sequence of 20 tosses .015.

The *cum* (= cumulative) column gives the probability of r or more successes in the total number of trials (or at least r successes). So, for example, on the left side of the table, the chances of getting 6 or more (or at least 6) heads in 10 consecutive tosses of a coin is .377; in the right side of the table, the probability of getting 5 or more (or at least 5) heads in 20 tosses of the coin is .994. Look on the right side of the table at the r value 4. As you would expect, the probability of getting exactly four heads in a 20-toss sequence (the *ind* value) is very low (.005), but the chances of getting at least 4 heads in a 20-toss sequence (the *cum* value) is very high (.999, close to certainty).

Table: Binomial Distribution for a Six-Sided Die

p = .1667
n = 24

r	*ind*	*cum*
0	.013	1-
1	.060	.987
2	.139	.927
3	.204	.788
4	.214	.584
5	.171	.371
6	.108	.200
7	.056	.091
8	.024	.035
9	.008	.012
10	.003	.003
11	.001	.001

The other *ind* and *cum* values from r =12 to r = 24 are 0

Some Observations on Binomial Distributions

The binomial distribution, as illustrated in Tables above, is appropriate only when the events in the sequence are independent of each other (i.e., when the outcome of one event does not affect the outcome of another). For example, in any coin toss the probability of a head is always .5; it is not affected by the result of any outcome before it. In the roll of a single die, the probability of a 6 showing as the outcome is always the same ($p = .1667$), no matter what the results of previous rolls might have been. In other words in sequences of coin tosses or die rolls, the outcome of any single event is independent of the outcome of any other event.

This independence does not always apply in probability situations. For example, when one draws a single card from a full, well-shuffled deck, the probability of the outcome producing a red card is .5, because 26 of the 52 cards are red (26 divided by 52 equals .5). However, if we draw a red card and do not return it to the deck, then the probability of a second draw producing a second red card is no longer .5. Since one red card has been removed from the deck on the first draw, there are now only 51 cards in the deck, 26 black cards and 25 red ones. Therefore, the probability of drawing a red card on the second draw is now 25/51 ($p = .4902$); whereas, the chance of drawing a black card on the second draw with one red card removed is now 26/51 ($p = .5098$).

In this example, the probabilities in the second draw have been affected by the result of the first draw (one card less in the deck, one red card less in the deck). What this means is that the probability of a red card in the second draw is not independent of the first draw. Thus, the sort of binomial distribution which we have been considering in the coin toss or roll of the die (where the probability remains the same in each toss or roll) would not be appropriate.

In any card game where the deck is not full and shuffled before each event, the probabilities are constantly changing as cards are played and removed from the deck. This is the case, most famously, with blackjack, where after every hand the played cards are normally collected and placed on the bottom and a new hand dealt from the remaining cards. A careful player who keeps close track of the cards which have been played and not returned to the deck and who understands probabilities very well can calculate when the probabilities of a favourable outcome shift in her favour (i.e., when p for success becomes greater than .5). This is the basis for a number of well-known and successful legal systems for beating the casino dealer, since this is the only casino game in which the probabilities occasionally favour the player.

Here, for example, is a simple illustration. Suppose in playing blackjack at the casino (in a game between you and the dealer only), you keep careful track of all the cards which have been played and put on the bottom of the deck. Suppose, as a result of this careful observation, you know that there are only five cards left unplayed and, further, that these five are 3 eights and 2 sevens. In such a situation you pawn everything you have, sell your car, bet everything, and stand pat with the two cards dealt to you. Since in routine casino blackjack the dealer cannot stand pat with 16 or under (and you know that from the remaining cards he must have 16, 15, or 14), it is certain that the dealer will have to draw and that his hand will be over 21. In other words, the probability for your success here is 1 (certainty).

That is a very simple example. Sophisticated mathematicians have developed systems of keeping track of the shifting probabilities in blackjack, betting low when the probability is less than .5 and high when the probability for success shifts to above .5. Over time, given the fairness of the game, such players will win (and some have won enormous amounts). Casinos routinely cope with such gamblers, when they recognize them, by shuffling the deck before every hand, thus restoring the favourable probabilities to the dealer in every hand. At this point, the sophisticated gambler with a system moves on down the strip to another casino.

BERNOULLI THEOREM

The (historically) original form of the (weak) law of large numbers. The theorem appeared in the fourth part of Jacob Bernoulli's book Ars conjectandi (The art of conjecturing). This part may be considered as the first serious study ever of probability theory.

The theorem deals with sequences of independent trials, in each one of which the probability of occurrence of some event ("success") is P. Let n be the number of trials and let m be the random variable equal to the number of successful events. The Bernoulli theorem states that, whatever the value of the positive numbers ε and η, the probability $\supset$ of the inequality

$$-\varepsilon \leq \frac{m}{n} - p \leq \varepsilon$$

will be higher than $1-\eta$ for all sufficiently large $\eta\left(n \geq n_0\right)$. The proof of this theorem, which was given by Bernoulli and which was exclusively based on a study of the decrease of probabilities in the binomial distribution as one moves away from the most probable value, was accompanied by an inequality which made it possible to point out a certain bound for the given $n_0$if and were given. Thus, it was found by Bernoulli that if p =2/5, the probability of the inequality

$$-\frac{1}{50} \leq \frac{m}{n} - \frac{2}{5} \leq \frac{1}{50}$$

will be higher than 0.999 if $n \geq 25{,}550$. By introducing a slight improvement in the original reasoning of Bernoulli, it is possible to conclude that it is sufficient to select a value of n obeying the condition

$$n > \frac{1+\varepsilon}{\varepsilon^2} \log \frac{1}{\eta} + \frac{1}{\varepsilon},$$

which gives in turn, for the probability $1-\mathrm{P}$ of the inequality

$$\frac{m}{n} - p\Big| > \varepsilon,$$

an estimate of the form

$$2\exp\{-\frac{1}{2}n\varepsilon^2\}$$

The condition obtained for the above example is $n \geq 17,665$ (more sophisticated estimates show that it is sufficient to take $n \geq 6502$; one may note, for the sake of comparison, that the de Moivre–Laplace theorem yields 6498 as the approximate value of n_0). Other estimates for $1 - P$ may be obtained using the Bernstein inequality and its analogues.

An Investigation of Bernoulli's Theorem

The rest of this module describes an experiment to illustrate and test Bernoulli's Theorem, to see whether our observed results get closer and closer to the calculated probabilities as we increase the number of trials.

Equipment

For this experiment you need to collect together the following items:

1. Ten coins, preferably all the same size and in good condition;
2. A can or cup in which to shake the coins, which must be able to tumble freely in the container when you hold your hand over the opening and shake the coins;
3. A tally sheet to keep track of the results (number of heads) of head throw (the tally sheet in given below)

Procedure

Find a quiet place at a table onto which you can spill the coins without bothering your family, friends, neighbours, or the cat. Then go through the following steps:

1. Carefully shake out the ten coins so that they are well mixed and then spill them out on a table or similar surface. Count the number of heads, and enter the number in the appropriate place on the tally sheet on the next page (using a vertical stroke in the appropriate box for each head; for example, if on the first trial you get six heads, then put a vertical line on the six line in the table, under the Trials 1-25 column; then on the second trial, if you get three heads, then put a line in the three box in the same column, and so on).
2. Collect the data for the first 25 tosses (each with 10 coins) in the first column, so that at the end of the 25 trials there is a record of how many tosses produced no heads, one head, two heads, three heads, and so on. Then do the same for the next 25 trials, entering the results in the second column. Enter the data for the third group of 25 tosses in the third column and for the fourth group of 25 trials in the fourth column. Now you have a record of the number of heads in 100 tosses of 10 coins.
3. Calculate the average number of heads for each of the 25 toss sequences, and enter that figure in the bottom row (labelled Mean) under the appropriate column. Calculate the average by adding up the numbers in the column (to get the total number of heads) and dividing by 25 (i.e., the number of trials). Note that you add up the number of heads in any one square by multiplying the number of vertical strokes times the number which the square represents. For example, if you have four vertical strokes in the 7 box (indicating that on 4 occasions there were 7 heads in the 10 coin result), then the number of heads for that box is 4 x 7 or 28.

Thus, you will have four figures for the average number of heads in each of the four 25-toss sequences.

4. Add up the totals for the first two 25-trial sequences, and enter them in the appropriate boxes in Column 5 (headed Total 1-50). Then add up the totals for the first three 25-trial sequences, and enter them in the appropriate boxes in Column 6 (headed Total 1-75). Then do the same for all the trials, entering the results in the boxes under the heading Total 1-100.
5. Then calculate the average number of heads (the Mean) for each column, once again adding up the total number of heads and dividing by the number of trials. Thus, you should produce an average number of heads for the first 50 trials, the first 75 trials, and for the total 100 trials.

LAW OF LARGE NUMBERS

A general principle according to which under certain very general conditions the simultaneous action of random factors leads to a result which is practically non-random. That the frequency of occurrence of a random event tends to become equal to its probability as the number of trials increases (a phenomenon which was probably first noted for games of chance) may serve as the first example of this principle.

At the turn of the 17th century J. Bernoulli demonstrated a theorem stating that, in a sequence of independent trials, in each of which the probability of occurrence of a certain A event has the same value $p, 0 < p < 1$, the relationship

$$\mathrm{P}\left\{\left|\frac{\mu_n}{n} - \mathrm{p}\right| > \varepsilon\right\} \to 0$$

is valid for any $\varepsilon > 0$ if $n \to \infty$; here, μ_n is the number of occurrences of the event in the first n trials and *un/n* is the frequency of the occurrence. This Bernoulli theorem was extended by S. Poisson to the case of a sequence of independent trials in which the probability of the occurrence of an event Avaries with the number of the trial. Let this probability in the k-th trial be $pk, k = 1, 2, ...,$ and let

$$\overline{P}n = \frac{p_1 + ... + p_n}{n}$$

Then, according to the Poisson theorem,

$$\mathrm{P}\left\{\left|\frac{\mu_n}{n} - \overline{p}\right| > \varepsilon\right\} \to 0$$

for any $\varepsilon > 0$ if $n \to \infty$. A rigorous proof of this theorem was first given in 1846 by P.L. Chebyshev, whose method was quite different from that of Poisson and was based on certain extremal considerations, whereas Poisson deduced from an approximate formula for the above probability, based on a law of Gauss which at that time had not yet been proved rigorously.

Poisson was the first to use the term "law of large numbers", by which he denoted his own generalization of the Bernoulli theorem.

A further natural extension of the Bernoulli and Poisson theorems is a consequence of the fact that the random variables may be represented as the sum

$$\mu_n = X_1 + ... + X_n$$

of independent random variables, where $X_K = 1$ if A occurs in the k-th trial and $X_K = 0$ otherwise. The mathematical expectation $E(\mu_n/n)$ (which is identical with the average of the mathematical expectations $EX_K, 1 \le k \le n$) is P in the case of Bernoulli and is in $\overline{P}n$ the case of Poisson. In other words, in both cases one deals with a deviation of the arithmetical average of the variables X_K from the arithmetical average of their mathematical expectations.

In his work "On average quantities", which appeared in 1867, Chebyshev established that the relationship

$$p\left\{\left|\frac{X_1 + ... + X_n}{n} - \frac{EX_1 + ... + EX_n}{n}\right| > \varepsilon\right\} \to 0$$

is valid, for any if $\varepsilon > 0, n \to \infty$ for independent random variables $X_1, ..., X_n, ...,$ under fairly general assumptions. Chebyshev assumed that the mathematical expectations EX_K^2 are all bounded by the same constant, even though his proof makes it clear that the requirement of boundedness of the variances of $X_k, DX_k = EX_K^2 - (EX_K)^2$, or even the requirement

$$B_n^2 = DX_1 + ... + DX_n = o(n^2), \qquad n \to \infty$$

is sufficient. It was thus shown by Chebyshev that the Bernoulli theorem is susceptible of extensive generalizations. A.A. Markov noted the possibility of further extensions and proposed to apply the term "law of large numbers" to all extensions of the Bernoulli theorem (and, in particular, to). Chebyshev's method is based on a rigorous formulation of all the properties of the mathematical expectations and on the use of the so- called Chebyshev inequality in probability theory. For the probability it yields an estimate of the form

$$n^{-2}\varepsilon^{-2}\sum_{K=1}^{n} DX_k$$

This estimate may be replaced by a more exact one — but under more restrictive conditions, Subsequent proofs of the law of large numbers are all, to varying extents, developments of Chebyshev's method. Using an appropriate "truncation" of the random variables X_k(replacement of these variables by auxiliary variables $X_n^{'}k$, viz. $X_{n,k}^{'} = X_k$ if $X_k - EX_k| \le L_n$ and $X_{n,K}^{'} = 0$ if $X_k - EX_k| > L_n$, where L_n is a constant), Markov extended the law of large numbers to the case when the variance of the terms does not exist. He showed, for instance, that is valid if, for certain constants $\delta > 0$ and $L > 0$ for all n,

$$E\left|X_n - EX_n\right|^{1+\delta} < L$$

Khinchin's theorem (1929) is proved in a similar manner: If X_n have the same distribution and if EX_n exists, then the law of large numbers is valid.

It is possible to formulate more or less final versions of the law of large numbers for sums of independent random variables. To do this, it is expedient to adopt a more general point of view, involving the concept of a sequence of asymptotically-constant random variables. The random variables of a sequence $Y_1,...,Y_n,...,$ are said to be asymptotically constant if there exists a sequence of constants $C_1,...,C_n,...,$ such that for any $\varepsilon > 0$, if $n > \infty$,

$$\mathrm{P}\{|Y_n - C_n| > \varepsilon\} \to 0$$

(i.e. $Y_n - C_n$ converges to zero in probability). If is valid for C_n some, it is also valid for $C_n' = mYn$, where mY is the median of the random variable Y. Furthermore, instead of the $X_1,...,X_n,...,$ sequence of independent random variables, one may also take a so-called triangular array:

$$X_{1,1},...,X_{1,k_1},$$
$$X_{2,1},...,X_{2,k_2},$$
$$\square.......,$$
$$X_{n,1},...,X_{n,k_n}:$$
$$\square.......$$

of random variables (the first subscript is the row number, while the second indicates the number of the variable within the row). The random variables of each individual row are assumed to be mutually independent. A sequence of partial sums is readily reduced to a triangular array if one assumes that $k_1 = 1, k_2 = 2,...,$ $X_{n,k} = X_k / n$,

Let

$$Y_n = X_{n,1+...+}X_{n,k_n}.$$

The general problem of the applicability of the law of large numbers to sums of independent random variables may then be stated as follows: Under what conditions are the sums Y_n asymptotically constant?

This question was answered in 1928 by A.N. Kolmogorov. Assume, without loss of generality, that the medians of the variables $X_{n,k}$ are zero. Let $X_{n,k} = X_{n,k}$ if $|X_{n,}k| \leq 1$ and $X_{n,k} = 0$ if $X_{n,k}| > 1$. The simultaneous fulfillment of the two conditions

$$\sum_{k=1}^{kn} \mathrm{P}\{|X_{n,k}| > 1\} \to 0 \quad if\, n \to \infty$$

and

$$\sum_{k=1}^{k_n} \mathrm{E}X\square^2_{n,k} \to 0 \quad \text{if } n \to \infty,$$

is necessary and sufficient for the sums Y_n to be asymptotically constant. The sum $\sum_k^{k_n} 1 \mathrm{E}X_{n,k}$ may be taken as C_n. That these conditions are sufficient is readily demonstrated by Chebyshev's method. If the mathematical expectations $EX_{n,k}$ exist, it is easy to find supplementary conditions under which it is permissible to choose $C_n = \mathrm{E}Y_n$, which yields necessary and sufficient conditions for the law of large numbers in the classical formulation. For a sequence of independent identically-distributed variables $\{X_n\}$ these conditions are reduced — in accordance with Khinchin's theorem quoted above — to the existence of the mathematical expectation. At the same time, the condition

$$n\mathrm{P}\{|X_1| > n\} \to 0 \text{ if } n \to \infty$$

is necessary and sufficient for the arithmetical averages Y_n to be asymptotically constant. Examples of cases in which condition is not met are easily found. Thus, the condition is not met if all X_n have a Cauchy distribution with density $1/\pi(1+x^2)$ (to which corresponds the characteristic function $\exp\{-|t|\}$). Here, the arithmetical averages $Y_n = (X_1 + ... + X_n)/n$ have the characteristic function $\exp\{-n|t/n|\} = \exp\{-|t|\}$ and, as a result, have the same distribution for any n as do the individual terms.

The most important cases to which the law of large numbers does not apply involve the times of return to the starting point in a random walk. Thus, in a symmetric Bernoulli random walk the time T_n passing prior to the n-th return to the starting point is the sum of the n independent random variables $X_1,...,X_n$ where X_1 is the time passed prior to the first return, is the time between the first and the second return, etc. The distribution of the variable $2T_n/\pi n^2$ converges as $n \to \infty$ to a non-degenerate limit distribution with density

$$P(x) = \frac{1}{\sqrt{2\pi}} e^{-1/2x} x^{-3/2} \text{ for } x > 0$$

and zero for $x \le 0$. Accordingly, in this case the distribution of the arithmetical average of X_i, i.e. T_n/n, is located, roughly speaking, on a segment of length of order n(whereas, in cases where the law of large numbers is applicable, it is located on segments of length) $o(1)$.

The applicability of the law of large numbers to sums of dependent variables (both in its classical and in its more general formulations) is connected, first and foremost, with the asymptotic independence of the random variables X_i and X_j as the difference between their subscripts $|i-j|$ increases. The corresponding theorems were first proved by Markov in 1907 for variables connected in a Markov chain. In fact, let be a homogeneous Markov chain with finite number of states, and let $X_1,...,X_n,...,$ all one-step transition probabilities be positive. Here, the asymptotic independence of X_j and X_i as $|i-j| \to \infty$ follows from the fact that the conditional distribution of X_j under a fixed value of X_i tends, as $n \to \infty$, to a limit which does not depend on the chosen value of X_i (Markov's ergodic theorem). The law of large numbers is deduced from this theorem: It is first established that, as $n \to \infty$,

$$\mathrm{E}\left(\frac{X_1 + \ldots + X_n}{n} - \alpha\right)^2 \to 0,$$

where $a = \lim_{k\to\infty} EX_k$; hence it follows that as $n \to \infty$,

$$\mathrm{P}\left\{\frac{X_1 + \ldots + X_n}{n} - \alpha\right| > \varepsilon\right\} \to 0,$$

The following, more general case, is governed by Bernstein's conditions: If $DX_j < L|R(X_i, X_j)| \leq \phi(|i - j|)$, , where L is a constant, R is the correlation coefficient and $\phi(n)$ is a function which tends to zero as $n \to \infty$, then the law of large numbers is applicable to the variables $\{X_n\}$. For sequences $\{X_n\}$ which are stationary in the wide sense, the correlation condition may be weakened somewhat, by replacing it by

$$\lim_{n\to\infty} \frac{1}{n} \sum_{j=1}^{n} R_j = 0,$$

where

$$R_j = R(X_i, X_{i+j})$$

The above-mentioned results may be extended in various ways. First, only convergence in probability was considered so far. Other types of convergence may also be considered, including convergence with probability one, mean-square convergence, etc. (in fact, many of the above conditions ensure mean-square convergence, which implies convergence in probability). The case of convergence with probability one is important, and for this reason has been especially named the strong law of large numbers.

Furthermore, many of these theorems can be applied, with suitable modifications, to random vectors having values in Euclidean spaces of any dimension, in a Hilbert space or in certain Banach spaces. For instance, if $\{X_n\}$ is a sequence of independent identically-distributed random vectors with values in a separable Banach space, and if $\mathrm{E}\|X_n\|$ (where $\|x\|$ is the norm of x) is finite, one has

$$\mathrm{P}\left\{\frac{X_1 + \ldots + X_n}{n} - \mathrm{EX}_1\right\| > \varepsilon\right\} \to 0$$

for any $\varepsilon > 0$ if $n \to \infty$.

The law of large numbers, when considered in its most general form, is closely related to ergodic theorems (cf. Ergodic theorem). Clearly, many theorems are also applicable to the case of the average $(1/T)\int_0^T X(t)dt$, where $X(t)$ is a random process depending on a continuous parameter.

Finally, instead of the sums of random variables one may consider other symmetric functions of them. This was in fact done by A.Ya. Khinchin (1951–1955) to prove certain conclusions in statistical mechanics. The result obtained by him may be illustrated by the following example. Let $X_n,1 ,..., X_{n,n}$ be the coordinates of a point uniformly distributed on the surface of the sphere

$$X_{n,1}^2 + ... + X_{n,n}^2 = n\mu, \qquad \mu > 0$$

The law of large numbers then applies to a wide class of symmetric functions $f(X_{n,1},..., X_{n,n})$ in the sense that as $n \to \infty$, their values are asymptotically constant (this is similar to the observation made in 1925 by P. Lıvy to the effect that sufficiently regular functions of a very large number of variables are almost constant in a large part of their domain of definition).

BERNSTEIN INEQUALITY

Bernstein's inequality in probability theory is a more precise formulation of the classical Chebyshev inequality in probability theory, proposed by S.N. Bernshtein in 1911; it permits one to estimate the probability of large deviations by a monotone decreasing exponential function. In fact, if the equations

$$\mathrm{E}X_j = 0, \quad \mathrm{E}X_j^2 = B_j, \quad j = 1,..., n,$$

hold for the independent random variables $X_1 ,..., X_n$ with

$$\mathrm{E}|X_j|^l \le \frac{b_j}{2} H^{l-2} l!$$

(where $l > 2$ and H is a constant independent of), then the following inequality of Bernstein (where $r > 0$) is valid for the sum $S_n = X_1 + ... + X_n$:

$$\mathrm{P}\{|S_n| > r\} \le 2\exp\left\{-\frac{r^2}{2(B_n + Hr)}\right\},$$

where $B_n = \sum b_j$. For identically-distributed bounded random variables $X_j(EX_j = 0,\ EX_j^2 = \sigma^2$, and $|X_j| \le L$,) inequality takes its simplest form:

$$\mathrm{P}\{|S_n| > t\sigma\sqrt{n}\} \le 2\exp\left\{-\frac{t^2}{2(1+\alpha/3)}\right\}$$

where $\alpha = Lt/\sqrt{n\sigma}$. A.N. Kolmogorov gave a lower estimate of the probability in. The Bernstein–Kolmogorov estimates are used, in particular, in proving the law of the iterated logarithm. Some idea of the accuracy of may be obtained by comparing it with the approximate value of the left-hand side of which is obtained by the central limit theorem in the form

$$\frac{2}{\sqrt{2\pi}}\int_t^{\infty} e^{-u^2/2} du = \frac{2}{\sqrt{2\pi t}}\left(1 - \frac{\theta}{t^2}\right)e^{-t^2/2},$$

where $0 < \theta < 1$. Subsequent to 1967, Bernstein's inequalities were extended to include multidimensional and infinite-dimensional cases.

Bernstein's inequality for the derivative of a trigonometric or algebraic polynomial gives an estimate of this derivative in terms of the polynomial itself. If $T_n(x)$ is a trigonometric polynomial of degree not n exceeding and if

$$M = \max_{0 \le x \le 2\pi} |T_n(x)|,$$

then the following inequalities are valid for all x:

$$|T_n^{(r)}(x)| \le Mn^r, \quad r = 1, 2, \ldots.$$

These estimates cannot be improved, since the number $M=1$ for

$$T_n(x) = \cos n(x - x_0)$$

is sharp:

$$\max_x |T_n^{(r)}(x)| = n^r.$$

Bernstein's inequality for trigonometric polynomials is a special case of the following theorem: If $f(x)$ is an entire function of order $\le \sigma$ and if

$$M = \sup_{-\infty < x < \infty} |f(x)|,$$

then one has

$$\sup_{-\infty < x < \infty} |f^{(r)}(x)| \le M\sigma^r \quad (r = 1, 2, \ldots).$$

Bernstein's inequality for an algebraic polynomial has the following form: If the polynomial

$$P_n(x) = \sum_{k=0}^{n} \alpha_k x^k$$

satifies the condition

$$|P_n(x)| \le M, \quad \alpha \le x \le b$$

then its derivative $P_n'(x)$ has the property

$$|P_n'(x)| \le \frac{Mn}{\sqrt{(x-a)(b-x)}}, \qquad \alpha < x < b,$$

which cannot be improved. As was noted by S.N. Bernshtein, this inequality is a consequence of the proof of the Markov inequality given by A.A. Markov.

Bernstein's inequalities are in fact employed in proving converse theorems in the theory of approximation of functions. There are a number of generalizations of Bernstein's inequality, in particular for entire functions in several variables.

SAMPLING METHODS

The need for accurate information in order to make informed decisions has been emphasized throughout this online publication (refer to the sections on Data, information and statistics). Statistics are an important type of information and statistical agencies play an important part in the production of such information. In order to do this, the agencies rely on censuses, sample surveys and the use of administration records. These three sources, along with their advantages and disadvantages, have been explained.

Sampling allows statisticians to draw conclusions about a whole by examining a part. It enables us to estimate characteristics of a population by directly observing a portion of the entire population. Researchers are not interested in the sample itself, but in what can be learned from the survey—and how this information can be applied to the entire population.

It is essential that a sample survey be correctly defined and organized. If the wrong questions are posed to the wrong people, statisticians will not receive information that will be useful when applied to the entire population.

In the context of a national statistical agency like Statistics Canada, the following steps are needed to select a sample and ensure that this sample will fulfill its goals.

Establish the Survey's Objectives

The first step in planning a useful and efficient survey is to specify the objectives with as much detail as possible. Without objectives, the survey is unlikely to generate usable results. Clarifying the aims of the survey is critical to its ultimate success. The initial users and uses of the data should be identified at this stage.

The pros and cons of a census versus a sample survey or the use of administrative records should be evaluated and a decision made as to the most appropriate method. (At this point, we will assume that a sample survey is the best way to proceed in order to obtain the information we need. This assumption will hold true for the remainder of the sample selection steps, even though many of the steps mentioned will also apply to the other methods.)

Define The Target Population

The target population is the total population for which the information is required. For example, if you were to conduct a survey about the most popular types of cars in Saskatchewan, then the target population would be every car in Saskatchewan. The units that make up the population must be described in terms of characteristics that clearly identify them. Specifically, the target population is defined by the following characteristics:

1. Nature of data required: about persons, hospitals, schools, etc.
2. Geographic location: the geographic boundaries of the population have to be determined, as well as the level of geographic detail required for the survey estimate (by province, by city, etc.).
3. Reference period: the time period covered by the survey.
4. Other characteristics, such as socio-demographic characteristics (interest in a particular age group, for example) or type of industry.

Decide on the Data to be Collected

The data requirements of the survey must be established. To ensure that the requirements are operationally sound, the necessary data terms and definitions also need to be determined.

Set The Level of Precision

As mentioned in the section on Sampling error, there is a level of uncertainty associated with estimates coming from a sample. For example, if you are trying to estimate the average distance between home and school for students in your class of 25 from a sample of 5 persons, your estimate will depend on who the 5 sampled students are. If the 5 sampled students also live close to the school, the results will not be able to represent the class accurately. This sample-to-sample variation is what causes the sampling error. Statisticians can estimate the sampling error associated with a particular sampling plan, and try to minimize it.

When designing a survey, the acceptable level of uncertainty in the survey estimates has to be established. This level depends on what the end use of the results will be and on the size of the overall budget. The bigger the budget, the more resources available, and thus, less chance for error. And if the end result is to serve a specific purpose, then the acceptable level of uncertainty would be smaller than an end result that is simply looking for general trends.

The level of uncertainty will also be determined by the sample size. Increasing the sample size will decrease the sampling error. (If you sample 24 out of 25 students in your class, there will not be as much sample-to-sample variation as there would be if you only sampled 5 students from among the 25 possible samples.)

THE SAMPLE DESIGN

Once the objectives, guidelines and definitions have been worked out, the statistician can work on the survey plan. The survey plan is divided into three parts:

1. Sample design: how the sample will be collected.
2. Estimation techniques: how the results from the sample will be extended to the whole population.
3. Measures of precision: how the sampling error will be measured.

The estimation techniques and measures of precision are discussed in a later section. For the moment, we will look at the sample design. The following steps lead to the complete determination of the sample design:

1. Determine what the survey population will be (e.g., students, men aged 20 to 35, newborn babies, etc.).
2. Choose the most appropriate survey time frame.
3. Define the survey units.
4. Establish the sample size (e.g., a sample of 100 from a population of 1,000).
5. Select a sampling method.

The Survey Population

The target population must be defined early in the survey-designing process. This is the population for which information is required. However, some members of the population have to be excluded because of operational constraints: the high cost of collecting data in some remote areas, the difficulty of identifying and contacting certain components of the target population, etc. For example, because it would be too difficult to locate and survey each car owned by every resident in Saskatchewan, a survey population of just the major cities and towns might be conducted instead. When some of the members of the target population are excluded, we call the included population the survey population or, what is sometimes called, an *observed population*. The target population is the population we want to observe while the survey population is the population we can observe.

The goal of this process is to have the survey population as close as possible to the target population. It is also very important that the users of the data be informed of the differences between the two populations, as the results of the survey will apply only to the survey population.

For example, a target population for a survey could be all Canadians aged 15 years and over (on a particular reference date), while the survey population could exclude residents of the Yukon, Nunavut and Northwest Territories, persons living on Aboriginal reserves, full-time members of the Canadian Armed Forces and residents of institutions. These Canadians might be excluded for various reasons: to survey people in the territories might prove to be difficult and expensive, military personnel may not be available for surveying if they are out on a mission, etc. Using this example, about 2% of the target population would be excluded from the survey population.

The Survey Frame

The survey frame, also called the sampling frame, is the tool used to gain access to the population. There are two types of frames: list frames and area frames. A list frame is just a list of names and addresses that provide direct access to 'individuals' (e.g., a list of hospitals, a list of restaurants, a list of students at a university). Area frames are a list of geographic areas that provide indirect access to individuals (e.g., the neighbourhoods in a city). This type of access is called indirect because first, a list of geographic areas must be selected and then, access to individuals within each selected area must be worked out.

For instance, suppose that you were surveying a rural town in Quebec to see what percentage of residents are farmers. If you were provided with an area frame, then you would be able to locate which roads to visit, but you would still have to find out the names and addresses of the residents on each road.

When there is no single frame that is appropriate, multiple frames can be used. Some sampling techniques using both types of frames will be discussed later.

A good frame should be complete and up-to-date; no member of the survey population should be excluded from the frame or duplicated on the frame (represented more than once);

and no unit that is not part of the population (e.g., deceased persons) should be on the frame. The frame chosen will impact the selected survey population. For instance, if a list of telephone numbers is used to select a sample of households, then all households without telephones are excluded from the survey population.

The Survey Units

There are three types of units that have to be accurately identified in order to avoid problems during the selection, data collection and data analysis stages. They are as follows:

1. The sampling unit is part of the frame and therefore subject to being selected.
2. The respondent unit or reporting unit provides the information needed by the survey.
3. The unit of reference or unit of analysis—the unit about which information is provided—is used to analyse the survey results.

For example, in a survey about newborns in Edmonton, the sampling unit might be a household, the reporting unit one of the parents or a legal guardian, and the unit of reference the baby.

The sampling units may differ depending on the frame used. This is why the survey population, survey frame and survey units are defined in conjunction with one another.

The Sample Size

The level of precision needed for the survey estimates will impact the sample size. However, it is not as easy to determine the sample size as one may think. Generally, the actual sample size of a survey is a compromise between the level of precision to be achieved, the survey budget and any other operational constraints, such as budget and time. In order to achieve a certain level of precision, the sample size will depend, among other things, on the following factors:

1. The variability of the characteristics being observed: If every person in a population had the same salary, then a sample of one person would be all you would need to estimate the average salary of the population. If the salaries are very different, then you would need a bigger sample in order to produce a reliable estimate.
2. The population size: To a certain extent, the bigger the population, the bigger the sample needed. But once you reach a certain level, an increase in population no longer affects the sample size. For instance, the necessary sample size to achieve a certain level of precision will be about the same for a population of one million as for a population twice that size.
3. The sampling and estimation methods: Not all sampling and estimation methods have the same level of efficiency. You will need a bigger sample if your method is not the most efficient. But because of operational constraints and the unavailability of an adequate frame, you cannot always use the most efficient technique.

The Sampling Method

There are two types of sampling methods: probability sampling and non-probability sampling. The difference between them is that in probability sampling, every unit has a 'chance' of being selected, and that chance can be quantified. This is not true for non-probability sampling; every item in a population does not have an equal chance of being selected. The next section will describe features of both types of sampling and detail some of the methods related to each type.

PROBABILITY SAMPLING

Probability sampling involves the selection of a sample from a population, based on the principle of randomization or chance. Probability sampling is more complex, more time-consuming and usually more costly than non-probability sampling. However, because units from the population are randomly selected and each unit's probability of inclusion can be calculated, reliable estimates can be produced along with estimates of the sampling error, and inferences can be made about the population.

There are several different ways in which a probability sample can be selected. The method chosen depends on a number of factors, such as the available sampling frame, how spread out the population is, how costly it is to survey members of the population and how users will analyse the data. When choosing a probability sample design, your goal should be to minimize the sampling error of the estimates for the most important survey variables, while simultaneously minimizing the time and cost of conducting the survey.

Simple Random Sampling

In *simple random sampling*, each member of a population has an equal chance of being included in the sample. Also, each combination of members of the population has an equal chance of composing the sample. Those two properties are what defines simple random sampling. To select a simple random sample, you need to list all of the units in the survey population.

Example: To draw a simple random sample from a telephone book, each entry would need to be numbered sequentially. If there were 10,000 entries in the telephone book and if the sample size were 2,000, then 2,000 numbers between 1 and 10,000 would need to be randomly generated by a computer. Each number will have the same chance of being generated by the computer (in order to fill the simple random sampling requirement of an equal chance for every unit). The 2,000 telephone entries corresponding to the 2,000 computer-generated random numbers would make up the sample.

Simple random sampling can be done with or without replacement. A sample with replacement means that there is a possibility that the sampled telephone entry may be selected twice or more. Usually, the simple random sampling approach is conducted without replacement because it is more convenient and gives more precise results. For the purpose of these descriptions, when we discuss simple random sampling, we will refer to sampling without replacement.

Simple random sampling is the easiest method of sampling and it is the most commonly used. Advantages of this technique are that it does not require any additional information on the frame (such as geographic areas) other than the complete list of members of the survey population along with information for contact. Also, since simple random sampling is a simple method and the theory behind it is well established, standard formulas exist to determine the sample size, the estimates and so on, and these formulas are easy to use.

On the other hand, this technique makes no use of auxiliary information present on the frame (i.e., number of employees in each business) that could make the design of the sample more efficient. And although it is easy to apply simple random sampling to small populations, it can be expensive and unfeasible for large populations because all elements must be identified and labeled prior to sampling. It can also be expensive if personal interviewers are required since the sample may be geographically spread out across the population.

A lottery draw is a good example of simple random sampling. For example, when a sample of 6 numbers is randomly generated from a population of 49, each number has an equal chance of being selected and each combination of 6 numbers has the same chance of being the winning combination. Even though people tend to avoid combinations such as 1-2-3-4-5-6, it has the same chance of being the winning set of numbers as the combination of 8-15-21-28-32-40.

Example: Suppose your school has 500 students and you need to conduct a short survey on the quality of the food served in the cafeteria. You decide that a sample of 10 students should be sufficient for your purposes. In order to get your sample, you assign a number from 1 to 500 to each student in your school. To select the sample, you use a table of randomly generated numbers. All you have to do is pick a starting point in the table (a row and column number) and look at the random numbers that appear there. In this case, since the data run into three digits, the random numbers would need to contain three digits as well. Ignore all random numbers after 500 because they do not correspond to any of the students in the school. Remember that the sample is without replacement, so if a number recurs, skip over it and use the next random number. The first 10 different numbers between 001 and 500 make up your sample.

Example: Imagine that you own a movie theatre and you are offering a special horror movie film festival next month. To decide which horror movies to show, you survey moviegoers asking them which of the listed movies are their favourites. To create the list of movies needed for your survey, you decide to sample 100 of the 1,000 best horror movies of all time. The horror movie population is divided evenly into classic movies (those filmed in or before 1969) and modern movies (those filmed in or later than 1970). One way of getting a sample would be to write out all of the movie titles on slips of paper and place them in an empty box. Then, draw out 100 titles and you will have your sample. By using this approach, you will have ensured that each movie had an equal chance of selection.

You can also calculate the probability of a given movie being selected. Since we know the sample size (n) and the total population (N), calculating the probability of being included in the sample becomes a simple matter of division:

Probability of being selected (same for each horror movie)

$= (n \div N) \times 100\%$

$= (100 \div 1{,}000) \times 100\%$

$= 10\%$

This means that every movie title on the list has a 10% or a 1 in 10 chance of being selected.

You can see that that one disadvantage of simple random sampling (not the only disadvantage, but an important one) is that even if you know that the population is made up of 500 classic movies and 500 modern movies and you know each movie's release date from the sampling frame, no use is made of this information. This sample might contain 77 classic movies and 23 modern movies, which would not be representative of the whole horror movie population.

There are ways to overcome this problem (these will be briefly discussed in the Estimation section), but there are also ways to account for this information. (This will also be discussed later, under the section on Stratified sampling.)

Systematic sampling

Sometimes called *interval sampling, systematic sampling* means that there is a gap, or interval, between each selected unit in the sample. In order to select a systematic sample, you need to follow these steps:

1. Number the units on your frame from 1 to N (where N is the total population size).
2. Determine the sampling interval (K) by dividing the number of units in the population by the desired sample size. For example, to select a sample of 100 from a population of 400, you would need a sampling interval of $400 \div 100 = 4$. Therefore, $K = 4$. You will need to select one unit out of every four units to end up with a total of 100 units in your sample.
3. Select a number between one and K at random. This number is called *the random start* and would be the first number included in your sample. Using the sample above, you would select a number between 1 and 4 from a table of random numbers. If you choose 3, the third unit on your frame would be the first unit included in your sample; if you choose 2, your sample would start with the second unit on your frame.
4. Select every K^{th} (in this case, every fourth) unit after that first number. For example, the sample might consist of the following units to make up a sample of 100: 3 (the random start), 7, 11, 15, 19...395, 399 (up to N, which is 400 in this case).

Using the example above, you can see that with a systematic sample approach there are only four possible samples that can be selected, corresponding to the four possible random starts:

1. 1, 5, 9, 13...393, 397
2. 2, 6, 10, 14...394, 398
3. 3, 7, 11, 15...395, 399
4. 4, 8, 12, 16...396, 400

Each member of the population belongs to only one of the four samples and each sample has the same chance of being selected. From that, we can see that each unit has a one in four chance of being selected in the sample. This is the same probability as if a simple random sampling of 100 units was selected. The main difference is that with simple random sampling, any combination of 100 units would have a chance of making up the sample, while with systematic sampling, there are only four possible samples. From that, we can see how precise systematic sampling is compared with simple random sampling. The population's order on the frame will determine the possible samples for systematic sampling. If the population is randomly distributed on the frame, then systematic sampling should yield results that are similar to simple random sampling.

This method is often used in industry, where an item is selected for testing from a production line to ensure that machines and equipment are of a standard quality. For example, a tester in a manufacturing plant might perform a quality check on every 20th product in an assembly line. The tester might choose a random start between the numbers 1 and 20. This will determine the first product to be tested; every 20th product will be tested thereafter.

Interviewers can use this sampling technique when questioning people for a sample survey. The market researcher might select, for example, every 10th person who enters a particular store, after selecting the first person at random. The surveyor may interview the occupants of every fifth house on a street, after randomly selecting one of the first five houses.

Example: Imagine you have to conduct a survey on student housing for your university or college. Your school has an enrolment of 10,000 students and you want to take a systematic sample of 500 students. In order to do this, you must first determine what your sampling interval (K) would be:

Total population χ sample size = sampling interval

$N \div n = K$

$= 10{,}000 \div 500$

$= 20$

To begin this systematic sample, all students would have to be assigned sequential numbers. The starting point would be chosen by selecting a random number between 1 and 20. If this number were 9, then the 9th student on the list would be selected along with every 20th student thereafter. The sample of students would be those corresponding to student numbers 9, 29, 49, 69...9,929, 9,949, 9,969 and 9,989.

In the examples used thus far, the sampling interval K was always a whole number, but this is not always the case. For example, if you want a sample of 30 from a population of 740, your sampling interval (or K) will be 24.7. In these cases, there are a few options to make the number easier to work with. You can round the number—either round it up to the nearest whole number or round it down. Rounding down will ensure that you select at least the number of units you originally wanted (and you can then delete some units to get the exact sample size you wanted). Techniques exist to adapt systematic sampling to the case where N (total population) is not a multiple of n (sample size), but still give a sample exactly the same as the n units. These techniques will not be discussed here.

The advantages of systematic sampling are that the sample selection cannot be easier (you only get one random number—the random start—and the rest of the sample automatically follows) and that the sample is distributed evenly over the listed population. The biggest drawback of the systematic sampling method is that if there is some cycle in the way the population is arranged on a list and if that cycle coincides in some way with the sampling interval, the possible samples may not be representative of the population. This can be seen in the following example:

Example: Suppose you run a large grocery store and have a list of the employees in each section. The grocery store is divided into the following 10 sections: deli counter, bakery, cashiers, stock, meat counter, produce, pharmacy, photo shop, flower shop and dry cleaning. Each section has 10 employees, including a manager (making 100 employees in total). Your list is ordered by section, with the manager listed first and then, the other employees by descending order of seniority.

If you wanted to survey your employees about their thoughts on their work environment, you might choose a small sample to answer your questions. If you use a systematic sampling approach and your sampling interval is 10, then you could end up selecting only managers or the newest employees in each section. This type of sample would not give you a complete or appropriate picture of your employees' thoughts.

Sampling with Probability Proportional to Size

Probability sampling requires that each member of the survey population have a chance of being included in the sample, but it does not require that this chance be the same for everyone. If there is information available on the frame about the size of each unit (e.g., number of employees for each business) and if those units vary in size, this information can be used in the sampling selection in order to increase the efficiency. This is known as *sampling with probability proportional to size* (PPS). With this method, the bigger the size of the unit, the higher the chance it has of being included in the sample. For this method to bring increased efficiency, the measure of size needs to be accurate. This is a more complex sampling method that will not be discussed in further detail here.

Stratified Sampling

Using *stratified sampling*, the population is divided into homogeneous, mutually exclusive groups called strata, and then independent samples are selected from each stratum. Any of the sampling methods mentioned in this section (and others that exist) can be used to sample within each stratum. The sampling method can vary from one stratum to another. When simple random sampling is used to select the sample within each stratum, the sample design is called *stratified simple random sampling*. A population can be stratified by any variable that is available for all units on the sampling frame prior to sampling (e.g., age, sex, province of residence, income, etc.).

Why do we need to create strata? There are many reasons, the main one being that it can make the sampling strategy more efficient. It was mentioned earlier that you need a larger sample to get a more accurate estimation of a characteristic that varies greatly from one unit to the other than for a characteristic that does not. For example, if every person in a population had the same salary, then a sample of one individual would be enough to get a precise estimate of the average salary.

This is the idea behind the efficiency gain obtained with stratification. If you create strata within which units share similar characteristics (e.g., income) and are considerably different from units in other strata (e.g., occupation, type of dwelling) then you would only need a small sample from each stratum to get a precise estimate of total income for that stratum. Then you could combine these estimates to get a precise estimate of total income for the whole population. If you were to use a simple random sampling approach in the whole population without stratification, the sample would need to be larger than the total of all stratum samples to get an estimate of total income with the same level of precision.

Stratified sampling ensures an adequate sample size for sub-groups in the population of interest. When a population is stratified, each stratum becomes an independent population and you will need to decide the sample size for each stratum.

Example: Suppose you want to estimate how many high school students have part-time jobs at the national level and also in each province. If you were to select a simple random sample of 25,000 people from a list of all high school students in Canada (assuming such a list was available for selection), you would end up on average with just a little over 100 people from Prince Edward Island, since they account for less than half of a percent of the whole Canadian population. This sample would probably not be large enough for the kind of detailed analysis you had in mind. Stratifying your list by province, again assuming that this information is available, and then selecting a sample size for each province would allow you to decide on the exact sample size needed for that specific province. Thus, in order to get good representation of Prince Edward Island, you would use a larger sample than the one allotted to it by the simple random sampling approach.

Example: An Ontario school board wanted to assess student opinion on dropping Grade 13 from the secondary school programme. They decided to survey students from Elmsview High School. To ensure a representative sample of students from all grade levels, the school board used a stratified sampling technique.

In this case, the strata were the five grade levels (grades 9 to 13). The school board then selected a sample within each stratum. The students selected in this sample were extracted using simple random or systematic sampling, making up a total sample of 100 students.

Stratification is most useful when the stratifying variables are

1. Simple to work with,
2. Easy to observe, and
3. Closely related to the topic of the survey.

Cluster Sampling

Sometimes it is too expensive to spread a sample across the population as a whole. Travel costs can become expensive if interviewers have to survey people from one end of the country to the other. To reduce costs, statisticians may choose a *cluster sampling* technique.

Cluster sampling divides the population into groups or clusters. A number of clusters are selected randomly to represent the total population, and then all units within selected clusters are included in the sample. No units from non-selected clusters are included in the sample—they are represented by those from selected clusters. This differs from stratified sampling, where some units are selected from each group.

Examples of clusters are factories, schools and geographic areas such as electoral subdivisions. The selected clusters are used to represent the population.

Example: Suppose you are a representative from an athletic organization wishing to find out which sports Grade 11 students are participating in across Canada. It would be too costly and lengthy to survey every Canadian in Grade 11, or even a couple of students from every Grade 11 class in Canada. Instead, 100 schools are randomly selected from all over Canada.

These schools provide clusters of samples. Then every Grade 11 student in all 100 clusters is surveyed. In effect, the students in these clusters represent all Grade 11 students in Canada.

Example: Imagine that the municipal council of a small city wants to investigate the use of health care services by residents.

First, the council requests from Statistics Canada electoral subdivision maps that identify and label each city block. From these maps, the council creates a list of all city blocks. This list will serve as the sampling frame.

Every household in that city belongs to a city block, and each city block represents a cluster of households. The council randomly picks a number of city blocks. Using the simple random sample approach, then the council creates a list of all households in the selected city blocks; these households make up the survey sample.

As mentioned, cost reduction is a reason for using cluster sampling. It creates 'pockets' of sampled units instead of spreading the sample over the whole territory. Another reason is that sometimes a list of all units in the population (a requirement when conducting simple random sample, systematic sample or sampling with probability proportional to size) is not available, while a list of all clusters is either available or easy to create.

In most cases, the main drawback is a loss of efficiency when compared with simple random sampling. It is usually better to survey a large number of small clusters instead of a small number of large clusters. This is because neighbouring units tend to be more alike, resulting in a sample that does not represent the whole spectrum of opinions or situations present in the overall population. In the two previous examples, students in the same school tend to participate in the same types of sports (depending on the facilities available at their school); similarly, elderly people have a tendency to live in specific neighbourhoods and to be heavy users of health services.

Another drawback to cluster sampling is that you do not have total control over the final sample size. Since not all schools have the same number of Grade 11 students and city blocks do not all have the same number of households, and you must interview every student or household in your sample, the final size may be larger or smaller than you expected.

Multi-stage Sampling

Multi-stage sampling is like the cluster method, except that it involves picking a sample from within each chosen cluster, rather than including all units in the cluster. This type of sampling requires at least two stages. In the first stage, large groups or clusters are identified and selected. These clusters contain more population units than are needed for the final sample.

In the second stage, population units are picked from within the selected clusters (using any of the possible probability sampling methods) for a final sample. If more than two stages are used, the process of choosing population units within clusters continues until there is a final sample.

Example: A cluster sample would choose 100 schools and then interview every Grade 11 student from those schools. Instead in multi-stage sampling, you could select more schools, get a list of all Grade 11 students from these selected schools and select a random sample (e.g., simple random sample) of students from each school. This would be a two-stage sampling design.

You could also get a list of all Grade 11 classes in the selected schools, pick a random sample of classes from each of those schools, get a list of all the students in the selected classes and finally select a random sample of students from each class. This would be a three-stage sampling design. Each time we add a stage, the process becomes more complex.

Now imagine that each school has on average 80 Grade 11 students. Cluster sampling would then give your organization a sample of about 8,000 students (100 schools x 80 Grade 11 students). If you wanted a bigger sample, you could select schools with more students; and for a smaller sample you could select schools with fewer students.

One way to control the sample size would be to stratify the schools into large, medium and small sizes (in terms of the number of Grade 11 students) and select a sample of schools from each stratum. This is called *stratified cluster sampling*.

With a three-stage design, you could select a sample of 400 schools, then select two Grade 11 classes per school (assuming that there are two or more Grade 11 classes per school). Finally, you could select 10 students per class. This way, you still end up with a sample of about 8,000 students (400 schools x 2 classes x 10 students), but the sample will be more spread out.

You can see from this example that with multi-stage sampling, you still have the benefit of a more concentrated sample for cost reduction. However, the sample is not as concentrated as other clusters and the sample size is still bigger than for a simple random sample size. Also, you do not need to have a list of all of the students in the population. All you need is a list of the classes from the 400 schools and a list of the students from the 800 classes. Admittedly, more information is needed in this type of sample than what is required in cluster sampling.

However, multi-stage sampling still saves a great amount of time and effort by not having to create a list of all of the units in a population.

Multi-phase Sampling

A *multi-phase sample* collects basic information from a large sample of units and then, for a subsample of these units, collects more detailed information. The most common form of multi-phase sampling is two-phase sampling (or double sampling), but three or more phases are also possible.

Multi-phase sampling is quite different from multi-stage sampling, despite the similarities in name. Although multi-phase sampling also involves taking two or more samples, all samples are drawn from the same frame and at each phase the units are structurally the same. However, as with multi-stage sampling, the more phases used, the more complex the sample design and estimation will become.

Multi-phase sampling is useful when the frame lacks auxiliary information that could be used to stratify the population or to screen out part of the population.

Example : Suppose that an organization needs information about cattle farmers in Alberta, but the survey frame lists all types of farms—cattle, dairy, grain, hog, poultry and produce. To complicate matters, the survey frame does not provide any auxiliary information for the farms listed there.

A simple survey could be conducted whose only question is "Is part or all of your farm devoted to cattle farming?" With only one question, this survey should have a low cost per interview (especially if done by telephone) and, consequently, the organization should be able to draw a large sample. Once the first sample has been drawn, a second, smaller sample can be extracted from among the cattle farmers and more detailed questions asked of these farmers. Using this method, the organization avoids the expense of surveying units that are not in this specific scope (i.e., non-cattle farmers).

Multi-phase sampling can be used when there is insufficient budget to collect information from the whole sample, or when doing so would create excessive burden on the respondent, or even when there are very different costs of collection for different questions on a survey.

Example : A health survey asks participants some basic questions about their diet, smoking habits, exercise routines and alcohol consumption. In addition, the survey requires that respondents subject themselves to some direct physical tests, such as running on a treadmill or having their blood pressure and cholesterol levels measured.

Filling out questionnaires or interviewing participants are relatively inexpensive procedures, but the medical tests require the supervision and assistance of a trained health practitioner, as well as the use of an equipped laboratory, both of which can be quite costly. The best way to conduct this survey would be to use a two-phase sample approach. In the first phase, the interviews are performed on an appropriately sized sample. From this sample, a smaller sample is drawn. This second sample will take part in the medical tests.

NON-PROBABILITY SAMPLING

The difference between *probability* and *non-probability sampling* has to do with a basic assumption about the nature of the population under study. In probability sampling, every item has a chance of being selected. In non-probability sampling, there is an assumption that there is an even distribution of characteristics within the population. This is what makes the researcher believe that any sample would be representative and because of that, results will be accurate. For probability sampling, randomization is a feature of the selection process, rather than an assumption about the structure of the population.

In non-probability sampling, since elements are chosen arbitrarily, there is no way to estimate the probability of any one element being included in the sample. Also, no assurance is given that each item has a chance of being included, making it impossible either to estimate sampling variability or to identify possible bias.

Reliability cannot be measured in non-probability sampling; the only way to address data quality is to compare some of the survey results with available information about the population. Still, there is no assurance that the estimates will meet an acceptable level of error. Statisticians are reluctant to use these methods because there is no way to measure the precision of the resulting sample.

Despite these drawbacks, non-probability sampling methods can be useful when descriptive comments about the sample itself are desired. Secondly, they are quick, inexpensive and convenient. There are also other circumstances, such as in applied social research, when it is unfeasible or impractical to conduct probability sampling. Statistics Canada uses probability sampling for almost all of its surveys, but uses non-probability sampling for questionnaire testing and some preliminary studies during the development stage of a survey.

Most non-sampling methods require some effort and organization to complete, but others, like convenience sampling, are done casually and do not need a formal plan of action.

Convenience or Haphazard Sampling

Convenience sampling is sometimes referred to as *haphazard* or *accidental sampling*. It is not normally representative of the target population because sample units are only selected if they can be accessed easily and conveniently.

There are times when the average person uses convenience sampling. A food critic, for example, may try several appetizers or entrees to judge the quality and variety of a menu. And television reporters often seek so-called 'people-on-the-street interviews' to find out how people view an issue. In both these examples, the sample is chosen randomly, without use of a specific survey method.

The obvious advantage is that the method is easy to use, but that advantage is greatly offset by the presence of bias. Although useful applications of the technique are limited, it can deliver accurate results when the population is homogeneous.

For example, a scientist could use this method to determine whether a lake is polluted.

Assuming that the lake water is well-mixed, any sample would yield similar information. A scientist could safely draw water anywhere on the lake without fretting about whether or not the sample is representative.

Examples of convenience sampling include:

- The female moviegoers sitting in the first row of a movie theatre
- The first 100 customers to enter a department store
- The first three callers in a radio contest.

Volunteer Sampling

As the term implies, this type of sampling occurs when people volunteer their services for the study. In psychological experiments or pharmaceutical trials (drug testing), for example, it would be difficult and unethical to enlist random participants from the general public. In these instances, the sample is taken from a group of volunteers. Sometimes, the researcher offers payment to entice respondents. In exchange, the volunteers accept the possibility of a lengthy, demanding or sometimes unpleasant process.

Sampling voluntary participants as opposed to the general population may introduce strong biases. Often in opinion polling, only the people who care strongly enough about the subject one way or another tend to respond. The silent majority does not typically respond, resulting in large selection bias.

Television and radio media often use call-in polls to informally query an audience on their views. The Much Music television channel uses this kind of survey in their CombatZone programme. The programme asks viewers to cast a vote for one of two music videos by telephone, e-mail or through their online website.

Oftentimes, there is no limit imposed on the frequency or number of calls one respondent can make. So, unfortunately, a person might be able to vote repeatedly. It should also be noted that the people who contribute to these surveys might have different views than those who do not.

Judgement Sampling

This approach is used when a sample is taken based on certain judgements about the overall population. The underlying assumption is that the investigator will select units that are characteristic of the population. The critical issue here is objectivity: how much can judgment be relied upon to arrive at a typical sample? *Judgement sampling* is subject to the researcher's biases and is perhaps even more biased than haphazard sampling. Since any preconceptions the researcher may have are reflected in the sample, large biases can be introduced if these preconceptions are inaccurate.

Statisticians often use this method in exploratory studies like pre-testing of questionnaires and focus groups. They also prefer to use this method in laboratory settings where the choice of experimental subjects (i.e., animal, human, vegetable) reflects the investigator's pre-existing beliefs about the population.

One advantage of judgement sampling is the reduced cost and time involved in acquiring the sample.

Quota Sampling

This is one of the most common forms of non-probability sampling. Sampling is done until a specific number of units (quotas) for various sub-populations have been selected. Since there are no rules as to how these quotas are to be filled, *quota sampling* is really a means for satisfying sample size objectives for certain sub-populations.

The quotas may be based on population proportions. For example, if there are 100 men and 100 women in a population and a sample of 20 are to be drawn to participate in a cola taste challenge, you may want to divide the sample evenly between the sexes—10 men and 10 women. Quota sampling can be considered preferable to other forms of non-probability sampling (e.g., judgement sampling) because it forces the inclusion of members of different sub-populations.

Quota sampling is somewhat similar to stratified sampling in that similar units are grouped together. However, it differs in how the units are selected. In probability sampling, the units are selected randomly while in quota sampling it is usually left up to the interviewer to decide who is sampled. This results in selection bias. Thus, quota sampling is often used by market researchers (particularly for telephone surveys) instead of stratified sampling, because it is relatively inexpensive and easy to administer and has the desirable property of satisfying population proportions. However, it disguises potentially significant bias.

As with all other non-probability sampling methods, in order to make inferences about the population, it is necessary to assume that persons selected are similar to those not selected. Such strong assumptions are rarely valid.

Example: The student council at Cedar Valley Public School wants to gauge student opinion on the quality of their extracurricular activities. They decide to survey 100 of 1,000 students using the grade levels (7 to 12) as the sub-population.

The table below gives the number of students in each grade level.

Table. Number of students enrolled at Cedar Valley Public School, by grade

Grade level of students	**Number of students (%)**	**Percentage**	**Quota of studentsin sample of 100**
7	150	15	15
8	220	22	22
9	160	16	16
10	150	15	15
11	200	20	20
12	120	12	12
Total	1,000	100	100

The student council wants to make sure that the percentage of students in each grade level is reflected in the sample. The formula is:

Percentage of students in Grade 10

= (number of students χ number of students) x 100%

$= (150 \div 1{,}000) \times 100$

= 15%

Since 15% of the school population is in Grade 10, 15% of the sample should contain Grade 10 students. Therefore, use the following formula to calculate the number of Grade 10 students that should be included in the sample:

Sample of Grade 10 students

$= (15\% \text{ of } 100) \times 100$

$= 0.15 \times 100$

= 15 students

The main difference between stratified sampling and quota sampling is that stratified sampling would select the students using a probability sampling method such as simple random sampling or systematic sampling. In quota sampling, no such technique is used. The 15 students might be selected by choosing the first 15 Grade 10 students to enter school on a certain day, or by choosing 15 students from the first two rows of a particular classroom. Keep in mind that those students who arrive late or sit at the back of the class may hold different opinions from those who arrived earlier or sat in the front. The main argument against quota sampling is that it does not meet the basic requirement of randomness. Some units may have no chance of selection or the chance of selection may be unknown. Therefore, the sample may be biased.

It is common, but not necessary, for quota samples to use random selection procedures at the beginning stages, much in the same way as probability sampling does. For instance, the first step in multi-stage sampling would be randomly selecting the geographic areas. The difference is in the selection of the units in the final stages of the process.

In multi-stage sampling, units are based on up-to-date lists for selected areas and a sample is selected according to a random process. In quota sampling, by contrast, each interviewer is instructed on how many of the respondents should be men and how many should be women, as well as how many people should represent the various age groups. The quotas are therefore calculated from available data for the population, so that the sexes, age groups or other demographic variables are represented in the correct proportions. But within each quota, interviewers may fail to secure a representative sample of respondents. For example, suppose that an organization wants to find out information about the occupations of men aged 20 to 25. An interviewer goes to a university campus and selects the first 50 men aged 20 to 25 that she comes across and who agree to participate in her organization's survey. However, this sample does not mean that these 50 men are representative of all men aged 20 to 25.

Quota sampling is generally less expensive than random sampling. It is also easy to administer, especially considering the tasks of listing the whole population, randomly selecting the sample and following-up on non-respondents can be omitted from the procedure. Quota sampling is an effective sampling method when information is urgently required and can be carried out independent of existing sampling frames. In many cases where the population has no suitable frame, quota sampling may be the only appropriate sampling method.

ESTIMATION

As we now know, the goal of conducting surveys is to obtain information about a particular population. When the sample has been selected and the information collected and processed, there still remains the task of linking the information gathered from the sample back to the overall population.

Estimation is the process of determining a likely value for a variable in the survey population, based on information collected from the sample. Researchers are usually interested in looking at estimates of many statistics—totals, averages and proportions being the most frequent—for different variables. For example, a sample survey could be used to produce any of the following statistics: estimates for the proportion of smokers among all people aged 15 to 24 in the population; the average earnings of men and women with a university degree; or the total number of cars possessed by the whole survey population.

Underpinning the estimation process is the sampling weight of a unit, which indicates the number of units in the population (including the sampling weight) that are represented by this sampled unit. The sampling weight is the inverse of the unit's probability of selection.

Example: Suppose that the city of Winnipeg has decided to award bus travellers with free one-year bus passes as a way of promoting its services. A simple random sample of 10 people is selected from the 30 passengers on a city bus. Since simple random sampling gives equal probability of selection to every member of the population (in this case, all passengers on the bus), each passenger had one chance out of three of being selected. This translates into a sampling weight of three for every selected unit. This means that each person in the sample represents three persons in the population—himself or herself, plus two other persons

To estimate this sampling weight, one could take the survey information for the 10 selected passengers and copy it three times to create an artificial population of 30. Totals, averages or proportions for the real population could then be estimated by the corresponding statistics computed using the artificial population. Instead of doing this, however, survey statisticians attach a sampling weight to each unit in the sample and take this weight into account when estimating.

If one person in a sample (with a sampling weight of 18) had blue eyes and brown hair, then it is as if a total of 18 people in the population had blue eyes and brown hair.

Example: You are conducting a survey to determine the total number of people living on your street and the average number of cars owned by each household. You decide to select a systematic sample of 5 households from the 20 on your street and intend to use that sample to estimate the totals you are looking for. The following table summarizes the information that you gathered during your interviews with the sampled households:

Table: Sample of households on Rdwood Street

Household number	Number of persons	Number of cars	Probability of selection	Sampling weight
1	1	0	1/4	4
2	4	2	1/4	4
3.	2	1	1/4	4
4	2	1	1/4	4
5	3	2	1/4	4

The selection probability of 1 in 4 comes from the fact that systematic sampling gives an equal chance of being selected to each household on your street. The sampling weight of 4 is just the inverse of that probability. When estimating, you have to look at the characteristics of each sampled household. In this case, it is decided that 4 households from the population of 20 on your street have the same characteristics.

In order to estimate the total number of persons living on your street, you have to multiply the number of persons in a household by the number of households in that sampling weight, then add up all the final numbers. For example, there are 4 one-person households (represented by Household number 1), 4 four-person households, 8 two-person households (four households represented by Household number 3 and four households represented by Household number 4) and 4 three-person households. The estimation of the total number of persons would then be:

Estimated number of persons living on your street

$= (4 \times 1) + (4 \times 4) + (8 \times 2) + (4 \times 3)$

= 48 people

To estimate the average number of cars per household, you proceed in the same manner. Get an estimate of the total number of cars owned by households on your street and then, divide the estimate by the actual number of households on the street. For example, there are 4 households without a car (represented by Household number 1), 8 households with two cars (represented by Household number 2 and Household number 5), 8 households with one car each (represented by Household number 3 and Household number 4).

Estimated number of cars

$= (4 \times 0) + (8 \times 2) + (8 \times 1)$

= 24 cars

Estimated average

$= 24 \div 20$

= 1.2 cars per household

Self-weighting Designs

It is not always the case that all sampled units had the same sampling weight. Some designs give unequal probability of selection to units, resulting in units within the same sample having different sampling weights. Answers from one household or business could represent the answers for 200 units of the population, while the answers from another could represent only 50 units in the population.

When every unit in the sample has the same sampling weight, the sampling design is said to be self-weighted. This kind of design is time-saving and operationally convenient, particularly for large samples. Because every unit has the same weight, those weights can be ignored when estimating averages and proportions. The average for the sample gives an appropriate estimate of the average for the whole population.

Simple random sampling and systematic sampling are examples of self-weighted designs. In that sense, calculations could have been made easier in Example 2. For instance, to estimate the average number of cars per household in the population, we could have used the same average as the one used in the sample. The 5 sampled households own a total of 6 cars, an average of 1.2 cars per household. This is the same result as that obtained using the sampling weight procedure.

Adjusting The Weights

Sometimes, the sampling weights are adjusted prior to estimation. There are basically two reasons for weight adjustment:

1. *Adjusting for non-response:* Using sampling weights for estimation works fine when you have been able to interview all selected units. In Example 2, if two of the five sampled households refused to answer or were unavailable at the time of the survey, you would only have answers for three households, thus representing only 12 of the 20 households on the street. The two non-responding units represent four households each. This means that we have no information on the number of persons or cars for 8 households on your street. In order to adjust for that, survey statisticians usually increase the weights of responding units to account for the loss of representativeness caused by non-response. The goal would be to use only the 3 units for which we have information, but still represent the 20 households on the street.
2. *Adjusting for external information:* Sometimes, we know the actual total for one or more variables measured in the sample. In Example of the Probability sampling section, a population of the 1,000 best horror movies was equally divided into 500 classic movies and 500 modern movies. Even though you knew this prior to sampling, you decided to select a simple random sample of 100 movies and ended up with 77 classic movies and 23 modern movies. Each of these movies has a weight of 10 (because you selected 1 movie out of every 10 titles). Using the answers from the survey and the sampling weight, your sample would represent a population of 770 classic movies and 230 modern movies. This could lead to inaccurate estimates. One solution would be to decrease the weight of every sampled classic and increase the weight of every sampled modern movie

so that your sample gives an estimate of 500 classics and 500 modern films in the population. This should reduce the distortion caused by a 'bad' sample.

Of course, stratifying by release date prior to sampling would have solved this problem. However, in a lot of cases, we have totals at the population level, but we don't know the attribute of each unit on the sampling frame. For example, from the Census of Population, we know how many men and women there are in a specific city, but all we have for sampling is a list of households. Thus, stratifying our population by sex would not be possible. Demographic projections by age and sex for each province are often used in social surveys to adjust sampling weights.

The weights adjusted for non-response and/or external counts are used for estimation, in the same way as the sampling weight was used in Example.

Other Estimation Methods

Using the weights to inflate the sample results is not the only estimation method that exists, but it is the simplest one and the only one that we will cover. Nevertheless, it is important to know that there exist some other methods that could lead to more precise estimates (e.g., using auxiliary information). The estimation process has to take into account the sampling design that was used. Otherwise, the resulting estimates could be severely biased.

Estimating The Sampling Error

As mentioned before, any estimates derived from samples are subject to what is called the sampling error. This comes from the fact that only a part of the population was observed, instead of the whole. A different sample could have come up with different results. The amount of variation that exists among the estimates from the different possible samples is what makes the sampling error. (There are roughly 14 million different combinations of 6 numbers from 1 to 49, so imagine how many ways there are to select a sample of 25,000 Canadian households!) Of course, this sampling error is unknown, since we would need to know the answer for each unit of the population in order to calculate it. Nevertheless, it can be estimated by using the survey data. The extent of the sampling error depends on many things, including the sampling method, the estimation method, the sample size and the variability of the estimated characteristic. This is why each sample estimate has its own sampling error. This error should thus be approximated for each estimated total, average, proportion, etc. produced by the survey.

Examples of Estimation Using An Simple Random Sampling Design

Simple random sampling is the simplest of all sampling methods. Estimation using the simple random sampling method has been studied extensively. There are simple formulas to estimate the sampling error for many statistics when simple random sampling is used, especially since it is a self-weighting design. We present here the most common estimator for a population average (mean) and total, under simple random sampling.

Estimation of The Population Mean

In a simple random sample, the estimate of the population mean is identical to the mean of the sample:

$$\hat{x} = \frac{\sum x}{n}$$

where

x = an observed value

$\hat{x}$ = estimate of the population mean

$\sum$ x= sum of all observed x values in the sample

n = number of observations in the sample.

Note: Lowercase x and n should be used if you are referring to a sample survey and upper case X and N should be used when referring to a population.

If the sample results have been summarized in a frequency table, then the estimate for the population mean is the same as the sample. Thus,

$$\hat{x} = \frac{\sum xf}{\sum f}$$

where

x = an observed value

f = the frequency of the value (the number of times that this value have been observed in the sample)

$\hat{x}$ = estimate of the population mean

$\sum xf$ = sum of all observed xf values (the product of the observed values times its frequency) in the sample

$\sum f$ = sum of the frequencies in the sample.

Example: A farmer randomly selects 10 eggs from a gross of 12 dozen eggs (144 eggs) he finds in his hen house. He carefully weighs each egg.

The following weights were recorded in grams:

0.75, 0.70, 0.55, 0.50, 0.60, 0.65, 0.75, 0.65, 0.75, 0.50

What is the mean weight of the gross of eggs?

Using the above formula, we can determine the mean weight of all of the eggs:

$$\hat{x} = \frac{\sum x}{n}$$

$$= \frac{6.4}{10}$$

$$= 0.64 \text{ grammes}$$

Estimation of The Population Total

For a simple random sample, the estimation formula of a total for the population is

$$\hat{x} = N\frac{\sum x}{n}$$

where

x = an observed value

$\hat{x}$ = estimated population total

$\sum x$ = sum of all observed x values in the sample

n = number of observations in the sample

N = total number of observations in the population.

It is just the estimate for the mean value multiplied by the number of units in the population. In the previous example, the mean weight of an egg is 0.64 grams, so it is logical to think that the total weight of the 144 eggs would be 92.16 grams (144 x 0.64 = 92.16 grams).